Advances in Experimental Medicine and Biology

Volume 805

Advances in Experimental Medicine and Biology

For further volumes:
http://www.springer.com/series/5584

Ke-li Han • Xin Zhang • Ming-jun Yang
Editors

Protein Conformational Dynamics

Springer

Editors
Ke-li Han
Chinese Academy of Sciences
Dalian Institute of Chemical Physics
Dalian, People's Republic of China

Xin Zhang
The Scripps Research Institute
La Jolla, CA, USA

Ming-jun Yang
Chinese Academy of Sciences
Dalian Institute of Chemical Physics
Dalian, People's Republic of China

ISSN 0065-2598 ISSN 2214-8019 (electronic)
ISBN 978-3-319-02969-6 ISBN 978-3-319-02970-2 (eBook)
DOI 10.1007/978-3-319-02970-2
Springer Cham Heidelberg New York Dordrecht London

Library of Congress Control Number: 2014930165

Printed on acid-free paper

Springer is part of Springer Science+Business Media (www.springer.com)

Preface

Proteins exert a variety of fascinating functions to participate in virtually all cellular processes. Underlying these highly controlled activities are the ordered conformational changes, which lead to molecular events that drive efficient and precise regulation and control of these processes. Although it has been widely acknowledged that the conformational dynamics of proteins contributes enormously in these molecular events and their biological function, it remains a major challenge to unambiguously discern their working mechanism both experimentally and computationally. The difficulties primarily arise from the complexity and heterogeneity of protein assemblies, the ruggedness of free energy landscapes, and the largely varied scales of temporal and spatial changes in different events.

Complementary to experimental studies, computational simulation has become a powerful and unique tool to dissect the working mechanism of proteins and provide information otherwise inaccessible to other methods. In recent years, the persistent progress in methodologies (super-computational resources, multi-scale modeling, enhanced sampling methods, etc.) has demonstrated a number of important applications in biological processes, e.g. molecular recognition, enzyme catalysis, molecular transport, protein folding and aggregation, and signal transduction. In this book, we present an extensive review of the recent theoretical and computational advances as well as their applications to key biological questions.

In Chaps, 1, 2, and 3, different methods are described to generate the conformational ensembles during the folding and function of proteins. Chapter 1 describes the generalized-ensemble algorithms and their applications to the protein folding problem. Chapter 2 introduces another powerful tool in conformational sampling, the Markov State Models (MSMs), which can increase the simulation length to microseconds or even milliseconds. This chapter explains the general concepts of MSMs, the model construct procedure, and its application to the long time-scale molecular dynamics simulation of biological macromolecules. As the conformational dynamics can now be characterized by advanced experimental methods, Chap. 3 shows how these information can be combined with computational simulation to build the conformational ensembles.

The conformational ensembles in biological macromolecules contain useful information that describes the important features of the system. Methods to extract this information will be summarized in Chaps. 4, 5, 6, and 7. Chapter 4 reviews methods that produce generative models of conformational dynamics to manifest the hidden thermodynamic and kinetic properties from the ensembles generated during simulations. To study the large-scale motions of proteins, Chap. 5 describes various coarse-grain elastic network models (ENMs) that predict the magnitudes and directions of protein motion, focusing on the recently developed generalized spring tensor model. When the size of biological systems increase, high-resolution structure can only be obtained in fragments and the structural information of the system is often of low-resolution. Several flexible fitting methods are presented in Chap. 6 to extract deeper understanding from the combination of these experimental data. Along this line, various coarse-grained models are discussed in Chap. 7 to study the functional change of the polypeptide backbones around their native states.

Protein folding is a critical biological process in which proteins acquire their defined three-dimensional structures from linear polypeptides. Chapters 8 and 9 focus on this topic and review methods that unveil this dynamic process. Because the cellular environment plays a major role in regulating the dynamics of protein folding, Chap. 8 describes novel coarse-grained methods that go beyond traditional aqueous solvent conditions and study *in vivo* protein folding dynamics. Since various optical spectroscopic techniques have been used to probe protein folding/unfolding events, Chap. 9 addresses progress on modeling the unfolding dynamics of several model proteins using the combined theoretical spectroscopic and MSMs approaches.

Conformational dynamics plays a major role in dictating function of proteins. Chapters 10 and 11 explore the conformational flexibility in enzyme catalysis and drug designs. Chapter 12 addresses how the nuclear magnetic resonance spectroscopy and computational methods are combined to understand the receptor-ligand interaction. The structural dynamics of membrane proteins stands as a major challenge and Chap. 13 presents a survey of the current methods and technique issues for simulations of membrane proteins. Besides the well-structured proteins, the intrinsically disordered proteins (IDPs) play important roles in a range of biological processes through the distinct coupled folding and binding mechanism. Chapter 14 presents the free energy analysis of the IDPs with the enhanced sampling methods.

Chapters 15, 16, 17, and 18 summarize recent computational studies on several biological processes. In Chap. 15, two machineries that transform chemical energy to mechanical work are studied using atomistic molecular dynamics simulations, coarse-grained analyses, and stochastic modeling techniques. In Chap. 16, recent theoretical and experimental progresses are reviewed on the universally conserved signal recognition particle machinery that mediates co-translational protein targeting reaction. Chapter 17 discusses the detailed analyses of the ATP-driven rotary motor enzyme that can perform ATP synthesis/hydrolysis using reversible motor rotation. Among different membrane proteins, the G protein coupled receptor (GPCR) is one of the most important families, which constitutes the target of about one third drugs in the market. Chapter 18 summarizes recent computational studies on the chemo-sensorial GPCRs that are responsible to detect odorant and tasting molecules.

Finally, we would like to express our gratitude to all the authors who have contributed their excellent work to this book. We also acknowledge the editorial team at the Springer, in particular Thijs van Vlijmen, Sara Germans-Huisman, and Ilse Hensen, for their helpful guidance during the entire project.

Dalian, People's Republic of China Ke-li Han
La Jolla, CA, USA Xin Zhang
Dalian, People's Republic of China Ming-jun Yang

Contents

About the Editors

Professor **Ke-li Han** is working at Dalian Institute of Chemical Physics, Chinese Academy of Sciences (CAS). He received his Ph.D. degree in physical chemistry in 1990. His research group is mainly interested in the theoretical and computational studies of the interdisciplines among physics, chemistry, biology and materials, including protein dynamics, hydrogen-bond dynamics in inter-/intramolecular electron transfer, molecular dynamics with attosecond resolution, and nonadiabatic dynamics in chemical processes etc. To date, he has published more than 330 scientific papers with a high citation over 8,000.

He receives the support of the National Outstanding Youth Foundation. In 1999, he received a first class Natural Science Award from CAS and the Young Chemist Award from the Chinese Chemical Society (CCS). He is the chair of the Virtual Laboratory for Computational Chemistry (VLCC), Supercomputing Center, and Computer Network Information Center at CAS. He chaired many international meetings and presented many invited talks. He is a member of the editorial board of the *Journal of Physical Chemistry*. He is also a member of the advisory editorial board of the *Journal of Theoretical & Computational Chemistry*, *Chinese Journal of Chemical Physics*, *Acta Physico-Chimica Sinica*, *Progress in Natural Science*, *Chinese Science Bulletin*, and *Journal of Atomic and Molecular Physics*.

Dr. **Xin Zhang** received his Ph.D. degree in 2010 from California Institute of Technology. His research interests focus on (i) molecular mechanism of the efficiency and fidelity of the co-translational protein targeting, (ii) developing fluorescence tag and biosensor by chemoselective approaches, and (iii) rendering protein evolution more efficient by adapting the proteostasis network. He has published 26 peer-reviewed papers and a book *Multistate GTPases Control Co-translational Protein Targeting* by The Springer Publishing Company IIC as 'Springer Thesis' in 2012. Due to his excellent work, he was awarded the American Chemical Society National Award – Noble Laureate Signature Award for Graduate Education in Chemistry in 2012, Helen Hay Whitney Postdoctoral Fellowship (2011–2013), Herbert Newby McCoy in 2010, and Chinese Government Award for Outstanding Self-financed Students Abroad in 2009.

Dr. **Ming-jun Yang** received his Ph.D. degree in Dalian Institute of Chemical Physics, Chinese Academy of Sciences in 2011. During Ph.D. studies, he mainly worked on the computational simulation of the conformational dynamics of signal recognition particle (SRP) GTPases to reveal their mechanism in protein targeting events. His current research interests focus on the development of new enhanced sampling methods in study of the protein conformational dynamics.

Chapter 1
Protein Folding Simulations by Generalized-Ensemble Algorithms

Takao Yoda, Yuji Sugita, and Yuko Okamoto

Abstract In the protein folding problem, conventional simulations in physical statistical mechanical ensembles, such as the canonical ensemble with fixed temperature, face a great difficulty. This is because there exist a huge number of local-minimum-energy states in the system and the conventional simulations tend to get trapped in these states, giving wrong results. Generalized-ensemble algorithms are based on artificial unphysical ensembles and overcome the above difficulty by performing random walks in potential energy, volume, and other physical quantities or their corresponding conjugate parameters such as temperature, pressure, etc. The advantage of generalized-ensemble simulations lies in the fact that they not only avoid getting trapped in states of energy local minima but also allows the

T. Yoda
Nagahama Institute of Bio-Science and Technology, Tamura, Nagahama, Shiga 526-0829, Japan

Y. Sugita
RIKEN Theoretical Molecular Science Laboratory, 2-1 Hirosawa, Wako, Saitama 351-0198, Japan

RIKEN Quantitative Biology Center, 7-1-26 Minatojima-minamimachi, Chuo-ku, Kobe, Hyogo 650-0047, Japan

RIKEN Advanced Science Institute for Computational Science, 7-1-26 Minatojima-minamimachi, Chuo-ku, Kobe, Hyogo 650-0047, Japan

Y. Okamoto (✉)
Department of Physics, Graduate School of Science, Nagoya University, Furo-cho, Chikusa-ku, Nagoya, Aichi 464-8602, Japan

Structural Biology Research Center, Graduate School of Science, Nagoya University, Furo-cho, Chikusa-ku, Nagoya, Aichi 464-8602, Japan

Center for Computational Science, Graduate School of Engineering, Nagoya University, Furo-cho, Chikusa-ku, Nagoya, Aichi 464-8603, Japan

Information Technology Center, Nagoya University, Furo-cho, Chikusa-ku, Nagoya, Aichi 464-8601, Japan
e-mail: okamoto@phys.nagoya-u.ac.jp

K. Han et al. (eds.), *Protein Conformational Dynamics*, Advances in Experimental Medicine and Biology 805, DOI 10.1007/978-3-319-02970-2_1,

calculations of physical quantities as functions of temperature or other parameters from a single simulation run. In this article we review the generalized-ensemble algorithms. Four examples, multicanonical algorithm, replica-exchange method, replica-exchange multicanonical algorithm, and multicanonical replica-exchange method, are described in detail. Examples of their applications to the protein folding problem are presented.

Keywords Generalized-ensemble algorithm • Multicanonical algorithm • Replica-exchange molecular dynamics • Replica-exchange multicanonical algorithm • Multicanonical replica-exchange method • Protein folding

1.1 Introduction

In order to study the protein folding problem, molecular simulation methods such as Monte Carlo (MC) and molecular dynamics (MD) are often used. However, conventional canonical simulations at physically relevant temperatures tend to get trapped in states of energy-local-minima, giving wrong results. A class of simulation methods, which are referred to as the *generalized-ensemble algorithms*, overcome this difficulty (for reviews see, e.g., Refs. [1–5]). In the generalized-ensemble algorithm, each state is weighted by an artificial, non-Boltzmann probability weight factor so that random walks in potential energy, volume, and other physical quantities or their corresponding conjugate parameters such as temperature, pressure, etc. may be realized. The random walks allow the simulation to escape from any energy barrier and to sample much wider conformational space than by conventional methods.

One of effective generalized-ensemble algorithms for molecular simulations is the multicanonical algorithm (MUCA) [6, 7], which was first applied to the protein folding problem in Ref. [8]. In this method, the weight factor is defined to be inversely proportional to the density of states and a free random walk in potential energy space is realized. Another effective generalized-ensemble algorithm is the *replica-exchange method* (REM) [9] (the method is also referred to as *parallel tempering* [10]), which was first applied to the protein folding problem in Ref. [11]. In this method, a number of non-interacting copies (or, replicas) of the original system at different temperatures are simulated independently and exchanged with a specified transition probability. The details of molecular dynamics algorithm for REM, which is referred to as the *replica-exchange molecular dynamics* (REMD), have been worked out in Ref. [12], and this led to a wide application of REMD in the protein and other biomolecular systems. One is naturally led to combine MUCA and REM, and two methods, *replica-exchange multicanonical algorithm* (REMUCA) and *multicanonical replica-exchange method* (MUCAREM), have been developed [13–15]. MUCAREM can be considered to be a special case of the multidimensional (or, multivariable) extension of REM, which we refer to as the *multidimensional replica-exchange method* (MREM) [16]. MREM is now widely used and often referred to as *Hamiltonian replica-exchange method* [17].

In this article, we describe the generalized-ensemble algorithms mentioned above. Namely, we review the four methods: MUCA, REM, REUMCA, and MUCAREM. Examples of the results in which these methods were applied to the protein folding problem are then presented.

1.2 Methods

1.2.1 Multicanonical Algorithm

Let us consider a system of N atoms of mass m_k ($k = 1, \ldots, N$) with their coordinate vectors and momentum vectors denoted by $q = (\boldsymbol{q}_1, \ldots, \boldsymbol{q}_N)$ and $p = (\boldsymbol{p}_1, \ldots, \boldsymbol{p}_N)$, respectively. The Hamiltonian $H(q,p)$ of the system is the sum of the kinetic energy $K(p)$ and the potential energy $E(q)$:

$$H\,(q, p) = K(p) + E(q), \tag{1.1}$$

where

$$K(p) = \sum_{k=1}^{N} \frac{\boldsymbol{p}_k{}^2}{2m_k}. \tag{1.2}$$

In the canonical ensemble at temperature T each state $x \equiv (q,p)$ with the Hamiltonian $H(q,p)$ is weighted by the Boltzmann factor:

$$W_{\rm B}\,(x;T) = \exp\left(-\beta H\,(q,p)\right), \tag{1.3}$$

where the inverse temperature β is defined by $\beta = 1/k_{\rm B}T$ ($k_{\rm B}$ is the Boltzmann constant). The average kinetic energy at temperature T is then given by

$$\left\langle K(p)\right\rangle_T = \left\langle \sum_{k=1}^{N} \frac{\boldsymbol{p}_k{}^2}{2m_k} \right\rangle_T = \frac{3}{2} N k_{\rm B} T. \tag{1.4}$$

Because the coordinates q and momenta p are decoupled in Eq. (1.1), we can suppress the kinetic energy part and can write the Boltzmann factor as

$$W_{\rm B}\,(x;T) = W_{\rm B}\,(E;T) = \exp\left(-\beta E\right). \tag{1.5}$$

The canonical probability distribution of potential energy $P_{\rm NVT}(E;T)$ is then given by the product of the density of states $n(E)$ and the Boltzmann weight factor $W_{\rm B}(E;T)$:

$$P_{\rm NVT}\,(E;T) \propto n(E) W_{\rm B}\,(E;T). \tag{1.6}$$

Because $n(E)$ is a rapidly increasing function and the Boltzmann factor decreases exponentially, the canonical ensemble yields a bell-shaped distribution of potential energy which has a maximum around the average energy at temperature T. The conventional MC or MD simulations at constant temperature are expected to yield $P_{\mathrm{NVT}}(E;T)$. A MC simulation based on the Metropolis algorithm [18] is performed with the following transition probability from a state x of potential energy E to a state x' of potential energy E':

$$w\left(x \to x'\right) = \min\left(1, \frac{W_{\mathrm{B}}\left(E'; T\right)}{W_{\mathrm{B}}\left(E; T\right)}\right) = \min\left(1, \exp\left(-\beta \Delta E\right)\right), \qquad (1.7)$$

where

$$\Delta E = E' - E. \qquad (1.8)$$

A MD simulation, on the other hand, is based on the following Newton equations of motion:

$$\dot{\boldsymbol{q}}_k = \frac{\boldsymbol{p}_k}{m_k}, \qquad (1.9)$$

$$\dot{\boldsymbol{p}}_k = -\frac{\partial E}{\partial \boldsymbol{q}_k} = \boldsymbol{f}_k, \qquad (1.10)$$

where $\boldsymbol{f}_k$ is the force acting on the k-th atom ($k = 1, \ldots, N$). This set of equations actually yield the microcanonical ensemble, however, and we have to add a thermostat in order to obtain the canonical ensemble at temperature T. Here, we just follow Nosé's prescription [19, 20], and we have

$$\dot{\boldsymbol{q}}_k = \frac{\boldsymbol{p}_k}{m_k}, \qquad (1.11)$$

$$\dot{\boldsymbol{p}}_k = -\frac{\partial E}{\partial \boldsymbol{q}_k} - \frac{\dot{s}}{s}\boldsymbol{p}_k = \boldsymbol{f}_k - \frac{\dot{s}}{s}\boldsymbol{p}_k, \qquad (1.12)$$

$$\dot{s} = s\frac{P_s}{Q}, \qquad (1.13)$$

$$\dot{P}_s = \sum_{k=1}^{N} \frac{{\boldsymbol{p}_k}^2}{m_k} - 3Nk_{\mathrm{B}}T = 3Nk_{\mathrm{B}}\left(T(t) - T\right), \qquad (1.14)$$

where s is Nosé's scaling parameter, P_s is its conjugate momentum, Q is its mass, and the "instantaneous temperature" $T(t)$ is defined by

$$T(t) = \frac{1}{3Nk_{\mathrm{B}}} \sum_{k=1}^{N} \frac{\boldsymbol{p}_k(t)^2}{m_k}. \qquad (1.15)$$

However, in practice, it is very difficult to obtain accurate canonical distributions of complex systems at low temperatures by conventional MC or MD simulation methods. This is because simulations at low temperatures tend to get trapped in one or a few of local-minimum-energy states. This difficulty is overcome by, for instance, the generalized-ensemble algorithms, which greatly enhance conformational sampling.

In the multicanonical ensemble [6, 7], on the other hand, each state is weighted by a non-Boltzmann weight factor $W_{\rm MUCA}(E)$ (which we refer to as the *multicanonical weight factor*) so that a uniform potential energy distribution $P_{\rm MUCA}(E)$ is obtained:

$$P_{\rm MUCA}(E) \propto n(E) W_{\rm MUCA}(E) \equiv \text{const.} \tag{1.16}$$

The flat distribution implies that a free random walk in the potential energy space is realized in this ensemble. This allows the simulation to escape from any local minimum-energy states and to sample the configurational space much more widely than the conventional canonical MC or MD methods.

The definition in Eq. (1.16) implies that the multicanonical weight factor is inversely proportional to the density of states, and we can write it as follows:

$$W_{\rm MUCA}(E) \equiv \exp\left[-\beta_0 E_{\rm MUCA}\left(E; T_0\right)\right] = \frac{1}{n(E)}, \tag{1.17}$$

where we have chosen an arbitrary reference temperature, $T_0 = 1/k_{\rm B}\beta_0$, and the "*multicanonical potential energy*" is defined by

$$E_{\rm MUCA}\left(E; T_0\right) \equiv k_{\rm B} T_0 \ln n(E) = T_0 S(E). \tag{1.18}$$

Here, $S(E)$ is the entropy in the microcanonical ensemble. Because the density of states of the system is usually unknown, the multicanonical weight factor has to be determined numerically by iterations of short preliminary runs [6, 7].

A multicanonical MC simulation is performed, for instance, with the usual Metropolis criterion [18]: The transition probability of state x with potential energy E to state x' with potential energy E' is given by

$$\begin{aligned} w\left(x \to x'\right) &= \min\left(1, \frac{W_{\rm MUCA}\left(E'\right)}{W_{\rm MUCA}(E)}\right) = \min\left(1, \frac{n(E)}{n\left(E'\right)}\right) \\ &= \min\left(1, \exp\left(-\beta_0 \Delta E_{\rm MUCA}\right)\right), \end{aligned} \tag{1.19}$$

where

$$\Delta E_{\rm MUCA} = E_{\rm MUCA}\left(E'; T_0\right) - E_{\rm MUCA}\left(E; T_0\right). \tag{1.20}$$

The MD algorithm in the multicanonical ensemble also naturally follows from Eq. (1.17), in which the regular constant temperature MD simulation (with $T = T_0$) is performed by replacing E by E_{MUCA} in Eq. (1.12) [21, 22]:

$$\dot{\boldsymbol{p}}_k = -\frac{\partial E_{\mathrm{MUCA}}(E; T_0)}{\partial \boldsymbol{q}_k} - \frac{\dot{s}}{s}\boldsymbol{p}_k = \frac{\partial E_{\mathrm{MUCA}}(E; T_0)}{\partial E}\boldsymbol{f}_k - \frac{\dot{s}}{s}\boldsymbol{p}_k. \tag{1.21}$$

From Eq. (1.18) this equation can be rewritten as

$$\dot{\boldsymbol{p}}_k = \frac{T_0}{T(E)}\boldsymbol{f}_k - \frac{\dot{s}}{s}\boldsymbol{p}_k. \tag{1.22}$$

where the following thermodynamic relation gives the definition of the "effective temperature" $T(E)$:

$$\left.\frac{\partial S(E)}{\partial E}\right|_{E=E_a} = \frac{1}{T(E_a)}, \tag{1.23}$$

with

$$E_a = \langle E \rangle_{T(E_a)}. \tag{1.24}$$

If the exact multicanonical weight factor $W_{\mathrm{MUCA}}(E)$ is known, one can calculate the ensemble averages of any physical quantity A at any temperature T ($= 1/k_B\beta$) as follows:

$$\langle A \rangle_T = \frac{\sum_E A(E) P_{\mathrm{NVT}}(E; T)}{\sum_E P_{\mathrm{NVT}}(E; T)} = \frac{\sum_E A(E) n(E) \exp(-\beta E)}{\sum_E n(E) \exp(-\beta E)}, \tag{1.25}$$

where the density of states is given by (see Eq. (1.17))

$$n(E) = \frac{1}{W_{\mathrm{MUCA}}(E)}. \tag{1.26}$$

The summation instead of integration is used in Eq. (1.25), because we often discretize the potential energy E with step size ε ($E = E_{\mathrm{i}}$; $\mathrm{i} = 1, 2, \ldots$). Here, the explicit form of the physical quantity A should be known as a function of potential energy E. For instance, $A(E) = E$ gives the average potential energy $\langle E \rangle_T$ as a function of temperature, and $A(E) = \beta^2 (E - \langle E \rangle_{\mathrm{T}})^2$ gives specific heat.

In general, the multicanonical weight factor $W_{\mathrm{MUCA}}(E)$, or the density of states $n(E)$, is not *a priori* known, and one needs its estimator for a numerical simulation.

This estimator is usually obtained from iterations of short trial multicanonical simulations. However, the iterative process can be non-trivial and very tedious for complex systems.

In practice, it is impossible to obtain the ideal multicanonical weight factor with completely uniform potential energy distribution. The question is when to stop the iteration for the weight factor determination. Our criterion for a satisfactory weight factor is that as long as we do get a random walk in potential energy space, the probability distribution $P_{\text{MUCA}}(E)$ does not have to be completely flat with a tolerance of, say, an order of magnitude deviation. In such a case, we usually perform with this weight factor a multicanonical simulation with high statistics (production run) in order to get even better estimate of the density of states. Let $N_{\text{MUCA}}(E)$ be the histogram of potential energy distribution $P_{\text{MUCA}}(E)$ obtained by this production run. The best estimate of the density of states can then be given by the single-histogram reweighting techniques [23] as follows (see the proportionality relation in Eq. (1.16)):

$$n(E) = \frac{N_{\text{MUCA}}(E)}{W_{\text{MUCA}}(E)}. \tag{1.27}$$

By substituting this quantity into Eq. (1.25), one can calculate ensemble averages of physical quantity $A(E)$ as a function of temperature. Moreover, ensemble averages of any physical quantity A (including those that cannot be expressed as functions of potential energy) at any temperature T $(=1/k_{\text{B}}\beta)$ can now be obtained as long as one stores the "trajectory" of configurations (and A) from the production run. Namely, we have

$$<A>_T = \frac{\sum_{k=1}^{n_0} A\left(x(k)\right) W_{\text{MUCA}}^{-1}\left(E\left(x(k)\right)\right) \exp\left[-\beta E\left(x(k)\right)\right]}{\sum_{k=1}^{n_0} W_{\text{MUCA}}^{-1}\left(E\left(x(k)\right)\right) \exp\left[-\beta E\left(x(k)\right)\right]}, \tag{1.28}$$

where $x(k)$ is the configuration at the k-th MC (or MD) step and n_0 is the total number of configurations stored. Note that when A is a function of E, Eq. (1.28) reduces to Eq. (1.25) where the density of states is given by Eq. (1.27).

Equations (1.25) and (1.28) or any other equations which involve summations of exponential functions often encounter with numerical difficulties such as overflows. These can be overcome by using, for instance, the following equation [24, 25]: For $C = A + B$ (with $A > 0$ and $B > 0$) we have

$$\begin{aligned} \ln C &= \ln\left[\max(A, B)\left(1 + \frac{\min(A, B)}{\max(A, B)}\right)\right] \\ &= \max(\ln A, \ln B) + \ln\{1 + \exp[\min(\ln A, \ln B) - \max(\ln A, \ln B)]\}. \end{aligned} \tag{1.29}$$

1.2.2 Replica-Exchange Method

The *replica-exchange method* (REM) is another effective generalized-ensemble algorithm. The system for REM consists of M *non-interacting* copies (or, replicas) of the original system in the canonical ensemble at M different temperatures $T_m (m = 1, \ldots, M)$. We arrange the replicas so that there is always exactly one replica at each temperature. Then there exists a one-to-one correspondence between replicas and temperatures; the label $i(=1, \ldots, M)$ for replicas is a permutation of the label $m(=1, \ldots, M)$ for temperatures, and vice versa:

$$\begin{cases} i = i(m) \equiv f(m), \\ m = m(i) \equiv f^{-1}(i), \end{cases} \tag{1.30}$$

where $f(m)$ is a permutation function of m and $f^{-1}(i)$ is its inverse.

Let $X = \{x_1^{[i(1)]}, \ldots, x_M^{[i(M)]}\} = \{x_{m(1)}^{[1]}, \ldots, x_{m(M)}^{[M]}\}$ stand for a "state" in this generalized ensemble. Each "substate" $x_m^{[i]}$ is specified by the coordinates $q^{[i]}$ and momenta $p^{[i]}$ of N atoms in replica i at temperature T_m:

$$x_m^{[i]} \equiv \left(q^{[i]}, p^{[i]}\right)_m. \tag{1.31}$$

Because the replicas are non-interacting, the weight factor for the state X in this generalized ensemble is given by the product of Boltzmann factors for each replica (or at each temperature):

$$\begin{aligned} W_{\mathrm{REM}}(X) &= \prod_{i=1}^{M} \exp\left\{-\beta_{m(i)} H\left(q^{[i]}, p^{[i]}\right)\right\} = \prod_{m=1}^{M} \exp\left\{-\beta_m H\left(q^{[i(m)]}, p^{[i(m)]}\right)\right\} \\ &= \exp\left\{-\sum_{i=1}^{M} \beta_{m(i)} H\left(q^{[i]}, p^{[i]}\right)\right\} = \exp\left\{-\sum_{m=1}^{M} \beta_m H\left(q^{[i(m)]}, p^{[i(m)]}\right)\right\}, \end{aligned} \tag{1.32}$$

where $i(m)$ and $m(i)$ are the permutation functions in Eq. (1.30).

We now consider exchanging a pair of replicas in this ensemble. Suppose we exchange replicas i and j which are at temperatures T_m and T_n, respectively:

$$X = \left\{\ldots, x_m^{[i]}, \ldots, x_n^{[j]}, \ldots\right\} \to X' = \left\{\ldots, x_m^{[j]'}, \ldots, x_n^{[i]'}, \ldots\right\}. \tag{1.33}$$

The exchange of replicas can be written in more detail as

$$\begin{cases} x_m^{[i]} \equiv \left(q^{[i]}, p^{[i]}\right)_m \to x_m^{[j]'} \equiv \left(q^{[j]}, p^{[j]'}\right)_m, \\ x_n^{[j]} \equiv \left(q^{[j]}, p^{[j]}\right)_n \to x_n^{[i]'} \equiv \left(q^{[i]}, p^{[i]'}\right)_n, \end{cases} \tag{1.34}$$

where the definitions for $p^{[i]'}$ and $p^{[j]'}$ will be given below.

In the original implementation of the *replica-exchange method* (REM) [9], Monte Carlo algorithm was used, and only the coordinates q (and the potential energy function $E(q)$) had to be taken into account. In molecular dynamics algorithm, on the other hand, we also have to deal with the momenta p. We proposed the following momentum assignment in Eq. (1.34) [12]:

$$\begin{cases} p^{[i]'} \equiv \sqrt{\frac{T_n}{T_m}}\, p^{[i]}, \\ p^{[j]'} \equiv \sqrt{\frac{T_m}{T_n}}\, p^{[j]}, \end{cases} \tag{1.35}$$

which we believe is the simplest and the most natural. This assignment means that we just rescale uniformly the velocities of all the atoms in the replicas by the square root of the ratio of the two temperatures so that the temperature condition in Eq. (1.4) may be satisfied immediately after replica exchange is accepted. We remark that similar momentum rescaling formulae for various constant-temperature algorithms have been worked out in Ref. [26].

The transition probability of this replica-exchange process is given by the usual Metropolis criterion:

$$w\left(X \rightarrow X'\right) \equiv w\left(x_m^{[i]} \middle| x_n^{[j]}\right) = \min\left(1, \frac{W_{\mathrm{REM}}\left(X'\right)}{W_{\mathrm{REM}}(X)}\right) = \min\left(1, \exp\left(-\Delta\right)\right), \tag{1.36}$$

where in the second expression (i.e., $w(x_m^{[i]}|x_n^{[j]})$) we explicitly wrote the pair of replicas (and temperatures) to be exchanged. From Eqs. (1.1), (1.2), (1.32), and (1.35), we have

$$\Delta = \beta_m \left(E\left(q^{[j]}\right) - E\left(q^{[i]}\right)\right) - \beta_n \left(E\left(q^{[j]}\right) - E\left(q^{[i]}\right)\right) \tag{1.37}$$

$$= \left(\beta_m - \beta_n\right)\left(E\left(q^{[j]}\right) - E\left(q^{[i]}\right)\right). \tag{1.38}$$

Note that after introducing the momentum rescaling in Eq. (1.35), we have the same Metropolis criterion for replica exchanges, i.e., Eqs. (1.36) and (1.38), for both MC and MD versions.

Without loss of generality we can assume that $T_1 < T_2 < \ldots < T_M$. The lowest temperature T_1 should be sufficiently low so that the simulation can explore the experimentally relevant temperature region, and the highest temperature T_M should be sufficiently high so that no trapping in an energy-local-minimum state occurs. A REM simulation is then realized by alternately performing the following two steps:

1. Each replica in canonical ensemble of the fixed temperature is simulated *simultaneously* and *independently* for a certain MC or MD steps.

2. A pair of replicas at neighboring temperatures, say, $x_m^{[i]}$ and $x_{m+1}^{[j]}$, are exchanged with the probability $w(x_m^{[i]}|x_{m+1}^{[j]})$ in Eq. (1.36).

A random walk in "temperature space" is realized for each replica, which in turn induces a random walk in potential energy space. This alleviates the problem of getting trapped in states of energy local minima.

After a long production run of a replica-exchange simulation, the canonical expectation value of a physical quantity A at temperature $T_m(m = 1, \ldots, M)$ can be calculated by the usual arithmetic mean:

$$\langle A \rangle_{T_m} = \frac{1}{n_m} \sum_{k=1}^{n_m} A\left(x_m(k)\right), \tag{1.39}$$

where $x_m(k)(k = 1, \ldots, n_m)$ are the configurations obtained at temperature T_m and n_m is the total number of measurements made at $T = T_m$. The expectation value at any intermediate temperature T $(= 1/k_B\beta)$ can also be obtained from Eq. (1.25), where the density of states $n(E)$ in Eq. (1.25) is now given by the multiple-histogram reweighting techniques, or, the weighted histogram analysis method (WHAM) [27, 28] as follows. Let $N_m(E)$ and n_m be respectively the potential-energy histogram and the total number of samples obtained at temperature $T_m = 1/k_B\beta_m(m = 1, \ldots, M)$. The best estimate of the density of states is then given by

$$n(E) = \frac{\sum_{m=1}^{M} N_m(E)}{\sum_{m=1}^{M} n_m \exp\left(f_m - \beta_m E\right)}, \tag{1.40}$$

where we have for each $m(=1, \ldots, M)$

$$\exp\left(-f_m\right) = \sum_{E} n(E) \exp\left(-\beta_m E\right). \tag{1.41}$$

Note that Eqs. (1.40) and (1.41) are solved self-consistently by iteration [27, 28] to obtain the density of states $n(E)$ and the dimensionless Helmholtz free energy f_m. Namely, we can set all the $f_m(m = 1, \ldots, M)$ to, e.g., zero initially. We then use Eq. (1.40) to obtain $n(E)$, which is substituted into Eq. (1.41) to obtain next values of f_m, and so on.

Moreover, ensemble averages of any physical quantity A (including those that cannot be expressed as functions of potential energy) at any temperature T $(= 1/k_B\beta)$ can now be obtained from the "trajectory" of configurations of the production run. Namely, we first obtain $f_m(m = 1, \ldots, M)$ by solving Eqs. (1.40) and (1.41) self-consistently, and then we have [14]

$$<A>_T = \frac{\displaystyle\sum_{m=1}^{M}\sum_{k=1}^{n_m} A\left(x_m(k)\right) \frac{1}{\displaystyle\sum_{l=1}^{M} n_l \exp\left[f_l - \beta_l E\left(x_m(k)\right)\right]} \exp\left[-\beta E\left(x_m(k)\right)\right]}{\displaystyle\sum_{m=1}^{M}\sum_{k=1}^{n_m} \frac{1}{\displaystyle\sum_{l=1}^{M} n_l \exp\left[f_l - \beta_l E\left(x_m(k)\right)\right]} \exp\left[-\beta E\left(x_m(k)\right)\right]}, \tag{1.42}$$

where $x_m(k)(k = 1, \ldots, n_m)$ are the configurations obtained at temperature T_m.

1.2.3 Replica-Exchange Multicanonical Algorithm and Multicanonical Replica-Exchange Method

The *replica-exchange multicanonical algorithm* (REMUCA) [13–15] overcomes both the difficulties of MUCA (the multicanonical weight factor determination is non-trivial) and REM (a lot of replicas, or computation time, is required). In REMUCA we first perform a short REM simulation (with M replicas) to determine the multicanonical weight factor and then perform with this weight factor a regular multicanonical simulation with high statistics. The first step is accomplished by the multiple-histogram reweighting techniques. Let $N_m(E)$ and n_m be respectively the potential-energy histogram and the total number of samples obtained at temperature T_m $(= 1/k_B\beta_m)$ of the REM run. The density of states $n(E)$ is then given by solving Eqs. (1.40) and (1.41) self-consistently by iteration.

Once the estimate of the density of states is obtained, the multicanonical weight factor can be directly determined from Eq. (1.17) (see also Eq. (1.18)). Actually, the density of states $n(E)$ and the multicanonical potential energy, $E_{\text{MUCA}}(E;T_0)$, thus determined are only reliable in the following range:

$$E_1 \leq E \leq E_M, \tag{1.43}$$

where

$$\begin{cases} E_1 = <E>_{T_1}, \\ E_M = <E>_{T_M}, \end{cases} \tag{1.44}$$

and T_1 and T_M are respectively the lowest and the highest temperatures used in the REM run. Outside this range we extrapolate the multicanonical potential energy linearly [13]:

$$\mathcal{E}_{\rm MUCA}^{\{0\}}(E) \equiv \begin{cases} \left.\frac{\partial E_{\rm MUCA}(E;T_0)}{\partial E}\right|_{E=E_1}(E-E_1)+E_{\rm MUCA}(E_1;T_0), & \text{for } E<E_1, \\ E_{\rm MUCA}(E;T_0), & \text{for } E_1 \le E \le E_M, \\ \left.\frac{\partial E_{\rm MUCA}(E;T_0)}{\partial E}\right|_{E=E_M}(E-E_M)+E_{\rm MUCA}(E_M;T_0), & \text{for } E>E_M. \end{cases} \tag{1.45}$$

The multicanonical MC and MD runs are then performed respectively with the Metropolis criterion of Eq. (1.19) and with the modified Newton equation in Eq. (1.21), in which $\mathcal{E}_{\rm MUCA}^{\{0\}}(E)$ in Eq. (1.45) is substituted into $E_{\rm MUCA}(E;T_0)$. We expect to obtain a flat potential energy distribution in the range of Eq. (1.43). Finally, the results are analyzed by the single-histogram reweighting techniques as described in Eq. (1.27) (and Eq. (1.25)).

Some remarks are now in order. From Eqs. (1.18), (1.23), (1.24), and (1.44), Eq. (1.45) becomes

$$\mathcal{E}_{\rm MUCA}^{\{0\}}(E) \equiv \begin{cases} \frac{T_0}{T_1}(E-E_1)+T_0S(E_1)=\frac{T_0}{T_1}E+\text{const}, & \text{for } E<E_1, \\ T_0S(E), & \text{for } E_1 \le E \le E_M, \\ \frac{T_0}{T_M}(E-E_M)+T_0S(E_M)=\frac{T_0}{T_M}E+\text{const}, & \text{for } E>E_M. \end{cases} \tag{1.46}$$

The Newton equation in Eq. (1.21) is then written as (see Eqs. (1.22), (1.23), and (1.24))

$$\dot{\boldsymbol{p}}_k = \begin{cases} \frac{T_0}{T_1}\boldsymbol{f}_k - \frac{\dot{s}}{s}\boldsymbol{p}_k, & \text{for } E<E_1, \\ \frac{T_0}{T(E)}\boldsymbol{f}_k - \frac{\dot{s}}{s}\boldsymbol{p}_k, & \text{for } E_1 \le E \le E_M, \\ \frac{T_0}{T_M}\boldsymbol{f}_k - \frac{\dot{s}}{s}\boldsymbol{p}_k, & \text{for } E>E_M. \end{cases} \tag{1.47}$$

Because only the product of inverse temperature β and potential energy E enters in the Boltzmann factor (see Eq. (1.5)), a rescaling of the potential energy (or force) by a constant, say α, can be considered as the rescaling of the temperature by $1/\alpha$ [21]. Hence, our choice of $\mathcal{E}_{\rm MUCA}^{\{0\}}(E)$ in Eq. (1.45) results in a canonical simulation at $T=T_1$ for $E<E_1$, a multicanonical simulation for $E_1 \le E \le E_M$, and a canonical simulation at $T=T_M$ for $E>E_M$. Note also that the above arguments are independent of the value of T_0, and we will get the same results, regardless of its value.

For Monte Carlo method, the above statement follows directly from the following equation. Namely, our choice of the multicanonical potential energy in Eq. (1.45) gives (by substituting Eq. (1.46) into Eq. (1.17))

$$W_{\mathrm{MUCA}}(E) \equiv \exp\left[-\beta_0 \mathcal{E}_{\mathrm{MUCA}}^{\{0\}}(E)\right] = \begin{cases} \exp(-\beta_1 E + \mathrm{const}), & \text{for } E < E_1, \\ \frac{1}{n(E)}, & \text{for } E_1 \le E \le E_M, \\ \exp(-\beta_M E + \mathrm{const}), & \text{for } E > E_M. \end{cases} \tag{1.48}$$

We now present the *multicanonical replica-exchange method* (MUCAREM) [13–15]. In MUCAREM the production run is a REM simulation with a few replicas not in the canonical ensemble but in the multicanonical ensemble, i.e., different replicas perform MUCA simulations with different energy ranges. While MUCA simulations are usually based on local updates, a replica-exchange process can be considered to be a global update, and global updates enhance the sampling further.

Let $\mathcal{M}$ be the number of replicas for a MUCAREM simulation. Here, each replica is in one-to-one correspondence not with temperature but with multicanonical weight factors of different energy range. Note that because multicanonical simulations cover much wider energy ranges than regular canonical simulations, the number of required replicas for the production run of MUCAREM is much less than that for the regular REM ($\mathcal{M} \ll M$). The weight factor for this generalized ensemble is now given by (see Eq. (1.32))

$$W_{\mathrm{MUCAREM}}(X) = \prod_{i=1}^{\mathcal{M}} W_{\mathrm{MUCA}}^{\{m(i)\}}\left(E\left(x_{m(i)}^{[i]}\right)\right) = \prod_{m=1}^{\mathcal{M}} W_{\mathrm{MUCA}}^{\{m\}}\left(E\left(x_m^{[i(m)]}\right)\right), \tag{1.49}$$

where we prepare the multicanonical weight factor (and the density of states) separately for $\mathcal{M}$ regions (see Eq. (1.17)):

$$W_{\mathrm{MUCA}}^{\{m\}}\left(E\left(x_m^{[i]}\right)\right) = \exp\left[-\beta_m \mathcal{E}_{\mathrm{MUCA}}^{\{m\}}\left(E\left(x_m^{[i]}\right)\right)\right] \equiv \frac{1}{n^{\{m\}}\left(E\left(x_m^{[i]}\right)\right)}. \tag{1.50}$$

Here, we have introduced $\mathcal{M}$ arbitrary reference temperatures T_m $(= 1/k_\mathrm{B}\beta_m)$ $(m = 1, \ldots, \mathcal{M})$, but the final results will be independent of the values of T_m, as one can see from the second equality in Eq. (1.50) (these arbitrary temperatures are necessary only for MD simulations).

Each multicanonical weight factor $W_{MUCA}^{\{m\}}(E)$, or the density of states $n^{\{m\}}(E)$, is defined as follows. For each m $(m = 1, \ldots, \mathcal{M})$, we assign a pair of temperatures $(T_L^{\{m\}}, T_H^{\{m\}})$. Here, we assume that $T_L^{\{m\}} < T_H^{\{m\}}$ and arrange the temperatures so that the neighboring regions covered by the pairs have sufficient overlaps. Without loss of generality we can assume $T_\mathrm{L}^{\{1\}} < \cdots < T_\mathrm{L}^{\{\mathcal{M}\}}$ and $T_\mathrm{H}^{\{1\}} < \cdots < T_\mathrm{H}^{\{\mathcal{M}\}}$. We define the following quantities:

$$\begin{cases} E_\mathrm{L}^{\{m\}} = \langle E \rangle_{T_\mathrm{L}^{\{m\}}}, \\ E_\mathrm{H}^{\{m\}} = \langle E \rangle_{T_\mathrm{H}^{\{m\}}}, \ (m = 1, \ldots, \mathcal{M}). \end{cases} \tag{1.51}$$

Suppose that the multicanonical weight factor $W_{\text{MUCA}}(E)$ (or equivalently, the multicanonical potential energy $E_{\text{MUCA}}(E;T_0)$ in Eq. (1.18)) has been obtained as in REMUCA or by any other methods in the entire energy range of interest ($E_{\text{L}}^{\{1\}} < E < E_{\text{H}}^{\{\mathcal{M}\}}$). We then have for each m ($m = 1, \ldots, \mathcal{M}$) the following multicanonical potential energies (see Eq. (1.45)) [13]:

$$\mathcal{E}_{\text{MUCA}}^{\{m\}}(E) \equiv \begin{cases} \left.\frac{\partial E_{\text{MUCA}}(E;T_m)}{\partial E}\right|_{E=E_{\text{L}}^{\{m\}}} \left(E - E_{\text{L}}^{\{m\}}\right) + E_{\text{MUCA}}\left(E_{\text{L}}^{\{m\}}; T_m\right), & \text{for } E < E_{\text{L}}^{\{m\}}, \\ E_{\text{MUCA}}\left(E; T_m\right) & \text{for } E_{\text{L}}^{\{m\}} \leq E \leq E_{\text{H}}^{\{m\}}, \\ \left.\frac{\partial E_{\text{MUCA}}(E;T_m)}{\partial E}\right|_{E=E_{\text{H}}^{\{m\}}} \left(E - E_{\text{H}}^{\{m\}}\right) + E_{\text{MUCA}}\left(E_{\text{H}}^{\{m\}}; T_m\right), & \text{for } E > E_{\text{H}}^{\{m\}} \end{cases} \tag{1.52}$$

Finally, a MUCAREM simulation is realized by alternately performing the following two steps.

1. Each replica of the fixed multicanonical ensemble is simulated *simultaneously* and *independently* for a certain MC or MD steps.
2. A pair of replicas, say i and j, which are in neighboring multicanonical ensembles, say m-th and $(m+1)$-th, respectively, are exchanged:

$$X = \left\{\ldots, x_m^{[i]}, \ldots, x_{m+1}^{[j]}, \ldots\right\} \rightarrow X' = \left\{\ldots, x_m^{[j]}, \ldots, x_{m+1}^{[i]}, \ldots\right\}. \tag{1.53}$$

The transition probability of this replica exchange is given by the Metropolis criterion:

$$w\left(X \rightarrow X'\right) = \min\left(1, \exp\left(-\Delta\right)\right), \tag{1.54}$$

where we now have (see Eq. (1.37)) [13]

$$\begin{aligned} \Delta = \beta_m &\left\{\mathcal{E}_{\text{MUCA}}^{\{m\}}\left(E\left(q^{[j]}\right)\right) - \mathcal{E}_{\text{MUCA}}^{\{m\}}\left(E\left(q^{[i]}\right)\right)\right\} \\ - \beta_{m+1} &\left\{\mathcal{E}_{\text{MUCA}}^{\{m+1\}}\left(E\left(q^{[j]}\right)\right) - \mathcal{E}_{\text{MUCA}}^{\{m+1\}}\left(E\left(q^{[i]}\right)\right)\right\}. \end{aligned} \tag{1.55}$$

Here, $E(q^{[i]})$ and $E(q^{[j]})$ are the potential energy of the i-th replica and the j-th replica, respectively.

Note that in Eq. (1.55) we need to newly evaluate the multicanonical potential energy, $\mathcal{E}_{\text{MUCA}}^{\{m\}}\left(E\left(q^{[j]}\right)\right)$ and $\mathcal{E}_{\text{MUCA}}^{\{m+1\}}\left(E\left(q^{[i]}\right)\right)$, because $\mathcal{E}_{\text{MUCA}}^{\{m\}}(E)$ and $\mathcal{E}_{\text{MUCA}}^{\{n\}}(E)$ are, in general, different functions for $m \neq n$.

In this algorithm, the m-th multicanonical ensemble actually results in a canonical simulation at $T = T_L^{\{m\}}$ for $E < E_L^{\{m\}}$, a multicanonical simulation for $E_L^{\{m\}} \leq E \leq E_H^{\{m\}}$, and a canonical simulation at $T = T_H^{\{m\}}$ for $E > E_H^{\{m\}}$, while the replica-exchange process samples states of the whole energy range ($E_{\text{L}}^{\{1\}} \leq E \leq E_{\text{H}}^{\{\mathcal{M}\}}$).

For obtaining the canonical distributions at any intermediate temperature T, the multiple-histogram reweighting techniques are again used. Let $N_m(E)$ and n_m be respectively the potential-energy histogram and the total number of samples obtained with the multicanonical weight factor $W_{MUCA}^{\{m\}}(E)$ $(m = 1, \ldots, \mathcal{M})$. The expectation value of a physical quantity A at any temperature T $(= 1/k_B\beta)$ is then obtained from Eq. (1.25), where the best estimate of the density of states is obtained by solving the WHAM equations, which now read [13]

$$n(E) = \frac{\sum_{m=1}^{\mathcal{M}} N_m(E)}{\sum_{m=1}^{\mathcal{M}} n_m \exp(f_m) W_{\mathrm{MUCA}}^{\{m\}}(E)} = \frac{\sum_{m=1}^{\mathcal{M}} N_m(E)}{\sum_{m=1}^{\mathcal{M}} n_m \exp\left(f_m - \beta_m \mathcal{E}_{\mathrm{MUCA}}^{\{m\}}(E)\right)}, \tag{1.56}$$

where we have for each m $(= 1, \ldots, \mathcal{M})$

$$\exp(-f_m) = \sum_E n(E) W_{\mathrm{MUCA}}^{\{m\}}(E) = \sum_E n(E) \exp\left(-\beta_m \mathcal{E}_{\mathrm{MUCA}}^{\{m\}}(E)\right). \tag{1.57}$$

Note that $W_{MUCA}^{\{m\}}(E)$ is used instead of the Boltzmann factor $\exp(-\beta_m E)$ in Eqs. (1.40) and (1.41).

Moreover, ensemble averages of any physical quantity A (including those that cannot be expressed as functions of potential energy) at any temperature T $(= 1/k_B\beta)$ can now be obtained from the "trajectory" of configurations of the production run. Namely, we first obtain f_m $(m = 1, \ldots, \mathcal{M})$ by solving Eqs. (1.56) and (1.57) self-consistently, and then we have [14]

$$<A>_T = \frac{\sum_{m=1}^{\mathcal{M}} \sum_{k=1}^{n_m} A(x_m(k)) \frac{1}{\sum_{l=1}^{\mathcal{M}} n_l \exp(f_l) W_{\mathrm{MUCA}}^{\{l\}}(E(x_m(k)))} \exp[-\beta E(x_m(k))]}{\sum_{m=1}^{\mathcal{M}} \sum_{k=1}^{n_m} \frac{1}{\sum_{l=1}^{\mathcal{M}} n_l \exp(f_l) W_{\mathrm{MUCA}}^{\{l\}}(E(x_m(k)))} \exp[-\beta E(x_m(k))]}, \tag{1.58}$$

where the trajectories $x_m(k) (k = 1, \ldots, n_m)$ are taken from each multicanonical simulation with the multicanonical weight factor $W_{MUCA}^{\{m\}}(E)$ $(m = 1, \ldots, \mathcal{M})$ separately.

As seen above, both REMUCA and MUCAREM can be used to obtain the multicanonical weight factor, or the density of states, for the entire potential energy range of interest. For complex systems, however, a single REMUCA or MUCAREM simulation is often insufficient. In such cases we can iterate MUCA (in REMUCA) and/or MUCAREM simulations in which the estimate of the multicanonical weight factor is updated by the single- and/or multiple-histogram reweighting techniques, respectively.

To be more specific, this iterative process can be summarized as follows. The REMUCA production run corresponds to a MUCA simulation with the weight factor $W_{\mathrm{MUCA}}(E)$. The new estimate of the density of states can be obtained by the single-histogram reweighting techniques of Eq. (1.27). On the other hand, from the MUCAREM production run, the improved density of states can be obtained by the multiple-histogram reweighting techniques of Eqs. (1.56) and (1.57).

The improved density of states thus obtained leads to a new multicanonical weight factor (see Eq. (1.17)). The next iteration can be either a MUCA production run (as in REMUCA) or MUCAREM production run. The results of this production run may yield an optimal multicanonical weight factor that yields a sufficiently flat energy distribution for the entire energy range of interest. If not, we can repeat the above process by obtaining the third estimate of the multicanonical weight factor either by a MUCA production run (as in REMUCA) or by a MUCAREM production run, and so on.

We remark that as the estimate of the multicanonical weight factor becomes more accurate, one is required to have a less number of replicas for a successful MUCAREM simulation, because each replica will have a flat energy distribution for a wider energy range. Hence, for a large, complex system, it is often more efficient to first try MUCAREM and iteratively reduce the number of replicas so that eventually one needs only one or a few replicas (instead of trying REMUCA directly from the beginning and iterating MUCA simulations).

1.3 Simulation Results

We now present some examples of the simulation results by the algorithms described in the previous section. The computer code developed in Refs. [12, 13, 29, 30], which is based on the version 2 of PRESTO [31], was used after modifications that were necessary for each calculation.

The first example is the C-peptide of ribonuclease A in explicit water [32]. The N-terminus and the C-terminus of the C-peptide analogue were blocked with the acetyl group and the N-methyl group, respectively. The number of amino acids is 13 and the amino-acid sequence is: Ace-Ala-Glu$^-$-Thr-Ala-Ala-Ala-Lys$^+$-Phe-Leu-Arg$^+$-Ala-His$^+$-Ala-Nme [33, 34]. It is known by experiments that this peptide forms α-helix structures [33, 34]. The initial configuration of our simulation was first generated by a high temperature molecular dynamics simulation (at $T = 1{,}000$ K) in gas phase, starting from a fully extended conformation. We randomly selected one of the structures that do not have any secondary structures such as α-helix and β-sheet. The peptide was then solvated in a sphere of radius 22 Å, in which 1,387 water

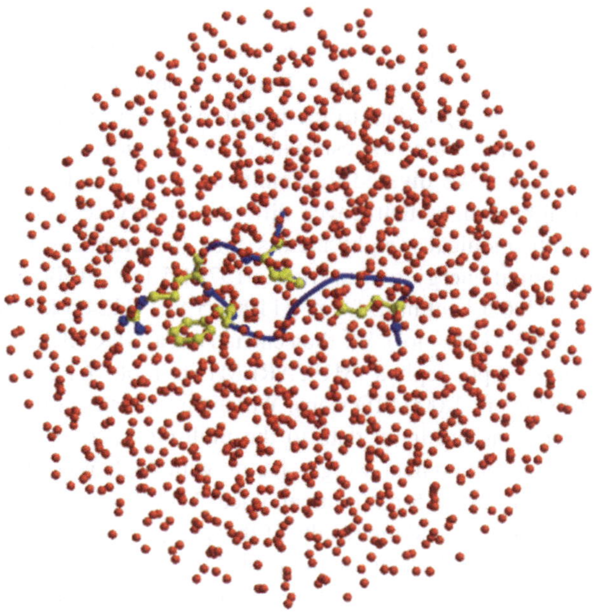

Fig. 1.1 The initial configuration of C-peptide in explicit water, which was used in all of the 32 replicas of the first REMD simulation (REMD1 in Table 1.1). The *red filled circles* stand for the oxygen atoms of water molecules. The number of water molecules is 1,387, and they are placed in a sphere of radius 22 Å. As for the peptide, besides the backbone structure (in *blue*), side chains of only Glu^{-}-2, Phe-8, Arg^{+}-10, and His^{+}-12 are shown (in *yellow*) (Reprinted from Ref. [32] with kind permission of Cell Press (2005))

Table 1.1 Summary of parameters in REMD, MUCAREM, and REMUCA simulations of C-peptide in explicit water[a]

Simulation	Number of replicas, M	Temperature, T_m (K) ($m = 1, \ldots, M$)	MD steps per replica
REMD1[b]	32	250, 258, 267, 276, 286, 295, 305, 315, 326, 337, 348, 360, 372, 385, 398, 411, 425, 440, 455, 470, 486, 502, 519, 537, 555, 574, 593, 613, 634, 655, 677, 700	2.0×10^5
MUCAREM1	4	360, 440, 555, 700	2.0×10^6
REMUCA1	1	700	3.0×10^7

[a]Reprinted from Ref. [32] with kind permission of Cell Press (2005)
[b]REMD1 stands for the replica-exchange molecular dynamics simulation, MUCAREM1 stands for the multicanonical replica-exchange molecular dynamics simulation, and REMUCA1 stands for the final multicanonical molecular dynamics simulation (the production run) of REMUCA. The results of REMD1 were used to determine the multicanonical weight factors for MUCAREM1, and those of MUCAREM1 were used to determine the multicanonical weight factor for REMUCA1

molecules were included (see Fig. 1.1). Harmonic restraint was applied to prevent the water molecules from going out of the sphere. The total number of atoms is 4,365. The dielectric constant was set equal to 1.0. The force-field parameters for protein were taken from the all-atom version of AMBER parm99 [35], which was found to be suitable for studying helical peptides [36, 37], and TIP3P model [38] was used for water molecules. The unit time step, Δt, was set to 0.5 fs. In Table 1.1 the parameter values in the simulations performed are summarized.

We first performed a REMD simulation with 32 replicas for 100 ps per replica (REMD1 in Table 1.1). During this REMD simulation, replica exchange was tried every 200 MD steps. Using the obtained potential-energy histogram of each replica as input data to the multiple-histogram analysis in Eqs. (1.40) and (1.41), we obtained the first estimate of the multicanonical weight factor, or the density of states. We divided this multicanonical weight factor into four multicanonical weight factors that cover different energy regions [13–15] and assigned these multicanonical weight factors into four replicas (the weight factors cover the potential energy ranges from −13791.5 to −11900.5 kcal/mol, from −12962.5 to −10796.5 kcal/mol, from −11900.5 to −9524.5 kcal/mol, and from −10796.5 to −8293.5 kcal/mol). We then carried out a MUCAREM simulation with four replicas for 1 ns per replica (MUCAREM1 in Table 1.1), in which replica exchange was tried every 1,000 MD steps. We again used the potential-energy histogram of each replica as the input data to the multiple-histogram analysis and finally obtained the multicanonical weight factor with high precision. As a production run, we carried out a 15-ns multicanonical MD simulation with one replica (REMUCA1 in Table 1.1) and the results of this production run were analyzed in detail.

In Fig. 1.2 we show the probability distributions of potential energy that were obtained from the above three generalized-ensemble simulations, namely, REMD1, MUCAREM1, and REMUCA1. We see in Fig. 1.2a that there are enough overlaps between all pairs of neighboring canonical distributions, suggesting that there were sufficient numbers of replica exchange in REMD1. We see in Fig. 1.2b that there are good overlaps between all pairs of neighboring multicanonical distributions, implying that MUCAREM1 also performed properly. Finally, the multicanonical distribution in Fig. 1.2c is completely flat between around −13,000 kcal/mol and around −8,000 kcal/mol. The results suggest that a free random walk was realized in this energy range.

In Fig. 1.3a we show the time series of potential energy from REMUCA1. We indeed observe a random walk covering as much as 5,000 kcal/mol of energy range. We show in Fig. 1.3b the average potential energy as a function of temperature, which was obtained from the trajectory of REMUCA1 by the reweighting techniques. The average potential energy monotonically increases as the temperature increases.

The accuracy of average quantities calculated depend on the "quality" of the random walk in the potential energy space, and the measure for this quality can be given by the number of tunneling events [7, 15]. One tunneling event is defined by a trajectory that goes from $E_{\rm H}$ to $E_{\rm L}$ and back, where $E_{\rm H}$ and $E_{\rm L}$ are the values near the highest energy and the lowest energy, respectively, which the random walk can reach. If $E_{\rm H}$ is sufficiently high, the trajectory gets completely uncorrelated when it reaches $E_{\rm H}$. On the other hand, when the trajectory reaches near $E_{\rm L}$, it tends to get trapped in local-minimum states. We thus consider that the more tunneling events we observe during a fixed number of MC/MD steps, the more efficient the method is as a generalized-ensemble algorithm (or, the average quantities obtained by the reweighting techniques are more reliable). Here, we took $E_{\rm H} = -8{,}250$ kcal/mol and $E_{\rm L} = -12{,}850$ kcal/mol for the measurement of the tunneling events. The

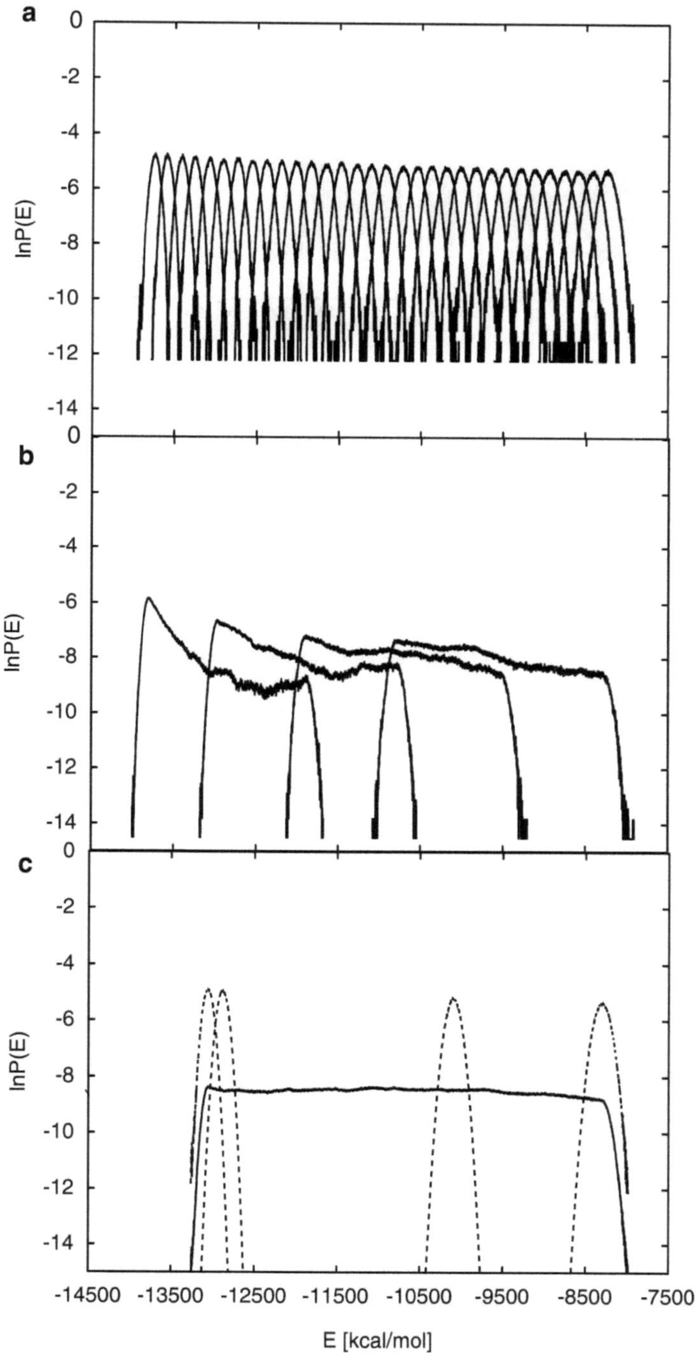

Fig. 1.2 Probability distributions of potential energy of the C-peptide system obtained from (**a**) REMD1, (**b**) MUCAREM1, and (**c**) REMUCA1. See Table 1.1 for the parameters of the simulations. *Dashed curves* in (**c**) are the reweighted canonical distributions at 290, 300, 500, and 700 K (from *left* to *right*) (Reprinted from Ref. [32] with kind permission of Cell Press (2005))

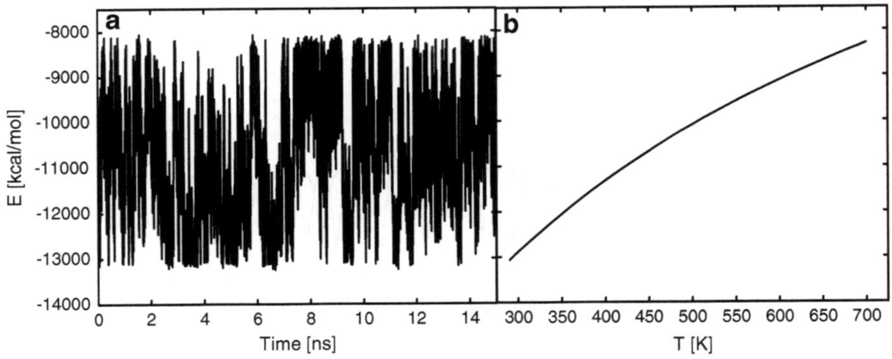

Fig. 1.3 Time series of potential energy of the C-peptide system from the REMUCA production run (REMUCA1 in Table 1.1) (**a**) and the average potential energy as a function of temperature (**b**). The latter was obtained from the trajectory of REMUCA1 by the single-histogram reweighting techniques (Reprinted from Ref. [32] with kind permission of Cell Press (2005))

random walk in REMUCA1 yielded as many as 55 tunneling events in 15 ns. The corresponding numbers of tunneling events for REMD1 and for MUCAREM1 were 0 in 3.2 ns and 5 in 4 ns, respectively. Hence, REMUCA is the most efficient and reliable among the three generalized-ensemble algorithms.

In Fig. 1.4 the potential of mean force along the first two principal component axes at 300 K is shown. There exist three distinct minima in the free-energy landscape, which correspond to three local-minimum-energy states. We show representative conformations at these minima in Fig. 1.5. The structure of the global-minimum free-energy state (GM) has a partially distorted α-helix with the salt bridge between Glu^{-}-2 and Arg^{+}-10. The structure is in good agreement with the experimental structure obtained by both NMR and X-ray experiments. In this structure there also exists a contact between Phe-8 and His^{+}-12. This contact is again observed in the corresponding residues of the X-ray structure. At LM1 the structure has a contact between Phe-8 and His^{+}-12, but the salt bridge between Glu^{-}-2 and Arg^{+}-10 is not formed. On the other hand, the structure at LM2 has this salt bridge, but it does not have a contact between Phe-8 and His^{+}-12. Thus, only the structures at GM satisfy all of the interactions that have been observed by the X-ray and other experimental studies.

The next example is the C-terminal β-hairpin of streptococcal protein G B1 domain [39]. This peptide is sometimes referred to as G-peptide [40] and is known by experiments to form β-hairpin structures in aqueous solution [41, 42]. The number of amino acids is 16 and the amino-acid sequence is: Gly-Glu^{-}-Trp-Thr-Tyr-Asp^{-}-Asp^{-}-Ala-Thr-Lys^{+}-Thr-Phe-Thr-Val-Thr-Glu^{-}. The N-terminus and C-terminus were set to be in the zwitter ionic form ($NH_3{}^{+}$ and COO^{-}), following the conditions in the experiments. GROMOS96 (43a1) force field [43] was used for the solute molecule. SPC model [44] was employed for solvent water molecules according to the GROMOS prescription. We first performed a REMD simulation of

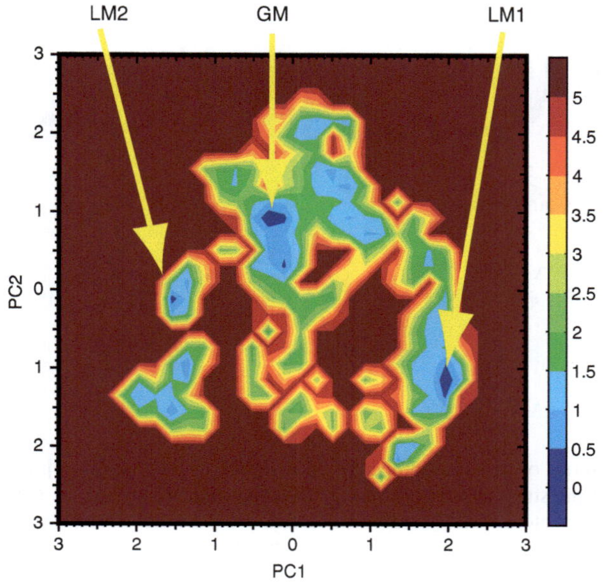

Fig. 1.4 Potential of mean force (kcal/mol) of the C-peptide system along the first two principal components at 300 K. The free energy was calculated from the results of REMUCA production run (REMUCA1 in Table 1.1) by the single-histogram reweighting techniques and normalized so that the global-minimum state (GM) has the value zero. GM, LM1, and LM2 represent three distinct minimum free-energy states (Reprinted from Ref. [32] with kind permission of Cell Press (2005))

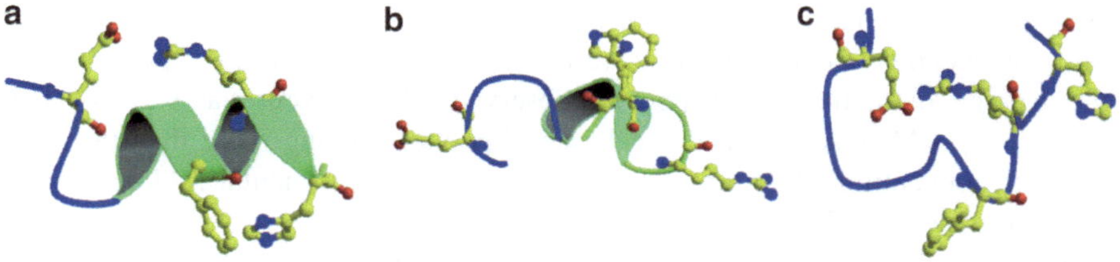

Fig. 1.5 The representative structures at the global-minimum free-energy state ((**a**) GM) and the two local-minimum states ((**b**) LM1 and (**c**) LM2). As for the peptide structures, besides the backbone structure, side chains of only Glu^{-}-2, Phe-8, Arg^{+}-10, and His^{+}-12 are shown in ball-and-stick model (Reprinted from Ref. [32] with kind permission of Cell Press (2005))

G-peptide without explicit solvents from a fully extended polypeptide conformation. In the simulation, we used the distance-dependent dielectric constant. We then selected the final conformation in the replica that was simulated at the highest temperature at the end of the simulation. This conformation was soaked in a water cap whose radius was 26 Å. Before starting the MUCAREM simulation, we performed a 100-ps REMD simulation with 64 replicas twice. (One of them was done for optimization of temperature table for the second REMD.) Using the results of the second REMD, we determined the initial multicanonical weight

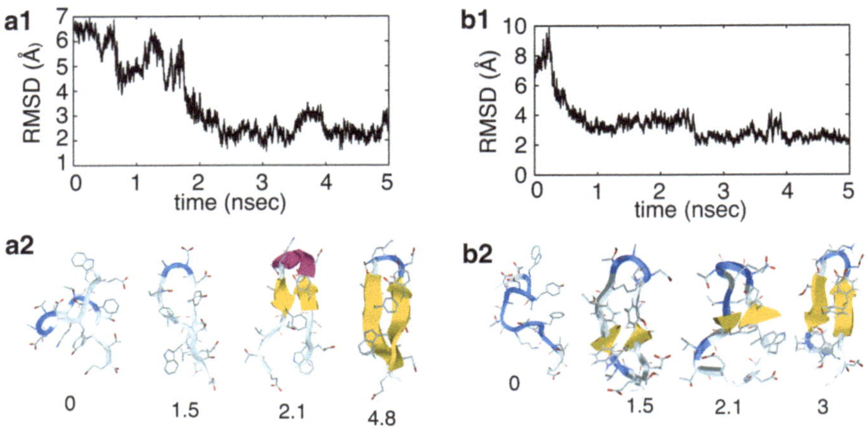

Fig. 1.6 Time series of structural properties of two folding events of G-peptide, resulting in the native-like β-hairpin structures. Those during 5-ns time windows are shown. (**a1**) and (**b1**) are the time series of heavy-atom RMSD values for Replica 4 and Replica 8, respectively. Likewise, (**a2**) and (**b2**) are representative snapshot structures observed in these 5-ns time windows. The numbers written under the snapshot structures represent the time when it was observed (Reprinted from Ref. [39] with kind permission of Wiley (2007))

factor. By iterating cycles of a short MUCAREM with 8 replicas and an update to a new weight factor [15], we refined the multicanonical weight factor. After that we performed a MUCAREM MD with 8 replicas for 34.75 ns (per replica) as a production run. Thus, the total production MD length was 278 ns. In total, three independent folding events were observed in three different replicas. Thus, the average simulation length per one observed folding event was 92.7 ns. This suggests that MUCAREM can accelerate G-peptide folding more than 60 times than the conventional MD simulations, because the experimental folding time of G-peptide is 6 μs [45].

Figure 1.6 shows the time series of the heavy-atom Root Mean Square Deviation (RMSD) from the native configuration (coordinates in the PDB entry 2GB1) and representative snapshot structures observed in the folding events are shown for two replicas. They indeed folded into native-like conformations.

We also evaluated the canonical expectation values of secondary-structure contents (β-bridge contents) of each residue at 320 K using the multiple-histogram reweighting techniques in Eqs. (1.56), (1.57), and (1.58). The results are shown in Fig. 1.7. These results are qualitatively similar to the previous ones that were derived from shorter MUCAREM simulations [36, 37]. They clearly imply that the β-hairpin structures are formed at this temperature.

The third example is the chicken villin headpiece subdomain in explicit water [46]. The number of amino acids is 36. The force field CHARMM22 [47] with CMAP [48, 49] and TIP3P water model [38, 47] were used. The number of water molecules was 3,513. The MD time step was 1.0 fs. We made two production runs of about 1 μs, each of which was a MUCAREM simulation with eight replicas.

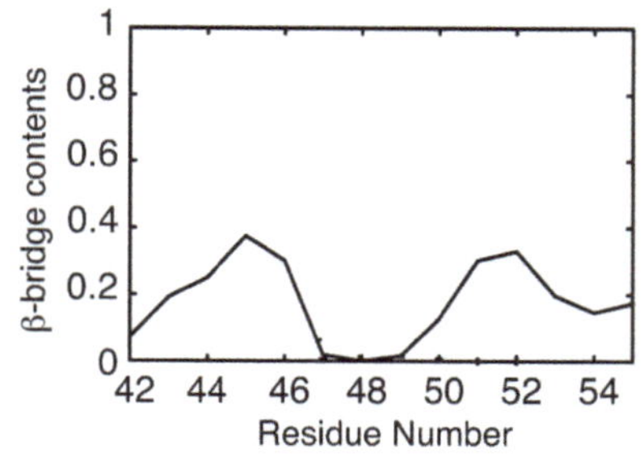

Fig. 1.7 Canonical expectation values of the β-bridge contents of G-peptide at 320 K as a function of the residue number. Values are evaluated by the multiple-histogram reweighting techniques (Reprinted from Ref. [39] with kind permission of Wiley (2007))

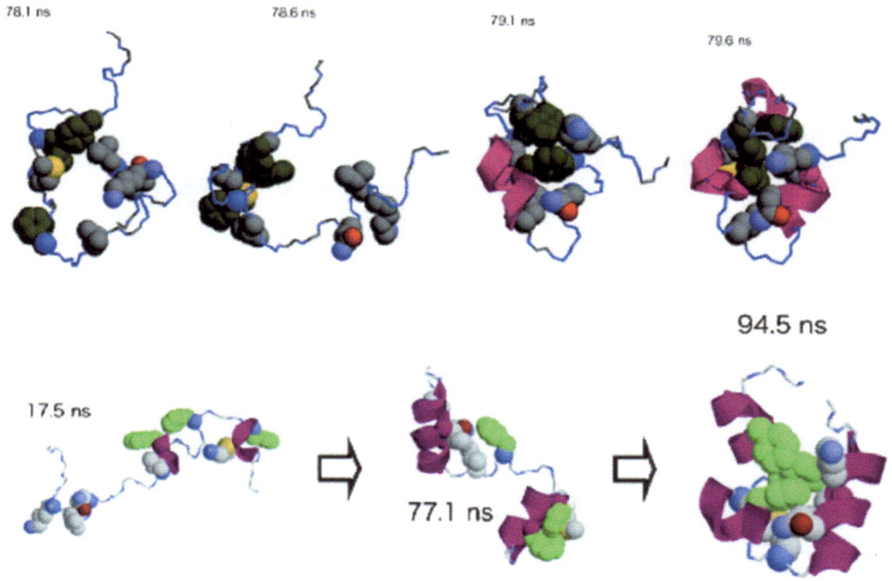

Fig. 1.8 Snapshots of villin headpiece during the MUCAREM production runs that folded into native-like conformations: MUCAREM1 (*above*) and MUCAREM2 (*below*)

They are referred to as MUCAREM1 and MUCAREM2. The former consisted of 1.127 μs covering the temperature range between 269 and 699 K, and the latter 1.157 μs covering the temperature range between 289 and 699 K.

We consider that the backbone folded into the native structure from unfolded ones if the mainchain RMSD becomes less than or equal to 3.0 Å. The folding event is counted separately if it goes through an unfolded structure (with the backbone RMSD greater than or equal to 6.5 Å). With this criterion, we observed 11 folding events in seven different replicas (namely, Replicas 5, 7, and 8 in MUCAREM1 and Replicas 1, 2, 4, and 5 in MUCAREM2). In Fig. 1.8 we show the snapshots of the

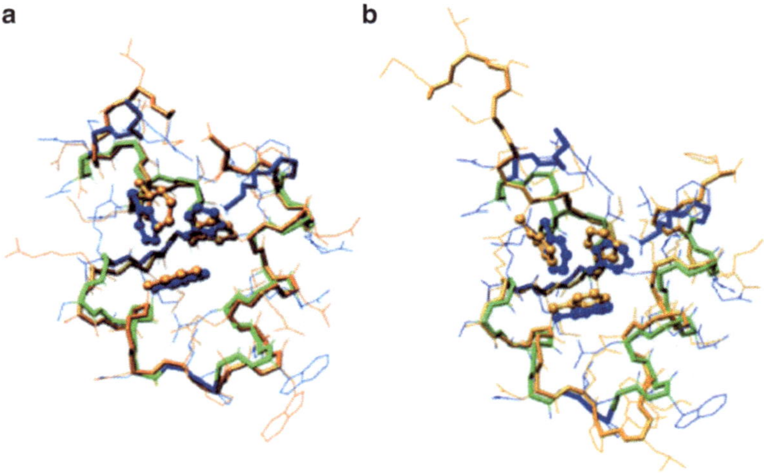

Fig. 1.9 Low-RMSD conformations of villin headpiece subdomain HP36 obtained in MUCAREM1 and MUCAREM2 (colored in *orange*). The X-ray structure (PDB ID: 1YRF) is also superimposed (colored in *blue* and *green*). Here, the α-helices in the X-ray structure are colored in *green* and the rest in *blue*. Three phenylalanine side chains (Phe7, Phe11, and Phe18), which form a hydrophobic core, are shown in *ball*-and-*stick* representation. (**a**) The lowest-bakcbone-RMSD conformation observed in the two MUCAREM production runs (Replica 5 of MUCAREM2). The backbone RMSD value is 1.1 Å (for non-terminal 34 residues). (**b**) A low-RMSD conformation observed in MUCAREM1 (Replica 8). The RMSD value is 1.0 Å for residues 9–32 and 3.3 Å for non-terminal 34 residues (Reprinted from Ref. [46] with kind permission of Cell Press (2010))

replicas folding into native-like conformations for the two MUCAREM production runs. In Fig. 1.9 we compare the obtained low-RMSD conformations and the native structure. They are indeed very close to the native structure.

1.4 Conclusions

In this article we introduced four powerful generalized-ensemble algorithms, namely, multicanonical algorithm (MUCA), replica-exchange method (REM), replica-exchange multicanonical algorithm (REMUCA), and multicanonical replica-exchange method (MUCAREM), which can greatly enhance conformational sampling of biomolecular systems. The results of protein folding simulations by these methods were presented. Because it is very difficult to determine the multicanonical weight factors for very large systems, MUCAREM is the most promising method among the four methods for large biomolecular systems.

Acknowledgments This work was supported, in part, by Grants-in-Aid for Scientific Research on Innovative Areas, "Transient Macromolecular Complex" (Y.S.) and "Fluctuations and Biological Functions" (Y.O.), for Computational Materials Science Initiative (Y.O.), and for High Performance Computing Infrastructure (HPCI) (Y.S. and Y.O.) from MEXT, Japan

References

1. Hansmann UHE, Okamoto Y (1999) New Monte Carlo algorithms for protein folding. Curr Opin Struct Biol 9:177–183
2. Mitsutake A, Sugita Y, Okamoto Y (2001) Generalized-ensemble algorithms for molecular simulations of biopolymers. Biopolymers 60:96–123
3. Sugita Y, Okamoto Y (2002) Free-energy calculations in protein folding by generalized-ensemble algorithms. In: Schlick T, Gan HH (eds) Lecture notes in computational science and engineering. Springer, Berlin, pp 304–332; e-print: cond-mat/0102296
4. Okumura H, Itoh SG, Okamoto Y (2012) Generalized-ensemble algorithms for simulations of complex molecular systems. In: Leszczynski J, Shukla MK (eds) Practical aspects of computational chemistry II: an overview of the last two decades and current trends. Springer, Dordrecht, pp 69–101
5. Sugita Y, Miyashita N, Li P-C, Yoda T, Okamoto Y (2012) Recent applications of replica-exchange molecular dynamics simulations of biomolecules. Curr Phys Chem 2:401–412
6. Berg BA, Neuhaus T (1991) Multicanonical algorithms for 1st order phase transitions. Phys Lett B267:249–253
7. Berg BA, Neuhaus T (1992) Multicanonical ensemble: a new approach to simulate first-order phase transitions. Phys Rev Lett 68:9–12
8. Hansmann UHE, Okamoto Y (1993) Prediction of peptide conformation by multicanonical algorithm – new approach to the multiple-minima problem. J Comput Chem 14:1333–1338
9. Hukushima K, Nemoto K (1996) Exchange Monte Carlo method and application to spin glass simulations. J Phys Soc Jpn 65:1604–1608
10. Marinari E, Parisi G, Ruiz-Lorenzo JJ (1997) Numerical simulations of spin glass systems. In: Young AP (ed) Spin glasses and random fields. World Scientific, Singapore, pp 59–98
11. Hansmann UHE (1997) Parallel tempering algorithm for conformational studies of biological molecules. Chem Phys Lett 281:140–150
12. Sugita Y, Okamoto Y (1999) Replica-exchange molecular dynamics method for protein folding. Chem Phys Lett 314:141–151
13. Sugita Y, Okamoto Y (2000) Replica-exchange multicanonical algorithm and multicanonical replica-exchange method for simulating systems with rough energy landscape. Chem Phys Lett 329:261–270
14. Mitsutake A, Sugita Y, Okamoto Y (2003) Replica-exchange multicanonical and multicanonical replica-exchange Monte Carlo simulations of peptides. I. Formulation and benchmark test. J Chem Phys 118:6664–6675
15. Mitsutake A, Sugita Y, Okamoto Y (2003) Replica-exchange multicanonical and multicanonical replica-exchange Monte Carlo simulations of peptides. II. Application to a more complex system. J Chem Phys 118:6676–6688
16. Sugita Y, Kitao A, Okamoto Y (2000) Multidimensional replica-exchange method for free-energy calculations. J Chem Phys 113:6042–6051
17. Fukunishi F, Watanabe O, Takada S (2002) On the Hamiltonian replica exchange method for efficient sampling of biomolecular systems: application to protein structure prediction. J Chem Phys 116:9058–9067
18. Metropolis N, Rosenbluth AW, Rosenbluth MN, Teller AH, Teller E (1953) Equation of state calculations by fast computing machines. J Chem Phys 21:1087–1092
19. Nosé S (1984) A molecular dynamics method for simulations in the canonical ensemble. Mol Phys 52:255–268
20. Nosé S (1984) A unified formulation of the constant temperature molecular dynamics methods. J Chem Phys 81:511–519
21. Hansmann UHE, Okamoto Y, Eisenmenger F (1996) Molecular dynamics, Langevin and hybrid Monte Carlo simulations in a multicanonical ensemble. Chem Phys Lett 259:321–330
22. Nakajima N, Nakamura H, Kidera A (1997) Multicanonical ensemble generated by molecular dynamics simulation for enhanced conformational sampling of peptides. J Phys Chem B 101:817–824

23. Ferrenberg AM, Swendsen RH (1988) New Monte Carlo technique for studying phase transitions. Phys Rev Lett 61:2635–2638; (1989). *ibid., 63*, 1658
24. Berg BA (2004) Markov chain Monte Carlo simulations and their statistical analysis. World Scientific, Singapore, p 253
25. Berg BA (2003) Multicanonical simulations step by step. Comput Phys Commun 153:397–406
26. Mori Y, Okamoto Y (2010) Replica-exchange molecular dynamics simulations for various constant temperature algorithms. J Phys Soc Jpn 79:074001
27. Ferrenberg AM, Swendsen RH (1989) Optimized Monte Carlo data analysis. Phys Rev Lett 63:1195–1198
28. Kumar S, Bouzida D, Swendsen RH, Kollman PA, Rosenberg JM (1992) The weighted histogram analysis method for free-energy calculations on biomolecules. 1. The method. J Comput Chem 13:1011–1021
29. Sugita Y, Kitao A (1998) Improved protein free energy calculation by more accurate treatment of nonbonded energy: application to chymotrypsin inhibitor 2, V57A. Proteins 30:388–400
30. Kitao A, Hayward S, Go N (1998) Energy landscape of a native protein: jumping-among-minima model. Proteins 33:496–517
31. Morikami K, Nakai T, Kidera A, Saito M, Nakamura H (1992) PRESTO (protein engineering simulator): a vectorized molecular dynamics program for biopolymers. Comput Chem 16:243–248
32. Sugita Y, Okamoto Y (2005) Molecular mechanism for stabilizing a short helical peptide studied by generalized-ensemble simulations with explicit solvent. Biophys J 88:3180–3190
33. Shoemaker KR, Kim PS, York EJ, Stewart JM, Baldwin RL (1987) Tests of the helix dipole model for stabilization of alpha-helices. Nature 326:563–567
34. Shoemaker KR, Faiman R, Schultz DA, Robertson AD, York EJ, Stewart JM, Baldwin RL (1990) Side-chain interactions in the C-peptide helix: Phe 8 ... His 12^{+}. Biopolymers 29:1–11
35. Wang J, Cieplak P, Kollman PA (2000) How well does a restrained electrostatic potential (RESP) model perform in calculating conformational energies of organic and biological molecules. J Comput Chem 21:1049–1074
36. Yoda T, Sugita Y, Okamoto Y (2004) Comparisons of force fields for proteins by generalized-ensemble simulations. Chem Phys Lett 386:460–467
37. Yoda T, Sugita Y, Okamoto Y (2004) Secondary-structure preferences of force fields for proteins evaluated by generalized-ensemble simulations. Chem Phys 307:269–283
38. Jorgensen WL, Chandrasekhar J, Madura JD, Impey RW, Klein ML (1983) Comparison of simple potential functions for simulating liquid water. J Chem Phys 79:926–935
39. Yoda T, Sugita Y, Okamoto Y (2007) Cooperative folding mechanism of a β-hairpin peptide studied by a multicanonical replica-exchange molecular dynamics simulation. Proteins 66:846–859
40. Honda S, Kobayashi N, Munekata E (2000) Thermodynamics of a β-hairpin structure: evidence for cooperative formation of folding nucleus. J Mol Biol 295:846–859
41. Blanco FJ, Rivas G, Serrano L (1994) A short linear peptide that folds into a native stable β-hairpin in aqueous solution. Nat Struct Biol 1:584–589
42. Kobayashi N, Honda S, Yoshii H, Uedaira H, Munekata E (1995) Complement assembly of two fragments of the streptococcal protein G B1 domain in aqueous solution. FEBS Lett 366:99–103
43. van Gunsteren WF, Billeter SR, Eising AA, Hunenberger PH, Kruger P, Mark AE, Scott WRP, Tironi IG (1996) Biomolecular simulation: the GROMOS96 manual and user guide. Vdf Hochschulverlag AG an der ETH, Zurich
44. Berendsen HJC, Postma JPM, van Gunsteren WF, Hermans J (1981) Interaction models for water in relation to protein hydration. In: Pullman B (ed) Intermolecular forces. Reidel, Dordrecht, pp 331–342
45. Munoz V, Thompson PA, Hofrichter J, Eaton WA (1997) Folding dynamics and mechanism of β-hairpin formation. Nature 390:196–199

46. Yoda T, Sugita Y, Okamoto Y (2010) Hydrophobic core formation and dehydration in protein folding studied by generalized-ensemble simulations. Biophys J 99:1637–1644
47. MacKerell AD Jr, Bashford D, Bellott M, Dunbrack RL Jr, Evanseck JD, Field MJ, Fischer S, Gao J, Guo H, Ha S, Joseph-McCarthy D, Kuchnir L, Kuczera K, Lau FTK, Mattos C, Michnick S, Ngo T, Nguyen DT, Prodhom B, Reiher WEIII, Roux B, Schlenkrich M, Smith JC, Stote R, Straub J, Watanabe M, Wiorkiewicz-Kuczera J, Yin D, Karplus M (1998) All-atom empirical potential for molecular modeling and dynamics studies of proteins. J Phys Chem B 102:3586–3616
48. MacKerell AD Jr, Feig M, Brooks CL III (2004) Improved treatment of the protein backbone in empirical force fields. J Am Chem Soc 126:698–699
49. Mackerell AD Jr, Feig M, Brooks CL III (2004) Extending the treatment of backbone energetics in protein force fields: limitations of gas phase quantum mechanics in reproducing protein conformational distributions in molecular dynamics simulations. J Comput Chem 25:1400–1415

Chapter 2
Application of Markov State Models to Simulate Long Timescale Dynamics of Biological Macromolecules

Lin-Tai Da*, Fu Kit Sheong*, Daniel-Adriano Silva*, and Xuhui Huang

Abstract Conformational changes of proteins are an essential part of many biological processes such as: protein folding, ligand binding, signal transduction, allostery, and enzymatic catalysis. Molecular dynamics (MD) simulations can describe the dynamics of molecules at atomic detail, therefore providing a much higher temporal and spatial resolution than most experimental techniques. Although MD simulations

*Author contributed equally with all other contributors.

L.-T. Da
Department of Chemistry, The Hong Kong University of Science and Technology, Clear Water Bay, Kowloon, Hong Kong

Skaggs School of Pharmacy and Pharmaceutical Sciences, University of California, San Diego, La Jolla, CA 92093, USA

F.K. Sheong
Department of Chemistry, The Hong Kong University of Science and Technology, Clear Water Bay, Kowloon, Hong Kong

D.-A. Silva
Department of Chemistry, The Hong Kong University of Science and Technology, Clear Water Bay, Kowloon, Hong Kong

Department of Biochemistry, University of Washington, Seattle, WA 98105, USA

X. Huang (✉)
Department of Chemistry, The Hong Kong University of Science and Technology, Clear Water Bay, Kowloon, Hong Kong

Division of Biomedical Engineering, The Hong Kong University of Science and Technology, Clear Water Bay, Kowloon, Hong Kong

Center of Systems Biology and Human Health, School of Science and Institute for Advance Study, The Hong Kong University of Science and Technology, Clear Water Bay, Kowloon, Hong Kong
e-mail: xuhuihuang@ust.hk

K. Han et al. (eds.), *Protein Conformational Dynamics*, Advances in Experimental Medicine and Biology 805, DOI 10.1007/978-3-319-02970-2_2,

have been widely applied to study protein dynamics, the timescales accessible by conventional MD methods are usually limited to timescales that are orders of magnitude shorter than the conformational changes relevant for most biological functions. During the past decades great effort has been devoted to the development of theoretical methods that may enhance the conformational sampling. In recent years, it has been shown that the statistical mechanics framework provided by discrete-state and -time Markov State Models (MSMs) can predict long timescale dynamics from a pool of short MD simulations. In this chapter we provide the readers an account of the basic theory and selected applications of MSMs. We will first introduce the general concepts behind MSMs, and then describe the existing procedures for the construction of MSMs. This will be followed by the discussions of the challenges of constructing and validating MSMs, Finally, we will employ two biologically-relevant systems, the RNA polymerase and the LAO-protein, to illustrate the application of Markov State Models to elucidate the molecular mechanisms of complex conformational changes at biologically relevant timescales.

Keywords Markov State Models • Molecular dynamics simulations • Free energy landscape • Molecular recognition • Biological macromolecules • Proteins • RNA polymerase

2.1 Introduction

Conformational changes are known to be critical in many biological processes such as protein folding, ligand binding, signal transduction, allostery, and enzymatic catalysis [1–4]. Due to its significance in biological field, a great amount of research attention has been drawn to investigate conformational changes in biological macromolecules. Throughout the decades of conformational studies in biomolecular systems, X-ray crystallography [5] has evolved and thus has already revolutionized our understanding on the atomic-level structural details as well as the functions of protein, DNA and RNA. More recently, the emergence of Cryo-electron microscopy [6] and small-angle X-ray scattering techniques [7] has further boosted the advance in the understanding of complex biomolecular structures. Despite their remarkable success in the field, these experimental methods can only provide static snapshots of the molecules under study but not the details of the conformational dynamics. To overcome this limitation, alternative experimental methods, including nuclear magnetic resonance spectroscopy (NMR) [8] and several different fluorescence spectroscopic techniques [9–13], are routinely used to study the dynamics of protein ensembles in real time. Even so, the atomic-level details of conformational changes in biological macromolecules are still hard to capture, mainly due to the fast-dynamics and microscopic nature of these systems.

Molecular dynamics (MD) simulations are a computational technique that can complement experiments and address the aforementioned issues. MD is a simulation technique based on Newton's equations of motion and in the recent years it has attracted great attention due to its ability to simulate dynamics of biological

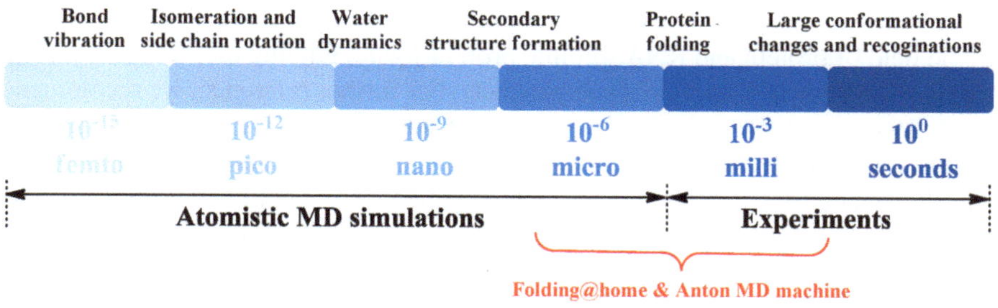

Fig. 2.1 The timescales gap between the conformational changes accessible by conventional MD simulations and the relevant biological functions observed experimentally for biomolecules. The picture illustrates the notion of timescale gap between the theory (MD simulations) and experiments, however, (*red color key*) the Folding@home project (using massive crowdsourcing computing) [22] and the specialized MD simulator machine Anton [23], have been able to MD simulate experimentally relevant timescales (μs-ms) by using its massive computational resources. However, these kinds of resources are not accessible to everyone and currently its capacity is indeed restricted to relatively small biomolecules

macromolecules [14, 15]. MD can describe the molecules dynamics in atomic detail, which is of a much higher spatial resolution than most experimental techniques. In the past decades great progress has been made in the development of force fields used for MD simulations of biologically relevant macromolecules, which had led to a more accurate description of the dynamics of protein, DNA and RNA. Furthermore, the exponential increase of computing power [16] and the development of crowd-sourcing computing [17], has allowed in recent years the simulation of biological macromolecules at timescales ranging from nanoseconds (ns) to microseconds (μs). In a limited number of cases, it has even been possible to simulate the millisecond timescale of small proteins with the aid of specialized MD computers [18], an unprecedented timescale [19, 20] that, for the first time, has permitted comparisons with experimental data and to elaborate hypothesis about the mechanisms of protein's function.

Although MD simulations have been widely applied to study protein dynamics at an atomic-level of detail, the timescales accessible by conventional MD methods are usually limited to the timescale that is orders of magnitude shorter than the conformational changes relevant to the biological function. In order to bridge this "timescale gap" [21] (see Fig. 2.1), many research efforts have been devoted to the development of theoretical methods that aim at faster sampling of the conformational space [24], some examples include: steered [25], targeted [26], accelerated [27, 28], replica exchange MD simulations [29] and metadynamics [30].

In contrast to these enhanced sampling techniques, a mathematical framework known as Markov State Model (MSM) has recently caught researchers' attention due to its potential to bridge the timescale gap. With the application of such framework, system dynamics at long timescales can be predicted by performing only short (time conventional) MD simulations [31–36]. The emergence of this promising method has thus opened the door of extracting at-equilibrium models of the complete energy landscape of biomolecules.

A number of successful examples of applying MSM have already been reported in the field of protein and RNA folding, protein-ligand binding mechanisms and the release of enzymatic reaction's sub-products. For further reference, we recommend the following notable recent examples: the Pande group has described the protein folding process of the Villin's headpiece [37], lambda repressor [38], NTL9 [20] and showed that for some protein the folded native-states are kinetic-hubs [39]; The Huang group used MSM to study the folding of small RNA hairpins [40], ligand binding mechanism of a periplasmic binding protein (PBP) [41] and the release of pyrophosphate ion from the active site of the yeast RNA polymerase II (Pol II) [42] and bacterial RNA polymerase [43]; the Noé group has used MSM to understand the folding mechanism of the PinWW protein [44]; while Bowman and Geissler have described a novel method to identify hidden allosteric sites in proteins based mainly on MSMs [45]. All these studies have demonstrated a good agreement with the available experimental observations.

In view of this widespread interest in the application of MSM to biomolecular studies, we will hereby give a general account of the theories as well as applications in the chapter. We will first introduce the basic concepts behind MSM and describe the detailed procedures for its construction, we will then illustrate the challenges of generating and validating a MSM, and finally we will employ two biological systems, the RNA polymerase (bacterial and eukaryotic) and the Lysine-, Arginine-, Ornithine-binding protein (LAO protein), as examples to explain the practical details of how MSM can be used to extract relevant kinetic and at-equilibrium information from an ensemble of short MD simulations.

2.2 Modeling the Dynamics of Biomolecules

Macromolecules, due to their high degrees of freedom and complicated molecular interaction, have numerous free energy minima in their conformational free energy landscape. In general, relatively low free energy barriers (within the order of several kcal/mol) separate these free energy minima. Because molecules are dynamic in nature at any temperature above absolute zero, and the amplitude of these motions increases with temperature, thermal fluctuation at biologically relevant temperatures are usually sufficient for the system to overcome these low conformational free energy barriers. Therefore, in most cases (if not all), at biologically relevant temperatures, what is called the native state of a protein is actually composed by a collection of protein conformations in dynamic equilibrium. To model the kinetics of such systems, a common approach is to divide the conformational space into a set of discrete states that are kinetically metastable in nature [31, 33], each corresponding to a free energy minima (or a grouped set of connected free energy minima). Therefore, the transitions between these metastable states can be approximated as the transitions between states in a kinetic scheme.

If the probability distribution of any future state $X(t+\Delta t)$ depends only on the present state $X(t)$ the transition process is known as Markovian, also sometimes dubbed as "memoryless". Such Markovian process in a kinetic scheme can be described by the memoryless Master equation:

$$\frac{dX(t)}{dt} = X(t)K \tag{2.1}$$

with $X(t)$ being an n-dimensional row vector describing the probability for the n-states to be occupied at time t. K is the rate matrix, where K_{ij} is the rate constant for the transition from state i to j. The diagonal elements of K are defined such that $K_{ii} = -\sum_{i \neq j} K_{ij}$ in order to have conservation of mass. This memoryless approximation is the underlying reason that allows modeling of macromolecular dynamics with MSM, which will be discussed in detail in the following section.

2.3 Markov Chain

To aid readers' understanding in the application of MSMs, some basic knowledge of Markov process will be first presented. A Markov Model (named in honor to Andrey Markov, who develop the theory of stochastic processes) defines mathematically a finite system (described by states) with transitions from one state to another. In this stochastic model, the fundamental assumption is that the population distribution $X(t)$ is sufficient to determine any later distribution $X(t+\Delta t)$ where $\Delta t > 0$. Under this model the states evolves over time in a probabilistic manner, and the distribution of states $X(t+\Delta t)$ after each Δt (namely propagation) depends only on its previous distribution $X(t)$, but not on any state before that. This is consistent with the "memoryless" approximation mentioned in the previous section, and thus MSM can be applied in the description of kinetics of macromolecular systems. Currently the prevailing type of Markov Model applied in macromolecular studies is known as Markov Chain, which considers an autonomous (no external contribution) process with fully observable states (occupancy of every states in the model are transparent to the observers). In a time-continuous Markov chain, the interval of propagation steps Δt is infinitesimally small such that the stochastic process can be represented as a continuous propagator. However, in the case of the applications of Markov Models for the analysis of data from MD simulations, due to the fact that MD simulations are intrinsically discrete in nature, the model most frequently employed is a discrete-time homogeneous Markov chain model in which the propagation only occurs as discrete steps. More details of this kind of MSM will be illustrated in detail in the following example.

2.4 The Transition Probability Matrix

Let us consider a simple case, suppose that a protein has three metastable states, namely: state 1, 2 and 3. If transitions only occur stochastically at some discrete time, the system can be modeled as a discrete-time Markov chain. In our example, we assume the propagation steps are equally spaced at an interval τ and the transition probabilities per propagation step are time-homogeneous (i.e. the transition probabilities depend only on Δt but not on t). Now, assume that the transition probabilities per propagation time step τ (also known as lag time) between any pair of these three states are known to be those listed in the following 2D matrix:

State	*1*	*2*	*3*
1	0.65	0.28	0.07
2	0.15	0.67	0.18
3	0.12	0.36	0.52

The previous matrix is known as the transition probability matrix (TPM). In a TPM we use the symbol p_{ij} to represent the transition probability from state i to state j. Hence, the probability of the transition from state 2 to 3 is represented as p_{23}. From the TPM above we know that $p_{23} = 0.18$. In terms of an MSM, the transition probability matrix ($\boldsymbol{P}$) is a row-normalized matrix ($\sum_j P_{ij} = 1, \forall\, i$), because the elements p_{ij} in each row represent the probability of a state i to transition to different states j and the summation $\sum_j P_{ij}$ is the probability to have a transition originating from state i to any state j. Please note that some literature uses a column-normalized matrix P^t instead of the row-normalized matrix, but in this chapter we will conform to the row-normalized matrix definition.

2.5 Propagation of the Markov Chain

As mentioned before, a Markov chain must meet the requirement that the probability of any state after the chain propagation is independent of all but the previous state. For the previous example, suppose that the initial distribution probabilities of the states follow the row vector $X(0) = [0.21, 0.68, 0.11]$ (i.e. 0.21 of the population is in state 1 and so forth). Because p_{ij} refers to the conditional probability of the transition from state i to j, the distribution after one chain propagation (Δt) can be calculated by: $X(0)P(\tau) \approx [0.25, 0.55, 0.19]$. In a similar way, the distribution after two chain propagations is determined as $\mathrm{X}(0)\mathrm{P}(\tau)^2$. Therefore, we can write the distribution vector after the time $n\tau$ as:

$$X\left(n\tau\right) = X(0)\left[P\left(\tau\right)\right]^n \tag{2.2}$$

Given Eq. (2.2) and a vector of initial population distributions for the system states, it is possible to predict the evolution of the system on a longer timescale by the simple exponentiation of probability matrix. The Eq. (2.2) is actually equivalent to Eq. (2.1), but at discrete times, and they are related by the expression: $P(\tau) = e^{\tau K}$ [32, 43].

2.6 Constructing a TPM from MD Simulations

TPM is the fundamental part of a discrete-time homogeneous MSM, because a vector of state probabilities can be propagated forward in time by simply multiplying it to the transition probability matrix. The construction of TPM is therefore the most important process in the construction of a discrete-time MSM. To construct the TPM in practice from MD trajectories, one has to first perform space discretization to group conformations in the trajectories together (details of the technique will be discussed in later sections of the chapter), because only if we consider a group of conformations as oppose to individual conformation we can empirically determine the observed conditional probability for a transition event to occur. With the conformational space properly discretized (in either microstate or macrostate level), a transition count matrix (TCM) N is then constructed by counting the total number of transitions (n_{ij}) from state i to j observed in all MD trajectories within a certain lag time τ. From the principle of detailed balance, the TCM obtained should be symmetric because all elementary transitions should be reversible under equilibrium condition. Yet due to the fact that equilibrium sampling is almost never reached in simulations (thus the need of MSM for equilibrium studies), the TCM is usually not strictly symmetric. In these cases, the TCM can be symmetrized by:

$$N^{symm} = \frac{N + N^{\mathrm{T}}}{2} \tag{2.3}$$

The TPM is then formulated by normalizing each row of the symmetrized TCM by:

$$\mathrm{P_{ij}} = \frac{\mathrm{N}_{\mathrm{ij}}^{symm}}{\sum\left(\mathrm{N}_{\mathrm{ij}}^{symm}\right)} \tag{2.4}$$

Simply symmetrizing the transition count matrix is the most trivial way to impose the detailed balance condition, but may introduce errors when the number of inter-state count is small or rather un-symmetric. Noé has introduced an algorithm to approximate the transition probability matrix induced by the observed count matrix. Under the framework of the Bayesian Inference, a distribution of transition probability matrices (posterior distributions) can be obtained using a Metropolis Monte Carlo scheme, subject to the constraint of the detailed balance with the observed transition count matrix as the maximum likelihood [32].

2.7 Considerations of TPM Construction

With a properly constructed TPM, it is possible to derive a MSM that can be used to understand the time-evolution of system. However, in spite of this attractive feature of MSM, it is necessary to understand that such property is founded on the Markovian assumption, but the assumption does not necessarily hold true for a kinetic scheme obtained from MD simulation. This is due to the fact that upon space discretization during the clustering step, one will introduce discretization error in the model [46]. This error is mainly produced by the existence of small internal free energy barriers inside of the discrete states, which will give rise to differences in the dynamics among the conformations existent within the state (due to the inertial effects of molecule dynamics at short timescales). In order to reduce such error, a finer discretization can be performed or a longer lag time could be chosen. None of these approaches is perfect, a finer discretization could reduce the differences in dynamics among conformations within a state, but at the same time the statistical significance of each cluster is reduced. On the other hand, by coarse-graining the simulations time into long time steps, the system can have more time to "lost its memory" so as to achieve better Markovianity. For example, if we consider the limit, theoretically any model will be Markovian at the infinity limit, despite the fact that this scenario is unfeasible in any time-dependent simulation. Yet if we could achieve perfect Markovianity under such condition, the dynamic information of the system will be completely lost. Therefore, in order to generate an insightful Markov model using MD data, apart from choosing a lag-time long enough to give a good approximation of the Markovian condition, it is equally necessary to choose a lag-time that is short enough to be useful (few ps-ns in most cases). In other words, when building an MSM from MD simulations, one always has to face the tradeoff among Markovianity, spatial resolution and preserving certain timescale-resolution of the dynamics. If we try to take into account these constraints as well as our aim of representing the energy landscape with a Markovian kinetic scheme, we can deduce one possible balance between these factors could be that the state defined in the model should be metastable (thus correspond the minima of the energy landscape), the intra-state relaxation times (the time a state takes for a conformation inside the state to transit to other conformations within the state and lost the memory) is minimal (thus all conformations can have similar kinetic behavior via fast interconversion) and the interstate transition times (the time that takes for a conformation to transit to a conformation in other state) is maximal (or in other words, high barriers lie between states), this implies the generation of metastable states without internal high free energy barriers [33, 34, 40].

If we examine closely the protocol presented above, one can discover that under the Markovian assumption the construction of a TPM does not require individual MD simulations to visit all the metastable regions in the free energy landscape [34, 44, 47]. Instead, only probabilities of local interstate transitions are necessary for constructing TPM and the corresponding MSM. This actually lessens the burden of computational cost to strive for a converged sampling with long trajectories, as many short sampling-trajectories (just long enough to sample interstate transitions) are all that is required. Furthermore, apart from standard MD simulation, there are other sampling methods that can be applied in order to generating descriptions of the energy landscape that can be used in a MSM construction (e.g. Monte Carlo simulations), but those are beyond the scope of this chapter.

Now, the only remaining question for MSM construction is how to define the micro- and macro-states that are used to calculate the TPM.

2.8 Free Energy Landscapes of Biomolecules and Its Relation to Microstates, Macrostates and MSM

From the notion above, it can be understood that a proper partitioning of the conformational space such that the metastable states correspond to distinctive energy minima is necessary for generating a TPM that can give a valid MSM. Usually such partitioning is achieved in two stages, namely the microstate clustering and the macrostate lumping. The microstate clustering aims to generate clusters of conformations fulfilling the criteria that conformations within each cluster have similar kinetic behavior, while the macrostate lumping aims at putting the clusters generated in the previous stage together, in order to give place to larger groups, each composed of several microstates, such that the major energy barriers in the system lie between macrostates. In the case of the microstates, if we consider that an ensemble of converged MD simulations represents an "energy landscape" [44, 47], then we can assume that it is possible to generate a partition of the system just by grouping conformations that are related kinetically at short time-intervals. If the partition of the microstates is fine enough (see Fig. 2.2), a group of the microstates will correspond/minimize to the same energy-minimum basin [31, 33, 34].

Therefore, at the macrostate lumping stage (see Fig. 2.2), one can look at the transition/kinetics between the microstates in order to connect those microstates

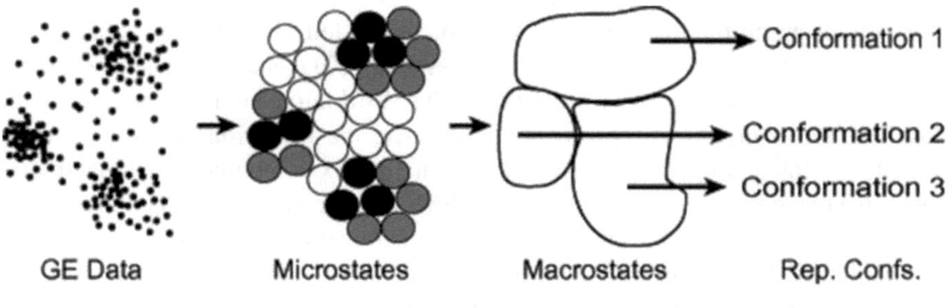

Fig. 2.2 The steps required when building a MSM. The conformations (GE data, represented by *points*) obtained from the MD simulations are firstly grouped into microstates; next, the structures are clustered in microstates based on its degree of geometric similarity. Next, the microstates are further lumped into several kinetically related macrostates (Figure adapted from reference [34])

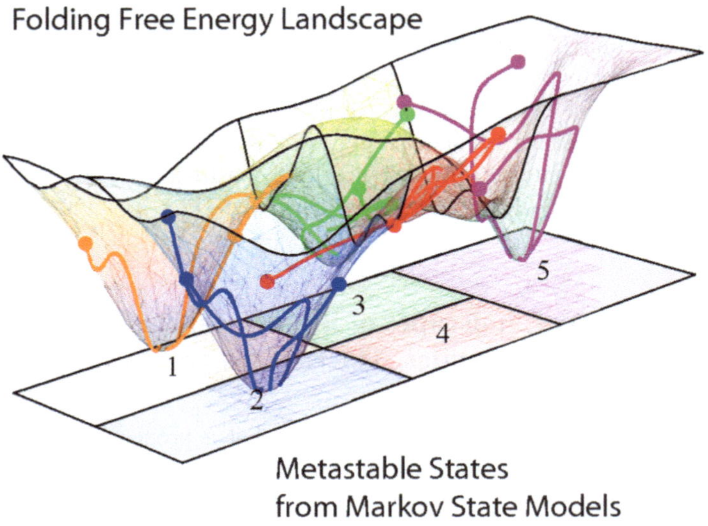

Fig. 2.3 Relationship between energy landscape, MD trajectories and metastable states. The schematic 3D free-energy landscape, comprised by 5 energy minimums, represents the conformational space of a certain protein. The different lines illustrate the behavior of 3 hypothetical MD trajectories started from different energy minima. It can be appreciated that some trajectories are able to overcome the energy barrier, escaping from its starting energy minimum to enter into another minimum, but none of the independent trajectories is able to visit the complete energy landscape (i.e. all the relevant protein conformations). Nevertheless, if considered as a conjunct the MD simulations had in fact visited all the energy landscape. The 2D projection in the bottom shows an idealized discretization of this energy landscape into the corresponding 5 metastable states (Figure adapted from reference [47])

separated by low-energy barriers (i.e. those with fast inter-microstates transitions) into a single metastable state (macrostate). In this way, by first partitioning the conformations into microstates and then lumping the microstates into macrostates the complete energy landscape can be partitioned into a small set of metastable states (see Fig. 2.3) [34], such macrostate division of the energy landscape is not only representing the underlying kinetics of the system, but also by reducing the number of states in the system (usually to less than 100) it is easier to analyze and in many cases it is also possible to extrapolate the MSM into a human-comprehensible fashion (e.g. a graphical representation showing the transitions between states). Finally, by calculating the transition probability matrix at the micro and macrostate level, we can construct and validate the MSM, which can be used to extract useful thermodynamic and kinetic properties of the dynamic process that we are interested. As explained before, any of the two models: micro- and macro-state level can be valid, however one should choice between the finer and the coarser model based. Nevertheless, usually the macrostate model is the common choice, since it is simpler, easier to analyze and its statistical certainty is intrinsically higher.

2.9 Microstate Clustering of MD Conformations

Although there is a lack of consensus about the best method to cluster kinetically related conformations, the most usual methods are based on some sort of geometrical clustering. The assumption behind structural clustering is that structures closely related in the geometrical space should also be closely related in the kinetic space, hence, grouping structures that are close in geometry will approximately give structures that are close in dynamics. Several structure-based clustering methods are already available, with the most fundamental and widely used are: K-centers, K-means and K-medoids clustering. The common goal for these three methods is to partition a set of n conformations into k mutually exclusive partitions C_1, $C_2, \ldots, C_k$. These k partitions are then used for macrostate lumping in the later stage.

K-centers clustering aims at find k "centers" (see Fig. 2.4A), which is defined by a subset S from the set of points V such that $|S| = k$ and minimizing the expression:

$$\max_{v \in V} \min_{s \in S} (v, s) \tag{2.5}$$

or, in simple words, find k points from the dataset such that the longest distance between any point to its closest corresponding center is minimized. The k partitions can then be obtained by assigning all points into their closest corresponding centers to form k mutually exclusive groups.

The k-centers problem is actually NP-hard, which implies that solving the exact solution is computationally expensive. In real practice though, k-centers clustering algorithm usually refers to an approximate algorithm shown below:

1. Randomly select one conformation as the center of the first microstate k_1.
2. Calculate the distance $d(x_i, k_1)$ between each of the conformations x_i in the dataset and k_1.
3. Choose the conformation with the largest $d(x_i, k_1)$ value as the second microstate center k_2.
4. Reassign the conformations in the dataset to the new cluster if the distance to the new cluster center is shorter than the distance to any other cluster centers (i.e. for a new cluster center k_2, conformation x_i is assigned to C_2 if $d(x_i, k_2)$ is shorter than $d(x_i, k_1)$.
5. Then choose the next cluster center that is furthest from the all previous centers and repeat step 4.
6. Repeat the same procedure until the desired number of microstates is obtained.

The k-centers clustering method can create clusters with an approximately equal geometric volume. Moreover, the clustering speed can be greatly improved by applying triangle inequality in the step of cluster assignment, which has been currently implemented in the MSMBuilder package [34, 50, 51].

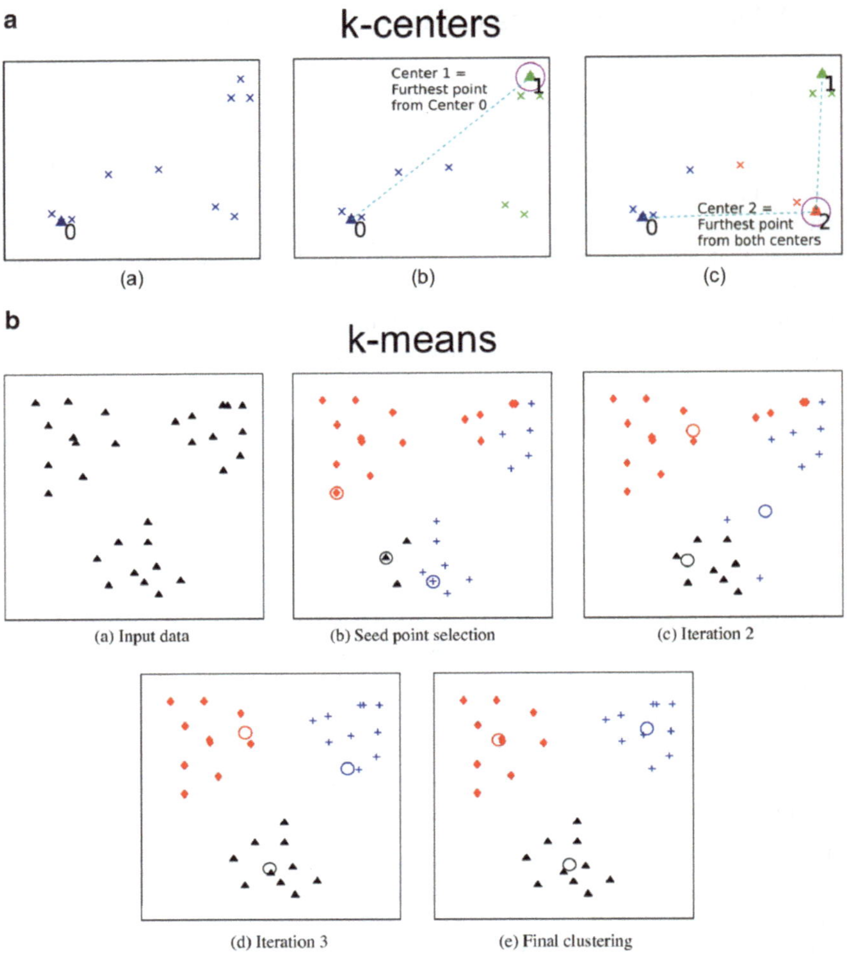

Fig. 2.4 (**A**) **Illustration of an approximate k-centers clustering algorithm.** The process of generating k geometric groups from a given dataset with the approximate k-centers algorithm is illustrated as follows: (from *left* to *right*) (*a*) From the given data points, choose a random point as the first cluster center. (*b*) Measure the distance of all points against the first center and choose the one with the furthest distance as the second cluster center. Assign all points to their closest cluster center such that all points are divided into two clusters ("partitions" in the mathematical sense), illustrated here with two different colors. (*c*) Measure the distance of all points against their assigned cluster centers, find the point with the maximum distance (i.e. furthest from all existing center) as the next cluster center. Re-assign all points to their closest centers into partitions. Repeat until the desired number of clusters k is obtained. The final partitioning is used as the geometrical grouping of the points (Figure adapted from reference [48]). (**B**) **Illustration of an approximate k-means clustering algorithm.** K-means algorithm attempts to divide the given dataset (*a*) into k geometric partitioning in the following way (From *left* to *right*, *top* to *bottom*). (*b*) From the data points, randomly choose k points as initial centers (*circled*). Assign all points to their closest corresponding centers into k partitions, shown here in different colors. (*c–d*) For each partition, take the "mean" position of the points within the group as the updated center position (*circles*). Re-assign all the points again to the new centers. (*e*) Repeat the process until no change in the cluster assignment is observed. The final cluster assignment is taken as the geometrical partitioning of the points (Figure adapted from reference [49])

K-means clustering refers to something very different from k-centers clustering (see Fig. 2.4B). Instead of aiming solely at points to be centers, the k-means clustering attempts to find k partitions so as to minimize:

$$\sum_{K=1}^{k} \sum_{x_i \in C_k} \sqrt{\left(x_i - \overline{k}_j\right)^T \left(x_i - \overline{k}_j\right)} \tag{2.6}$$

where $\overline{k}_j = \frac{1}{N_k} \sum_{x_i \in C_k} x_i$. In other words, the points are put into k partitions so that the sum of distances of all points to the partition average of their assigned partitions is minimized.

Just like k-centers, k-means is also an NP-hard problem, and so an approximation is also needed. A commonly used approximate k-means clustering protocol is illustrated here:

1. Instead of using one conformation as the first microstate center, k conformations are (randomly) chosen as the initial centers for the k microstates.
2. Calculate the Euclidean distance between every conformations in the dataset to each centers defined in step 1.
3. Assign the conformation to the microstate with the minimum distance.
4. Determine the mean vectors by averaging the distance vector for all the conformations within each microstate, and using the mean vector as the new center.
5. Repeat step 2, 3 and 4 until the clustering process is converged. That is, the new round of iteration does not change the assignments of any conformations from the previous iteration.

Despite the popularity of k-means cluster, this clustering technique is actually sensitive to the low density regions and tends to lump the points from the low density regions into the clusters from high density regions, which in the context of microstate clustering leads to the incorrect description of the some interesting states such as the transition states. A clustering algorithm that closely resembles k-means, which is known as k-medoids clustering, can overcome the above drawbacks of k-means by taking actual data points ("medoids") instead of the means of the partitions as centers. Unfortunately, k-medoids also has its own limitations, such as being inefficient for large data sets and offer a poor control of the cluster size.

2.10 Implied Timescales and Number of Macrostates

With the microstates generated from the first stage of clustering, we can construct the TPM based on the transitions between these states. Then if we attempt to do eigenvector decomposition of the TPM:

$$X_i P(\tau) = \mu_i X_i \tag{2.7}$$

where each eigenvector X_i actually corresponds to a certain state distribution that can give rise to a sustainable mode of transitions between groups of states, with the signed structure of the eigenvector indicating the two groups between which the transition occurs. The eigenvalue of each mode can be interpreted as reflecting the decay of the occupancy of the mode (N_i). If initially, the occupancy of the mode i is taken as:

$$N_i(0) = 1 \tag{2.8}$$

At $t = \tau$, the occupancy of the mode i then become:

$$N_i(\tau) = \mu_i \tag{2.9}$$

where: $\mu_i \leq 1$.

From the decaying property, the time dependence of the occupancy of each mode can be modeled as an exponential decay with decay rate constant $\frac{1}{\tau_i}$ (or τ_i as time constant), such that:

$$N_i(t) = e^{-\frac{t}{\tau_i}} \tag{2.10}$$

If we put $t = \tau$ in Eq. (2.10), and combine that with Eq. (2.9), we have:

$$N_i\,(\tau) = e^{-\frac{\tau}{\tau_i}} = \mu_i \tag{2.11}$$

We can then express the time constant τ_i of the decay of transition mode as:

$$\tau_i = \frac{-\tau}{\ln\,(\mu_i)} \tag{2.12}$$

This time constant τ_i is also known in the literature as the "implied timescale" of the transition mode. Due the fact that τ_i reflects the lifetime of a particular transition mode, it can also be used in the assessment of the timescale of the dynamics of the system and the identification of modes. A slow τ indicates a persistent transition mode, which can correspond to slow dynamics, and such modes are usually of particular interest in MSM construction. When the implied timescales are used in the determination of the Markovian time of the system, multiple transition matrices are built at different lag time τ and the corresponding sets of implied timescales τ_i are then determined (see Fig. 2.5). If all the microstates generated are ideally Markovian, all the implied timescales should remain constant regardless of the choice of lag time, but this is usually not the case in practice. It is instead expected that the implied time scale will first quickly rise and then flatten off. In such cases, the time at which all implied timescales have plateaued will be treated as the Markovian time of the system.

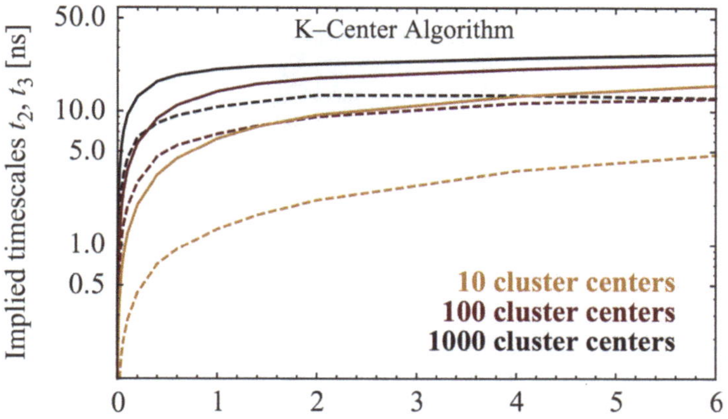

Fig. 2.5 Implied timescales convergence at different discretization levels. The plot shows the lag-time dependent implied timescales of the two slowest processes (t_2, *solid lines*) and (t_3 *dashed lines*), computed from different MSM of the MR121-GSGSW peptide. The different models (*different line colors*) correspond to k-centers clusterings of the same MD simulations data, but using different number of clusters (microstates). As it has been explained in the text, a model start to be Markovian at the shortest lag-time in which the slowest implied time scales converge, it is to say when the plot flattens. It can be seen that, as explained in the text, increasing the number of clusters (finer discretization) enhances a faster convergence of the implied timescales; hence, models with more microstates are Markovian at shorter lag-times (Figure adapted from reference [46])

2.11 Lumping Microstates into Macrostates

Under the protocol that we discussed, the second stage of MSM building involves lumping the microstates generated in the first stage into larger macrostates. The number of macrostates of the system can be chosen based on the major gap(s) between two consecutive modes in the implied timescales (see Figs. 2.5 and 2.19), as such gap indicates the slower modes and the faster modes have a significant separation in timescale (i.e. the slower modes will be significantly more sustainable than the faster ones). By choosing the number of macrostates in such way, theoretically the slow dynamics can all be properly preserved in the final model (i.e. no mixing of fast and slow dynamics in a single state), which in turn allows the model to fulfill the basic MSM requirements: (1) states are metastable, (2) intrastate transitions are fast, and (3) interstate transitions are distinct and slow.

The actual macrostate lumping can be performed using several methods, two of the most commonly used are: Perron cluster cluster analysis (PCCA), and its improved version (PCCA+). Basically, PCCA utilizes the properties of the eigenvectors and eigenvalues of the TPM to split the set of microstates into groups (see Fig. 2.6). As stated in the previous section, each eigenvector of the TPM corresponds to a certain mode of sustainable transitions between groups of states

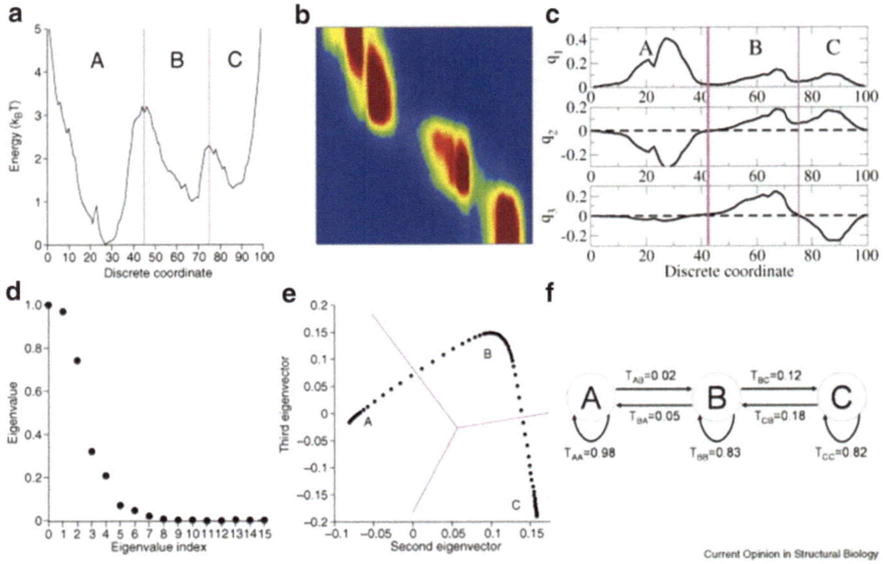

Fig. 2.6 PCCA lumping of microstates into macrostates. (**a**). Projection of the energy potential for 100 microstates onto one discrete coordinate, three energy bins were identified. (**b**). Transition matrix **T** for the 100 microstates. (**c**). Left eigenvectors of **T** indicating the transition information between different microstates. Each eigenvector corresponds to a certain mode of sustainable transitions between groups of states with the signed structure indicating the two groups. Except the first eigenvector provides the stationary distribution. (**d**). The eigenvalue spectrum of **T**. (**e**) Projections of the 100 microstates onto the second and third right eigenvectors of **T**. (**f**) Transition information for the macrostates *A*, *B* and *C* (Figure adapted from reference [33])

with the signed structure indicating the two groups. With the first eigenvector neglected (as it has an eigenvalue of 1 or implied timescale which represents the equilibrium), the set of microstates can thus be split into N groups by successively choosing the first N_i eigenvectors and partition the microstates into two groups according to their sign structure (see Fig. 2.6). After lumping we can recalculate the TCM on macrostate level, from which we can calculate the stationary distribution for the different metastable states. The next section treats a different algorithm that aims to generate multiple timescale-resolution MSMs at the macrostate level, which is based in the use of hierarchical clustering method.

2.12 Hierarchical Lumping of Microstates in Macrostates

In PCCA, microstates are lumped together based solely on the feature of TPM. Despite the mathematical correctness of the previous method, practically the method could suffer from sampling noise and cause errors in the resulted lumping. This is especially true because of the multi-resolution nature of energy landscapes [40] and

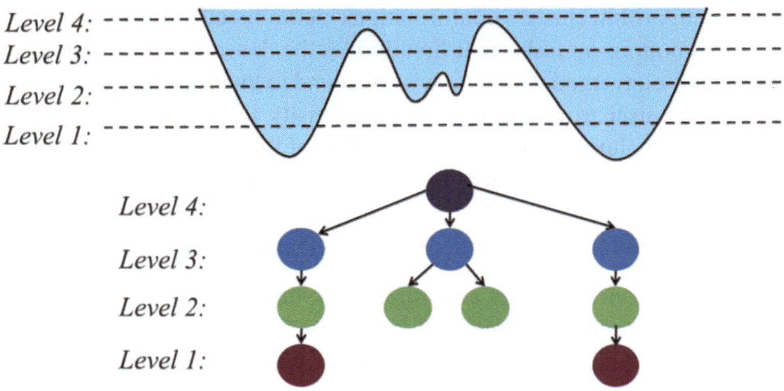

Fig. 2.7 Hierarchical clustering of microstates into macrostates. The kinetic clustering is done separately on different density levels. At first level (highest density), two separate states are identified, which are represented by the two nodes in *red*. At the next level, four separate states are identified. After the identification of states at all density levels, the connectivity between states at different level of hierarchy is identified, as shown in the figure below. The number of macrostate and their corresponding lumping can be identified from the leaf nodes of the graph (Figure adapted from reference [52])

also because the sampling quality of low populated microstates (e.g. near to high energy barriers) can be very different to the highly-populated microstates (i.e. those in the bottom of energy minima). The Super-density-level Hierarchical Clustering (SHC) introduced by Huang et al. [40] attempts to address this issue by treating the energy landscape in an hierarchical way instead of simply extracting features from the TPM (see Fig. 2.7). As stated in the previous discussions, to achieve a Markovian model the macrostates should be defined in a way that large internal free energy barriers are avoided and conformations within the same macrostate can interconvert quickly (within one lag time). Therefore, at smaller lag-times an MSM will require more macrostates to ensure that each state is small enough such that its dynamics per lag-time are memoryless. Intrinsically, shorter lag times result in higher resolution MSM that capture more free energy minima separated by small energy barriers. In the other hand, longer lag times result in a lower resolution MSM, with only a few macrostates separated by high-energy barriers. From other point of view, in a lower resolution MSM each macrostate is composed of multiple local free energy minima. SHC attempts to do lumping at different resolution by considering subsets of conformations with different densities successively, thus improves the accuracy of the kinetic lumping by treating poorly sampled states differently from states with better statistics [40]. Furthermore, for the issue that popular kinetic lumping algorithms such as PCCA and PCCA+ tend to identify poorly sampled states as being kinetically distinct from the others [40] and preserve them as metastable state in the resultant macrostate model, despite further investigations can easily show that they are likely just due to sampling noise rather than representing the true free energy minima, SHC can handle these states with very small populations separately and so the resultant model will not be skewed by these poorly sampled states.

The SHC algorithm clusters conformations hierarchically, by using super density level sets in a bottom-up fashion. It first divides the densest regions of phase space, which in a well-converged system may correspond to the free energy minima. Then, by allowing the user to fine-tune the super density level sets, this algorithm can generate multi-resolution models. From the best of the authors' knowledge, SHC is the first algorithm known to address the construction of MSM at different resolutions (see Fig. 2.7).

The SHC algorithm lump the microstate together in the following way:

1. Partition the conformations into a large number of microstates. K-centers clustering has been recommended by the authors, since it gives states with approximately uniform size, which has been proposed that this might result in a correlation between the population of each state and its density [40].
2. Split the microstates into n-density levels $L = \{L_1, \ldots L_n\}$.
3. Calculate the super density level sets $S_i = \mathrm{L}_1 \cup \mathrm{L}_2 \ldots \cup \mathrm{L}_{i-1} \cup \mathrm{L}_i$, and then each super density level also contains all previous levels $S_1 \subseteq S_2 \subseteq S_{i-1} \subseteq S_i$.
4. Perform spectral clustering, in each super density level, to group kinetically related microstates.
5. Build a graph of the states connectivity across super density levels. Then, generate a directed gradient flows along the edges of the graph from low to high-density levels. In SHC, it is denominated an attraction node (or attractive basin) where the gradient flow ends. Each attraction node is assigned to a new metastable state.
6. Assign every microstate not belonging to an attraction node, to the metastable state that it has the largest transition probability to (see Fig. 2.7).

In the SHC algorithm, the populations of microstates obtained from the K-centers clustering are used to approximate the conformation density, since K-centers algorithm can generate clusters with approximately equal radii in RMSD. However, it is extremely challenging to accurately estimate the conformational densities in high dimensional spaces, since small variances in the cluster RMSD radius may cause large differences in volume.

To address the above issue, Huang and coworkers have developed a new algorithm, which is based on the Nyström method and its multilevel extensions (HNEG) [52]. The HNEG algorithm allows us to approximate the transition probability matrix (P) with its dominant submatrix (A). Using the Nyström approximation, it can be shown that the leading eigenvectors of the submatrix A containing the most populated states (i.e. the entries in A are significantly larger than those in B and C) have the same sign structure as those of the original transition probability matrix P:

$$P = \begin{pmatrix} \mathrm{A} & \mathrm{B} \\ \mathrm{B}^{\mathrm{T}} & C \end{pmatrix} \tag{2.13}$$

Therefore, one can perform the kinetic lumping based on the eigenvector components of the submatrix A using either PCCA or PCCA+. In order to define the boundary between A and C, the same multi-level procedure as laid out in the SHC algorithm has been adopted.

In both SHC and HNEG algorithms, there are many possible choices of the super level sets (S) and each could result in different lumpings. The authors recommend trying many of them and then use Bayesian model comparison to choose the best model [52]. One additional advantage of SHC and HNEG is that they can automatically determine the number of macrostates, whereas many other methods (like PCCA and PCCA+) require the state number to be known in prior.

2.13 MSM Validation

With the microstates properly lumped into macrostates the TPM can be constructed at the macrostate level and the corresponding MSM can be built. Yet before using the MSM for any analysis, an assessment of the model accuracy is necessary. Apart from the aforementioned plateauing of the implied timescales, there also exist other tests for assessing the validity of the model. A notable method is known as Chapman-Kolmogorov Test [40, 46], which assesses the validity of the model based on the premise that the Markovian assumption:

$$P(t + s) = P(t)P(s) \tag{2.14}$$

must hold within the error margin (see Fig. 2.8). The actual testing procedures vary between implementations, but the general idea lies on testing the transition matrix propagated at the chosen Markovian time against the transitions counted from populations bootstrapped from simulations. Only if the transitions predicted by the MSM match with those observed in the simulation, the model is deemed valid. An implementation example of the test is shown in Fig. 2.8. Alternative approaches of model validation include application of Bayesian factors or information entropy [36, 53, 54], which will not be discussed in detail here. Apart from these pure theoretical validations, assessments can also be done via comparison with experimental data [19, 40, 41, 55, 56], which will be illustrated later in this chapter (see Sect. 2.18).

2.14 Mean First Passage Time

Having a properly built and validated MSM, kinetic information of the system can then be harvested from the model. Apart from simply propagating the TPM so as to obtain equilibrium populations, timescale information of the transition can also be

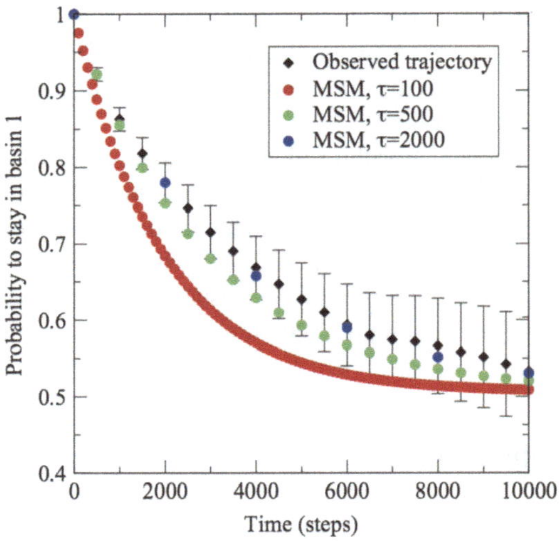

Fig. 2.8 Example of Chapman-Kolmogorov test. A representative implementation of Chapman-Kolmogorov test is illustrated here on one partitioned state. In this implementation the initial population is first solely populated on the state to be tested (i.e. $X_i(t=0)=1$, $X_{j\neq i}(t=0)=0$), the probability for this state to be occupied in the subsequent steps (which can be considered as the "self-transition" probability at the test step) are calculated via repeated propagation with TCM following Eq. (2.3) in the main text. The calculated self-transition probability is then compared against the actual occupation probability counted from MD trajectories for all propagated steps. In this example, at $\tau=100$ the self-transition probabilities propagated to different time steps clearly deviates from the ones directly observed from MD simulations, thus the MSM constructed at $\tau=100$ cannot represent the kinetics of the model properly. For the MSM constructed at $\tau=500$, the propagated self-transition probability marginally lies within the error bar, and thus can better reflect the dynamics of the state than the one constructed at $\tau=100$. The MSM constructed at $\tau=2{,}000$ gives self-transition probability that is close to the one observed from MD, and thus is considered to be the model that best conform to the Chapman-Kolmogorov equation under the test and should best represent the dynamics of the system (Figure adapted from reference [46])

obtained from the model. One of the commonly used timescale for interstate transition is known as mean first passage time (MFPT), which is defined as the average time taken for the transitions starting at state i until reaching state f for the first time, including both the direct transitions from states i to f and transitions through other intermediate states. MFPT of the transitions from i to f can be written as:

$$F_{if} = \sum_{j} P_{ij}\left(\tau + F_{jf}\right) \quad \text{or} \quad F_{if} = \tau + \sum_{j\neq f} P_{ij}F_{jf} \tag{2.15}$$

where τ is the lag time used to construct the transition matrix $P(\tau)$. Thus the MFPT for all transitions in the model can be determined by solving a set of linear equations defined by Eq. (2.11) with the boundary condition $F_{ff}=0$. Therefore, we

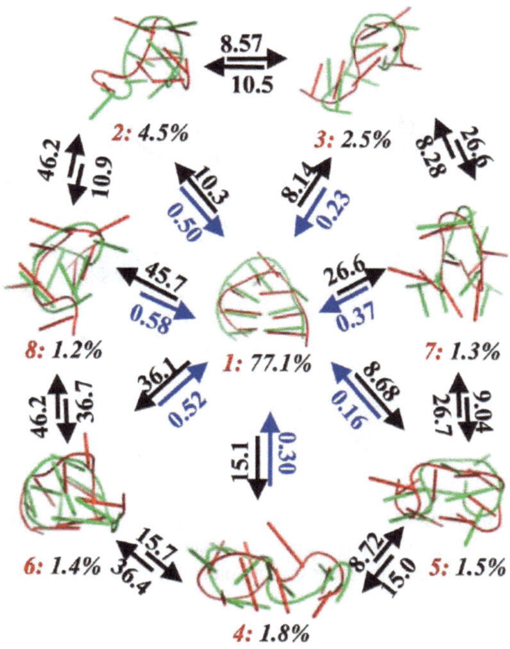

Fig. 2.9 Application of MFPT to the study of small-RNA folding mechanism. The figure shows the folding mechanism of a small RNA-hairpin tetraloop (5′-GCGGCAGC-3′). Next to the *arrows* are the corresponding Mean First Passage Times (MFPTs units: μs) between the eight most populated states in the MSM. States are labeled in *red* from 1 to 8 and the state populations are shown in *black* (Figure adapted from reference [40])

can understand that while the implied timescales describe the aggregated timescales for transitions between groups of states, the MFPT are average transition times between specific pairs or groups of metastable conformational states, and thus the MFPT calculation can provide detailed information about system's kinetics (see Fig. 2.9).

2.15 Transition Path Theory

Apart from the overall dynamical behaviors of the system, one might also be interested in some particular states or transitions. For example, in the case of a protein-ligand interaction, one might be interested more in the transitions from a unliganded protein state to a liganded protein state than the transitions between different unliganded states. Because multiple possible pathways between the two states are likely to coexist in an MSM, simple network analysis might not be adequate for such purpose. In this case, a framework known as transition path theory could be applied in order to study the relative likelihood of transitions between a particular state A to another state B [57]. Under this framework, by recursively solving for the interstate transitions between states, the pathway with highest flux, defined as highest number of transitions per unit time, can be identified. Such pathway can thus be understood as the "dominant pathway" for state A to B

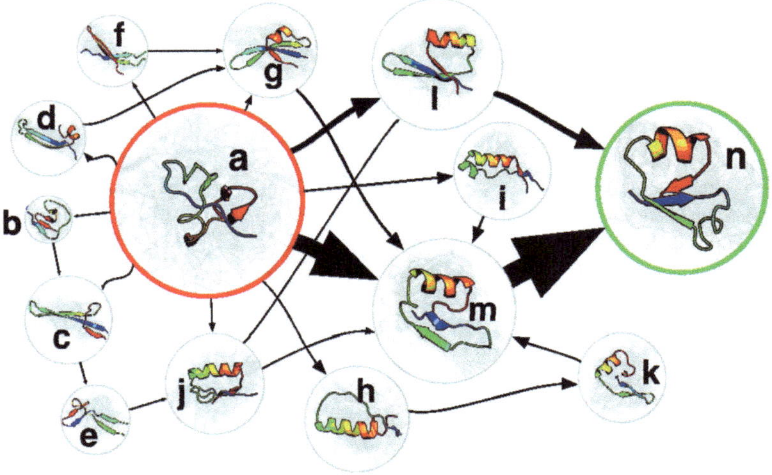

Fig. 2.10 Determination of the top 15 folding pathways of NTL9 using TPT. The transitions between the macrostates are connected with *sized arrows*. The pathway along the *larger arrows* indicate the dominant folding path (Figure adapted from reference [20])

transitions (see Fig. 2.10), and second or third pathways can thereby be identified by following this scheme. Detailed derivations are beyond the scope of this chapter and interested readers are advised to consult the relevant literature referenced for further information [33, 46, 54, 57, 58].

2.16 Visualization of MSM

As with many parts of research, on top of the quantitative analysis, visualization is also a very important part of understanding and appreciating a MSM. An intuitive approach of illustrating MSM would be to present them as graphs with macrostates as vertices and connectivity as weighted edges. Yet it is common to come across MSM with more than just a few macrostates, and the connectivity of these states may form a high dimensional network, which could make visualization difficult. Depending on the systems, solutions of visualization issue might vary considerably. Apart from plotting out the macrostate connectivity in whole or in part, if some specific representative geometric parameters exist in the system that can be used to describe the progress of the dynamics (e.g. a dihedral angle can be used to describe the rotation of a bond), conformations could be projected on such parameters and use them to describe the most prominent geometric differences between different macrostates. This is a commonly used approach for simple system as represented by the well-known system in the field: the terminally blocked alanine

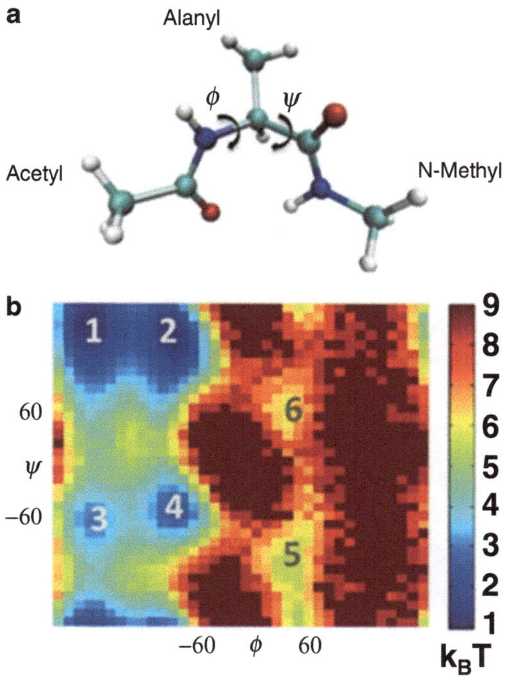

Fig. 2.11 Projections of the conformational space of the terminally blocked alanine peptide. (**a**) Illustration of the Φ and Ψ dihedral angles of the alanine in the terminally blocked alanine peptide. These two dihedral angles represent the major possible conformational changes in the system. (**b**) Energy landscape of the peptide projected onto the Φ-Ψ plane. The "energy" (shown here as the "potential of mean force") of the bins in the grid is determined from the density of the projected points in each bin. High density regions of the projection are shown as the minima of the energy landscape. With a proper choice of geometric parameters (in this case Φ and Ψ dihedral angle), geometric differences between different minima (which also correspond to six macrostates in this case) can become prominent and intuitive visualization of the states is thus possible (Figure adapted from reference [52])

peptide (NMe-Ala-Ace, also known as alanine dipeptide in some literature), which represents the peptide motion with the **Φ** and **Ψ** angle of alanine (see Fig. 2.11). If such parameters are not available, an alternative approach could be to project the conformations of interest on the first one or more principle components of the system in order to have a general understanding of the spatial distribution of the macrostates.

Recently a new program, the MSMExplorer [59], has become freely available. This Java suite allows interactive visualization and analysis of MSM built using the MSMBuilder package [34, 50, 51] (see Fig. 2.12). The representations of a MSM that it can generate includes graph of states connections, scatter plots of user definable data, visualization of hierarchical MSM and transition paths.

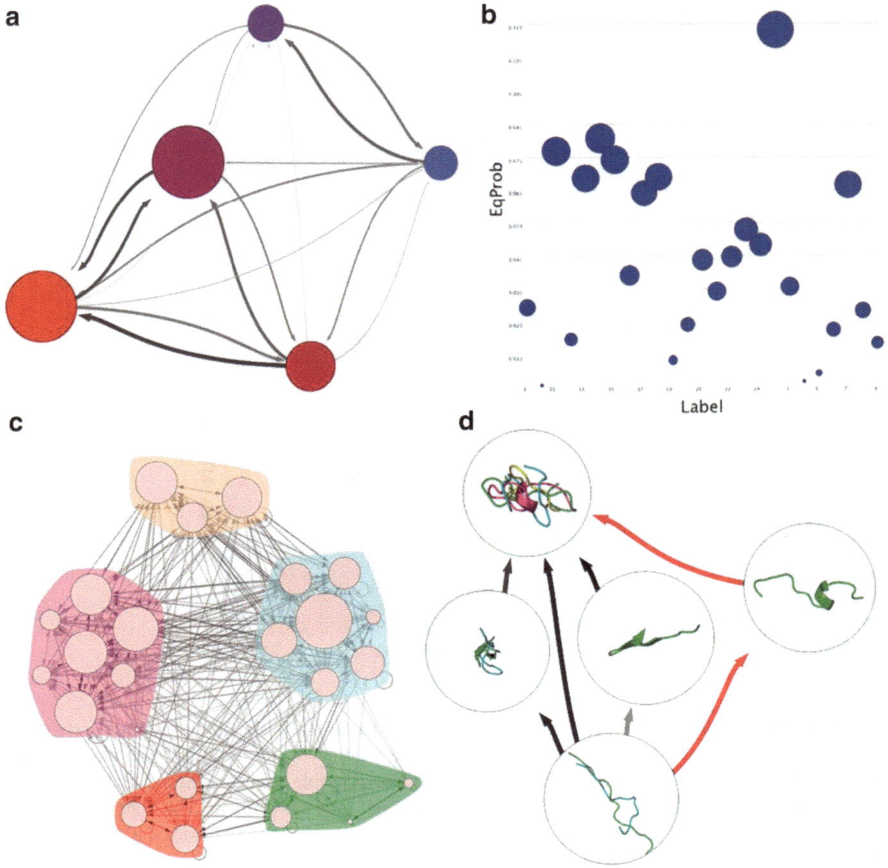

Fig. 2.12 Visualizations of MSM using the program MSMExplorer. The recently released MSMExplorer program allows to easily perform several visual analysis of MSM built using the MSMbuilder-2. For example: (**a**) Network representations of the states in the model and its connections to other states. (**b**) 2D scatter plots of arbitrary RAW data vs. state number, with visual representation of the state size. (**c**) Visualization of hierarchical MSM, in which the membership of a finer-grained model can be overlaid with a coarser-grained model, this allows the visualization of multiple resolutions of the MSM in a single plot. (**d**) TPT diagrams of the highest flux paths between two macrostates, with the advantage that images depicting conformations in each state can be overlaid on each node (Figure adapted from reference [59])

2.17 Mining Data from MSM

Constructing and validating an MSM can be challenging, but extracting relevant data from the model could be even more difficult. Some MSM construction packages have therefore offered an all-in-one solution for application of MSM in MD studies. For example, the MSMbuilder 2 package [50] offers several general tools coded in python for analysis of MSM, such as: (1) Extracting random conformations from the micro and macrostates, which allows rapid visual identification of the structural properties of the system in each state. When one is dealing with large dataset, it is often convenient to measure physical properties (e.g. distances, SASA, RMSF, correlations, among others) in a reduced set of representative conformation of each state, which can be accomplished by randomly extracting an statistically significant number of random conformations; (2) Calculating cluster radii, which allows rapid assessment of the structural diversity among the clusters; (3) Calculating cluster RMSD to a reference structure, which can be used to identify known states of interest (e.g. folded and unfolded states, bound state or intermediates); (4) Calculating the aforementioned transition path between two states, and generate a plain text graph in a DOT file that can be visualized with a number of open source software widely available for several operating systems.

Nevertheless, as in any scientific research problem, each studied system can pose unique challenges and the users will frequently find themselves without the necessary tools to perform a particular analysis. In such scenario, in many cases it is possible to use simple Linux-like command line scripting (BASH, CSH, TCSH, etc.) to combine existing analysis tools in order to perform more complex analyses. However, the most powerful approach is to code custom analysis tools. Open-source packages that provide a framework to interact with MD trajectories are valuable aids for such customized programming, such as: MDAnalysis (in python language) [60] which can be advantageously combined with the open source NumPy and SciPy suites, VMD (Tcl/Tk) [61] and Gromacs (C) [62, 63]. Anyhow, there is no unique or simple answer of how to perform a certain analysis, we compel the readers to make an incursion in programming their own analysis tools. One can, in most cases, find that by just following the existent online literature and asking advice from more experienced programmers (in the many-existent Internet communities), it takes little time to get used to programming analysis tools.

2.18 Practical Examples of MSMs Construction

With all the basic theories and tools discussed above, readers should have already get hold of the essential techniques for applying MSM in practical biomolecular studies. In the following sections, two of our works are presented here as practical examples to illustrate how all these aforementioned techniques work in practice.

2.18.1 MSM Example #1. PP_i Release Mechanism in the Yeast RNA Polymerase II and Bacterial RNA Polymerases

RNA polymerase is a critical biological machine that is responsible for transferring the genetic information from the DNA template to the messenger RNA (mRNA) [64–66]. The nucleic addition cycle (NAC) of the RNA polymerase consists of several steps: (1) The post-state of the polymerase contains an empty active site at register +1 site that can accommodate the incoming NTP. (2) The binding of the NTP that can form several important contacts with a critical domain of the polymerase named Trigger Loop (TL) and then fix it in a closed state. (3) The catalytic reaction which forms the phosphodiester bond. (4) The release of the produced pyrophosphate ion (PP_i) from the active site. (5) The opening of TL domain accompanied by the forward shifting template DNA by one register site, which creates a new active site and new NAC starts. Extensive experimental and theoretical studies have been devoted to understand the specific steps during the NAC, including NTP binding, TL motion, catalytic reactions and translocation.

The interplays between the PP_i release, TL opening motion and translocation have attracted extensive attentions [67, 68]. The crystallography studies indicated that the PP_i release in T7 RNA polymerase is the driving force to trigger the opening of the adjacent O-helix that allows the translocation [69]. However, the *E. coil* single molecular study did not observe the coupling between PP_i release and translocation [68]. Recent fluorescence studies suggest that translocation process proceeds shortly after or concurrently with the PP_i release in *E. coil* system [67]. Although these experimental studies shed light on the roles of the PP_i release on the translocation, the detailed mechanism of the PP_i release process as well as its role on the TL opening motion has been elusive. We have used MSM to address these questions [42, 43].

People have obtained the crystal structures of the RNA polymerase in both eukaryote and bacterial systems [70, 71]. Based on the NTP-bound RNA polymerase complexes in yeast and *T. thermophilus* (termed as Pol II and RNAP respectively afterwards), we build the PP_i-bound RNA polymerase complexes by directly cleaving the P_α-O bond to form the phosphodiester bond and the PP_i group (see Fig. 2.13). The comparison of the structures of these two complexes shows different features in the secondary channel and TL domains. These structural differences suggest that the PP_i release mechanism and its effects on the TL domains are likely to be different.

In order to obtain the initial release pathways of the PP_i group in both systems, we adopted steered MD (SMD) simulations to pull the PP_i group out of the active site. The pulling simulations were conducted along different directions with the aim of considering all the possible PP_i release pathways. Then, representative structures from the SMD simulations were chosen for the following unbiased MD simulations to erase the biases introduced by the SMD. These MD simulations were then used to build the MSM. At first, we divided all the conformations from the seeding MD simulations into hundreds of microstates by employing the K-center clustering algorithm. The distance between a pair of conformations was set to be the RMSD value of three PP_i atoms (the bridge oxygen and two phosphorus atoms). To compute RMSD, the structure was aligned to the modeled PP_i-bound RNAP complex according to the C_α atoms of the BH residues. The microstates are small, and the average RMSD values to its central conformation in each state are only $\sim$2 Å in both systems.

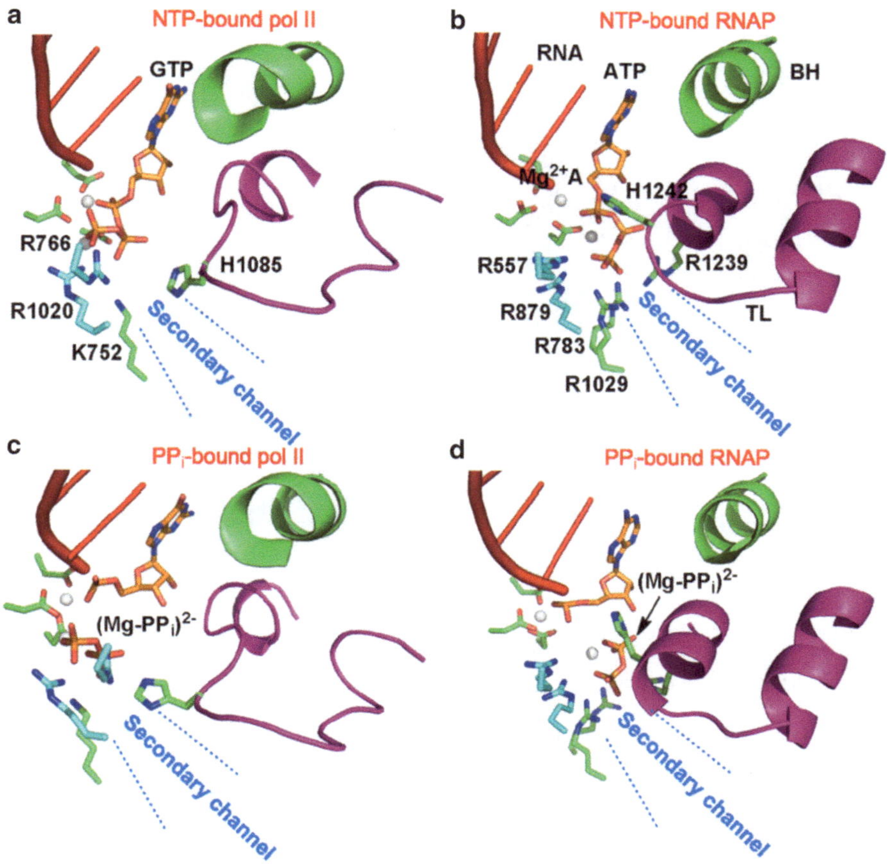

Fig. 2.13 Comparisons of the binding modes between the ligands and two RNA polymerases. (**a**) and (**b**) are the structures of the NTP-bound RNA Pol II and RNAP complexes respectively. (**c**) and (**d**) are the corresponding PP_i-bound models (Figure adapted from reference [43])

Next, we applied the Robust Perron Cluster Cluster Analysis (PCCA+) algorithm to lump these microstates obtained above into several macrostates. The number of the macrostates was determined from the major gap captured in the implied timescale plot on the microstates level. The macrostates number was finally determined to be 4 and 2 for the Pol II and RNAP system respectively.

Our results suggest that the PP_i release in the Pol II adopts a hopping mode in which four metastable states were well defined and several positively charged residues were observed to form favorable interactions with the PP_i group in each metastable state [42] (see Fig. 2.14). Furthermore, mutant MD simulations were individually performed to elucidate the specific roles of these residues on the PP_i release.

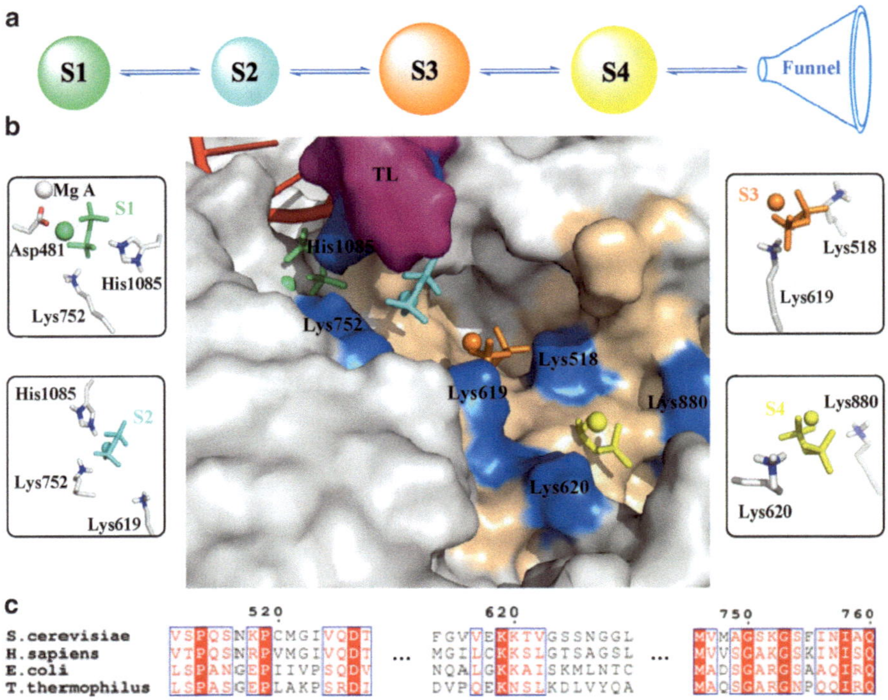

Fig. 2.14 PP_i release in Pol II adopts a hopping mode identified by MSM. (**a**). Four metastable states (S1–S4) on the releasing pathway are displayed in *sized circles* proportional to their equilibrium populations. (**b**). Key interactions between the PP_i group and the Pol II residues in each state. (**c**). Multiple sequence alignment of these positive residues in the secondary channel among different species (Figure adapted from reference [42])

However, a simpler two-state model was observed for the PP_i release in the RNAP [43] (see Fig. 2.15). We found that the difference in the number of metastable states in the release of the PPi between these systems is due to the different layout of the positive residues in the secondary channel. Specifically, in Pol II, the four residues, K619, K620, K518 and K880 are located at relatively separated sites. However, the positively charged residues in RNAP: K908, K912, K780 and K1369 are close to each other in a continuous region.

From the kinetic point of view, our MFPT calculation indicates that the PP_i release in bacterial RNAP is ~3 fold faster than that in Pol II (500 ns versus 1.5 μs), which is consistent with the faster elongation rates observed for RNAP. More strikingly, because of the higher stabilities of the TL domain in RNAP compared to that in Pol II, the PP_i release in RNAP cannot induce the backbone unfolding of the TL domain. Instead, the TL residue R1239 was observed to greatly facilitate the PP_i release in RNAP by rotating its long side chain (see Fig. 2.16). Further control MD simulations indicate that the TL domain must be exposed to the solvent before its secondary structures can be fully unfolded. And the full opening motion of the TL is likely to occur at a timescale longer than the timescale of PP_i release.

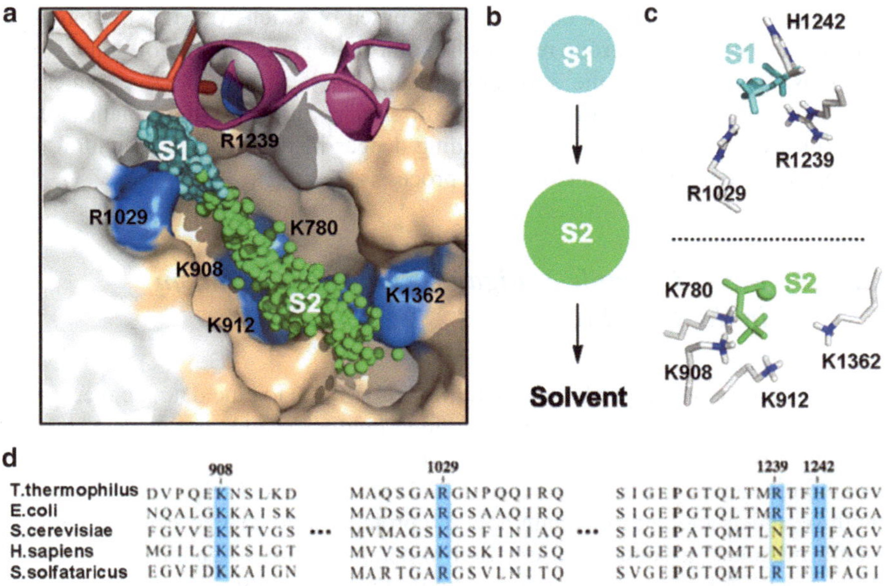

Fig. 2.15 Two-state model for the PP_i release in RNAP identified by MSM. (**a**). The distributions of the two macrostates. Each *sphere* represents the center of mass of the PP_i group. (**b**). *Sized circles* proportional to their equilibrium populations. (**c**). Key interactions between the PP_i group and the Pol II residues in each state. (**d**). Multiple sequence alignment of these positive residues in the secondary channel among different species (Figure adapted from reference [43])

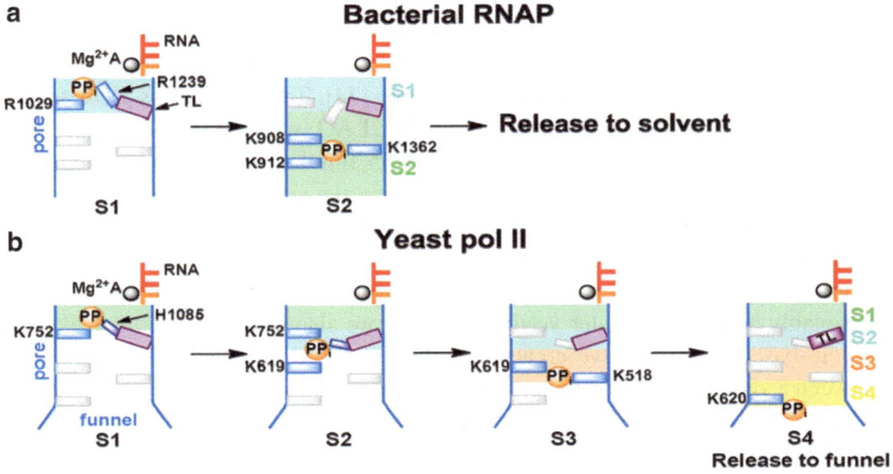

Fig. 2.16 Structural differences lead to distinct PP_i release mechanisms in RNAP (a) and Pol II (b) (Figure adapted from reference [43])

Taken together, we have built the MSM based on extensive MD simulations to investigate how the PP_i group releases from the active site in both Pol II and RNAP systems. By comparing the structural features of these two RNA polymerases, we have addressed how the structural differences influence the kinetics of the PP_i release from the active site, providing deeper insights on the structural basis underlying the transcription elongation process.

2.18.2 MSM Example 2. Ligand-Binding Mechanism in the LAO Protein

In this example, we used Markov State Models (MSM) to elucidate the mechanism by which the Lysine-, Arginine-, Ornithine-binding (LAO) periplasmic binding protein (PBP) binds to its ligand [41]. Two models of protein-ligand binding have been proposed for PBPs, the induced fit and conformational selection mechanisms, both of which attempt to explain how the protein could change from an unbound conformation to a bound conformation in complex with a ligand. In the induced fit model [72] the ligand first binds to the protein in its unbound conformation and this binding event induces the protein to go to the bound state. On the contrary, in the conformational selection model [73], the protein can access the protein-bound conformation even in the absence of the ligand, therefore the ligand can diffuse directly to the bound conformation and displace the equilibrium towards it. Using MSM, we directly monitored the mechanism of LAO binding to assess the role of conformational selection and induced fit.

We used the aforementioned MSMBuilder and SHC programs and algorithms to construct the state decomposition for our MSM of LAO's binding. We first performed 65 molecular dynamics simulations using the program GROMACS [62, 63], each 200 ns long, of the LAO protein from the organism *Salmonella typhimurium* and one of its ligands, L-arginine [74]. Ten simulations were started from the open protein conformation (PDB ID: 2LAO) with the ligand at more than 25 Å away from the binding site. The other 55 simulations were initialized from conformations randomly selected from those first ten simulations. To construct the microstate partition, we first used the k-centers algorithm in MSMBuilder to cluster our data into a large number of microstates. The objective of this clustering was to group together conformations that are so geometrically similar that one can reasonably assume (and later verify) that they are also kinetically similar. For the protein-based clustering, we created 50 clusters based on the Euclidean distance between a vector containing the protein opening and twisting angles (see Fig. 2.17). Then for the ligand-based clustering, we created 5,000 clusters using the Euclidean distance between all heavy-atoms of the ligand.

We then had to modify our clustering to account for the fact that the ligand dynamics fall into two different regimes (see Fig. 2.18): one where the ligand moves slowly due to interactions with the protein and one where the ligand is freely diffusing in solution.

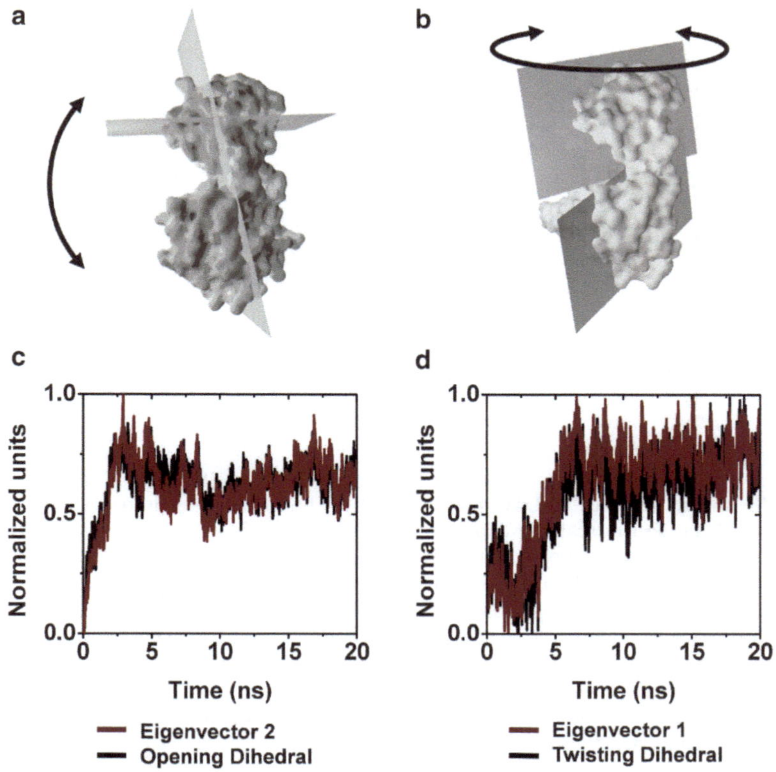

Fig. 2.17 Ad-hoc reaction coordinates used to describe the energy landscape of the LAO protein. (**a**) Opening and (**b**) twisting angles used to describe the motion of the protein. (**c**). The projection of conformations on the second eigenvector from Principle Component Analysis (PCA), and protein opening dihedral angle as a function of time are shown in *red* and *black* respectively. The 20 ns simulation is started from protein in the closed state (PDB ID: 1LAF), but ligand was not included in the simulation. (**d**) Same as (**c**) except that the projection of conformations on the first eigenvector from PCA and protein twisting dihedral angle are plotted. In this system, the twisting and opening angles are correlated well with the first and second eigenvectors from PCA (Figure adapted from reference [41])

The clusters described previously are adequate for describing the first regime, when the ligand interacts with the protein. However, when the ligand is freely diffusing (at more than ∼5 Å from the protein) the procedure outlined above results in a large number of clusters with poor statistics (less than ten transitions to other states). Better sampling of these states would be a waste of computational resources as there are analytical theories for diffusing molecules and a detailed MSM would provide little new insight. Instead, we chose to re-cluster these states using the Euclidean distance between the ligand's center of mass (as opposed to the Euclidean distance between all ligand heavy-atoms). At this stage, we created 10 new protein clusters and 100 new ligand clusters. After dropping empty clusters, this procedure yielded 3,730 microstates, of which 3,290 microstates came from the initial high

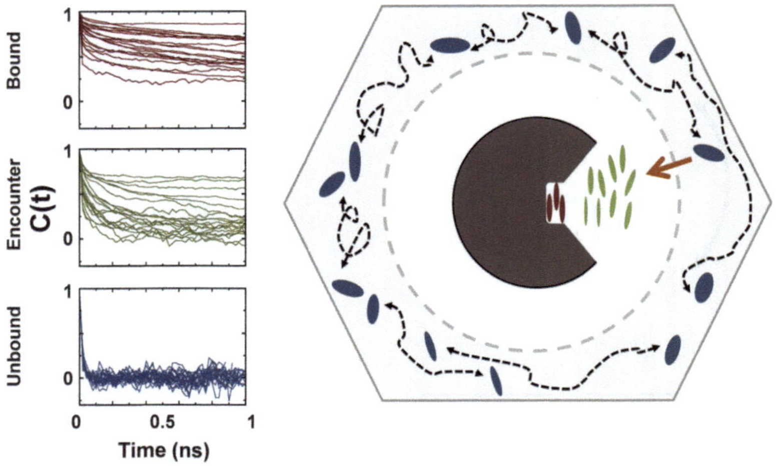

Fig. 2.18 In the presence of LAO the diffusion of the ligand L-arginine experiences two different relaxation-timescales. We found that the ligand of the LAO protein experience two very different timescales. As exemplified in the schematic figure, it can be seen that the ligand rotates quickly when it is far away from the protein but its rotation is restrained when it interacts with the protein. Thus, when constructing MSM, we only consider the ligand center of mass motion when the ligand does not have strong interactions with the protein (*blue color*) but we consider motion of all the ligand heavy atoms when the ligand is strongly interacting with the protein (*green* and *red color*). The graphs show the difference in the relaxation time of the ligand, which was assessed by analyzing its rotational autocorrelation in many independent MD trajectories, for: the unbound states (*blue*), the encounter complex state (*green*), and the bound state (*red*) (Figure adapted from reference [41])

resolution clustering and 440 came from the data that was clustered again at low resolution. To verify that the final microstate model is valid (i.e. Markovian) we plotted the implied timescales and found that they level off at a lag time between 2 and 6 ns (see Fig. 2.19), implying that the model is Markovian for lag times in this range.

We then lumped kinetically related microstates into macrostates using the SHC algorithm with density levels $L_{high} = [0.50, 0.55, 0.60, 0.65, 0.70, 0.75, 0.80, 0.85, 0.90, 0.95, 0.99]$ and $L_{low} = [0.4, 0.95]$, for the high and low-density regions respectively. The low and high resolution states were lumped separately because the states in each set have different sizes, so it is difficult to compare their densities. We then combined these two sets of macrostates to construct an MSM with 54 macrostates. Once again, we used the implied timescales test to verify that the model is Markovian and found that a 6 ns lag time yields Markovian behavior (see Fig. 2.19).

To generate the transition matrix using the above state decomposition, we have used a sliding window of the lag time on each 200 ns trajectory with a 20 ps interval between stored conformations (i.e. each trajectory contains 10,000 conformations)

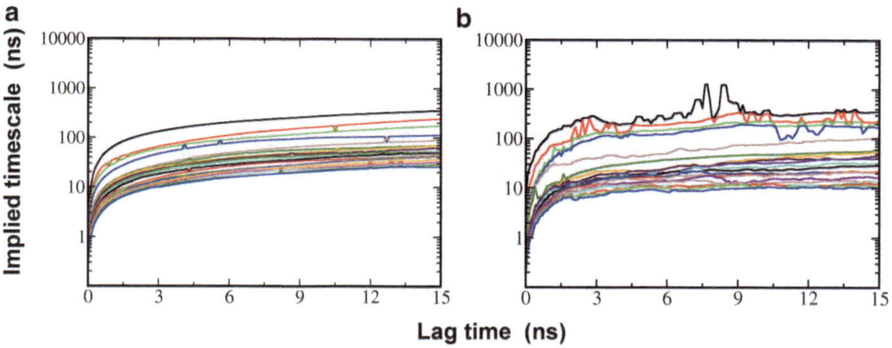

Fig. 2.19 Validating the MSM by analyzing the micro- and macro-states implied timescales. To validate the Markovianity of our model, we examined the 20 slowest implied timescales as a function of the lag time computed from: (**a**) MSM containing 3,730 microstates and (**b**) MSM containing 54 macrostates. It can be appreciated that both plots level off at a lag time of ~4 ns, hence from this point the model can be consider Markovian. Thus we choose a lag-time of 6 ns to construct our MSM. Furthermore, it can be seen that the implied timescales in the micro and macrostate models has good correspondence, meaning that both models (micro and macro) give a similar representation of the system (Figure adapted from reference [41])

to count the transitions. Because we used a hard cutoff between states, simulations at the top of the barriers between states can quickly oscillate from one state to the other, leading to an over-estimate of the transition rate between such states. To mitigate the effect of these recrossing events, we only counted the transitions from state x to state y if the protein remained in state y for at least 300 ps before transitioning to a new state. To generate the transition probability matrix we normalized each row of the transition count matrix.

To further assess the validity of our model we also verified that the system could reproduce known experimental observables. First we confirmed that the state with the largest population closely resembled the bound conformation observed in crystals (see Fig. 2.20); we also confirmed that the model is also in reasonable agreement with the experimentally measured binding free energy and association rates. From the MFPT from all unbound states to the bound state, our model predicts an association timescale of 0.258 ± 0.045 μs, in reasonable agreement with the experimental value of ~2.0 μs found in the highly homologous HisJ protein. Also, by using the algorithm introduced by van Gunsteren and co-workers [75] in conjunction with the equilibrium populations derived from our model, we estimate a binding free energy of −8.46 kcal/mol, in reasonable agreement with the experimental value of −9.95 kcal/mol. Together, this agreement between theory and experiment suggests that our model is in good reflection of reality.

Our final model suggests that three dominant-states need to be considered to adequately describe LAO's binding mechanism and that both: conformational selection and induced fit, play important roles in the transitions between these states (see Fig. 2.21). The third dominant state in our model—besides the previously

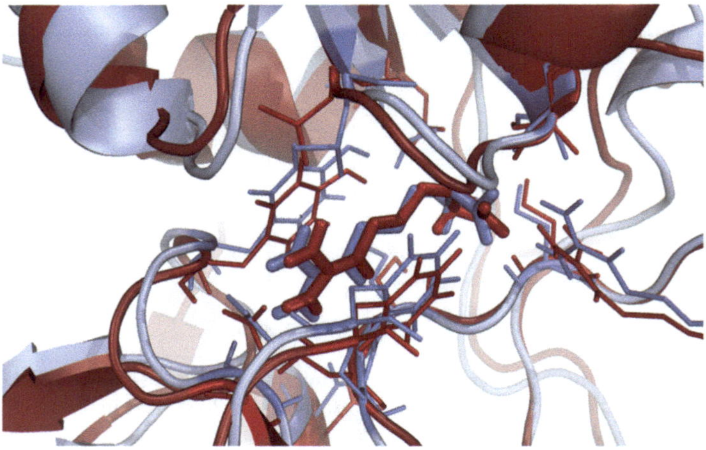

Fig. 2.20 Validating the MSM by structural comparison of the bound state. In our MSM, as in the reality, the bound state is the most populated state in the system, (in our model having an equilibrium population of 74.9 %). We used a snapshot from our simulations (structure in *red color*), to verify that the bound state is equivalent to the known crystal structure. We found that a snapshot in our model achieves a Cα-atoms RMSD of 1.2 Å (within 8 Å of the ligand C.O.M.) to the crystal structure of the bound state (*blue structure*, PDB ID: 1LAF), which confirms that the structures contained in our model are in good agreement with the experimental information (Figure adapted from reference [41])

known open and closed states—is only partially closed and weakly bound to the ligand, thus representing an encounter complex state. The ligand can induce the protein to have transition from the open state to the encounter complex (induced fit); however, the ligand-free protein can also go directly to the encounter complex state, indicating also an important role for the conformational selection mechanism (see Fig. 2.21).

2.19 Remarks and Future Perspectives

In this chapter we have reviewed the fundamental theories underlying the construction and applications of MSM, we have also highlighted that the main advantage of this method is to access timescales that are usually unreachable through conventional MD simulations. Finally, we presented two applications to illustrate the ability of MSM to investigate protein dynamics (at biologically relevant timescales) and to extract information about biological mechanisms. In conjunction with the increasing computing power, MSMs hold a great potential to address many more important problems related to the dynamics of complex biological macromolecules,

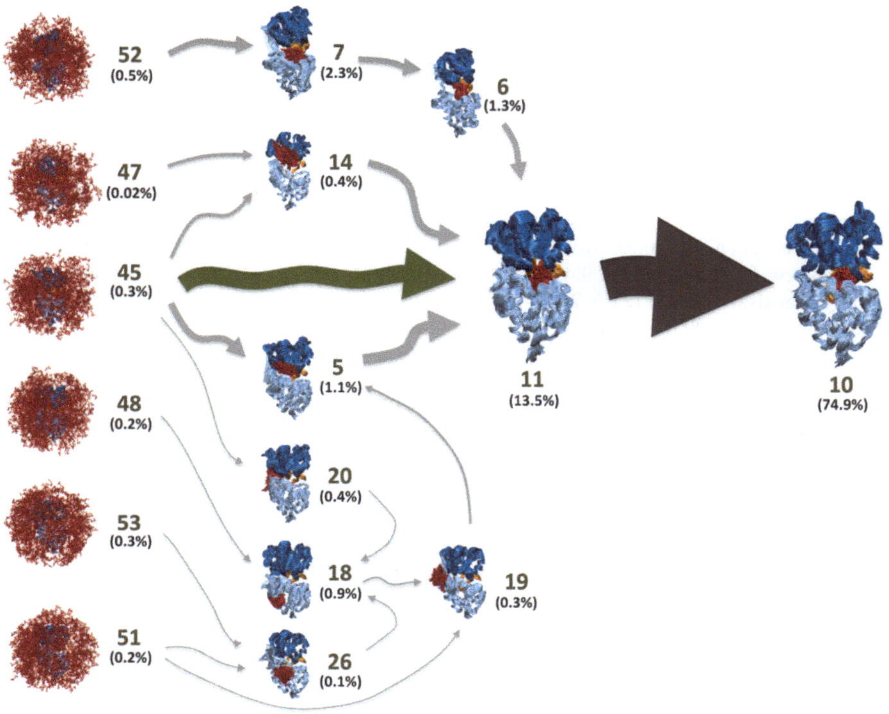

Fig. 2.21 **The mechanism of LAO's binding revealed by TPT.** The figure shows the superposition of the 10 highest flux pathways from the unbound macrostates to the bound macrostate. These pathways account for 35 % of the total flux from unbound states to the bound state. The conformational selection and induced fit pathways from the unbound states to the encounter complex state is shown in *green* and *grey arrows* respectively; it can be seen that the two mechanisms coexist. The arrow sizes are proportional to the interstate flux. State numbers and their equilibrium population calculated from MSM are also shown. The flux was calculated using a greedy backtracking algorithm applied to our 54-states MSM (Figure adapted from reference [41])

including problems that were impossible to attack just few years ago, mainly due to their prohibitive computational cost and the intrinsic complexity of analyzing complete free energy landscapes from a MD trajectory perspective. We envision that MSMs will be widely applied to elucidate molecular mechanisms of functional conformational changes in the near future.

Acknowledgements XH acknowledges support from the National Basic Research Program of China (973 Program 2013CB834703), National Science Foundation of China: 21273188, and Hong Kong Research Grants Council GRF 661011 and HKUST2/CRF/10. DAS acknowledges support from the PEW Charitable Trusts as postdoctoral fellow in the Biomedical Sciences. FKS acknowledges support from Hong Kong PhD Fellowship Scheme (2012/13).

References

1. Parak FG (2003) Proteins in action: the physics of structural fluctuations and conformational changes. Curr Opin Struct Biol 13:552
2. Reichmann D, Rahat O, Cohen M, Neuvirth H, Schreiber G (2007) The molecular architecture of protein-protein binding sites. Curr Opin Struct Biol 17:67
3. Mackerell AD Jr, Nilsson L (2008) Molecular dynamics simulations of nucleic acid-protein complexes. Curr Opin Struct Biol 18:194
4. Warshel A et al (2006) Electrostatic basis for enzyme catalysis. Chem Rev 106:3210
5. Kendrew JC et al (1958) A three-dimensional model of the myoglobin molecule obtained by x-ray analysis. Nature 181:662
6. Frank J et al (1995) A model of protein synthesis based on cryo-electron microscopy of the E. coli ribosome. Nature 376:441
7. Svergun DI (1999) Restoring low resolution structure of biological macromolecules from solution scattering using simulated annealing. Biophys J 76:2879
8. Wuthrich K (1986) NMR of proteins and nucleic acids. Wiley, New York
9. Callender R, Dyer RB (2002) Probing protein dynamics using temperature jump relaxation spectroscopy. Curr Opin Struct Biol 12:628
10. Ha T et al (1999) Single-molecule fluorescence spectroscopy of enzyme conformational dynamics and cleavage mechanism. Proc Natl Acad Sci USA 96:893
11. Lippincott-Schwartz J, Snapp E, Kenworthy A (2001) Studying protein dynamics in living cells. Nat Rev Mol Cell Biol 2:444
12. Michalet X, Weiss S, Jager M (2006) Single-molecule fluorescence studies of protein folding and conformational dynamics. Chem Rev 106:1785
13. Misteli T (2001) Protein dynamics: implications for nuclear architecture and gene expression. Science 291:843
14. Levitt M (1983) Protein folding by restrained energy minimization and molecular dynamics. J Mol Biol 170:723
15. Karplus M, McCammon JA (2002) Molecular dynamics simulations of biomolecules. Nat Struct Biol 9:646
16. Schaller RR (1997) Moore's law: past, present and future. Spectr IEEE 34:52
17. Larson SM, Snow CD, Shirts M (2002) Folding@ Home and Genome@ Home: using distributed computing to tackle previously intractable problems in computational biology
18. Shaw DE et al (2007) Anton, a special-purpose machine for molecular dynamics simulation. SIGARCH Comput Archit News 35:1
19. Snow CD, Nguyen H, Pande VS, Gruebele M (2002) Absolute comparison of simulated and experimental protein-folding dynamics. Nature 420:102
20. Voelz VA, Bowman GR, Beauchamp K, Pande VS (2010) Molecular simulation of ab initio protein folding for a millisecond folder NTL9(1–39). J Am Chem Soc 132:1526
21. Schlick T, Barth E, Mandziuk M (1997) Biomolecular dynamics at long timesteps: bridging the timescale gap between simulation and experimentation. Annu Rev Biophys Biomol Struct 26:181
22. Beberg AL, Ensign DL, Jayachandran G, Khaliq S, Pande VS (2009) IEEE international symposium on Parallel & Distributed Processing, 2009 (IPDPS 2009), Italy, pp 1–8
23. Shaw DE et al (2008) Anton, a special-purpose machine for molecular dynamics simulation. Commun ACM 51:91
24. Voter AF (1997) A method for accelerating the molecular dynamics simulation of infrequent events. J Chem Phys 106:4665
25. Isralewitz B, Gao M, Schulten K (2001) Steered molecular dynamics and mechanical functions of proteins. Curr Opin Struct Biol 11:224
26. Schlitter J, Engels M, Kruger P (1994) Targeted molecular dynamics: a new approach for searching pathways of conformational transitions. J Mol Graph 12:84

27. Hamelberg D, Mongan J, McCammon JA (2004) Accelerated molecular dynamics: a promising and efficient simulation method for biomolecules. J Chem Phys 120:11919
28. Voter AF (1997) Hyperdynamics: accelerated molecular dynamics of infrequent events. Phys Rev Lett 78:3908
29. Zhou R (2007) Replica exchange molecular dynamics method for protein folding simulation. Methods Mol Biol 350:205
30. Laio A, Parrinello M (2002) Escaping free-energy minima. Proc Natl Acad Sci USA 99:12562
31. Chodera JD, Singhal N, Pande VS, Dill KA, Swope WC (2007) Automatic discovery of metastable states for the construction of Markov models of macromolecular conformational dynamics. J Chem Phys 126:155101
32. Noe F (2008) Probability distributions of molecular observables computed from Markov models. J Chem Phys 128:244103
33. Noe F, Fischer S (2008) Transition networks for modeling the kinetics of conformational change in macromolecules. Curr Opin Struct Biol 18:154
34. Bowman GR, Huang X, Pande VS (2009) Using generalized ensemble simulations and Markov state models to identify conformational states. Methods 49:197
35. Singhal N, Snow CD, Pande VS (2004) Using path sampling to build better Markovian state models: predicting the folding rate and mechanism of a tryptophan zipper beta hairpin. J Chem Phys 121:415
36. Park S, Pande VS (2006) Validation of Markov state models using Shannon's entropy. J Chem Phys 124:054118
37. Ensign DL, Kasson PM, Pande VS (2007) Heterogeneity even at the speed limit of folding: large-scale molecular dynamics study of a fast-folding variant of the villin headpiece. J Mol Biol 374:806
38. Bowman GR, Voelz VA, Pande VS (2011) Atomistic folding simulations of the five-helix bundle protein lambda(6–85). J Am Chem Soc 133:664
39. Bowman GR, Pande VS (2010) Protein folded states are kinetic hubs. Proc Natl Acad Sci USA 107:10890
40. Huang X et al (2010) Constructing multi-resolution Markov State Models (MSMs) to elucidate RNA hairpin folding mechanisms. Pac Symp Biocomput 15:228
41. Silva DA, Bowman GR, Sosa-Peinado A, Huang X (2011) A role for both conformational selection and induced fit in ligand binding by the LAO protein. PLoS Comput Biol 7:e1002054
42. Da LT, Wang D, Huang X (2012) Dynamics of pyrophosphate ion release and its coupled trigger loop motion from closed to open state in RNA polymerase II. J Am Chem Soc 134:2399
43. Da LT, Pardo Avila F, Wang D, Huang X (2013) A two-state model for the dynamics of the pyrophosphate ion release in bacterial RNA polymerase. PLoS Comput Biol 9:e1003020
44. Noe F, Schutte C, Vanden-Eijnden E, Reich L, Weikl TR (2009) Constructing the equilibrium ensemble of folding pathways from short off-equilibrium simulations. Proc Natl Acad Sci USA 106:19011
45. Bowman GR, Geissler PL (2012) Equilibrium fluctuations of a single folded protein reveal a multitude of potential cryptic allosteric sites. Proc Natl Acad Sci USA 109:11681
46. Prinz JH et al (2011) Markov models of molecular kinetics: generation and validation. J Chem Phys 134:174105
47. Huang X, Bowman GR, Bacallado S, Pande VS (2009) Rapid equilibrium sampling initiated from nonequilibrium data. Proc Natl Acad Sci USA 106:19765
48. Zhao Y, Sheong FK, Sun J, Sander P, Huang X (2013) A fast parallel clustering algorithm for molecular simulation trajectories. J Comput Chem 34:95
49. Jain AK (2010) Data clustering: 50 years beyond K-means. Pattern Recognit Lett 31:651
50. Beauchamp KA et al (2011) MSMBuilder2: modeling conformational dynamics at the picosecond to millisecond scale. J Chem Theory Comput 7:3412
51. Bowman GR, Beauchamp KA, Boxer G, Pande VS (2009) Progress and challenges in the automated construction of Markov state models for full protein systems. J Chem Phys 131:124101

52. Yao Y et al (2013) Hierarchical Nystrom methods for constructing Markov state models for conformational dynamics. J Chem Phys 138:174106
53. Bacallado S, Chodera JD, Pande V (2009) Bayesian comparison of Markov models of molecular dynamics with detailed balance constraint. J Chem Phys 131:045106
54. Pande VS, Beauchamp K, Bowman GR (2010) Everything you wanted to know about Markov State Models but were afraid to ask. Methods 52:99
55. Beauchamp KA, Ensign DL, Das R, Pande VS (2011) Quantitative comparison of villin headpiece subdomain simulations and triplet-triplet energy transfer experiments. Proc Natl Acad Sci USA 108:12734
56. Zhuang W, Cui RZ, Silva DA, Huang X (2011) Simulating the T-jump-triggered unfolding dynamics of trpzip2 peptide and its time-resolved IR and two-dimensional IR signals using the Markov state model approach. J Phys Chem 115:5415
57. Weinan E, Vanden-Eijnden E (2010) Transition-path theory and path-finding algorithms for the study of rare events. Annu Rev Phys Chem 61:391
58. Bowman GR, Voelz VA, Pande VS (2011) Taming the complexity of protein folding. Curr Opin Struct Biol 21:4
59. Cronkite-Ratcliff B, Pande V (2013) MSMExplorer: visualizing Markov state models for biomolecule folding simulations. Bioinformatics 29:950
60. Michaud-Agrawal N, Denning EJ, Woolf TB, Beckstein O (2011) MDAnalysis: a toolkit for the analysis of molecular dynamics simulations. J Comput Chem 32:2319–2327
61. Humphrey W, Dalke A, Schulten K (1996) VMD: visual molecular dynamics. J Mol Graph 14:33
62. Lindahl E, Hess B, van der Spoel D (2001) GROMACS 3.0: a package for molecular simulation and trajectory analysis. J Mol Model 7:306
63. Hess B, Kutzner C, van der Spoel D, Lindahl E (2008) GROMACS 4: algorithms for highly efficient, load-balanced, and scalable molecular simulation. J Chem Theory Comput 4:435
64. Kornberg R (2007) The molecular basis of eukaryotic transcription (Nobel Lecture). Angew Chem 46:6956
65. Lee TI, Young RA (2000) Transcription of eukaryotic protein-coding genes. Annu Rev Genet 34:77
66. Shilatifard A, Conaway RC, Conaway JW (2003) The RNA polymerase II elongation complex. Annu Rev Biochem 72:693
67. Malinen AM et al (2012) Active site opening and closure control translocation of multisubunit RNA polymerase. Nucleic Acids Res 40:7442
68. Abbondanzieri EA, Greenleaf WJ, Shaevitz JW, Landick R, Block SM (2005) Direct observation of base-pair stepping by RNA polymerase. Nature 438:460
69. Yin YW, Steitz TA (2004) The structural mechanism of translocation and helicase activity in T7 RNA polymerase. Cell 116:393
70. Wang D, Bushnell DA, Westover KD, Kaplan CD, Kornberg RD (2006) Structural basis of transcription: role of the trigger loop in substrate specificity and catalysis. Cell 127:941
71. Vassylyev DG et al (2007) Structural basis for substrate loading in bacterial RNA polymerase. Nature 448:163
72. Koshland DE (1958) Application of a theory of enzyme specificity to protein synthesis. Proc Natl Acad Sci USA 44:98
73. Tsai CJ, Kumar S, Ma B, Nussinov R (1999) Folding funnels, binding funnels, and protein function. Protein Sci Publ Protein Soc 8:1181
74. Oh BH et al (1993) Three-dimensional structures of the periplasmic lysine/arginine/ornithine-binding protein with and without a ligand. J Biol Chem 268:11348
75. Hünenberger PH et al (1997) Experimental and theoretical approach to hydrogen-bonded diastereomeric interactions in a model complex. J Am Chem Soc 119:7533

Chapter 3
Understanding Protein Dynamics Using Conformational Ensembles

X. Salvatella

Abstract Conformational ensembles are powerful tools to represent the range of conformations that can be sampled by proteins. They can be generated by using purely theoretical methods or, as is most often the case, by fitting ensembles of conformations to experimental data that report on the amplitude of protein dynamics. Conformational ensembles have been useful instruments to study fundamental properties of proteins such as the mechanism of molecular recognition, the early stages of protein folding and the mechanism by which structural information propagates through the structures of globular proteins structures *via* correlated backbone motions. In this chapter I will review the various approaches that have been put forward in the literature to generate conformation ensembles for proteins and present a selection of examples of how such representations of the structural heterogeneity of proteins have been used to explore the fundamental properties of these macromolecules. Finally, I will look ahead at likely future developments in this area, which is important for structural and chemical biology as well as for biophysics.

Keywords Conformational ensembles • Nuclear magnetic resonance • Correlated motions • Conformational selection • Induced fit • Allostery

3.1 Protein Dynamics

The structures of proteins fluctuate in various timescales and with various amplitudes [1]. Since thesefluctuations play important roles in biological function it is desirable to complement the structural information contained in protein structures

X. Salvatella (✉)
Joint BSC-IRB Research Programme in Computational Biology,
Institute for Research in Biomedicine (IRB Barcelona), Barcelona, Spain
e-mail: xavier.salvatella@irbbarcelona.org

K. Han et al. (eds.), *Protein Conformational Dynamics*, Advances in Experimental Medicine and Biology 805, DOI 10.1007/978-3-319-02970-2_3,

with an account of how these fluctuate [2]. Various methodological approaches have been put forward to reach this goal, ranging from purely theoretical methods that predict the fate of protein structures from first principles [3] to experimental methods that provide very detailed equilibrium distributions of well-defined structural properties such as specific inter-atomic distances.

Protein conformational ensembles aim at representing the range of conformations that a given protein samples at equilibrium [4]. Several methods have been put forward for their generation and most of these rely on the fitting of experimental data sensitive to structural fluctuations to ensembles of conformations that model the structural heterogeneity of proteins [2]. It is important to state early on that conformational ensembles do not in general report on protein dynamics in the sense that they do not provide information about the rate of inter-conversion between conformers. They can however report on the amplitude of the dynamics and this property has found wide use in the analysis of the behavior of proteins [4].

3.2 Generating Conformational Ensembles

As previously mentioned a wide range of tools is available for the generation of conformational ensembles. Although they vary quite widely in how the ensembles are built they share one important feature, which is that they extract information on the amplitude of protein dynamics from experimental data sensitive to structural fluctuations. I will now describe the different conceptual approaches that have been used in the field to generate such representations of the structural heterogeneity of proteins.

3.2.1 Using Molecular Dynamics and Advanced Sampling Methods

Molecular dynamics (MD) [3] is a simulation technique that can in principle provide an extremely detailed description of protein dynamics. It is based on modeling interatomic interactions by using empiric potentials called force fields and in the prediction of the time evolution of experimental structures by integration of Newton's equations of motion. The accuracy of these trajectories relies of course on the quality of such force fields and on the ability of computer hardware to simulate biologically relevant timescales, which is still a challenge, especially for large proteins and multi-protein complexes.

Although the technique was developed long ago it has experienced an extraordinary surge in recent years (Fig. 3.1) thanks to the availability of hardware designed specifically to carry out MD simulations, such as the supercomputer Anton [9], built by D. E. Shaw Research, and hardware designed to carry out other tasks but

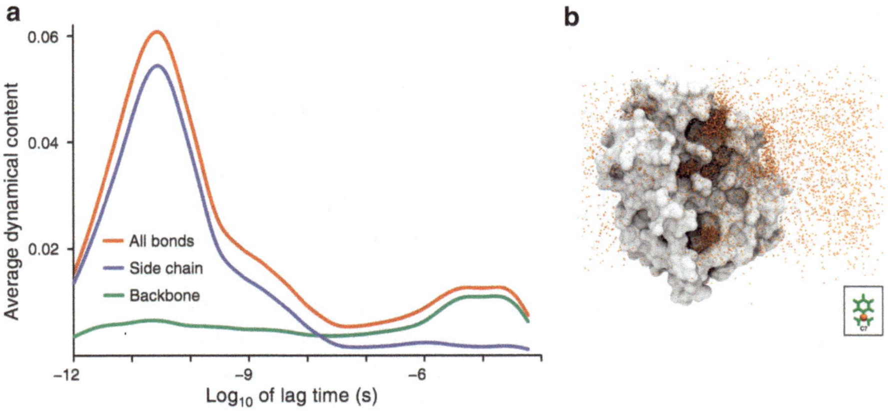

Fig. 3.1 MD is a very powerful simulation tool to characterize the structural heterogeneity and the dynamics of macromolecules [5] as well as their interactions will small molecules [6, 7]. (**a**) Analysis of a 1 ms simulation of BPTI run using the supercomputer Anton built by D. E. Shaw research [5] which illustrates how the dynamics of side chains are in general much faster than those of the backbone (**b**) snapshots of a simulation of the binding of benzamidine to trypsin carried out by using the GPUGRID distributed computing network [8], where the *orange dots* represent the various positions adopted by the C7 atom of benzamidine, shown in the *inset*, illustrating that the small molecule can bind to several sites on the surface of the enzyme before forming a stable complex

that performs MD particularly efficiently, such as graphics processing units (GPUs) [8, 10]. Mainly thanks to these improvements in hardware it has recently become possible to produce trajectories that are sufficiently long to sample the averaging time of experimental observables [11]. This is an important development because it allows the direct comparison of the simulated trajectories with experimental data, which is necessary for iteratively improving force field parameters to render the simulated behavior increasingly realistic.

One illustrative example of how increases in simulation time can lead to improvements in the quality of force fields is provided by a recent report of the D. E. Shaw Research team [12]. To improve the force field parameters that describe the conformational preferences of amino acid side chains the authors first compared the rotamer distributions obtained in very long (720 ns) MD simulations of model α-helices using a state of the art force field (Amber ff99SB) [13] with those derived from a statistical analysis of experimental structures deposited in the protein data bank (PDB).

After identifying four side amino acid types for which the agreement was poor (Ile, Leu, Asp, Asn) they optimized the force field parameters that govern the conformational properties of their side chains against quantum chemical calculations. Finally, and crucially, they validated the force field thus obtained (Amber ff99SB – ILDN) by predicting the NMR parameters (scalar and dipolar couplings, see Table 3.1) of a number of globular proteins that have been well studied using this technique such as hen egg white lysozyme, bovine pancreatic trypsin inhibitor, ubiquitin, and the B3 domain of Protein G. That the authors were able to generate trajectories with a length (1.2 μs) that matches the averaging time of the NMR parameters was key for proving that the new force field parameters, that provide a better validation that the old ones, represent a substantial improvement [12].

In cases where the size of the system and the timescale of the dynamics of interest allow investigation by MD this is undoubtedly the most informative technique that is currently available for characterizing the fluctuations of the structure of proteins. Even in cases where MD can be used it is nevertheless necessary to validate the resulting trajectories either by predicting experimental data sensitive to dynamics such as nuclear magnetic resonance (NMR) parameters (see below) or by predicting the outcome of perturbations of the system such as point mutations. Only in cases when these validations are successful is it advisable to consider the trajectories provided by MD a realistic model of the behavior of the protein [14].

An illustrative example of how it is possible to use MD trajectories validated by experiments to analyze very subtle but important dynamical properties of proteins is provided by the work of the Bruschweiler group on the protein ubiquitin. Ubiquitin is small protein that is used as a model system for this type of studies because it is of a size that allows the simulation of relative long timescales, is stable in most force fields and because its spectroscopic properties render its characterization using NMR relatively straightforward.

In an important study published in 2007 Showalter and Bruschweiler showed that simulating the dynamics of this protein using MD and a state of the art force field (Amber ff99SB) for 50 ns lead to a trajectory that agreed with NMR data sensitive to dynamics (RDCs) better than the X-ray structure (1UBQ) and only slightly worse than the NMR structure (1D3Z) refined against the NMR data used for validation [15]. These results indicate that, at least for ubiquitin, proper consideration of the contribution of motional averaging to the measured NMR data by using MD can lead to much improved representations of the structural properties of proteins (Fig. 3.2).

Armed with this validation Bruschweiler and co-workers analyzed the degree of correlation of the motions of the backbone torsions in this protein [16]. They recently found that there is a weak but certain degree of correlation of the motions of the ϕ and ψ torsion angles of residues facing one another across the β-sheet of ubiquitin, especially when they are hydrogen bonded, but that this decays very quickly as the distance between two given residues increases. It is important to emphasize that this type of analysis, that relies on a very accurate representation of the dynamics of the protein, is warranted due to the notable ability of the trajectory obtained by these researchers to validate against experiments [15].

Although MD has made spectacular progress in the last few years there are still many biological processes that cannot yet be routinely simulated using this technique. Processes that fall under this category include those involving intrinsically disordered proteins (IDPs), that have important biological functions but that fail to fold into conventional structures that can be characterized using the tools of structural biology such as X-ray crystallography, conventional NMR and cryo-

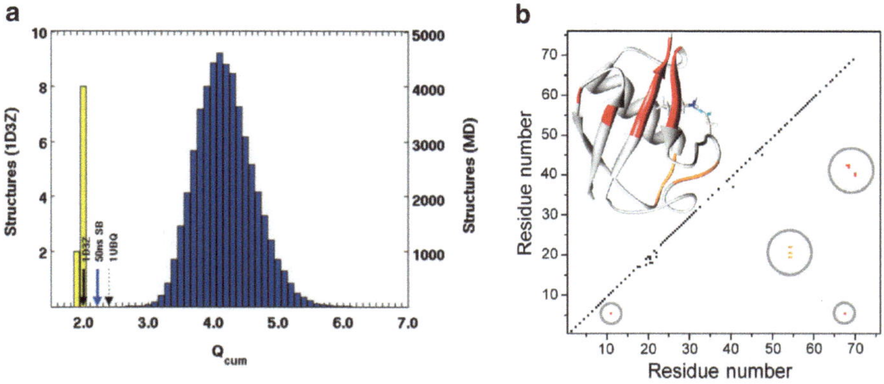

Fig. 3.2 MD trajectories validated against experimental data that reports on the amplitude of protein dynamics can be used to understand fundamental properties of proteins such as the presence of correlated backbone motions. (**a**) Histogram of the quality factors, with $Q = rms(D^{exp} - D^{calc})/rms(D^{exp})$, of the various conformations sampled during a 50 ns trajectory of ubiquitin (in *blue*) and of the conformations of the conventional NMR ensemble (in *yellow*, pdb code 1D3Z); quality factors of the various structural representations (with 1UBQ representing the highest resolution X-structure) (**b**) matrix representation of the pairwise correlation coefficient between backbone torsion angles of ubiquitin with R^2 larger than 0.1, which shows the presence of correlated motions across the β-sheet of the protein as indicated in *red* in the structure as well as in other hydrogen bonded residues

electron microscopy [17]. These proteins present an extreme degree of structural heterogeneity and play important roles in molecular mechanisms of enormous importance for biology, such as transcription [18, 19], and biomedicine [20].

Due to the challenges involved in sampling the particularly vast conformational space explored by IDPs and to the possibility that current force fields are not optimized to accurately describe the weak inter-atomic interactions that dominate the behavior of this type of proteins the use of conventional MD to study such systems is still in its infancy [21]. For this reason there is substantial interest in the development of approaches that allow determining conformational ensembles for IDPs by combining molecular simulations with the information contained in experimental observables reporting on the amplitude of protein dynamics such as SAXS and NMR [22] (Sect. 3.2.2).

In cases where the size of the protein is too large, the conformational space too vast [23] and the dynamics often too slow to be sampled by conventional MD it is possible to use advanced sampling methods such as replica exchange MD [24], umbrella sampling [25], accelerated MD [26], local elevation [27] and metadynamics [28], among others available, to explore the conformational space sampled by proteins. In these methods various strategies are used to overcome the free energy barriers present in the energy landscape, that prevent efficient sampling in conventional MD. A detailed theoretical and technical description of these techniques is beyond the scope of this chapter and can be found elsewhere [29, 30]. Advanced sampling tools have been important to show that appropriate

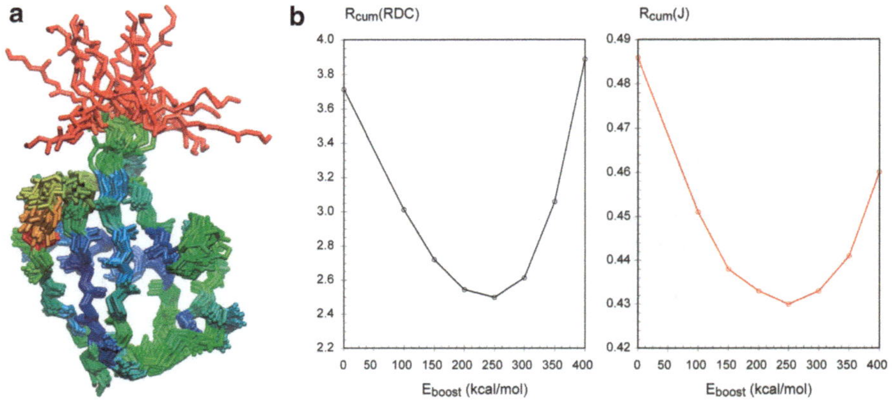

Fig. 3.3 Advanced sampling methods can lead to accurate representations of the structural heterogeneity of proteins. (**a**) Ubiquitin ensemble obtained by the McCammon and Blackledge groups by AMD with a degree of boost optimized by validation against NMR data [31], where the residues are colored according to their flexibility (*blue*: rigid, *red*: flexible). (**b**) Level of agreement of AMD trajectories, expressed as $R_{cum} = Q_{cum}/\sqrt{2}$, obtained with increasing degrees of boost with an indication that both the scalar and the dipolar couplings, that average on the same timescale, validate best when $E_{boost} = 250$ kcal.mol^{-1}, strongly indicating that this level of boost allows sampling all conformations that contribute to the average, experimental NMR parameter [31]

consideration of the averaging implicit in the measurement of nuclear magnetic resonance (NMR) parameters can lead to very accurate descriptions of the structural heterogeneity of proteins and peptides [31].

In a particularly powerful illustration of how such methods can be used to report on the amplitude of protein dynamics McCammon, Blackledge and co-workers use accelerated molecular dynamics (AMD) [26] to analyze the dynamics of the protein ubiquitin [31]. In AMD sampling is accelerated by adding an energy term to the potential energy, that depends on the difference between the potential energy and a reference energy called the boost energy (E_{boost}), that effectively decreases the height of free energy barriers encountered by the simulation [26]. The correct value of the E_{boost} that needs to be used to reach a particular timescale is however, in principle, not known *a priori*.

To generate a trajectory describing all conformations contributing to the average NMR parameters measured experimentally for the small model protein ubiquitin (scalar and residual dipolar couplings, that average in the ms timescale) these authors carried out AMD simulations of this protein with increasing values of E_{boost}. An analysis of how well the simulated trajectories agreed with the NMR experiments showed that conventional MD simulations did not validate well, that moderate degrees of acceleration lead to very significant improvements in the agreement against experiment and that it was possible to use the experimental data to determine the optimal value of E_{boost}, one that samples all conformations that contribute to the time-averaged NMR parameters (Fig. 3.3).

3.2.2 *From Experimental Data*

In the methods discussed in the previous section experimental data plays a modest role. It is merely used as a validation tool to reassure the researcher that the trajectory provided by MD or by an advanced sampling method is a realistic representation of the dynamics of the system. There are however scenarios where it is desirable that experimental data plays a much more important role because the conformational space sampled by the system is too vast to be sampled by MD, for example, when there are reasons to think that a particular property of the system may not we well-described by force fields or when, for a variety of possible reasons, it is desirable to minimize the role played by the force field in determining the structural and dynamical properties of the system.

There are a number of experimental methods that can provide structural and dynamical information to determine conformational ensembles for proteins. These include Förster resonance energy transfers (FRET) measured using fluorescence, that provide information about r^{-6} averaged inter-dye distances [32] and, in single molecule mode, distributions of distances as well as small angle x-ray scattering (SAXS), that provides information about the hydrodynamic properties of proteins [33]. The most powerful experimental method is however undoubtedly NMR [34] because it provides information at atomic resolution, unlike SAXS, and because it does not require, unlike FRET, labeling the protein with fluorescent groups that can alter the structure, the dynamics and the interactions of the protein and therefore require performing extensive experimental controls.

It is possible to measure a number of parameters by NMR (chemical shifts, scalar couplings, nuclear Overhauser effects, residual dipolar couplings and chemical shift anisotropy in aligned samples, cross-correlated relaxation rates) and these can be related to quantities that can be in principle computed from structures, trajectories and ensembles (distances, angles, torsion angles) as shown in Table 3.1. For the purposes of validating and, especially, generating conformational ensembles it is of course important to take into appropriate consideration the range of validity of the equations used to back-calculate NMR parameters, the accuracy with which NMR parameters can be measured experimentally [35] and the accuracy with which they can be back-calculated [36]. Another factor to take into account is the way these equations were parameterized i.e. whether the parameters were determined from first principles or whether they were instead determined by fitting to known crystallographic structures, a procedure that leads to equations that under-estimate the contribution of conformational averaging [37].

3.2.2.1 Selection Methods

Various approaches are available for the generation of conformational ensembles from experimental but I will start will selection methods because they these conceptually related to the use of conventional MD and advanced sampling methods

Table 3.1 NMR parameters that can be used for generating conformational ensembles and their relationship to protein structure

Symbol	Description	Structural interpretation
NOE_{ij}	Nuclear Overhauser effect	Distance between ^1H nuclei i and j (<6 Å)
PRE_{ij}	Paramagnetic relaxation enhancement	Distance between an unpaired electron attached to site i and nucleus j (<30 Å)
$^3J_{ij}$	Three-bond scalar coupling	Dihedral angle between bond vectors i and j
$^{3h}J_{ij}$	Trans-hydrogen bond scalar coupling	Geometry of hydrogen bond linking heavy atoms i and j
RDC or D_{ij}	Residual dipolar coupling	Angle between bond vectors i and j and the molecular frame defined by macromolecular alignment
CS_i	Chemical shifts	Convolution of a large number of structural properties in the vicinity of nucleus i
$S_i{}^2$	Order parameter	Rigidity of bond vector i in the molecular frame defined by macromolecular tumbling ($0 \le S_i{}^2 \le 1$)

described inSect. 3.2.1. Selection methods use experimental data to, as their name suggests, select conformations from a pre-defined pool of conformations generated *a priori*. The pool is meant to contain all physically possible conformations that the protein can sample in a defined timescale with some probability but that these are not present with their correct statistical weights. Since any experimental (NMR or otherwise) parameter contains information about the distribution of conformations contributing to the average, it can be in principle be used to optimize the statistical weights.

Several algorithms have been used to select the conformations from the pool. These range from Monte Carlo algorithms, such as the ENSEMBLE method developed by the Forman-Kay laboratory to generate ensembles for IDPs [38], to genetic algorithms, such as the OED method of the Svergun laboratory to generate ensembles from SAXS data [39] and the ASTEROIDS method (Fig. 3.4a) developed by the Blackledge laboratory to generate ensembles for IDPs from NMR data [41]. These methods differ in the nature of the pool and in the technical details of the selection method but are all based on the same idea and make the same key assumption, which is that all possible conformations are present in the pool.

Of course whether a selection method performs well depends fundamentally on the properties of the pool. This must be an accurate representation of the range of conformations that can be sampled by the protein and its size must be representative of the size of the conformational space available. If the quality of the pool is low, that is if the conformations present in the pool do not represent the conformations that the protein can adopt, the selection method will generate a conformational

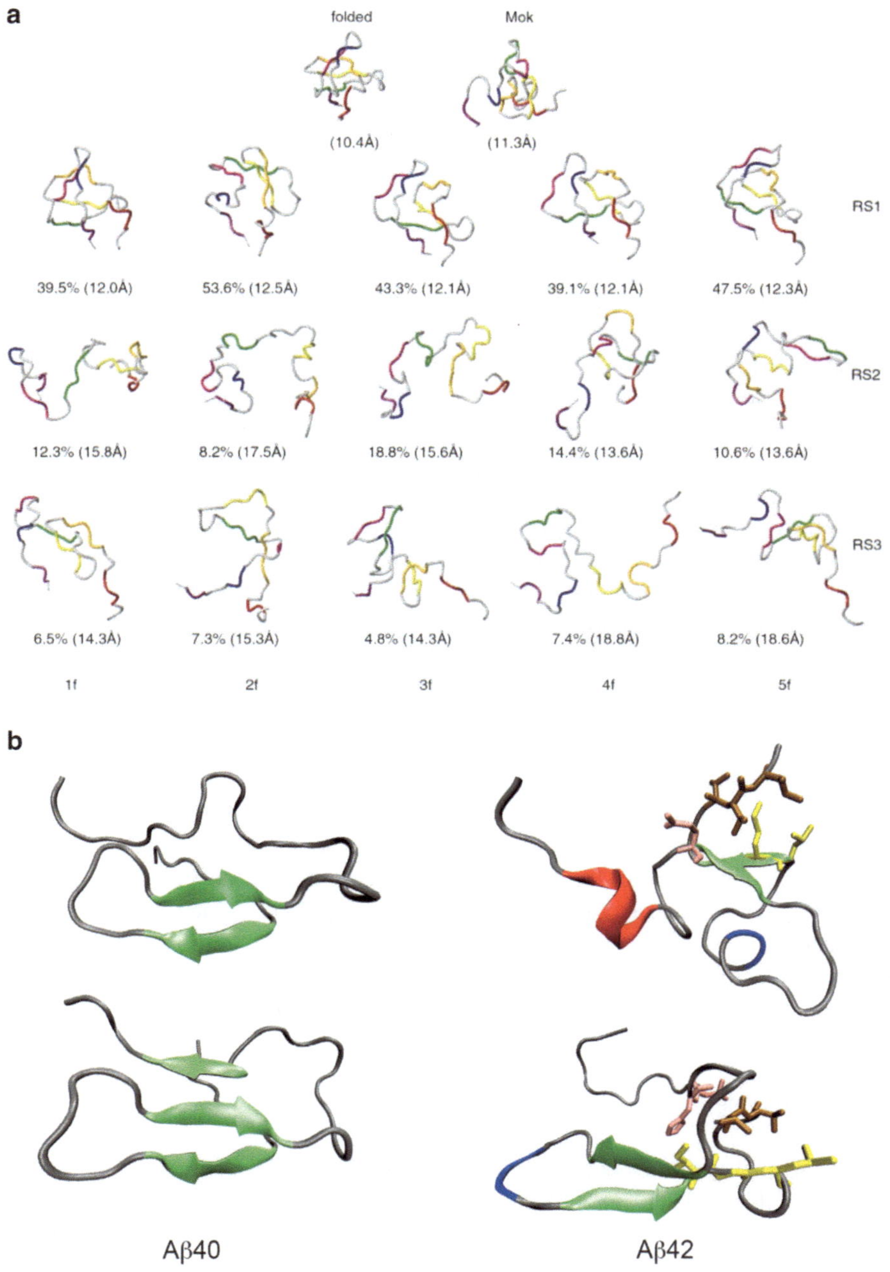

Fig. 3.4 (**a**) Representative structures belonging to the three main clusters (RS1, RS2, RS3) defining the denatured state of a drkN SH3 domain by selecting conformations from a pool generated by a simulation of the thermal denaturation of the native state of this protein [38] (**b**) Comparison of the dominant conformations of Aβ40 and Aβ42 as determined by selecting structures from a pool derived from trajectories of these peptides computed using MD or REMD. It can be observed that the C-terminal residues of Aβ42, that are not present in Aβ40, form long-range transient contacts [40]

ensemble that will be in agreement with experiment but that will be not be a realistic description of the structural properties of the protein. If the pool is of good quality but it contains too few conformations, that is if conformations actually sampled by the protein are absent from the pool, the selection algorithm will also fail to produce a useful result.

From Statistical Coils

As already mentioned selection methods have used quite extensively for the generation of conformational ensembles describing the properties of chemically denatured proteins and IDPs. The pioneering work of Dobson and co-workers [42], recently followed up by the Sosnick [23] and Blackledge [43] groups, showed that it is possible to produce reasonable representations for the range of structures sampled by these systems by generating conformational ensembles where the distributions of backbone torsion angles of the different residues of the protein match those of the same residues in loops and termini of structures deposited in the PDB.

These conformational ensembles aim at representing the structural properties of polypeptides where these are dominated by the local structural preferences [42], that is in the absence of the long-range interactions that play an important role in stabilizing the tertiary structure of globular proteins. These statistical coils have, in spite of the simplicity of the approach used to generate them, structural properties that match those determined experimentally for disordered proteins and that validate reasonably well with SAXS and NMR experimental parameters such as backbone scalar and residual dipolar couplings [23, 42, 43].

There is considerable experimental evidence, mainly from FRET, chemical shifts and paramagnetic relaxation enhancement (PRE) experiments measured using NMR (Table 3.1) that IDPs can form transient long range interactions that are important for their physiological and physiopathological roles [44–47]. Since these long range interactions cannot be definition be described by statistical coils [42] efforts have been made in improving the description of IDPs by using selection methods where statistical coils are used as pools.

From Ensembles Determined Using Simulations

Since the quality of the pool is key for the performance of selection methods efforts have been directed at using molecular simulations for constructing pools that better reproduce the structural properties of IDPs. These were pioneered by the Forman-Kay group, that used thermal unfolding trajectories to generate the pool in a study of the denatured state of an SH3 domain [38] in which they used ^{1}H-^{1}H NOEs, scalar couplings, ^{13}C chemical shifts, among other parameters, to optimize the statistical weights of the conformations of the pool (Fig. 3.4b). This [38], as well as other studies by the same group [48, 49], in which they also explored the use of statistical coils as pools, indicated that the denatured state is substantially collapsed, with a fraction of the secondary structure of the native state, and that it is stabilized by native as well as non-native long range interactions.

This pioneering work has been followed up by the Head-Gordon group, that has used instead ensembles derived from MD or ERMD trajectories to generate ensembles for the peptides Aβ40 and Aβ42 [40]. Although these two peptides have very similar sequences the latter is much more prone than the former to form neurotoxic oligomeric species thought to trigger Alzheimer's disease [50, 51]. There has been much discussion about how can the addition of only two residues have such a profound effect on the structural properties of an IDP. The recent study of the Head-Gordon group, in which the authors used ^{1}H-^{1}H NOEs as well as scalar and residual dipolar couplings to bias a selection algorithm, is an important contribution to this topic. It shows that Aβ42 has a substantially different contact map due to the propensity of the two additional residues to form long-range contacts with hydrophobic residues in the rest of the sequence (Fig. 3.4b).

3.2.2.2 Restrained Simulations

Both selection methods and, especially, MD rely heavily on an accurate description of the conformational space available to proteins either by using motional models, such as statistical coils, or molecular simulations force fields. These approaches are therefore unsuitable when these descriptions are not available or when it is thought that the conformations that they provide are not correct. In these cases it is possible to carry out restrained molecular simulations in which empirical potentials are added to the potential energy of the protein provided by the force field to penalize configurations with back-calculated experimental parameters that are in disagreement with those measured experimentally.

Since these methods bias the sampling they have the potential to generate conformations that would otherwise not be sampled in an unrestrained MD simulation *i.e.* they use the experimental data as a protein-specific force field correction [52]. Restrained simulations have of course a long history in structure determination and in fact lie at the heart of the ability of NMR to produce average structures for proteins by generating configurations with structural properties (bond lengths, angles, inter-atomic distances, etc.) that do not deviate too much from those considered optimal for molecular simulation force fields and are in addition in agreement with NMR parameters reporting on protein structure (Table 3.1) [53, 54].

Given that the NMR parameters cannot be measured or back-calculated from structures with infinite accuracy they usually do not define a unique conformation and it is customary to represent NMR structures as ensembles of conformations that fit the NMR data. The spread of these ensembles depends on the ability of the experimental data to define the average structure and can be considered equivalent to the resolution of crystallographic structures, with significant heterogeneity reflecting poor resolution. Even though the presence of significant dynamics can lead, like in X-ray crystallography, to poor resolution it is important to emphasize that the spread of conventional NMR ensembles does not directly report on protein dynamics [52].

For NMR ensembles to report on protein dynamics it is necessary to use restrained simulation protocols that fit the NMR parameters to an ensemble of conformations rather to a single average conformation [2, 55] In these protocols an energy penalty is applied when the ensemble–averaged back-calculated NMR parameters are in disagreement with experiment.

Ensemble Averaged Restrained Simulations

In ensemble-averaged restrained simulations an ensemble of conformations of the protein of interest is simulated simultaneously and the ensemble-averaged back-calculated NMR parameters are restrained by an empirical quadratic potential to be in agreement with the experimental value [2]. When there is a violation the energy penalty generates a force in all conformations that contribute to the average so that the NMR parameter is fulfilled. This leads to conformational ensembles where individual conformations may have NMR parameters that deviate from experiment but where the ensemble collectively does not and is, therefore, a representation of the range of conformations that is sampled by the protein at equilibrium.

As early as the 1990s there was a general awareness of the importance of conformational averaging in protein structure determination by NMR and, as a consequence, attempts at using this type of simulation protocols to generate conformational ensembles from the NMR parameters commonly used for determining structures [55, 56]. This was however a challenging task because NOEs, the main source of structural information available at the time, are not suitable restraints for ensemble simulations because they average non-linearly [57] and could only be measured semi-quantitatively due to the low signal to noise ratio of NMR and the presence of spin-diffusion [58, 59]. As a consequence it was not possible to cross-validate the resulting conformational ensembles, which were significantly under-restrained and therefore presented artifactual structural heterogeneity. It is worth mentioning that the measurement of very exact NOEs in per-deuterated proteins is to a large extent alleviating the problems of this NMR observable as restraint in ensemble simulations as illustrated by recent work of Riek and co-workers [58, 59].

This situation, however, changed quite significantly when Tjandra and Bax showed that it was possible to induce a small degree of anisotropy in the rotational diffusion on protein samples by using an external alignment medium [62]. This lead to the possibility of measuring residual dipolar couplings (RDCs) for pairs of nuclei, which otherwise average to zero when the inter-nuclear vector rotates isotropically around the magnetic field of the NMR apparatus. For proteins which do not experience important changes in alignment when their structures fluctuate [63] the value of the RDC of a given conformation is given by the degree of alignment, which depends on the overall shape of the protein, and on the orientation

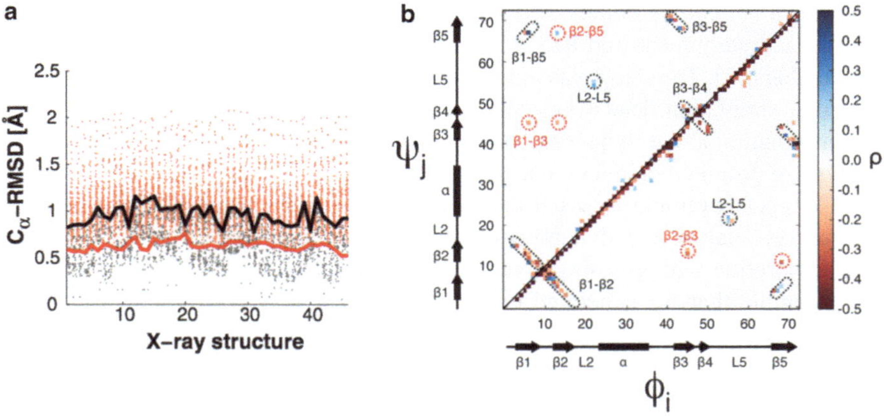

Fig. 3.5 Restrained ensemble simulations can be used to generate detailed descriptions of the structural heterogeneity of proteins that can be of great use to understand fundamental properties of their structures (**a**) A conformational ensemble determined for the protein ubiquitin using RDCs in its free state contains the structures that the protein adopts upon binding other proteins, indicating that the molecular recognition of this protein occurs by conformational selection. Plot of the Calpha-RMSD between each bound structure of ubiquitin (*x-axis*) and the ensemble members of free (*red dots*) and bound (*black dots*) ubiquitin [60] (**b**) Using a similar approach we showed that the motions of residues in the β-sheet of this protein are correlated, which suggests that the collective motions that these correlations underlie could play role in molecular recognition. The matrix represents the circular correlation coefficient between two backbone torsion angles of ubiquitin [61]

of the inter-nuclear vector in the molecular structure [62]. Unlike NOEs RDCs can be measured with quite high accuracy and, in addition, average linearly; they are therefore much better suited to ensemble simulation protocols that aim at generating conformational ensembles and have been used quite extensively in the last few years [60, 61, 64, 65].

Parallel to these developments methods were also set up to use S^2 order parameters, that are derived from ^{15}N relaxation rates and that report on the amplitude of the motions of individual backbone NH bond vectors, as restraints for ensemble restrained simulations [66]. Vendruscolo, Dobson an co-workers showed that using these in combination with NOEs allowed the generation of conformational ensembles that validated against ensemble-averaged experimental RDCs better than single structures [67]. In further developments Vendruscolo and co-workers showed how implementing changes to the simulation protocol, particularly to the way in which the averaging of the various observables was carried out, lead to very substantial increases in accuracy of the ensembles [68].

The use of RDCs as restraints in ensemble simulations was pioneered by Clore and co-workers, which setup up the basic simulation protocol and put forward an ingenious strategy to simultaneously fit the coordinates of the ensemble members with the 5 independent elements of the alignment tensor [64]. This mathematical object describes the degree and direction of alignment or the protein in the molecular frame and can be represented by a symmetric traceless matrix that is in generally not known but can be obtained by single value decomposition from a set of more than 5 accurately measured RDCs if a reasonable structural model for the protein is available [69]. The setup was initially implemented to explore whether restrained ensemble simulations lead to ensembles that agreed with alternative ways of treating the information about dynamics contained in RDCs [64] but has found use in the analysis of protein dynamics since then [61, 65].

In a recent example we used the simulation setup proposed by Clore and co-workers to analyze the dynamics of ubiquitin from a very large set of RDCs in collaboration with Griesinger and co-workers [61]. One important property of the ensemble that we generated is that it is in very good agreement with NMR parameters that we did not use to restrain the ensemble simulation such as trans-hydrogen bond scalar couplings and cross-correlated relaxation rates. An analysis of the correlated motions present in this ensemble lead to the observation of weak but statistically significant correlated motions that connect the dynamics of residues that can be quite far apart in the structure of ubiquitin.

As previously mentioned the degree and directions of alignment of a protein structure expressed in the alignment tensor are generally not known and need to be fit to the experimental RDCs [69]. This is an important drawback of using RDCs for characterizing the structural heterogeneity of proteins because it decreases the information content of these NMR parameters and because it narrows the range of systems that can be studied to those for which the alignment is assumed not to change significantly during the dynamics. One possible way to alleviate this problem is to calculate the alignment tensor of the various conformations that contribute to the average RDC *on the fly* [70, 71], which is possible for mechanisms of external that are well understood such as steric [69] and electrostatic [72] alignment. This new approach to determining conformational ensembles is still under development [73, 74] but it is likely to be an important development because it will allow the characterization of the structural heterogeneity of proteins that experience large shape changes such as intrinsically disordered and multi-domain proteins.

In parallel to these developments in the characterization of the structural heterogeneity of globular proteins restrained ensemble simulations have been used extensively to generate conformational ensembles for chemically denatured and intrinsically disordered proteins. The main source of structural information for these proteins are paramagnetic relaxation enhancements (PREs). PREs are increases in the relaxation of NMR resonances caused by transient interactions of the corresponding nuclei with paramagnetic functional groups introduced by using protein engineering and can be used to probe long-range (up to 25 Å) transient interactions [77]. PREs can be used as restraints in ensemble simulations of disordered proteins [78] but suffer from the same averaging problems of NOEs *i.e.* they do not average linearly.

In order to clarify to what extent PREs are useful restraints for ensemble simulations of disordered proteins our laboratory recently carried out a detailed characterization of their information content. The conclusion that we reached is that PREs are indeed very useful probes of transient long-range interactions but that their averaging properties render them quite inadequate restraints for ensemble simulations because their average is to a large insensitive to the shape of the distribution of distances. We find that ensemble-averaging does not provide significant advantages for obtaining an accurate characterization of transient long-range interactions and that fitting the PREs to a small number of structures, that can be as small as one, provides the most accurate map for a disordered protein [79].

Time Averaged Restrained Simulations

An alternative to ensemble restrained simulations is time averaged restrained simulations. In this approach, developed by Van Gunsteren and co-workers, a single conformation of the protein is simulated and a quadratic empirical potential ensures that the time-averaged value of a given NMR parameter is equivalent to its experimental counterpart [75]. The key parameter of this simulation protocol is the averaging time *i.e.* the time after which the trajectory is expected to satisfy the experimental values, which needs to be determined *a priori*. Although this approach was an important conceptual development when it was proposed [75, 76] it seems that carrying our ensemble-averaged restrained simulations is a more common approach to the problem of determining conformational ensembles from NMR data.

3.3 Looking Ahead

Conformational ensembles represent an exciting new development in biophysics because they allow for an explicit represent of the dynamics of proteins. Although they do not contain information about the timescale of the dynamics these ensembles provide quite accurate representations of the amplitude of the motions. It is however the case that the determination of such ensembles from experiment is not a routine endeavor because it requires the measurement of a substantial number of NMR parameters such as RDCs. From this point of view it seems that an important priority should be to extract as much information about the amplitude of dynamics from NMR chemical shifts because this NMR parameter is easy to measure. It is therefore likely that we will see developments in this area soon [80].

References

1. Karplus M, Kuriyan J (2005) Molecular dynamics and protein function. Proc Natl Acad Sci USA 102:6679–6685
2. Vendruscolo M (2007) Determination of conformationally heterogeneous states of proteins. Curr Opin Struct Biol 17:15–20
3. McCammon JA, Gelin BR, Karplus M (1977) Dynamics of folded proteins. Nature 267: 585–590

4. Fenwick RB, Esteban-Martín S, Salvatella X (2011) Understanding biomolecular motion, recognition, and allostery by use of conformational ensembles. Eur Biophys J 40:1339–1355
5. Shaw DE, Maragakis P, Lindorff-Larsen K, Piana S, Dror RO et al (2010) Atomic-level characterization of the structural dynamics of proteins. Science 330:341–346
6. Shan Y, Kim ET, Eastwood MP, Dror RO, Seeliger MA et al (2011) How does a drug molecule find its target binding site? J Am Chem Soc 133:9181–9183
7. Buch I, Giorgino T, De Fabritiis G (2011) Complete reconstruction of an enzyme-inhibitor binding process by molecular dynamics simulations. Proc Natl Acad Sci USA 108: 10184–10189
8. Buch I, Harvey MJ, Giorgino T, Anderson DP, De Fabritiis G (2010) High-throughput all-atom molecular dynamics simulations using distributed computing. J Chem Inf Model 50:397–403
9. Shaw DE, Chao JC, Eastwood MP, Gagliardo J, Grossman JP et al (2008) Anton, a special-purpose machine for molecular dynamics simulation. Commun ACM 51:91
10. Harvey MJ, Giupponi G, Fabritiis GD (2009) ACEMD: accelerating biomolecular dynamics in the microsecond time scale. J Chem Theory Comput 5:1632–1639
11. Maragakis P, Lindorff-Larsen K, Eastwood MP, Dror RO, Klepeis JL et al (2008) Microsecond molecular dynamics simulation shows effect of slow loop dynamics on backbone amide order parameters of proteins. J Phys Chem B 112:6155–6158
12. Lindorff-Larsen K, Piana S, Palmo K, Maragakis P, Klepeis JL et al (2010) Improved side-chain torsion potentials for the Amber ff99SB protein force field. Proteins 78:1950–1958
13. Hornak V, Abel R, Okur A, Strockbine B, Roitberg A et al (2006) Comparison of multiple Amber force fields and development of improved protein backbone parameters. Proteins 65:712–725
14. van Gunsteren WF, Dolenc J, Mark A (2008) Molecular simulation as an aid to experimentalists. Curr Opin Struct Biol 18:149–153
15. Showalter SA, Brüschweiler R (2007) Quantitative molecular ensemble interpretation of NMR dipolar couplings without restraints. J Am Chem Soc 129:4158–4159
16. Li D-W, Meng D, Brüschweiler R (2009) Short-range coherence of internal protein dynamics revealed by high-precision in silico study. J Am Chem Soc 131:14610–14611
17. Dunker AK, Silman I, Uversky VN, Sussman JL (2008) Function and structure of inherently disordered proteins. Curr Opin Struct Biol 18:756–764
18. Fuxreiter M, Tompa P, Simon I, Uversky VN, Hansen JC et al (2008) Malleable machines take shape in eukaryotic transcriptional regulation. Nat Chem Biol 4:728–737
19. Garcia-Pino A, Balasubramanian S, Wyns L, Gazit E, De Greve H et al (2010) Allostery and intrinsic disorder mediate transcription regulation by conditional cooperativity. Cell 142: 101–111
20. Iakoucheva LM, Brown CJ, Lawson JD, Obradović Z, Dunker AK (2002) Intrinsic disorder in cell-signaling and cancer-associated proteins. J Mol Biol 323:573–584
21. Lindorff-Larsen K, Trbovic N, Maragakis P, Piana S, Shaw DE (2012) Structure and dynamics of an unfolded protein examined by molecular dynamics simulation. J Am Chem Soc 134:3787–3791
22. Jensen MR, Ruigrok RW, Blackledge M (2013) Describing intrinsically disordered proteins at atomic resolution by NMR. Curr Opin Struct Biol 23:426–435
23. Jha AK, Colubri A, Freed KF, Sosnick TR (2005) Statistical coil model of the unfolded state: resolving the reconciliation problem. Proc Natl Acad Sci USA 102:13099–13104
24. Sugita Y, Okamoto Y (1999) Replica-exchange molecular dynamics method for protein folding. Chem Phys Lett 314:141–151
25. Torrie GM, Valleau JP (1977) Nonphysical sampling distributions in Monte Carlo free-energy estimation: umbrella sampling. J Comput Phys 23:187–199
26. Hamelberg D, Mongan J, McCammon JA (2004) Accelerated molecular dynamics: a promising and efficient simulation method for biomolecules. J Chem Phys 120:11919
27. Huber T, Torda AE, van Gunsteren WF (1994) Local elevation: a method for improving the searching properties of molecular dynamics simulation. J Comput Aided Mol Des 8:695–708

28. Laio A, Parrinello M (2002) Escaping free-energy minima. Proc Natl Acad Sci USA 99: 12562–12566
29. Berne BJ, Straub JE (1997) Novel methods of sampling phase space in the simulation of biological systems. Curr Opin Struct Biol 7:181–189
30. Zuckerman DM (2011) Equilibrium sampling in biomolecular simulations. Annu Rev Biophys 40:41–62
31. Markwick PRL, Bouvignies G, Salmon L, McCammon JA, Nilges M et al (2009) Toward a unified representation of protein structural dynamics in solution. J Am Chem Soc 131: 16968–16975
32. Sisamakis E, Valeri A, Kalinin S, Rothwell PJ, Seidel CAM (2010) Accurate single-molecule FRET studies using multiparameter fluorescence detection. Methods Enzymol 475:455–514
33. Svergun D, Barberato C, Malfois M, Volkov V, Konarev P et al (1995) Evaluation of the solution scattering from macromolecules with known atomic structure and fitting to experimental data. J Appl Crystallogr 28:768–773
34. Dyson HJ, Wright PE (2004) Unfolded proteins and protein folding studied by NMR. Chem Rev 104:3607–3622
35. Fenwick RB, Esteban-Martín S, Salvatella X (2010) Influence of experimental uncertainties on the properties of ensembles derived from NMR residual dipolar couplings. J Phys Chem Lett 1:3438–3441
36. Robustelli P, Cavalli A, Dobson CM, Vendruscolo M, Salvatella X (2009) Folding of small proteins by Monte Carlo simulations with chemical shift restraints without the use of molecular fragment replacement or structural homology. J Phys Chem B 113:7890–7896
37. Case DA (2013) Chemical shifts in biomolecules. Curr Opin Struct Biol 23:172–176
38. Choy WY, Forman-Kay JD (2001) Calculation of ensembles of structures representing the unfolded state of an SH3 domain. J Mol Biol 308:1011–1032
39. Bernadó P, Mylonas E, Petoukhov MV, Blackledge M, Svergun DI (2007) Structural characterization of flexible proteins using small-angle X-ray scattering. J Am Chem Soc 129:5656–5664
40. Ball KA, Phillips AH, Wemmer DE, Head-Gordon T (2013) Differences in β-strand populations of monomeric Aβ40 and Aβ42. Biophys J 104:2714–2724
41. Nodet G, Salmon L, Ozenne V, Meier S, Jensen MR et al (2009) Quantitative description of backbone conformational sampling of unfolded proteins at amino acid resolution from NMR residual dipolar couplings. J Am Chem Soc 131:17908–17918
42. Smith LJ, Bolin KA, Schwalbe H, MacArthur MW, Thornton JM et al (1996) Analysis of main chain torsion angles in proteins: prediction of NMR coupling constants for native and random coil conformations. J Mol Biol 255:494–506
43. Bernadó P, Blanchard L, Timmins P, Marion D, Ruigrok RWH et al (2005) A structural model for unfolded proteins from residual dipolar couplings and small-angle x-ray scattering. Proc Natl Acad Sci USA 102:17002–17007
44. Fuxreiter M, Simon I, Friedrich P, Tompa P (2004) Preformed structural elements feature in partner recognition by intrinsically unstructured proteins. J Mol Biol 338:1015–1026
45. Song J, Guo L-W, Muradov H, Artemyev NO, Ruoho AE et al (2008) Intrinsically disordered gamma-subunit of cGMP phosphodiesterase encodes functionally relevant transient secondary and tertiary structure. Proc Natl Acad Sci USA 105:1505–1510
46. Baker JMR, Hudson RP, Kanelis V, Choy W-Y, Thibodeau PH et al (2007) CFTR regulatory region interacts with NBD1 predominantly via multiple transient helices. Nat Struct Mol Biol 14:738–745
47. Sugase K, Dyson HJ, Wright PE (2007) Mechanism of coupled folding and binding of an intrinsically disordered protein. Nature 447:1021–1025
48. Marsh JA, Neale C, Jack FE, Choy W-Y, Lee AY et al (2007) Improved structural characterizations of the drkN SH3 domain unfolded state suggest a compact ensemble with native-like and non-native structure. J Mol Biol 367:1494–1510
49. Marsh JA, Forman-Kay JD (2009) Structure and disorder in an unfolded state under nondenaturing conditions from ensemble models consistent with a large number of experimental restraints. J Mol Biol 391:359–374

50. Chiti F, Dobson CM (2006) Protein misfolding, functional amyloid, and human disease. Annu Rev Biochem 75:333–366
51. Roychaudhuri R, Yang M, Hoshi MM, Teplow DB (2009) Amyloid beta-protein assembly and Alzheimer disease. J Biol Chem 284:4749–4753
52. Esteban-Martín S, Bryn Fenwick R, Salvatella X (2012) Synergistic use of NMR and MD simulations to study the structural heterogeneity of proteins. WIREs Comput Mol Sci 2:466–478
53. Kaptein R, Zuiderweg ER, Scheek RM, Boelens R, van Gunsteren WF (1985) A protein structure from nuclear magnetic resonance data. lac repressor headpiece. J Mol Biol 182:179–182
54. Wüthrich K (2003) NMR studies of structure and function of biological macromolecules (Nobel Lecture). J Biomol NMR 27:13–39
55. Bonvin AMJJ, Brünger AT (1995) Conformational variability of solution nuclear magnetic resonance structures. J Mol Biol 250:80–93
56. Mierke DF, Kurz M, Kessler H (1994) Peptide flexibility and calculations of an ensemble of molecules. J Am Chem Soc 116:1042–1049
57. Bürgi R, Pitera J, van Gunsteren WF (2001) Assessing the effect of conformational averaging on the measured values of observables. J Biomol NMR 19:305–320
58. Vögeli B, Segawa TF, Leitz D, Sobol A, Choutko A et al (2009) Exact distances and internal dynamics of perdeuterated ubiquitin from NOE buildups. J Am Chem Soc 131:17215–17225
59. Vögeli B, Kazemi S, Güntert P, Riek R (2012) Spatial elucidation of motion in proteins by ensemble-based structure calculation using exact NOEs. Nat Struct Mol Biol 19:1-53-1057
60. Lange OF, Lakomek N-A, Farès C, Schröder GF, Walter KFA et al (2008) Recognition dynamics up to microseconds revealed from an RDC-derived ubiquitin ensemble in solution. Science 320:1471–1475
61. Fenwick RB, Esteban-Martín S, Richter B, Lee D, Walter KFA et al (2011) Weak long-range correlated motions in a surface patch of ubiquitin involved in molecular recognition. J Am Chem Soc 133:10336–10339
62. Tjandra N, Bax A (1997) Direct measurement of distances and angles in biomolecules by NMR in a dilute liquid crystalline medium. Science 278:1111–1114
63. Salvatella X, Richter B, Vendruscolo M (2008) Influence of the fluctuations of the alignment tensor on the analysis of the structure and dynamics of proteins using residual dipolar couplings. J Biomol NMR 40:71–81
64. Clore GM, Schwieters CD (2004) How much backbone motion in ubiquitin is required to account for dipolar coupling data measured in multiple alignment media as assessed by independent cross-validation? J Am Chem Soc 126:2923–2938
65. Clore GM, Schwieters CD (2004) Amplitudes of protein backbone dynamics and correlated motions in a small alpha/beta protein: correspondence of dipolar coupling and heteronuclear relaxation measurements. Biochemistry 43:10678–10691
66. Best RB, Vendruscolo M (2004) Determination of protein structures consistent with NMR order parameters. J Am Chem Soc 126:8090–8091
67. Lindorff-Larsen K, Best RB, Depristo MA, Dobson CM, Vendruscolo M (2005) Simultaneous determination of protein structure and dynamics. Nature 433:128–132
68. Richter B, Gsponer J, Várnai P, Salvatella X, Vendruscolo M (2007) The MUMO (minimal under-restraining minimal over-restraining) method for the determination of native state ensembles of proteins. J Biomol NMR 37:117–135
69. Zweckstetter M, Bax A (2000) Prediction of sterically induced alignment in a dilute liquid crystalline phase: aid to protein .... J Am Chem Soc 122:3791–3792
70. Huang J-R, Grzesiek S (2010) Ensemble calculations of unstructured proteins constrained by RDC and PRE data: a case study of urea-denatured ubiquitin. J Am Chem Soc 132:694–705
71. Esteban-Martín S, Fenwick RB, Salvatella X (2010) Refinement of ensembles describing unstructured proteins using NMR residual dipolar couplings. J Am Chem Soc 132:4626–4632
72. Zweckstetter M, Hummer G, Bax A (2004) Prediction of charge-induced molecular alignment of biomolecules dissolved in dilute liquid-crystalline phases. Biophys J 86:3444–3460

73. De Simone A, Montalvao RW, Vendruscolo M (2011) Determination of conformational equilibria in proteins using residual dipolar couplings. J Chem Theory Comput 7:4189–4195
74. Montalvao RW, De Simone A, Vendruscolo M (2012) Determination of structural fluctuations of proteins from structure-based calculations of residual dipolar couplings. J Biomol NMR 53:281–292
75. Torda AE, Scheek RM, van Gunsteren WF (1989) Time-dependent distance restraints in molecular dynamics simulations. Chem Phys Lett 157:289–294
76. Torda AE, Scheek RM, van Gunsteren WF (1990) Time-averaged nuclear Overhauser effect distance restraints applied to tendamistat. J Mol Biol 214:223–235
77. Clore GM, Iwahara J (2009) Theory, practice, and applications of paramagnetic relaxation enhancement for the characterization of transient low-population states of biological macromolecules and their complexes. Chem Rev 109:4108–4139
78. Dedmon MM, Lindorff-Larsen K, Christodoulou J, Vendruscolo M, Dobson CM (2005) Mapping long-range interactions in α-synuclein using spin-label NMR and ensemble molecular dynamics simulations. J Am Chem Soc 127:476–477
79. Silvestre-Ryan J, Bertoncini CW, Fenwick RB, Esteban-Martín S, Salvatella X (2013) Average conformations determined from PRE data provide high-resolution maps of transient tertiary interactions in disordered proteins. Biophys J 104:1740–1751
80. Camilloni C, Robustelli P, De Simone A, Cavalli A, Vendruscolo M (2012) Characterization of the conformational equilibrium between the two major substates of RNase A using NMR chemical shifts. J Am Chem Soc 134:3968–3971

Chapter 4
Generative Models of Conformational Dynamics

Christopher James Langmead, Ph.D.

Abstract Atomistic simulations of the conformational dynamics of proteins can be performed using either Molecular Dynamics or Monte Carlo procedures. The ensembles of three-dimensional structures produced during simulation can be analyzed in a number of ways to elucidate the thermodynamic and kinetic properties of the system. The goal of this chapter is to review both traditional and emerging methods for learning *generative models* from atomistic simulation data. Here, the term 'generative' refers to a model of the joint probability distribution over the behaviors of the constituent atoms. In the context of molecular modeling, generative models reveal the correlation structure between the atoms, and may be used to predict how the system will respond to structural perturbations. We begin by discussing traditional methods, which produce multivariate Gaussian models. We then discuss GAMELAN (GRAPHICAL MODELS OF ENERGY LANDSCAPES), which produces generative models of complex, non-Gaussian conformational dynamics (e.g., allostery, binding, folding, etc.) from long timescale simulation data.

Keywords Probabilistic graphical models • Generative models • Energy landscapes • Conformational ensembles • Molecular dynamics • Thermodynamics • Kinetics • Inference • Learning • Parametric • Semi-parametric • Non-parametric

4.1 Introduction

Atomistic simulations are widely used to investigate the conformational dynamics of proteins and other molecules (e.g., [22, 24]). The raw output from any simulation is an ensemble of three-dimensional conformations. These ensembles can be analyzed

C.J. Langmead (✉)
Carnegie Mellon University, Pittsburgh, PA, USA
e-mail: cjl@cs.cmu.edu

K. Han et al. (eds.), *Protein Conformational Dynamics*, Advances in Experimental Medicine and Biology 805, DOI 10.1007/978-3-319-02970-2_4,

using a variety of methods, ranging from simple descriptive statistics (e.g., average energies, radius of gyration, etc.) to generative models (e.g., normal mode analysis, quasi-harmonic analysis, etc.). Here, the term 'generative' refers to any model of the joint probability distribution, $P(X_1, \ldots, X_n)$, over a set of user-defined random variables, $\mathbf{X} = \{X_1, \ldots, X_n\}$, representing the system's degrees of freedom (e.g., distances, fluctuations, angles, etc.). In this chapter, we focus on techniques for learning generative models from conformational ensembles.

Generative models provide important insights into conformational dynamics. In particular, they elucidate the inter-atomic correlations that give rise to collective motions within and across dynamical domains. Consequently, generative models can be used to estimate important quantities, including the magnitudes of atomic fluctuations (e.g., [50]), configurational entropies (e.g., [21]), and free energies (e.g., [17, 18, 20]). They can also be used to predict how the system will respond to local structural changes (e.g., ligand binding) (e.g., [39]).

Many techniques exist for learning generative models from conformational ensembles. Well-known examples include: Normal Modes Analysis [6, 13, 25], Quasi Harmonic Analysis [21, 26], Essential Dynamics [1], and Elastic Network Models [50]. Ultimately, the differences between these methods amount to: (a) which variables are modeled, and (b) the mathematical form used to define $P(\mathbf{X})$. This chapter contrasts several strategies for specifying $P(\mathbf{X})$ (whether implicitly or explicitly), starting from simple harmonic models (where $P(\mathbf{X})$ takes the form of a multivariate Gaussian), and proceeding to more expressive models that are better suited for anharmonic (i.e., non-Gaussian) motions. This latter category is presented in the context of GAMELAN (GRAPHICAL MODELS OF ENERGY LANDSCAPES), which is a new framework for learning generative models from conformational ensembles.

GAMELAN is motivated by recent developments in atomistic simulation technologies. In particular, advances in hardware and software (e.g., [5, 15, 31, 34, 40, 47]) have dramatically increased the timescales accessible to simulation. Microsecond ($\mu s = 10^{-6}$ s) and millisecond (ms $= 10^{-3}$ s) simulations are increasingly common, but the resulting conformational ensembles pose significant challenges. First and foremost, the conformational dynamics observed on the μs and ms timescales are usually very complex. In particular, they are not well suited to harmonic approximations. GAMELAN addresses this problem by providing users the option of learning multi-modal, non-Gaussian, and even time-varying generative models from the ensemble. This is achieved through a combination of parametric, semi-parametric, and non-parametric models. The second challenge is the size of the ensemble, which naturally increases with both the size of the system and the timescale. GAMELAN addresses this challenge by using efficient, but provably optimal algorithms for estimating the parameters of the generative model.

4.2 Conformational Ensembles

As previously noted, atomistic simulations can be performed using Molecular Dynamics (MD) and/or Monte Carlo (MC) sampling. Molecular dynamics simulations involve numerically solving Newton's laws of motion for a system of atoms whose interactions are defined according to a given force field. Monte Carlo simulations involve iteratively modifying an existing structure. Each modification is either accepted or rejected, stochastically, according to its energy, as defined by a force field. The theory and practice behind MD and MC algorithms is beyond the scope of this chapter. Here, we will simply assume that each method produces an ensemble of m conformations. The ensemble will be denoted as $\mathbf{C} = \{C(1), \ldots, C(m)\}$, where $C(i)$ specifies the cartesian coordinates for each atom in the ith conformation.

In principle, generative models can be constructed from the raw ensemble, $\mathbf{C}$, but it is much more common to limit the analysis to a limited number of covariates. Most analyses operate on either: (a) the cartesian coordinates of a subset of the atoms (e.g., non-solvent molecules, or even just the alpha carbons); (b) atomic fluctuations (i.e., displacements from a reference conformation); (c) pairwise distances between atoms; or (d) dihedral angles. The methods discussed in this chapter can be applied to any set of covariates, and so we will not restrict them to any particular type. Let $\mathbf{X} = \{X_1, \ldots, X_n\}$ be a vector encoding the n covariates to be analyzed, and recall that a generative model encodes the joint probability distribution $P(\mathbf{X})$. The parameters of the model are estimated from a set of data, $\mathbf{D} = \{\mathbf{X}(1), \ldots, \mathbf{X}(m)\}$, where $\mathbf{X}(i)$ is a vector containing the values of the n covariates extracted from $C(i)$.

4.3 Learning Generative Models from Conformational Ensembles

This section presents several methods for learning generative models from a set of data, $\mathbf{D}$, starting with simple Gaussian models and progressing to non-Gaussian models.

4.3.1 *Simple Gaussian Models*

The most straightforward way to produce a model of the joint distribution, $P(\mathbf{X})$, is to fit a multivariate Gaussian distribution to the data. This can be accomplished, for example, by computing the n-dimensional empirical mean vector, $\mu = \frac{1}{m}\sum \mathbf{X}(i)$, and the $n \times n$ empirical covariance matrix $\Sigma = E[(\mathbf{X} - \mu)(\mathbf{X} - \mu)^T]$. Given these parameters, the probability density for any n-vector $\mathbf{x} = \{x_1, \ldots, x_n\}$ is:

$$P(\mathbf{x}) = \frac{1}{Z} \exp\left\{ -\frac{1}{2}(\mathbf{x} - \mu)^T \Sigma^{-1} (\mathbf{x} - \mu) \right\}, \tag{4.1}$$

where $Z = \sqrt{(2\pi)^n |\Sigma|}$ is the partition function and $|\Sigma|$ denotes the determinant of Σ.

Well-known methods for building harmonic models, including Normal Modes Analysis [6, 13, 25], Quasi Harmonic Analysis [21, 26], and Essential Dynamics [1], also produce multivariate Gaussian models, but not in the manner outlined above. Instead, they transform the data in some way. Quasi-Harmonic Analysis, for example, performs Principle Components Analysis (PCA) on a mass-weighted covariance matrix of atomic fluctuations. PCA diagonalizes the covariance matrix, producing a set of eigenvectors and their corresponding eigenvalues. Each eigenvector can be interpreted as one of the principal modes of vibration within the system or, equivalently, as a univariate Gaussian with zero mean and variance proportional to the corresponding eigenvalue. The eigenvectors are orthogonal by construction, and so the off-diagonal elements of the correlation matrix are zero.

Principal Components Analysis operates on covariance matrices, which capture pairwise relationships between variables. It is sometimes desirable to capture the relationships between tuples of variables (triples, quadruples, etc.). Here, Tensor Analysis may be used instead of PCA [35, 36]. The model produced via Tensor Analysis is also Gaussian.

4.3.1.1 Computing with Gaussian Models

When appropriate, multivariate Gaussian models have a number of attractive properties. For example, the Kullback-Leibler divergence[1] between two different models, M_0 and M_1 can be computed analytically:

$$KL(M_0||M_1) = \frac{1}{2}(\text{trace}(\Sigma_1^{-1}\Sigma_0) + (\boldsymbol{\mu}_1 - \boldsymbol{\mu}_0)^T \Sigma_1^{-1} (\boldsymbol{\mu}_1 - \boldsymbol{\mu}_0) - \ln(|\Sigma_0|/|\Sigma_1|) - n). \tag{4.2}$$

The ability to quantify the differences between two models has a number of practical uses. For example, the symmetric version of the Kullback-Leibler can be used to cluster a set of models, or to compare models learned from independent sources (e.g., different simulations of the same system).

[1]The Kullback-Leibler divergence is a non-symmetric measure of the difference between two distributions. It is non-negative and zero if and only if the two distributions are identical. The divergence can be symmetrized by taking the sum or average of $KL(M_0||M_1)$ and $KL(M_1||M_0)$.

The generative nature of the model means, among other things, that one can sample new conformations (i.e., those that weren't in **D**). In the case of a multivariate Gaussian, sampling can be accomplished in two steps. First, an n-dimensional vector of independent Gaussian random numbers is generated, $\mathbf{r} = [r_1, \ldots, r_n]$. Second, the random sample is produced by computing $\mathbf{x} = \mathbf{A}\mathbf{r} + \mu$, where $\mathbf{A}$ is the lower triangular matrix satisfying $\Sigma = \mathbf{A}\mathbf{A}^T$.

Finally, Gaussian models also make it easy to predict how the system will respond to local perturbations by computing conditional distributions. For example, let $\mathbf{V} \subset \mathbf{X}$ be an arbitrary subset of $\mathbf{X}$, and let $\mathbf{W} = \mathbf{X} \setminus \mathbf{V}$ be the complement set. Here, $\mathbf{V}$ might correspond to an allosteric binding site. We can simulate a local structural change (e.g., due to binding) by setting $\mathbf{V}$ to some particular value, say $\mathbf{v}$. Next, we can predict how the rest of the molecule will respond by conditioning the distribution on $\mathbf{v}$, and computing the conditional distribution $P(\mathbf{W}|\mathbf{v})$. This conditional distribution is also a multivariate Gaussian with parameters $(\mu_{W|v}, \Sigma_W)$ where:

$$\mu_{W|\mathbf{v}} = \mu_W + \Sigma_W^T \Sigma_{VV}^{-1}(\mathbf{v} - \mu_v) \tag{4.3}$$

$$\Sigma_W = \Sigma_{WW} - \Sigma_{WV}^T \Sigma_{VV}^{-1} \Sigma_{WV} \tag{4.4}$$

Here, $\Sigma = \begin{pmatrix} \Sigma_{WW} & \Sigma_{WV} \\ \Sigma_{WV}^T & \Sigma_{VV} \end{pmatrix}$. The vector $\mu^* = \mathbf{v} \cup \mu_{W|\mathbf{v}}$ is the mode of a new equilibrium distribution and is therefore the model's prediction for the most likely conformation, after the local perturbation. Significantly, this prediction is computed analytically via matrix operations. Alternatively, one might sample from the conditional distribution $P(\mathbf{W}|\mathbf{v}) \sim N(\mu_{W|\mathbf{v}}, \Sigma_W)$.

4.3.2 GAMELAN

The computational and analytical tractability of Gaussian models belies the fact that the conformational dynamics of proteins aren't normally distributed [1, 16]. Thus, while it is always possible to fit a Gaussian to a set of data, sometimes this approximation is valid, and sometimes it is not. In this section, we discuss a framework for creating generative models from conformational ensembles. This technique, called GAMELAN (GRAPHICAL MODELS OF ENERGY LANDSCAPES), is capable of producing a variety of generative models (including Gaussian), but it primarily intended for circumstances where the conformational dynamics are non-Gaussian.

GAMELAN produces generative models from a set of data by learning a *Probabilistic Graphical Model* (Fig. 4.1). Informally, a probabilistic graphical model is a factored encoding of a multivariate distribution, $P(\mathbf{X})$. It consists of a graph defined over the variables, and a set of functions defined over the nodes and edges in the

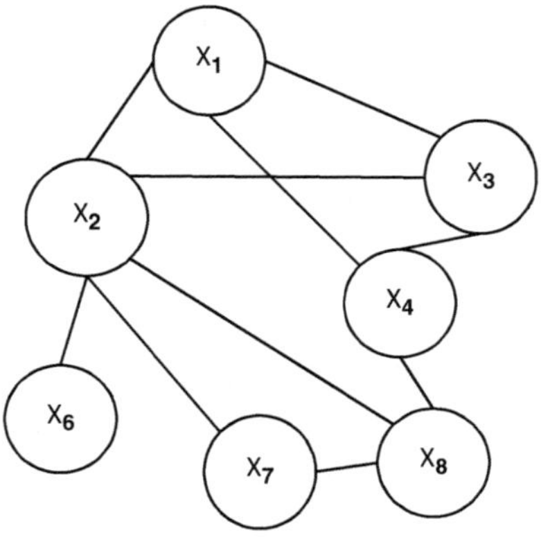

Fig. 4.1 A probabilistic graphical model over eight random variables. Nodes correspond to random variables. Edges reveal the conditional independencies among the variables. Each node and edge is associated with a function. When combined, the graph and the functions encode the joint probability distribution over the variables, $P(X_1, \ldots, X_8)$. Graphical models of protein structures may have hundreds or thousands of nodes, depending on which covariates are being modeled

graph. The topology of the graph distinguishes between indirect and direct couplings between random variables. The mathematical form of the functions determines the nature of the distribution. Here, there is a great amount of flexibility; depending on the choice of functions, GAMELAN can produce Gaussian distributions, multinomial distributions, circular distributions, and, most significantly, multi-modal distributions. Multi-modal distributions are essential if the system has more than one conformational substate [9, 10]. It is also possible to define the functions using molecular mechanics force-fields.

Like the simple Gaussian model discussed in Sect. 4.3.1, the models produced by GAMELAN can perform a variety of tasks efficiently, although not necessarily analytically. In particular, quantifying the difference between models, sampling, and computing conditional distributions are all possible using GAMELAN models. Additionally, if the node and edge functions are defined in terms of force-fields, the model can be used to estimate important quantities, like free energies [17, 18, 20].

4.3.2.1 Probabilistic Graphical Models

A probabilistic graphical model, $M = (G, \Theta)$, encodes a joint probability distribution $P(\mathbf{X})$ in terms of a graph, $G = (\mathbf{V}, \mathbf{E})$, and a set of functions $\Theta = (\theta_1, \ldots, \theta_p)$ defined over the nodes ($\mathbf{V} = (V_1, \ldots, V_n)$)—one for each random variable—and edges ($\mathbf{E} \subseteq \mathbf{V} \times \mathbf{V}$) of the graph. For the models that GAMELAN produces, G is undirected (Fig. 4.1). We note that undirected probabilistic graphical models are often called *Markov Random Fields* in the Machine Learning and Statistics literature.

The probability density encoded by a GAMELAN model is:

$$P(\mathbf{X}) = \frac{1}{Z(\Theta)} \exp\left(\sum_{c \in Cliques(G)} \theta_c(X_c) \right) \tag{4.5}$$

where the sum is over the fully connected subgraphs (i.e., cliques) of the graph and $X_C \subseteq \mathbf{X}$ is the subset of variables in clique c.

The topology of G defines the set of *conditional independencies* between the random variables. In particular, the absence of an edge between X_i and X_j means that the two variables are conditionally independent of each other. That is, given its neighbors in the graph, random variable X_i is independent of X_j (and visa-versa).

Informally, the notion of conditional independence means that any observed correlations between X_i and X_j can be explained in terms of the network of couplings in G. Note that the lack of an edge does *not* mean that two random variables are uncorrelated, only that the correlations are due to indirect couplings. By analogy, consider a mass-spring system. The motions of two masses may be correlated, even if they are not directly coupled by a spring. GAMELAN learns a minimal set of edges and the associated parameters that best explain the correlations observed in the data. We note, however, that the 'springs' in GAMELAN models are not necessarily harmonic.

There are two basic tasks associated with probabilistic graphical models: learning and inference. In general, both tasks are non-trivial, and there are a number of algorithms for solving these problems. Here, we will highlight some of the key concepts, and direct the reader to [23] for more information on these topics, and on graphical models, in general.

4.3.2.2 Learning

Learning refers to a procedure for estimating the parameters of the model from data. If the topology of the graph is given (i.e., imposed), then learning is generally performed maximizing the log-likelihood of the parameters, given the data: $\Theta^* = \underset{\Theta}{\operatorname{argmax}}\, ll(\Theta|\mathbf{D}; G)$. If the topology of the graph is not known, it can also be learned from the data. This is known as the structure learning problem. Finding the optimal topology and parameters simultaneously is much harder than finding the optimal parameters alone. This is because the number of undirected topologies over n variables is $2^{\binom{n}{2}}$. Practical algorithms for solving the structure learning problem place a prior over graph topologies, often in the form of a regularization penalty, which penalizes dense graphs. The intuition behind this penalty is that for every edge that is added to the graph, the parameters associated with the corresponding edge function must be estimated. Highly-parameterized models risk over-fitting the data, and so sparse models are preferred over dense models.

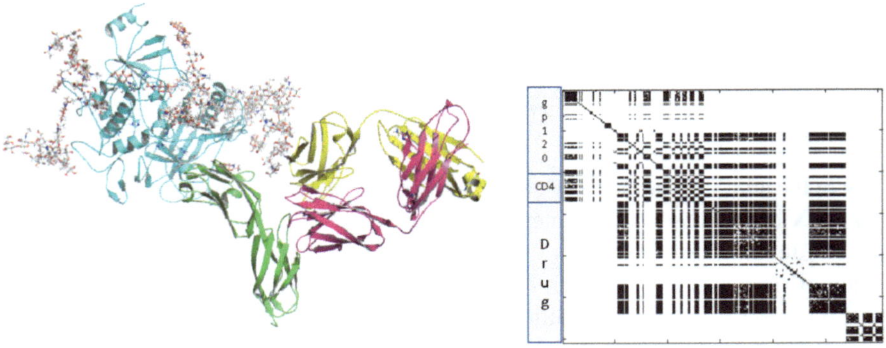

Fig. 4.2 *Left*: A complex consisting of gp120 (*cyan*), CD4 (*green*), and Ibalizumab (*magenta* and *yellow*). *Right*: The topology of the graphical model learned by GAMELAN. A *black dot* means there is an edge between residues i and j

When asked to solve the structure learning problem, GAMELAN optimizes a regularized version of a quantity known as the pseudo log-likelihood:

$$(G, \Theta)^* = \underset{G,\Theta}{\operatorname{argmax}}\, pll(G, \Theta|\mathbf{D}) - \lambda R(G, \Theta). \tag{4.6}$$

Here, pll is the pseudo log-likelihood of the graph and parameters, given the data, R is the regularization function, and λ is a parameter that controls the tradeoff between fitting the data (the first term) and having simple models. The pseudo log-likelihood is a consistent estimator (i.e., given enough data, it converges on the same solution as the exact log-likelihood), and is also much more efficient to compute [3]. The regularization penalty can be defined in a number of ways. GAMELAN uses an L_1 penalty, which encourages sparse graphs. $L1$ regularization also has desirable statistical properties. Specifically, it leads to consistent models (that is, given enough data our algorithm learns the true topology) while enjoying high efficiency (that is, the number of samples needed to achieve the true model is small). The regularization parameters, λ, can be set in a number of ways, including AIC and BIC, or through a simple permutation test that finds the value of λ that yields no edges on randomized versions of the data (where all correlations have been eliminated).

Figure 4.2 illustrates the topology of the network learned by GAMELAN from a 2ns simulation of a complex consisting of gp120 (a glycoprotein on the surface of the HIV envelope), the CD4 receptor (a glycoprotein expressed on the surface of T helper cells) and Ibalizumab, a humanized monoclonal antibody that binds to CD4 and inhibits the viral entry process. The set of edges includes intra- and inter molecular pairs.

4.3.2.3 Inference

Once the model has been learned, it can be used to perform a variety of tasks. For example, Gibbs sampling, and related procedures, can be used to generate new conformations. Approximate versions of the KL divergence can be computed by calculating an empirical estimate for $KL(M_0||M_1) = \sum P_{M_0}(i) \log \frac{P_{M_0}(i)}{P_{M_1}(i)}$. Marginal and conditional distributions can be computed using message-passing algorithms on the graph, such as Belief Propagation [33] and Expectation Propagation [29], depending on the nature of the functions (see [23] for a complete discussion).

The remainder of this section discusses the range of models that can be produced by GAMELAN. We start with parametric models, such as the Gaussian and von Mises distribution, and then move on to semi- and non parametric models, which are more expressive.

4.3.2.4 Parametric Models

The simplest graphical model for continuous-valued random variables is the *Gaussian Graphical Model*, which is defined as the pair $(\mathbf{h}, \Sigma^{-1})$. Here, Σ^{-1} is inverse of the covariance matrix (also known as the *precision* or *concentration* matrix) and h is an n-dimensional vector satisfying $\mu = h^T \Sigma$. Thus, a Gaussian Graphical Model can be constructed from the empirical mean and covariance (as in Sect. 4.3.1). Empirical estimates, however, are subject to over-fitting the data. Therefore, when asked to produce a Gaussian Graphical Model, GAMELAN computes a regularized estimate of the precision matrix (and hence a regularized version of its inverse, the covariance matrix). Specifically, GAMELAN learns a sparse precision matrix (i.e., one with many zeros among the off-diagonal elements) (see [39]). The non-zero elements of Σ^{-1} correspond to the edges in the graphical model.

Notice that unlike PCA-based methods, like Quasi-Harmonic Analysis, which produce Gaussian models after a change of basis, a Gaussian Graphical model is defined over the original variables, $\mathbf{X} = \{X_1, \ldots, X_n\}$. Thus while having a similar form, the resulting models are very different. In particular, PCA-based models encode the joint distribution in terms of *global* motions, since each eigenvector is a linear combination of the original variables. Gaussian Graphical Models, on the other hand, are defined in terms of a network of *local* couplings.

Some quantities are not well modeled using Gaussian variables. Angles, in particular, are defined on the circle, and so are best modeled using circular distributions. The circular analog to the Gaussian distribution is the von Mises distribution [8]. The univariate von Mises distribution over angle $\phi \in (-\pi, \pi]$ is defined as:

$$P(\phi) = \frac{e^{\kappa \cos(\phi - \mu)}}{2\pi I_0(\kappa)}$$

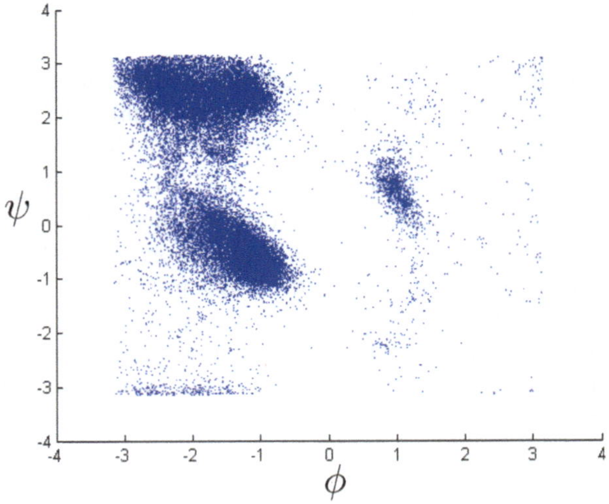

Fig. 4.3 Ramachandran plot. Axes are in radians

where $I_0(\kappa)$ is the modified Bessel function of order 0, and the parameters μ and $\frac{1}{\kappa}$ are analogous to μ and σ^2 (the mean and variance) in the Gaussian distribution. κ is known as the *concentration* of the variable, and so high concentration implies low variance. The bivariate von Mises distribution [28] over $\Phi = (\phi_1, \phi_2)$, can be defined as:

$$P(\Phi) = \frac{\exp\left\{\left[\sum_{i=1}^{2} \kappa_i \cos(\phi_i - \mu_i)\right] + \lambda g(\phi_1, \phi_2)\right\}}{Z(\mu_1, \mu_2, \kappa_1, \kappa_2, \lambda)},$$

where μ_1 and μ_2 are the means of ϕ_1 and ϕ_2, respectively, κ_1 and κ_2 are their corresponding concentrations, $g(\phi_1, \phi_2) = \sin(\phi_1 - \mu_1)\sin(\phi_2 - \mu_2)$, λ is a measure of the dependence between ϕ_1 and ϕ_2, and $Z(\cdot)$ is the normalization constant.

A *von Mises Graphical Model* can be defined using a combination of uni- and bivariate von Mises distributions as the node and edge functions, respectively. GAMELAN can produce von Mises graphical models using existing algorithms for regularized structure learning [38], and inference [37].

Gaussian and von Mises graphical models are most useful when the ensemble being analyzed is well-approximated by a unimodal distribution (e.g., fluctuations within a single conformational substate). For more complex ensembles, spanning more than one conformational substate, or exhibiting substantial asymmetry, these approximations will be poor. An obvious example of a complex distribution is the Ramachandran distribution (Fig. 4.3).

There are a number of strategies for addressing this problem of non-Gaussian distributed data. One simple-minded solution is to modify existing PCA-based methods (e.g., quasiharmonic analysis) so that they perform Independent Components Analysis, instead. Independent Components Analysis also performs a change of basis, but onto a set of statistically independent bases (as opposed to merely uncorrelated bases, which is what PCA produces). Such modifications are straightforward, but still define the joint distribution in terms of global motions. To address these same problem using graphical models, there are there are two basic options. The first is to discretize conformation space in some fashion, and then learn a multinomial model. For example, graphical models over discrete backbone or side-chains conformations (i.e., rotamers) have been developed (e.g., [17, 20]), and GAMELAN can construct such models. The second approach is to abandon parametric forms and utilize semi- or nonparametric graphical models. GAMELAN can produce these models too, as we discuss in the following sections.

4.3.2.5 Semi-parametric Models

If the data are not well-approximated via a Gaussian distribution, one option is to apply a function to the data so that the transformed data are (approximately) Gaussian. This is the central idea behind the *Nonparanormal* distribution [27]. Formally, random variable $\mathbf{X} = \{X_1, \ldots, X_n\}$ is distributed as a Nonparanormal, denoted by $\mathbf{X} \sim NPN(\mu, \Sigma, f)$, if there exist a set of functions $f = \{f_i, \ldots, f_n\}$ such that $f(\mathbf{X}) \sim N(\mu, \Sigma)$. Here, $f(\mathbf{X}) = \{f_1(X_1), \ldots, f_n(X_n)\}$. Under the constraints that the functions are monotone and differentiable, the probability density for any n-vector $\mathbf{x} = \{x_1, \ldots, x_n\}$ is:

$$P(\mathbf{x}) = \frac{1}{Z} \exp\left\{ -\frac{1}{2}(f(\mathbf{x}) - \mu)^T \Sigma^{-1} (f(\mathbf{x}) - \mu) \right\} \prod_i |f_i'(x_i)|, \tag{4.7}$$

where $Z = \sqrt{(2\pi)^n |\Sigma|}$.

GAMELAN learns the parameters of the Nonparanormal graphical model (i.e., μ, Σ, and f) from the data. Briefly, the fs are approximated as: $f_i = \mu_i + \sigma_i h_i(x)$, where μ_i and σ_i are the empirical mean and standard deviation for the ith variable, and h_i is the inverse cumulative distribution function applied to the marginal empirical cumulative distribution $F_i(t) = \frac{1}{m} \sum_j I(X_i(j) \leq t)$ for the ith variable. Here, I is the indicator function. Any operation that can be performed on a Gaussian (e.g., Sect. 4.3.1) can also be performed on the Nonparanormal. The functions, f, are invertible, so samples generated in the 'Nonparanormal space' can be projected into the real space. Similarly, predictions made by calculating conditional distributions can be inverted.

Figures 4.4 and 4.5 demonstrate the Nonparanormal. In Fig. 4.4 a scatter plot of two dimensional data is presented, along with histograms of the marginal distributions. The distribution is non-Gaussian. In Fig. 4.5 the red circles are the same points as in Fig. 4.4, after the Nonparanormal transformation. Notice that the marginals of the transformed data are now Gaussian.

Figure 4.6 illustrates the differences between making prediction using a Gaussian Graphical Model and a Nonparanormal Graphical Model. Models were fit to the

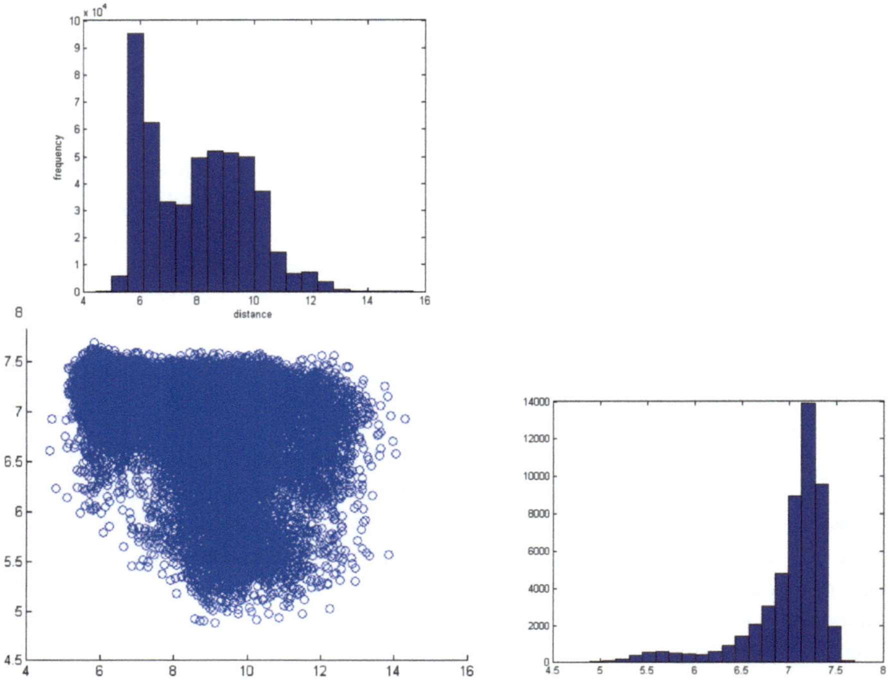

Fig. 4.4 Scatter plot of two dimensional data and histograms of the marginal distributions

same data as Fig. 4.4. The left-hand figure shows the predictions made for variable y, given different values of variable x under both models using Eq. 4.3. Similarly, the right-hand figure shows the predictions made for variable x, given different values of variable y. In each figure, the red line is the prediction made using a Gaussian model and the green line is the prediction made by the Nonparanormal model. Notice that while the Gaussian predictions form a line, the Nonparanormal predictions curve, better reflecting the distribution.

4.3.2.6 Non-parametric Models

The Nonparanormal Graphical Model is more expressive than a Gaussian Graphical Model, but there are models which are more expressive than the Nonparanormal. GAMELAN provides two options: (i) Hilbert-space embeddings of graphical models, and (ii) mixtures of graphical models.

A *Hilbert space* is a complete vector space endowed with an dot product operation. A *Reproducing kernel Hilbert space* (RKHS), H, is a Hilbert space of functions with kernel k satisfying the reproducing property:

$$f(x) = \langle f(\cdot), k(x, \cdot) \rangle,$$

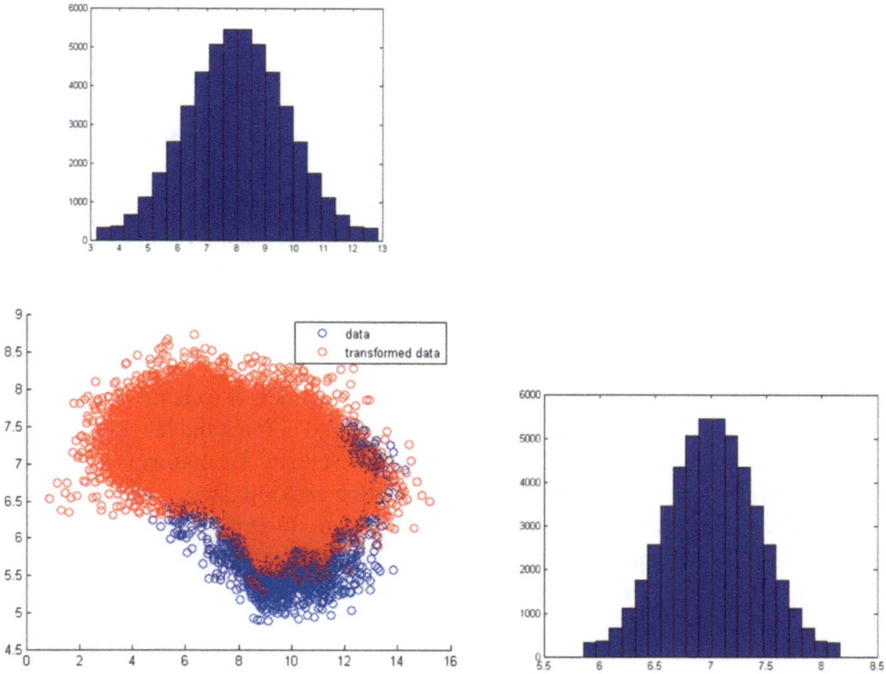

Fig. 4.5 The *red points* are the data from Fig. 4.4 in the 'Nonparanormal space'. Notice that the marginal distributions are now Gaussian

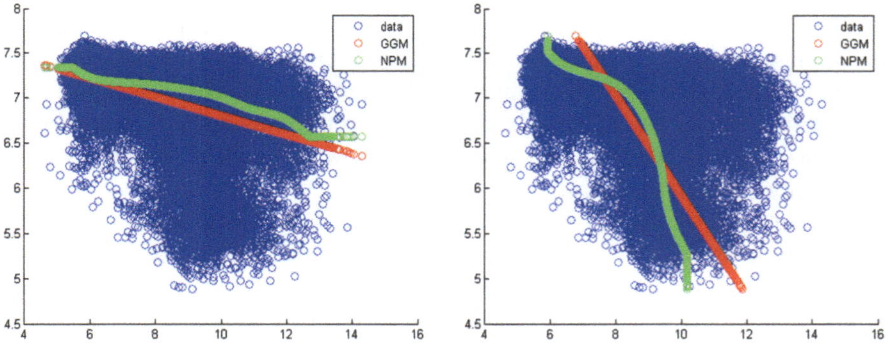

Fig. 4.6 *Left*: The *lines* show the maximum likelihood value for variable y, for different values of x. *Right*: The *lines* show the maximum likelihood value for variable x, for different values of y. The *red line* is computed using a Gaussian Graphical Model. The *green line* is computed using a Nonparanormal Graphical Model

and thus

$$k(x, x') = \langle k(x, \cdot), k(x', \cdot) \rangle.$$

The significance of RKHS's is that function evaluations can be performed via inner products.

Recently, it has been shown that joint and conditional probability distributions can be embedded into a suitable RKHS [42, 43]. That is, different distributions correspond to different points in the RKHS. The significance of these embeddings is that it becomes possible to efficiently learn non-parametric graphical models and perform inference [44,45]. The resulting distributions can be real-valued and multimodal.

The second approach to learning non-parametric models is to learn a mixture of graphical models. Here, the data can either be clustered beforehand into microstates, and a separate graphical model learned for each cluster (Gaussian, von Mises, Nonparanormal, or Hilbert-Space), or expectation-maximization can be used to identify the mixtures, and their parameters. Under a mixture model, each component is given a weight, w_i and the probability density for any n-vector $\mathbf{x} = \{x_1, \ldots, x_n\}$ is: $P(\mathbf{x}) = \prod_i w_i P(\mathbf{x}; G_i, \Theta_i)$, where $\sum_i w_i = 1$.

4.3.2.7 Time-Varying, Reaction-Coordinate Coupled, and Kinetic Models

GAMELAN can also be used to learn time-varying models from the data. In particular, if the conformational ensemble has been produced via Molecular Dynamics simulations, it is natural to wonder how the distribution changes over time. This is easily accomplished by learning models from (possibly overlapping) windows of the data. The width of the window is selected based on the timescale of interest. The resulting sequence of graphical models encodes a diffusion process and users may examine how the topology and parameters change over time. Similarly, GAMELAN may be applied to conformations obtained via umbrella sampling (or similar) along a reaction coordinate. The resulting models reveal how the distribution changes along the reaction coordinate (Fig. 4.7).

Alternatively, GAMELAN can be combined with Markov State Models [2, 30, 41, 46, 48] to produce a fully generative, kinetic model. Here, the data are clustered into microstates and the transition rates between states is estimated from the data. A graphical model is learned for each state, to facilitate sampling and inference. Unlike the time-varying model, Markov State Models are jump-processes.

4.4 Examples

To illustrate the predictive accuracy of different models, GAMELAN was applied to data from a 50 μs simulation of the engrailed homeodomain (Fig. 4.8-left). We extracted the $\theta - \tau$ angles from the data, which describes the configuration of

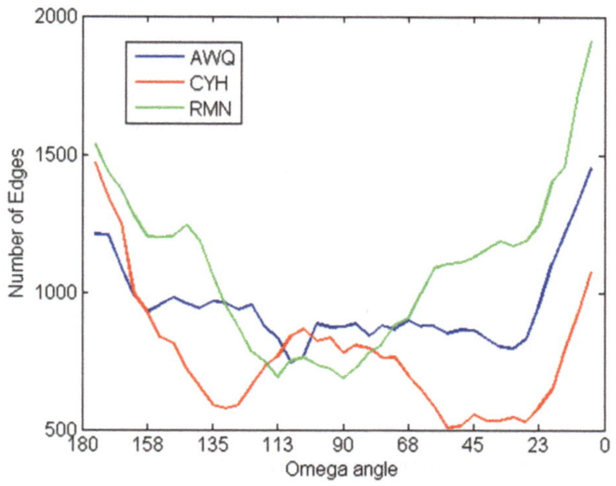

Fig. 4.7 The enzyme cyclophilin A isomerizes the omega angle of its substrate. Here, the number of edges learned by GAMELAN is plotted against the reaction coordinate for three substrates

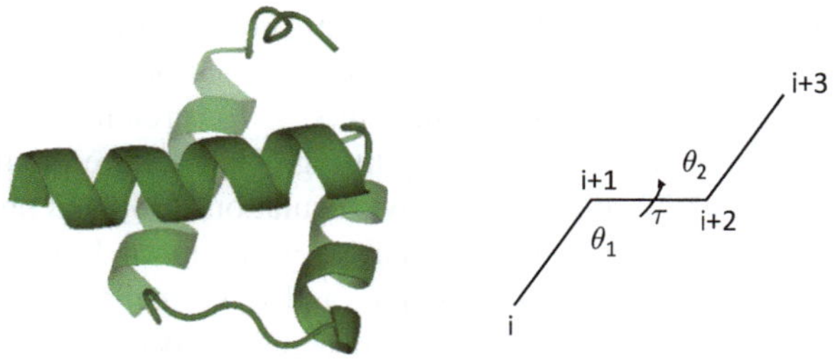

Fig. 4.8 *Left*: The engrailed homeodomain. *Right*: $\theta - \tau$ representation of the backbone is defined over the alpha carbons

Table 4.1 Predictive accuracy on angular data

Model	Root mean square error (°)
Gaussian	8.5
von Mises	5.7
Nonparanormal	7.2
Hilbert-Space	5.0
Mixture of Gaussian	7.2
Mixture of von Mises	4.9
Mixture of Nonparanormal	7.6

the alpha carbons (Fig. 4.8-right). The data was partitioned into training and test sets. The training data were used to learn a Gaussian, von Mises, Nonparanormal, Hilbert-Space, mixture of Gaussian, mixture of von Mises, and mixture of Nonparanormal Graphical models. Next, using the test data we conditioned each model on a random subset of the variables and predicted the values of the remaining variables. Table 4.1 demonstrates that the mixture of von Mises Graphical models gives the lowest errors.

Next, we extracted the pairwise distances between alpha carbons and partitioned the data into training and test sets. The training data were used to learn a Gaussian, Nonparanormal, Hilbert-Space Graphical, mixture of Gaussians ($k = 10$), and mixture of Nonparanormal models. Using the test data we conditioned each model

Table 4.2 Predictive accuracy on distances

Model	Root mean square error (Å)
Gaussian	3.9
Nonparanormal	4.9
Hilbert-Space	2.8
Mixture of Gaussian	1.6
Mixture of nonparanormal	3.1

on a random subset of the variables and predicted the values of the remaining variables. Table 4.2 demonstrates that the mixture of Gaussian Graphical models gives the lowest errors.

4.5 Discussion and Conclusion

Probabilistic Graphical Models of protein structures were first introduced in 2002 by Yanover and Weiss [51], who focused on predicting side chain configurations. Subsequent uses of graphical models for proteins have considered a wide range of problems, including density fitting [7], structure prediction [4, 14], protein design [3, 11, 12, 32, 49], free energy calculations [20], and predicting resistance mutations [19]. Their growing popularity in structural biology is due to their ability to represent complex distributions and solve challenging inference problems.

GAMELAN is the first graphical model specifically designed to aid in the analysis and modeling of conformational ensembles generated through simulation. The extreme complexity of the resulting distributions has necessitated the development of more expressive models. The Hilbert-Space embeddings presently represent the most powerful generative models of protein structure. Applying these models to application domains such as structure prediction and protein design is part of ongoing research. Other challenges include the development of graphical model based simulation algorithms where the model evolves in time along a reaction coordinate, generating conformations along the way.

Acknowledgements This work is supported in part by US NSF grant IIS-0905193, US NIH RC2GM093307, and US NIH P41 GM103712.

References

1. Amadei A, Linssen ABM, Berendsen HJC (1993) Essential dynamics of proteins. Proteins: Struct Funct Bioinformatics 17(4):412–425. doi:10.1002/prot.340170408. http://dx.doi.org/10.1002/prot.340170408
2. Andrec M, Felts AK, Gallicchio E, Levy RM (2005) Protein folding pathways from replica exchange simulations and a kinetic network model. Proc Natl Acad Sci USA 102(19):6801–6806. doi:10.1073/pnas.0408970102. http://www.pnas.org/content/102/19/6801.abstract

3. Balakrishnan S, Kamisetty H, Carbonell J, Lee S, Langmead CJ (2011) Learning generative models for protein fold families. Proteins: Struct Funct Bioinformatics 79(6):1061–1078
4. Boomsma W, Mardia KV, Taylor CC, Ferkinghoff-Borg J, Krogh A, Hamelryck T (2008) A generative, probabilistic model of local protein structure. Proc Natl Acad Sci USA 105(26):8932–8937
5. Bowers KJ, Chow E, Xu H, Dror RO, Eastwood MP, Gregersen BA, Klepeis JL, Kolossvary I, Moraes MA, Sacerdoti FD, Salmon JK, Shan Y, Shaw DE (2006) Scalable algorithms for molecular dynamics simulations on commodity clusters. In: SC'06: proceedings of the 2006 ACM/IEEE conference on supercomputing. ACM, New York, Tampa, Florida, USA, pp 84–96. doi:10.1145/1188455.1188544. http://dx.doi.org/10.1145/1188455.1188544
6. Brooks B, Karplus M (1983) Harmonic dynamics of proteins: normal modes and fluctuations in bovine pancreatic trypsin inhibitor. Proc Natl Acad Sci USA 80(21):6571–6575. http://www.pnas.org/content/80/21/6571.abstract
7. DiMaio F, Soni A, Phillips Jr. G, Shavlik J (2007) Creating all-atom protein models from electron-density maps using particle-filtering methods. Bioinformatics 23:2851–2858
8. Fisher N (1993) Statistical analysis of circular data. Cambridge University Press, Cambridge
9. Frauenfelder H, Petsko GA, Tsernoglou D (1979) Temperature-dependent x-ray diffraction as a probe of protein structural dynamics. Nature 280(5723):558–563
10. Frauenfelder H, Parak F, Young RD (1988) Conformational substates in proteins. Annu Rev Biophys Biophys Chem 17:451–479
11. Fromer M, Yanover C (2009) Accurate prediction for atomic-level protein design and its application in diversifying the near-optimal sequence space. Proteins: Struct Funct Bioinformatics 75(3):682–705. doi:10.1002/prot.22280
12. Fromer M, Yanover C (2008) A computational framework to empower probabilistic protein design. Bioinformatics 24(13):i214–222
13. Go N, Noguti T, Nishikawa T (1983) Dynamics of a small globular protein in terms of low-frequency vibrational modes. Proc Natl Acad Sci USA 80(12):3696–3700. http://www.pnas.org/content/80/12/3696.abstract
14. Harder T, Boomsma W, Paluszewski M, Frellsen J, Johansson K, Hamelryck T (2010) Beyond rotamers: a generative, probabilistic model of side chains in proteins. BMC Bioinformatics 11(1):306. doi:10.1186/1471-2105-11-306. http://www.biomedcentral.com/1471-2105/11/306
15. Harvey M, Giupponi G, Fabritiis G (2009) ACEMD: accelerating biomolecular dynamics in the microsecond time scale. J Chem Theory Comput 5(6):1632–1639
16. Hayward S, Kitao A, Go N (1995) Harmonicity and anharmonicity in protein dynamics: a normal mode analysis and principal component analysis. Proteins 23(2):177–186. doi:10.1002/prot.340230207. http://dx.doi.org/10.1002/prot.340230207
17. Kamisetty H, Xing E, Langmead C (2008) Free energy estimates of all-atom protein structures using generalized belief propagation. J Comp Bio 15(7):755–766
18. Kamisetty H, Bailey-Kellogg C, Langmead C (2009) A graphical model approach for predicting free energies of association for protein-protein interactions under backbone and side-chain flexibility
In: Proceedings structural bioinformatics and computational biophysics (3DSIG), Stockholm, pp 67–68
19. Kamisetty H, Xing E, Langmead C (2011) Approximating correlated equilibria using relaxations on the marginal polytope. In: Proceedings of the 28th international conference on machine learning (ICML), Helsinki, pp 1153–1160
20. Kamisetty H, Ramanathan A, Bailey-Kellogg C, Langmead C (2011) Accounting for conformational entropy in predicting bidning free energies of protein-protein interactions. Proteins 79(2):444–462
21. Karplus M, Kushick JN (1981) Method for estimating the configurational entropy of macromolecules. Macromolecules 14(2):325–332
22. Karplus M, McCammon JA (2002) Molecular dynamics simulations of biomolecules. Nat Struct Biol 9(9):646–652

23. Koller D, Friedman N (2009) Probabilistic graphical models: principles and techniques. MIT, Cambridge
24. Landau D, Binder K (2005) A guide to Monte Carlo simulations in statistical physics. Cambridge University Press, New York
25. Levitt M, Sander C, Stern PS (1983) The normal modes of a protein: native bovine pancreatic trypsin inhibitor. Intern J Quantum Chem 24(S10):181–199. doi:10.1002/qua.560240721. http://dx.doi.org/10.1002/qua.560240721
26. Levy RM, Srinivasan AR, Olson WK, McCammon JA (1984) Quasi-harmonic method for studying very low frequency modes in proteins. Biopolymers 23:1099–1112
27. Liu H, Lafferty J, Wasserman L (2009) The nonparanormal: semiparametric estimation of high dimensional undirected graphs. J Mach Learn Res 10:2295–2328. http://dl.acm.org/citation.cfm?id=1577069.1755863
28. Mardia KV (1975) Statistics of directional data. J R Stat Soc Ser B 37(3):349–393
29. Minka TP (2001) Expectation propagation for approximate bayesian inference. In: Breese JS, Koller D (eds) UAI'01: proceedings of the 17th conference in uncertainty in artificial intelligence, University of Washington, Seattle, pp 362–369
30. Pan AC, Roux B (2008) Building Markov state models along pathways to determine free energies and rates of transitions. J Chem Phys 129(6). doi:10.1063/1.2959573
31. Pande VS, Baker I, Chapman J, Elmer SP, Khaliq S, Larson SM, Rhee YM, Shirts MR, Snow C, Sorin EJ, Zagrovic B (2003) Atomistic protein folding simulations on the submillisecond time scale using worldwide distributed computing. Biopolymers 68(1):91–109
32. Parker AS, Griswold KE, Bailey-Kellogg C (2012) Structure-guided deimmunization of therapeutic proteins. In: Proceedings of the 16th annual international conference on research in computational molecular biology, Barcelona, pp 184–198
33. Pearl J (1986) Fusion, propagation, and structuring in belief networks. Artif Intell 29(3): 241–288
34. Phillips J, Braun R, Wang W, Gumbart J, Tajkhorshid E, Villa E, Chipot C, Skeel R, Kale L, Schulten K (2005) Scalable molecular dynamics with namd. J Comp Chem 26:1781–1802
35. Ramanathan A, Agarwal PK, Kurnikova M, Langmead C (2010) An online approach for mining collective behaviors from molecular dynamics simulations. J Comp Biol 17(3): 309–324
36. Ramanathan A, Yoo J, Langmead C (2011) On-the-fly identification of conformational substates from molecular dynamics simulations. J Chem Theory Comput 7(3):778–789
37. Razavian N, Kamisetty H, Langmead C (2011) The von mises graphical model: expectation propagation for inference. Technical report CMU-CS-11-108, Department of Computer Science, Carnegie Mellon University
38. Razavian N, Kamisetty H, Langmead C (2011) The von mises graphical model: regularized structure and parameter learning. Technical report CMU-CS-11-108, Department of Computer Science, Carnegie Mellon University
39. Razavian N, Kamisetty H, Langmead C (2012) Learning generative models of molecular dynamics. BMC Genomics 13(Suppl 1):S5. doi:10.1186/1471-2164-13-S1-S5
40. Shaw DE, Deneroff MM, Dror RO, Kuskin JS, Larson RH, Salmon JK, Young C, Batson B, Bowers KJ, Chao JC, Eastwood MP, Gagliardo J, Grossman JP, Ho CR, Ierardi DJ, Kolossváry I, Klepeis JL, Layman T, McLeavey C, Moraes MA, Mueller R, Priest EC, Shan Y, Spengler J, Theobald M, Towles B, Wang SC (2007) Anton, a special-purpose machine for molecular dynamics simulation. SIGARCH Comput Archit News 35:1–12
41. Singhal N, Snow CD, Pande VS (2004) Using path sampling to build better Markovian state models: predicting the folding rate and mechanism of a tryptophan zipper beta hairpin. J Chem Phys 121(1):415–425. doi:10.1063/1.1738647
42. Smola A, Gretton A, Song L, Schölkopf B (2007) A Hilbert space embedding for distributions. In: Hutter M, Servedio RA, Takimoto E (eds) Algorithmic learning theory. 18th International conference, ALT 2007, Sendai, Japan, October 1–4, 2007, proceedings. Lecture notes in computer science, vol 4754. Springer, New York, pp 13–31. ISBN: 978-3-540-75224-0

43. Song L, Huang J, Smola A, Fukumizu K (2009) Hilbert space embeddings of conditional distributions with applications to dynamical systems. In: Proceedings of the 26th annual international conference on machine learning, ICML'09, Montreal, pp 961–968. ACM, New York. doi:10.1145/1553374.1553497. http://doi.acm.org/10.1145/1553374.1553497
44. Song L, Gretton A, Guestrin C (2010) Nonparametric tree graphical models. In: Artificial intelligence and statistics (AISTATS), Sardinia
45. Song L, Gretton A, Bickson D, Low Y, Guestrin C (2011) Kernel belief propagation. In: International conference on artifical intelligence and statistics (AISTATS), Ft. Lauderdale
46. Sriraman S, Kevrekidis IG, Hummer G (2005) Coarse master equation from bayesian analysis of replica molecular dynamics simulations. J Phys Chem B 109(14):6479–6484. doi:10.1021/jp046448u. http://pubs.acs.org/doi/abs/10.1021/jp046448u. PMID:16851726
47. Stone J, Phillips JC, Freddolino PL, Hardy DJ, Trabuco LG, Schulten K (2007) Accelerating molecular modeling applications with graphics processors. J Comp Chem 28:2618–2640
48. Swope WC, Pitera JW, Suits F (2004) Describing protein folding kinetics by molecular dynamics simulations. 1. Theory. J Phys Chem B 108(21):6571–6581. doi:10.1021/jp037421y. http://pubs.acs.org/doi/abs/10.1021/jp037421y
49. Thomas J, Ramakrishnan N, Bailey-Kellogg C (2009) Protein design by sampling an undirected graphical model of residue constraints. IEEE/ACM Trans Comput Biol Bioinformatics 6(3):506–516. doi:10.1109/TCBB.2008.124
50. Tirion MM (1996) Large amplitude elastic motions in proteins from a single-parameter, atomic analysis. Phys Rev Lett 77:1905–1908. doi:10.1103/PhysRevLett.77.1905. http://link.aps.org/doi/10.1103/PhysRevLett.77.1905
51. Yanover C, Weiss Y (2002) Approximate inference and protein folding. In: Becker S, Thrun S, Obermayer K (eds) Advances in neural information processing systems (NIPS). MIT Press, Cambridge, pp 84–86

Chapter 5
Generalized Spring Tensor Models for Protein Fluctuation Dynamics and Conformation Changes

Hyuntae Na, Tu-Liang Lin, and Guang Song

Abstract **Background:** In the last decade, various coarse-grained elastic network models have been developed to study the large-scale motions of proteins and protein complexes where computer simulations using detailed all-atom models are not feasible. Among these models, the Gaussian Network Model (GNM) and Anisotropic Network Model (ANM) have been widely used. Both models have strengths and limitations. GNM can predict the relative magnitudes of protein fluctuations well, but due to its isotropy assumption, it cannot be applied to predict the directions of the fluctuations. In contrast, ANM adds the ability to do the latter, but loses a significant amount of precision in the prediction of the magnitudes.

Results: In this book chapter, we present a single model, called generalized spring tensor model (STeM), that is able to predict well both the magnitudes and the directions of the fluctuations. Specifically, STeM performs equally well in B-factor predictions as GNM and has the ability to predict the directions of fluctuations as

H. Na (✉)
Department of Computer Science, Iowa State University, 226 Atanasoff Hall,
Ames, IA 50011, USA
e-mail: htna@iastate.edu

T.-L. Lin (✉)
Department of Management Information Systems, National Chiayi University,
580 Sinmin Rd., Chiayi City 600, Taiwan
e-mail: tuliang@mail.ncyu.edu.tw

G. Song (✉)
Department of Computer Science, Iowa State University, 226 Atanasoff Hall,
Ames, IA 50011, USA

L.H. Baker Center for Bioinformatics and Biological Statistics, Iowa State University,
226 Atanasoff Hall, Ames, IA 50011, USA

Program of Bioinformatics and Computational Biology, Iowa State University,
226 Atanasoff Hall, Ames, IA 50011, USA
e-mail: gsong@iastate.edu

K. Han et al. (eds.), *Protein Conformational Dynamics*, Advances in Experimental Medicine and Biology 805, DOI 10.1007/978-3-319-02970-2_5,

ANM. This is achieved by employing a physically more realistic potential, the Gō-like potential. The potential, which is more sophisticated than that of either GNM or ANM, though adds complexity to the derivation process of the Hessian matrix (which fortunately has been done once for all and the MATLAB code is freely available electronically at http://www.cs.iastate.edu/~gsong/STeM), causes virtually no performance slowdown. In addition, we show that STeM can be further extended to an all-atom model and protein fluctuation dynamics computed by all-atom STeM matches closely with that by Normal Mode Analysis (NMA).

Conclusions: Derived from a physically more realistic potential, STeM proves to be a natural solution in which advantages that used to exist in two separate models, namely GNM and ANM, are achieved in one single model. It thus lightens the burden to work with two separate models and to relate the modes of GNM with those of ANM at times. By examining the contributions of different interaction terms in the Gō potential to the fluctuation dynamics, STeM reveals, (i) a physical explanation for why the distance-dependent, inverse distance square (i.e., $1/r^2$) spring constants perform better than the uniform ones, and (ii), the importance of three-body and four-body interactions to properly modeling protein dynamics.

STeM is not limited to coarse-grained protein models that use a single bead, usually the alpha carbon, to represent each residue. The core idea of STeM, deriving the Hessian matrix directly from a physically realistic potential, can be extended to all-atom models as well. We did this and discovered that all-atom STeM model represents a highly close approximation of NMA, yet without the need for energy minimization.

Keywords Normal mode analysis • Hessian matrix • Spring tensor model • Protein dynamics • Mean-square fluctuations

Abbreviations

ENM	Elastic Network Model
GNM	Gaussian Network Model
ANM	Anisotropic Network Model
STeM	Spring Tensor Model
NMA	Normal Model Analysis
ANMr2	ANM using $1/r^2$ as spring constant

5.1 Introduction

It is now well accepted that the functions of a protein are closely related to not only its structure but also its dynamics. With the advancement of the computational power and increasing availability of computational resources, function-related protein dynamics, such as large-scale conformation transitions, has been probed by various computational methods at multiple scales. Among these computational methods, coarse-grained models play an important role since many functional processes take place over time scales that are well beyond the capacity of all-atom simulations [1]. One type of coarse-grained models, the elastic network models (ENMs), have been particularly successful and widely used in studying protein dynamics and in relating the intrinsic motions of a protein with its function-related conformation changes over the last decade [2–5].

The reason why ENMs have been well received as compared to the conventional normal mode analysis (NMA) lies at its simplicity to use. ENMs do not require energy minimization and therefore can be applied directly to crystal structures to compute the modes of motions. In contrast, minimization is required for carrying out the conventional normal mode analysis (NMA). The problematic aspect of energy minimization is that it usually shifts the protein molecule away from its crystal conformation by about 2 Å. In addition, in ENMs analytical solutions to residue fluctuations and motion correlations can be easily derived. On the other hand, the simplicity of ENMs leaves much room for improvement and many new models have been proposed [6–12].

The two most widely used ENM models are Gaussian Network Model (GNM) and Anisotropic Network Model (ANM). They have been used to predict the magnitudes or directions of the residue fluctuations from a single structure and have been applied in many research areas [2, 5], such as domain decomposition [13] and allosteric communication [14–17]. Both models have their own advantages and disadvantages. GNM can predict the relative magnitudes of the fluctuations well, but due to its isotropy assumption, it cannot be applied to predict the directions of the fluctuations. In contrast, ANM adds the ability to do the latter, but it loses a significant amount of precision in the prediction of the magnitudes.

5.1.1 Gaussian Network Model

Gaussian Network Model (GNM) was first introduced in [3] under the assumption that the separation between a pair of residues in the folded protein is Gaussianly distributed. Given its simplicity, the model performs extremely well in predicting the experimental B-factors. The model represents a protein structure using its C_α atoms. The connectivity among the C_α's is expressed in Kirchhoff matrix $\mathbf{\Gamma}$ (see Eq. (5.1)). Two C_α's are considered to be in contact if their distance falls within a certain cutoff distance. The cutoff distance between a pair of residues is the only parameter in the model and is normally set to be 7–8 Å. Let $\Delta \boldsymbol{r}_i$ and $\Delta \boldsymbol{r}_j$ represent the instantaneous fluctuations from equilibrium positions of residues i and j and r_{ij} and $r_{0,ij}$ be the respective instantaneous and equilibrium distances between residues i and j. The Kirchhoff matrix $\mathbf{\Gamma}$ is:

$$\boldsymbol{\Gamma}_{ij} = \begin{cases} -1 & \text{if } i \neq j \cap r_{0,ij} \leq r_c \\ 0 & \text{if } i \neq j \cap r_{0,ij} > r_c \\ \sum_{j,j\neq i}^{N} \boldsymbol{\Gamma}_{ij} & \text{if } i = j \end{cases} \tag{5.1}$$

where i and j are the indices of the residues and r_c is the cutoff distance.

The simplicity of the Kirchhoff matrix formulation results from the assumption that the fluctuations of each residue are isotropic and Gaussianly distributed along the X, Y and Z directions. The expected value of residue fluctuations, $\langle \Delta \mathbf{r}_i^2 \rangle$, and correlations, $\langle \Delta \boldsymbol{r}_i \cdot \Delta \boldsymbol{r}_j \rangle$, can be easily obtained from the inverse of the Kirchhoff matrix:

$$\left\langle \Delta \boldsymbol{r}_i^2 \right\rangle = \frac{3k_B T}{\gamma} \left(\boldsymbol{\Gamma}^{-1} \right)_{ii}, \tag{5.2}$$

$$\left\langle \Delta \boldsymbol{r}_i \cdot \Delta \boldsymbol{r}_j \right\rangle = \frac{3k_B T}{\gamma} \left(\boldsymbol{\Gamma}^{-1} \right)_{ii}, \tag{5.3}$$

where k_B is the Boltzmann constant and T is the temperature. γ is the spring constant. The $\langle \varDelta \mathbf{r}_i^2 \rangle$ term is directly proportional to the crystallographic B-factors.

5.1.2 Anisotropic Network Model

GNM provides only the magnitudes of residue fluctuations. To study the motions of a protein in more details, especially to determine the directions of the fluctuations, normal mode analysis (NMA) is needed. Traditional NMA is all-atom based and requires a structure to be first energy-minimized before the Hessian matrix and normal modes can be computed, which was rather cumbersome. Even after the energy minimization, the derivation of the Hessian matrix is not easy due to the complicated all-atom potential. In Tirion's pioneering work [18], the energy minimization step was removed and a much simpler Hookean potential was used, and yet it was shown that the low frequency normal modes remained mostly accurate. Since then, the Hookean spring potentials have been favored in most coarse-grained C_α models [4, 19, 20]. One of such models is best known as Anisotropic Network Model (ANM) [4] since it has anisotropic, directional information of the fluctuations. The potential in ANM has the simplest harmonic form. Assuming that a given structure is at equilibrium, the Hessian matrix $3N \times 3N$ can be derived analytically from such a potential [4]. The $3N \times 3N$ Hessian matrix H_{ANM} can be repartitioned into $N \times N$ super elements and each super element is a 3×3 tensor.

$$\boldsymbol{H}_{\text{ANM}} = \begin{bmatrix} \boldsymbol{H}_{1,1} & \boldsymbol{H}_{1,2} & \dots & \boldsymbol{H}_{1,N} \\ \boldsymbol{H}_{2,1} & \boldsymbol{H}_{2,2} & \dots & \boldsymbol{H}_{2,N} \\ \vdots & \vdots & \ddots & \vdots \\ \boldsymbol{H}_{N,1} & \boldsymbol{H}_{N,2} & \dots & \boldsymbol{H}_{N,N} \end{bmatrix} \tag{5.4}$$

where $\boldsymbol{H}_{i,j}$ is the interaction tensor between residues i and j and can be expressed as:

$$\boldsymbol{H}_{i,j} = \begin{bmatrix} \frac{\partial^2 V}{\partial X_i \partial X_j} & \frac{\partial^2 V}{\partial X_i \partial Y_j} & \frac{\partial^2 V}{\partial X_i \partial Z_j} \\ \frac{\partial^2 V}{\partial Y_i \partial X_j} & \frac{\partial^2 V}{\partial Y_i \partial Y_j} & \frac{\partial^2 V}{\partial Y_i \partial Z_j} \\ \frac{\partial^2 V}{\partial Z_i \partial X_j} & \frac{\partial^2 V}{\partial Z_i \partial Y_j} & \frac{\partial^2 V}{\partial Z_i \partial Z_j} \end{bmatrix} \tag{5.5}$$

Let $\boldsymbol{H}^{+}$ be the pseudo inverse of Hessian matrix $\boldsymbol{H}_{\text{ANM}}$. The mean square fluctuation $\langle \Delta \boldsymbol{r}_i^2 \rangle$ and correlation $\langle \Delta \boldsymbol{r}_i \cdot \Delta \boldsymbol{r}_j \rangle$ can be calculated by summing over the X, Y and Z components:

$$\langle \Delta \mathbf{r}_i^2 \rangle = \frac{3k_B T}{\gamma} \left(\boldsymbol{H}^{+}_{3i-2,3i-2} + \boldsymbol{H}^{+}_{3i-1,3i-1} + \boldsymbol{H}^{+}_{3i,3i} \right) \tag{5.6}$$

$$\langle \Delta \mathbf{r}_i \cdot \Delta \mathbf{r}_j \rangle = \frac{3k_B T}{\gamma} \left(\boldsymbol{H}^{+}_{3i-2,3j-2} + \boldsymbol{H}^{+}_{3i-1,3j-1} + \boldsymbol{H}^{+}_{3i,3j} \right) \tag{5.7}$$

5.1.3 Strengths and Limitations of GNM and ANM

The advantages of ANM or GNM over the conventional NMA lie in several aspects: (i) it is a coarse-grained model and uses the C_a's to represent the residues in a structure; (ii) it does not require energy minimization and thus can be applied directly to crystal structures to compute the modes of motions; (iii) it provides analytical solutions to the mean square fluctuations and motion correlations.

The limitations of the GNM model. GNM provides only information on the magnitudes of residue fluctuations but no directional information. Therefore, the modes of GNM should not be interpreted as protein motions or components of the motions, since the potential in GNM is not rotationally invariant [21].

The limitations of the ANM model. In contrast to that in GNM, the potential in ANM is based on simple, harmonic Hookean springs and is rotationally invariant. And thus, the modes of ANM do represent the possible modes of protein motions. In doing this, however, ANM loses a significant amount of precision in predicting the magnitudes of the fluctuations. The reason is that, in GNM, the fluctuations in the separation between a pair of residues are assumed to be Gaussianly distributed and isotropic, while in ANM, because only a Hookean spring is attached between a pair of residues *i* and *j*, the fluctuation of residue *j* is constrained only longitudinally along the axis from *i* to *j*. The fluctuation is unconstrained transversely. The interaction spring tensor $\boldsymbol{H}_{i,j}^{\text{ANM}}$ between residues *i* and *j* in Eq. (5.5) becomes the following in the local frame (where the *Z* axis is along the direction from residues *i* to *j*):

$$\boldsymbol{H}_{i,j}^{\text{ANM}} = \begin{bmatrix} 0\,0\,0 \\ 0\,0\,0 \\ 0\,0\,1 \end{bmatrix} \tag{5.8}$$

Because the fluctuation of residue *j* is unconstrained transversely relative to residue *i*, the fluctuations given by ANM are less realistic than those by GNM, which are assumed to be isotropic. The isotropy in GNM is equivalent to an interaction spring tensor between residues *i* and *j* of the following form:

$$\boldsymbol{H}_{i,j}^{\text{GNM}} = \begin{bmatrix} 1\,0\,0 \\ 0\,1\,0 \\ 0\,0\,1 \end{bmatrix} \tag{5.9}$$

From the two tensors $\boldsymbol{H}_{i,j}^{\text{ANM}}$ and $\boldsymbol{H}_{i,j}^{\text{GNM}}$ given in Eqs. (5.8) and (5.9), the causes for the limitations in GNM and ANM are clearly displayed. The unrealistic-ness in ANM is an artifact resulting from its over-simplified potential. The isotropy assumption of GNM, on the other hand, does a better job than ANM in modeling the effect of residue interactions on the magnitudes of the fluctuations, but gives up completely on representing the anisotropic nature that is intrinsic to all physical forces and interactions, since only the magnitudes of the mean-square fluctuations and cross-correlations were of concern when GNM was first proposed. Therefore, to overcome the limitations of GNM and ANM, what is needed is a generalized interaction spring tensor that both is anisotropic and can exert more proper constraints on the fluctuations than the ANM tensor $\boldsymbol{H}_{i,j}^{\text{ANM}}$ does. This calls for a model that has a physically more realistic potential than that of ANM. Since potentials with only two-body interactions can provide only longitudinal constraints, it is necessary to include multi-body interactions in the potential in order to have transversal constraints as well. The multi-body interactions provide additional diagonal and off-diagonal terms to the interaction spring tensor between residues *i* and *j*. For example, by properly including three-body interactions, the interaction spring tensor may look like:

$$\boldsymbol{H}_{i,j}^{\text{STeM}} = \begin{bmatrix} 0 & 0 & 0 \\ 0 & 0 & 0 \\ 0 & 0 & \mathrm{T}(i,j) \end{bmatrix} + \sum_k \begin{bmatrix} \mathrm{s}_{11}(i,j,k) & \mathrm{s}_{12}(i,j,k) & \mathrm{s}_{13}(i,j,k) \\ \mathrm{s}_{21}(i,j,k) & \mathrm{s}_{22}(i,j,k) & \mathrm{s}_{23}(i,j,k) \\ \mathrm{s}_{31}(i,j,k) & \mathrm{s}_{32}(i,j,k) & \mathrm{s}_{33}(i,j,k) \end{bmatrix} \tag{5.10}$$

where *k* represent the indices of the residues that interact with both residues *i* and *j* through three-body interaction S. The first tensor on the right side of Eq. (5.10) represents the two-body interaction, which is similar to $\boldsymbol{H}_{i,j}^{\mathrm{ANM}}$, except that the interaction strength T(*i,j*) depends on residues *i* and *j*, and thus may be distance-dependent as well.

5.1.4 Our Contributions

To overcome the limitations of ANM and GNM, we have developed a generalized spring tensor model for studying protein fluctuation dynamics and conformation changes. It is called generalized spring tensor model, or STeM, for the reason that the interaction between a pair of residues *i* and *j* is no longer a Hookean spring that has the tensor form of Eq. (5.8), but takes a generalized tensor form (similar to that in Eq. (5.10)) that can provide both longitudinal and transversal constraints on a residue's fluctuations relative to its neighbors. We obtain the generalized tensor form by deriving the Hessian matrix from a physically more realistic Gō-like potential (Eq. (5.11)), which has been successfully used in many MD simulations to study protein folding processes and conformation changes [22–24]. In additional to the Hookean spring interactions, the potential includes bond bending and torsional interactions, both of which had been found to be helpful in removing the "tip effect" of the ANM model [9]. The inclusion of the bond bending and torsional interactions is reflected in the generalized tensor spring interaction between residues *i* and *j*, in such a way that the tensor now includes not only the two-body interaction between residues *i* and *j*, but also three-body and four-body interactions that involve residues *i* and *j* (see Eq. (5.10)).

In doing this, the STeM model is able to integrate all the aforementioned attractive features of ANM and GNM and overcome their limitations. Specifically, STeM performs equally well in B-factors predictions as GNM and has the ability to predict the directions of the fluctuations as ANM. This is accomplished with virtually no performance slowdown. The only potential drawback of this model is the significantly increased complexity in deriving the Hessian matrix. Fortunately, this has been done once for all and the derivation results are available electronically at http://www.cs.iastate.edu/~gsong/STeM.

STeM is physically more accurate by explicitly including the bond bending and torsional interactions since they capture the chain behavior of protein molecules, which are neglected in most elastic network models where a protein is treated as an elastic rubber. Therefore, we have reasons to expect this model will further distinguish itself in studying protein dynamics where a correct modeling of bond bending and/or torsional rotations is critical.

STeM is not limited to coarse-grained protein models that use a single bead, usually the alpha carbon, to represent each residue. The core idea of STeM, deriving the Hessian matrix directly from a physically realistic potential, can be extended to all-atom models as well. We did this and discovered that all-atom STeM model represents a closer approximation of NMA than most other models.

5.2 Results and Discussion

5.2.1 *Crystallographic B-Factor Prediction*

Table 5.1 shows the correlation coefficients between the experimental and calculated B-factors of the 111 proteins in the first dataset. The mean values of the correlation coefficients of ANM, GNM, and STeM are 0.53, 0.59, and 0.60 respectively. STeM provides the directional information of the residue fluctuations as ANM and has an accuracy even slightly better than GNM in B-factor predictions. Figure 5.1 shows the distributions of the correlation coefficients between the calculated B-factors and the experimental B-factors. STeM is the only model in which there are instances where the correlation coefficient is above 0.85 and no instances where the correlation coefficient is below 0.25. This implies that the performance of STeM is more steady than either ANM or GNM. The scatter plot of the correlation coefficients between ANM and STeM in Fig. 5.2 shows that STeM performs better than ANM for 80 % of the proteins in the dataset.

Table 5.1 The correlation coefficients between the experimental and calculated B-factors using different models

Protein	R(Å)	ANM	GNM	STeM	Protein	R(Å)	ANM	GNM	STeM
1AAC	1.31	0.7	0.71	0.76	1ADS	1.65	0.77	0.74	0.71
1AHC	2.00	0.79	0.68	0.61	1AKY	1.63	0.56	0.72	0.6
1AMM	1.20	0.56	0.72	0.55	1AMP	1.80	0.62	0.59	0.68
1ARB	1.20	0.78	0.76	0.83	1ARS	1.80	0.14	0.43	0.41
1ARU	1.60	0.7	0.78	0.79	1BKF	1.60	0.52	0.43	0.5
1BPI	1.09	0.43	0.56	0.57	1CDG	2.00	0.65	0.62	0.71
1CEM	1.65	0.51	0.63	0.76	1CNR	1.05	0.34	0.64	0.42
1CNV	1.65	0.69	0.62	0.68	1CPN	1.80	0.51	0.54	0.56
1CSH	1.65	0.44	0.41	0.57	1CTJ	1.10	0.47	0.39	0.62
1CUS	1.25	0.74	0.66	0.76	1DAD	1.60	0.28	0.5	0.42
1DDT	2.00	0.21	−0.01	0.49	1EDE	1.90	0.67	0.63	0.75
1EZM	1.50	0.56	0.6	0.58	1FNC	2.00	0.29	0.59	0.61
1FRD	1.70	0.54	0.83	0.77	1FUS	1.30	0.4	0.63	0.61
1FXD	1.70	0.58	0.56	0.7	1GIA	2.00	0.68	0.67	0.69
1GKY	2.00	0.36	0.55	0.44	1GOF	1.70	0.75	0.76	0.78
1GPR	1.90	0.65	0.62	0.66	1HFC	1.50	0.63	0.38	0.35
1IAB	1.79	0.36	0.42	0.53	1IAG	2.00	0.34	0.52	0.44
1IFC	1.19	0.61	0.67	0.53	1IGD	1.10	0.18	0.44	0.27
1IRO	1.10	0.82	0.51	0.85	1JBC	1.15	0.72	0.7	0.73
1KNB	1.70	0.63	0.66	0.54	1LAM	1.60	0.53	0.63	0.71
1LCT	2.00	0.52	0.57	0.61	1LIS	1.90	0.16	0.43	0.3
1LIT	1.55	0.65	0.62	0.76	1LST	1.80	0.39	0.72	0.73
1MJC	2.00	0.67	0.67	0.61	1MLA	1.50	0.59	0.57	0.54
1MRJ	1.60	0.66	0.49	0.5	1NAR	1.80	0.62	0.76	0.74
1NFP	1.60	0.23	0.48	0.41	1NIF	1.70	0.42	0.58	0.61

(continued)

Table 5.1 (continued)

Protein	R(Å)	ANM	GNM	STeM	Protein	R(Å)	ANM	GNM	STeM
1NPK	1.80	0.53	0.55	0.64	1OMP	1.80	0.61	0.63	0.65
1ONC	1.70	0.55	0.7	0.58	1OSA	1.68	0.36	0.42	0.55
1OYC	2.00	0.78	0.73	0.77	1PBE	1.90	0.53	0.61	0.63
1PDA	1.76	0.6	0.76	0.58	1PHB	1.60	0.56	0.52	0.59
1PHP	1.65	0.59	0.63	0.65	1PII	2.00	0.19	0.44	0.28
1PLC	1.33	0.41	0.47	0.42	1POA	1.50	0.54	0.66	0.42
1POC	2.00	0.46	0.52	0.39	1PPN	1.60	0.61	0.64	0.67
1PTF	1.60	0.47	0.6	0.54	1PTX	1.30	0.65	0.51	0.62
1RA9	2.00	0.48	0.61	0.53	1RCF	1.40	0.59	0.63	0.58
1REC	1.90	0.34	0.5	0.49	1RIE	1.50	0.71	0.25	0.52
1RIS	2.00	0.25	0.24	0.47	1RRO	1.30	0.08	0.31	0.36
1SBP	1.70	0.69	0.72	0.67	1SMD	1.60	0.5	0.62	0.67
1SNC	1.65	0.68	0.71	0.72	1THG	1.80	0.5	0.53	0.5
1TML	1.80	0.64	0.64	0.58	1UBI	1.80	0.56	0.69	0.61
1WHI	1.50	0.12	0.33	0.38	1XIC	1.60	0.29	0.4	0.47
2AYH	1.60	0.63	0.73	0.82	2CBA	1.54	0.67	0.75	0.8
2CMD	1.87	0.68	0.6	0.62	2CPL	1.63	0.61	0.6	0.72
2CTC	1.40	0.63	0.67	0.75	2CY3	1.70	0.51	0.5	0.67
2END	1.45	0.63	0.71	0.68	2ERL	1.00	0.74	0.73	0.85
2HFT	1.69	0.63	0.79	0.72	2IHL	1.40	0.62	0.69	0.72
2MCM	1.50	0.78	0.83	0.79	2MHR	1.30	0.65	0.52	0.64
2MNR	1.90	0.46	0.5	0.47	2PHY	1.40	0.54	0.55	0.68
2RAN	1.89	0.43	0.4	0.31	2RHE	1.60	0.28	0.38	0.33
2RN2	1.48	0.68	0.71	0.75	2SIL	1.60	0.43	0.5	0.51
2TGI	1.80	0.69	0.71	0.73	3CHY	1.66	0.61	0.75	0.68
3COX	1.80	0.71	0.71	0.72	3EBX	1.40	0.22	0.58	0.4
3GRS	1.54	0.44	0.57	0.59	3LZM	1.70	0.6	0.52	0.66
3PTE	1.60	0.68	0.83	0.77	4FGF	1.60	0.41	0.27	0.43
4GCR	1.47	0.73	0.81	0.75	4MT2	2.00	0.42	0.37	0.46
5P21	1.35	0.4	0.51	0.45	7RSA	1.26	0.42	0.63	0.59
8ABP	1.49	0.61	0.82	0.62	–	–	–	–	–

Column R (Å) gives the resolution of each structure

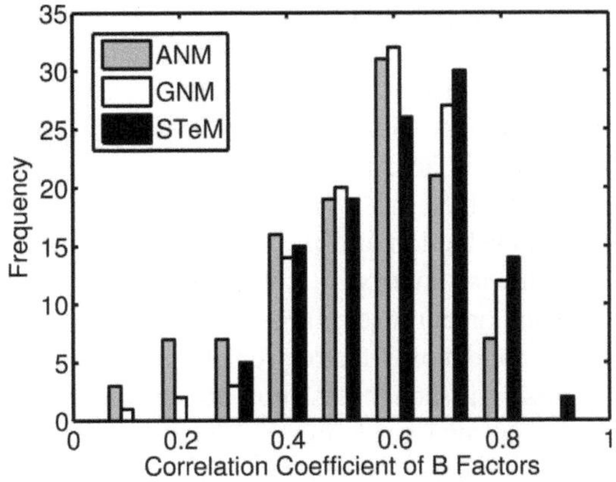

Fig. 5.1 The distributions of the correlation coefficients between the experimental and calculated B-factors

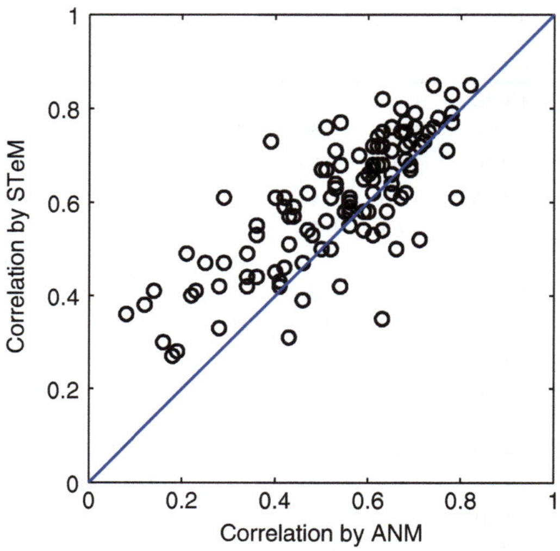

Fig. 5.2 The scatter plot of the correlation coefficients by ANM and those by STeM. For 80 % of the proteins listed in Table 5.1, STeM does better than ANM

Protein structures of higher resolution have more accurate data on atom coordinates and B-factors. We investigate whether our model's performance can be further improved when the dataset used is limited to structures with higher resolution. We select the 12 structures with resolution better than 1.3 Å from the first dataset. The mean values of the correlation coefficients of these 12 structures are 0.56, 0.62, and 0.63 for ANM, GNM, and STeM, respectively, which gives an improvement of about 5–6 % for all of the three models. Since the improvement is based on a relatively small set of 12 structures, a larger dataset is needed to further examine this potential dependence of B-factor prediction accuracy on structure quality.

5.2.2 The Contributions of Different Interaction Terms to the Fluctuations

The Gō-like potential in Eq. (5.11) has four different interaction terms, namely, bond stretching, bond bending, torsional interactions, and the non-bonded interactions. It is of great interest to investigate the relative contributions of these different terms to the agreement with experimental B-factors. Since only the non-bonded interaction term (V_4) is able to provide by itself enough constraints to ensure the Hessian matrix to have no more than six zero eigenvalues, V_4 is used as the base term for the evaluation of different terms' contributions to the mean-square fluctuations. The Hessian matrix of ANM, denoted by $\boldsymbol{H}_{\mathrm{ANM}}$, is used as another baseline for comparison purposes. Table 5.2 lists the contributions of these different terms to the improvement of B-factor predictions as they are added to the potential.

First, it is seen that the non-bonded interactions, as are present in $\boldsymbol{H}_{V_4}$ and $\boldsymbol{H}_{\mathrm{ANM}}$, play a dominant role in contributing to the B-factors. This is not surprising since the mean-square fluctuations of a residue are mostly constrained by its interactions

Table 5.2 The contributions of different interaction terms to the agreement with experimental B-factors H_{ANM}

Hessian matrices used	Correlation coefficient with B-factors	Improvement with respect to ANM
$\boldsymbol{H}_{\text{ANM}}$	0.53	0.00
$\boldsymbol{H}_{V_4}$	0.55	0.02
$\boldsymbol{H}_{V_4} + \boldsymbol{H}_{V_1}$	0.57	0.04
$\boldsymbol{H}_{V_4} + \boldsymbol{H}_{V_2}$	0.57	0.04
$\boldsymbol{H}_{V_4} + \boldsymbol{H}_{V_3}$	0.56	0.03
$\boldsymbol{H}_{V_4} + \boldsymbol{H}_{V_1} + \boldsymbol{H}_{V_2}$	0.59	0.06
$\boldsymbol{H}_{V_4} + \boldsymbol{H}_{V_1} + \boldsymbol{H}_{V_3}$	0.58	0.05
$\boldsymbol{H}_{V_4} + \boldsymbol{H}_{V_2} + \boldsymbol{H}_{V_3}$	0.57	0.04
$\boldsymbol{H}_{V_4} + \boldsymbol{H}_{V_1} + \boldsymbol{H}_{V_2} + \boldsymbol{H}_{V_3} (= \boldsymbol{H}_{\text{STeM}})$	0.60	0.07
$\boldsymbol{H}_{\text{ANM}} + \boldsymbol{H}_{V_1}$	0.54	0.01
$\boldsymbol{H}_{\text{ANM}} + \boldsymbol{H}_{V_2}$	0.54	0.01
$\boldsymbol{H}_{\text{ANM}} + \boldsymbol{H}_{V_3}$	0.54	0.01
$\boldsymbol{H}_{\text{ANM}} + \boldsymbol{H}_{V_1} + \boldsymbol{H}_{V_2} + \boldsymbol{H}_{V_3}$	0.56	0.03

$\boldsymbol{H}_{\text{ANM}}$ is the Hessian matrix of ANM. $\boldsymbol{H}_{V_1}$, $\boldsymbol{H}_{V_2}$, $\boldsymbol{H}_{V_3}$, and $\boldsymbol{H}_{V_4}$ are the Hessian matrices of the bond stretching (V_1), bond bending (V_2), torsional rotation (V_3), and non-local interaction (V_4) terms, respectively

with its spatial neighbors, most of which are through non-bonded interactions. What is more interesting is that $\boldsymbol{H}_{V_4}$ term alone performs better than $\boldsymbol{H}_{\text{ANM}}$. This is in agreement with recent results that the performance of B-factor predictions can be improved by using distance-dependent force constants [25, 26]. Particularly, the spring constants that take the form of inverse distance square have been shown to be superior in a recent exhaustive study that experimented with different distance-dependent spring constants on a large dataset [16]. The Taylor expansion of the non-bonded interaction term (V_4) shows that it has an equivalent spring constant of the form $\frac{120\varepsilon}{r_{0,ij}^2}$ (see Eq. (5.36)), which is exactly proportional to the inverse of the pairwise distance square. Thus, STeM provides a physics-based explanation for the choice of using inverse square distance spring constants.

The contribution to the improvement in B-factor predictions from each of the bonded interactions, such as that of bond stretching, is small, as had been pointed out by Bahar et al. when GNM was first proposed over a decade ago [3]. However, when the contributions of all of these four terms are added up, they together enable the STeM model to gain a significant improvement over ANM to reach the level of accuracy on a par with GNM.

5.2.3 *Conformational Change Evaluation*

It is known that the modes derived from the open form of a structure have better overlaps and correlations with the direction of a protein's conformation change than

Table 5.3 The overlaps and correlations between the observed conformation changes and the most involved modes using different models and the open conformations

Protein	Overlap in ANM	Correlation in ANM	Overlap in STeM	Correlation in STeM
Adenylate kinase	0.49(1)	0.62(1)	0.55(1)	0.63(1)
Alcohol dehydrogenase	0.69(3)	0.54(9)	0.73(2)	0.65(30)
Annexin V	0.33(1)	0.60(32)	0.33(1)	0.56(22)
Aspartate aminotransferase	0.56(9)	0.63(9)	0.68(6)	0.67(6)
Calmodulin	0.44(5)	0.62(77)	0.48(1)	0.62(16)
Che Y protein	0.46(1)	0.78(12)	0.40(1)	0.74(1)
Citrate synthase	0.48(7)	0.72(26)	0.49(5)	0.63(5)
Dihydrofolate reductase	0.71(1)	0.65(1)	0.73(1)	0.66(1)
Diphtheria toxin	0.43(1)	0.69(2)	0.50(2)	0.73(2)
Enolase	0.31(1)	0.45(34)	0.32(1)	0.49(53)
HIV-1 protease	0.67(1)	0.78(10)	0.85(1)	0.90(1)
Immunoglobulin	0.68(3)	0.57(3)	0.66(3)	0.58(3)
Lactoferrin	0.48(1)	0.64(24)	0.48(1)	0.70(36)
LAO binding protein	0.81(1)	0.74(1)	0.87(1)	0.80(1)
Maltodextrin binding protein	0.77(2)	0.66(2)	0.80(2)	0.70(2)
Seryl-tRNA synthetase	0.21(4)	0.59(10)	0.21(4)	0.60(37)
Thymidylate synthase	0.37(4)	0.69(9)	0.44(3)	0.68(9)
Triglyceride lipase	0.35(15)	0.50(25)	0.30(14)	0.56(24)
Triose phosphate isomerase	0.15(38)	0.28(11)	0.14(7)	0.30(8)
Tyrosine phosphatase	0.41(2)	0.57(27)	0.42(1)	0.59(25)

the ones derived from the closed form [20]. Here we apply the STeM model to study the conformation changes between the open and closed forms of 20 proteins. The open forms are used to calculate the normal modes. Table 5.3 lists the overlaps and correlations of the observed conformation changes and the indices of the modes that are most involved in the conformation changes. GNM is not considered since it cannot provide directional information. The mean values of the overlaps and correlation coefficients of ANM are 0.49 and 0.61 respectively, and 0.52 and 0.64 respectively for STeM. These amount to an improvement of about 5 % for STeM over ANM on both overlap and correlation. Since the results are obtained based on a relatively small set of 20 protein pairs, the significance of the improvement seen here needs to be further tested by conducting a more exhaustive analysis that uses a larger set of proteins and varying parameters, and preferably taking into account the effect of crystal packing as well. We will leave this for future work. It is also worth noting that, in both the overlap and correlation calculations, the modes that are most involved in the conformation change tend to have lower indices in STeM than in ANM (see Table 5.3), which may imply the modes of STeM be of higher quality than those of ANM.

5.2.4 *Protein Fluctuation Dynamics Predicted by All-Atom STeM Matches Closely with That of NMA*

In this section, we apply all-atom STeM model to a large number of proteins and show that the fluctuation dynamics produced by STeM matches closely with that of NMA. To avoid the uncertainties existing in experimental B-factors due to crystal packing and disorder, the atomic fluctuations computed from NMA and STeM are compared with each other and not with the experimental B-factors.

To compute the fluctuations, all the structures are first energetically minimized using the Tinker program [27] with the Charmm22 force field. The minimized structures are then used by NMA, STeM, and later on, by ANM and ANMr2 models, to compute the mean-square fluctuations. Some of the force field parameters from Charmm22 are used in computing the STeM Hessian matrix. Let $\boldsymbol{M}$ be the $N \times N$ diagonal mass matrix, I be the 3×3 identity matrix, and $\otimes$ be the operator of the Kronecker product. Let $\boldsymbol{b}^{\mathrm{NMA}}$ and $\boldsymbol{b}^{\mathrm{STeM}}$ denote the mean-square fluctuations from NMA and STeM, respectively. The following procedure details how they are determined:

1. Use Tinker [27] to determine the minimized conformation C whose potential energy as defined by Charmm22 is fully minimized;
2. Obtain $\boldsymbol{b}^{\mathrm{NMA}}$ using Tinker;
3. Compute the Hessian matrix $\boldsymbol{H}^{\mathrm{STeM}}$ of C using STeM;
4. Determine frequencies f_i and modes $\boldsymbol{m}_i$ of $\boldsymbol{H}^{\mathrm{STeM}}$ in the mass-weighted Cartesian coordinate as follows, where $i = 7, 8, \ldots, 3N$:
   (a) $\tilde{\boldsymbol{H}}^{\mathrm{STeM}} \leftarrow \left(\boldsymbol{M}^{1/2} \otimes I\right)^{-1} \boldsymbol{H}^{\mathrm{STeM}} \left(\boldsymbol{M}^{1/2} \otimes I\right)^{-1}$;
   (b) $\langle f_i, \tilde{\boldsymbol{m}}_i \rangle \leftarrow i$th eigenvalue and eigenvector of $\tilde{\boldsymbol{H}}^{\mathrm{STeM}}$;
   (c) $\boldsymbol{m}_i \leftarrow \left(\boldsymbol{M}^{1/2} \otimes I\right)^{-1} \tilde{\boldsymbol{m}}_{i,}$;
5. Compute the B-factor $\boldsymbol{b}^{\mathrm{STeM}}$ using f_i and $\boldsymbol{m}_i$;
6. Compute the correlation between $\boldsymbol{b}^{\mathrm{NMA}}$ and $\boldsymbol{b}^{\mathrm{STeM}}$.

The procedure is repeated for a dataset of 306 proteins.

5.2.4.1 STeM Outperforms ANM in Matching with NMA

Figure 5.3 compares the correlations between computed B-factors: $\boldsymbol{b}^{\mathrm{STeM}}$, $\boldsymbol{b}^{\mathrm{ANM}}$, or $\boldsymbol{b}^{\mathrm{ANMr2}}$, with $\boldsymbol{b}^{\mathrm{NMA}}$. 306 proteins, listed in Table 5.4, are used to compute these correlations. Denote corr($\boldsymbol{a}$,$\boldsymbol{b}$) by the correlation between two vectors $\boldsymbol{a}$ and $\boldsymbol{b}$. Figure 5.3a shows the scatter plot of corr($\boldsymbol{b}^{\mathrm{NMA}}$,$\boldsymbol{b}^{\mathrm{ANM}}$) and corr($\boldsymbol{b}^{\mathrm{NMA}}$,$\boldsymbol{b}^{\mathrm{STeM}}$),

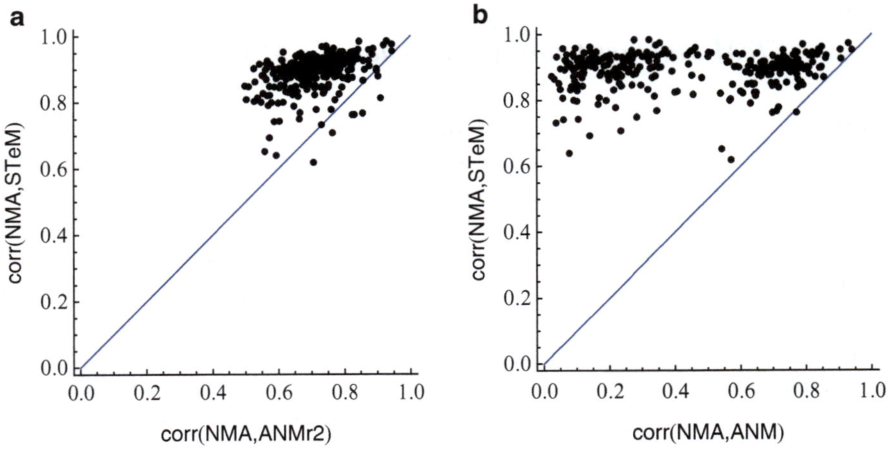

Fig. 5.3 The scatter plots of the correlation coefficients with NMA by different all-atom models. (**a**) the scatter plot of corr($\boldsymbol{b}^{\mathrm{NMA}}$,$\boldsymbol{b}^{\mathrm{ANM}}$) and corr($\boldsymbol{b}^{\mathrm{NMA}}$,$\boldsymbol{b}^{\mathrm{STeM}}$), and (**b**) the scatter plot of corr($\boldsymbol{b}^{\mathrm{NMA}}$,$\boldsymbol{b}^{\mathrm{ANMr2}}$) and corr($\boldsymbol{b}^{\mathrm{NMA}}$,$\boldsymbol{b}^{\mathrm{STeM}}$), where $\boldsymbol{b}$ denotes the mean-square fluctuations computed by the different models. 306 proteins as listed in Table 5.4 are used for the computation

Table 5.4 List of proteins used in the all-atom STeM study

1BKR	2A28	2R8U	3B0F	3JQU	3NAR	3R87	3TME	4BBD
1C5E	2AAJ	2RK5	3B7H	3JTN	3NBC	3RDM	3TOE	4D8D
1DBF	2BT9	2RKL	3BD1	3K6F	3NGP	3RDO	3TP5	4DCZ
1G2B	2CKK	2VE8	3BRI	3K6T	3NJK	3RDS	3TPX	4DRO
1G2R	2F5K	2VKL	3BRL	3KBL	3NJM	3RE6	3TSI	4DRP
1GK7	2FE5	2VZC	3BZY	3KIK	3NS6	3RGR	3TWE	4E34
1GU1	2FL4	2WJ5	3CCD	3KJL	3NTW	3RHB	3TXQ	4E35
1HG7	2GBJ	2WPU	3CNK	3KNG	3NXA	3RJS	3TXS	4E6I
1I2T	2GBN	2WQ0	3CPO	3KOV	3O48	3RKV	3U1C	4E6S
1IHR	2HIN	2X3D	3CX2	3KXY	3O5Z	3RL8	3U80	4E8O
1J8Q	2HO2	2X48	3D4W	3L1F	3OBL	3RNJ	3UD8	4EDL
1J9E	2I5C	2X5H	3DS4	3L1X	3OJB	3RNV	3UJ3	4EDM
1JCD	2IC6	2X5T	3E2B	3L7H	3OMT	3RQ9	3VA9	4ERR
1JO0	2IGD	2XDH	3E56	3LKY	3OV4	3RSW	3VBG	4ES3
1MFG	2IZX	2XEM	3FG7	3LLB	3P38	3RY2	3VEJ	4EWI
1MG4	2J5Y	2XF6	3FX7	3LLO	3P6J	3RZW	3VGN	4EZA
1MK5	2JDC	2XF7	3FZW	3LNQ	3P7J	3S02	3VI6	4F26
1MM9	2JDD	2XG3	3G21	3LNW	3PA7	3SEI	3VMX	4F55
1N9M	2JKU	2XRH	3G9R	3LRD	3PE9	3SFM	3VMY	4F8A
1NWW	2LIS	2XUS	3GZ2	3LRG	3PO8	3SGP	3ZR8	4FQN
1OOT	2O31	2XW6	3H00	3M0R	3PYJ	3SGR	3ZSK	4FYH

(continued)

Table 5.4 (continued)

1R29	2O37	2XX6	3HA4	3M0U	3Q47	3SHU	3ZSL	4GCN
1R6J	2O9V	2Y2T	3HFO	3M8J	3Q9Q	3SK4	3ZW2	4GCO
1T6F	2OEI	2Y3E	3HGM	3M9H	3QF3	3SK6	3ZZL	4GMQ
1TG0	2ON8	2Y3F	3HSH	3M9J	3QGL	3SK8	3ZZQ	4GS3
1U07	2ONQ	2Y4X	3HTU	3MAB	3QMQ	3SQF	4A1H	4GSW
1URR	2OVG	2Y9F	3I4O	3MBT	3QMX	3SSQ	4A6S	4HBX
1W53	2PMR	2Y9G	3ID1	3MCB	3QWG	3SSU	4A75	4HE6
1WM3	2PWO	2Y9R	3ID2	3MCE	3QWS	3SWY	4A9F	4HK2
1WYX	2PZV	2YEL	3ID3	3MHE	3R27	3T6F	4ABM	4HTH
1Y0M	2QCB	2YH5	3ID4	3MP9	3R3M	3T6L	4AEQ	4HTI
1YO3	2QCP	2YIZ	3IG9	3MSH	3R45	3T8N	4AGH	4HTJ
1Z96	2QJL	2ZWM	3IGE	3N27	3R69	3T8U	4B27	4HX8
2A26	2R6Q	3AXC	3IPT	3N4W	3R85	3TDM	4B6X	4IOG

while (b) the scatter plot of corr($\boldsymbol{b}^{\mathrm{NMA}}$,$\boldsymbol{b}^{\mathrm{ANMr2}}$) and corr($\mathbf{b}^{\mathrm{NMA}}$,$\mathbf{b}^{\mathrm{STeM}}$), respectively. The average correlation over all proteins is 0.44 for corr($\boldsymbol{b}^{\mathrm{NMA}}$,$\boldsymbol{b}^{\mathrm{ANM}}$), 0.71 for corr($\boldsymbol{b}^{\mathrm{NMA}}$,$\boldsymbol{b}^{\mathrm{ANMr2}}$), and 0.89 for corr($\boldsymbol{b}^{\mathrm{NMA}}$,$\boldsymbol{b}^{\mathrm{STeM}}$). STeM clearly outperforms both ANMr2 and ANM in matching with NMA, having a high average correlation in mean-square fluctuations with those of NMA. The results thus underscore the importance of including multi-body interactions for a finer portrait of protein fluctuation dynamics.

5.3 Conclusions

Protein mean-square fluctuations and conformation changes are two closely related aspects of protein dynamics. However, in the past, two separate groups of models were needed to best explain protein mean-square fluctuations or conformation changes. Specifically, the best models for predicting mean-square fluctuations cannot predict conformation changes, and the models that can predict conformation changes do not have the best performance in predicting mean-square fluctuations. There is thus an obvious gap between the models that work well in predicting one aspect of the dynamics and those in another.

Since protein mean-square fluctuations and conformation changes are two closely related dynamic phenomena and share a similar physical origin, we reasoned that models based on a physically more accurate potential should be able to bridge the gap and predict both aspects of the protein dynamics well. Indeed, by using a Gō-like potential, we have successfully developed a spring tensor model (STeM) that is able to singly predict well both mean-square fluctuations and conformation changes. Specifically, STeM performs equally well in B-factor predictions as GNM and has the ability to predict the directions of fluctuations as ANM.

The new STeM model does come with a cost. As is seen, the derivation process of the Hessian matrix in STeM is much more complex than models using only two-body Hookean potentials, such as those used in ANM. However, the introduced complexity in the potential is necessary in resolving the aforementioned gap that is mainly due to over-simplified potentials and in providing a single, unified model for protein dynamics. Moreover, the derivation process, though more complex, needs to be done only once.

Examining the different interaction terms in the Gō potential and their contributions to the agreement with experimental B-factors provides further benefits. Along the way, we have discovered a physical explanation for why the distance-dependent, inverse distance square (i.e., $\frac{1}{r^2}$) spring constants perform better than the uniform ones. The van der Waals interaction term in the potential naturally renders inverse distance square spring constants! By including the bond bending and torsional interactions and their contributions to the improvement in B-factor predictions, the STeM model confirms the importance of 3-body and 4-body potentials. The importance of multi-body potentials is made even more evident when their contribution to the interaction spring tensor is examined – the multi-body potentials are shown to be necessary in providing proper constraints on residue fluctuations, even transversely. In [28] we noted that the 3-body and 4-body potentials introduced through bond bending and torsional interactions in the coarse-grained STeM model only scratched the surface of the full extensity of the multi-body potentials. Indeed, results from all-atom STeM where the multi-body interactions are most accurately represented demonstrate that all-atom STeM has reached an even higher correlation with NMA in predicting mean-square fluctuations, yet without the need for energy minimization.

Finally, since STeM takes into account bond bending and torsional interactions, it is expected that it should further distinguish itself in studying protein dynamics where a correct modeling of bond bending or torsional rotations is critical, such as in predicting S^2 order parameters of NMR structures.

5.4 Methods

In this section we present the derivations of the Hessian matrix for a coarse-grained model from a Gō-like potential [23]. The derivations are mostly the same as what appeared in [28]. In addition, we show how the core idea of STeM can be extended to derive the STeM Hessian matrix for an all-atom model using an all-atom potential.

5.4.1 The Gō-Like Potential

The Gō-like potential in [23] takes the non-native and native (equilibrium) conformations as input and it can be divided into four terms. The first term of this Gō-like potential (defined as V_1 for later use) preserves the chain connectivity. The second (V_2) and third terms (V_3) define the bond angle and torsional interactions respectively and the last term (V_4) is the nonlocal interactions. The Gō-like potential has the following expression:

$$\begin{aligned}
V(X, X_0) &= \sum_{\text{bonds}} V_1(r, r_0) + \sum_{\text{angle}} V_2(\theta, \theta_0) + \sum_{\text{dihedral}} V_3(\phi, \phi_0) + \sum_{i<j-3} V_4\left(r_{ij}, r_{0,ij}\right) \\
&= \sum_{\text{bonds}} K_r (r - r_0)^2 + \sum_{\text{angle}} K_\theta (\theta - \theta_0)^2 \\
&\quad + \sum_{\text{dihedral}} \left\{ K_\phi^{(1)} [1 - \cos(\phi - \phi_0)] + K_\phi^{(3)} [1 - \cos 3(\phi - \phi_0)] \right\} \\
&\quad + \sum_{i<j-3} \varepsilon \left[5\left(\frac{r_{0,ij}}{r_{ij}}\right)^{12} - 6\left(\frac{r_{0,ij}}{r_{ij}}\right)^{10} \right]
\end{aligned} \tag{5.11}$$

In Eq. (5.11), r and r_0 represent respectively the instantaneous and equilibrium lengths of the virtual bonds between the C_α atoms of consecutive residues. Similarly, the θ (θ_0) and ϕ (ϕ_0) are respectively the instantaneous (equilibrium) virtual bond angles formed by three consecutive residues and the instantaneous (equilibrium) virtual dihedral angles formed by four consecutive residues. The r_{ij} and $r_{0,ij}$ represent respectively the instantaneous and equilibrium distances between two non-consecutive residues i and j.

The Gō-like potential in Eq. (5.11) includes several force parameters ($K_r, K_\theta, K_\phi^{(1)}, K_\phi^{(3)}$ *and* ε) and the values of these parameters are taken directly from [23] without any tuning. The values of these parameters are: $K_r = 100\varepsilon, K_\theta = 20\varepsilon$, $K_\phi^{(1)} = \varepsilon, K_\phi^{(3)} = 0.5\varepsilon$ *and* $\varepsilon = 0.36$.

5.4.2 Anisotropic Fluctuations from the Second Derivative of the Gō-Like Potential

Similar to ANM, STeM has a $3N \times 3N$ Hessian matrix that can be decomposed into $N \times N$ super-elements. Each super-element in STeM, $\boldsymbol{H}_{i,j}$, is a summation of four 3×3 matrices. The first 3×3 matrix is the contribution from bond stretching.

The second and third 3×3 matrices are the contributions from bond bending and torsional rotations respectively. The fourth 3×3 matrix is the contribution from nonlocal contacts.

$$
\boldsymbol{H}_{i,j} = \begin{bmatrix} \frac{\partial^2 V_1(r,r_0)}{\partial X_i \partial X_j} & \frac{\partial^2 V_1(r,r_0)}{\partial X_i \partial Y_j} & \frac{\partial^2 V_1(r,r_0)}{\partial X_i \partial Z_j} \\ \frac{\partial^2 V_1(r,r_0)}{\partial Y_i \partial X_j} & \frac{\partial^2 V_1(r,r_0)}{\partial Y_i \partial Y_j} & \frac{\partial^2 V_1(r,r_0)}{\partial Y_i \partial Z_j} \\ \frac{\partial^2 V_1(r,r_0)}{\partial Z_i \partial X_j} & \frac{\partial^2 V_1(r,r_0)}{\partial Z_i \partial Y_j} & \frac{\partial^2 V_1(r,r_0)}{\partial Z_i \partial Z_j} \end{bmatrix} +
\begin{bmatrix} \frac{\partial^2 V_2(\theta,\theta_0)}{\partial X_i \partial X_j} & \frac{\partial^2 V_2(\theta,\theta_0)}{\partial X_i \partial Y_j} & \frac{\partial^2 V_2(\theta,\theta_0)}{\partial X_i \partial Z_j} \\ \frac{\partial^2 V_2(\theta,\theta_0)}{\partial Y_i \partial X_j} & \frac{\partial^2 V_2(\theta,\theta_0)}{\partial Y_i \partial Y_j} & \frac{\partial^2 V_2(\theta,\theta_0)}{\partial Y_i \partial Z_j} \\ \frac{\partial^2 V_2(\theta,\theta_0)}{\partial Z_i \partial X_j} & \frac{\partial^2 V_2(\theta,\theta_0)}{\partial Z_i \partial Y_j} & \frac{\partial^2 V_2(\theta,\theta_0)}{\partial Z_i \partial Z_j} \end{bmatrix} +
\begin{bmatrix} \frac{\partial^2 V_3(\phi,\phi_0)}{\partial X_i \partial X_j} & \frac{\partial^2 V_3(\phi,\phi_0)}{\partial X_i \partial Y_j} & \frac{\partial^2 V_3(\phi,\phi_0)}{\partial X_i \partial Z_j} \\ \frac{\partial^2 V_3(\phi,\phi_0)}{\partial Y_i \partial X_j} & \frac{\partial^2 V_3(\phi,\phi_0)}{\partial Y_i \partial Y_j} & \frac{\partial^2 V_3(\phi,\phi_0)}{\partial Z_i \partial Z_j} \\ \frac{\partial^2 V_3(\phi,\phi_0)}{\partial Z_i \partial X_j} & \frac{\partial^2 V_3(\phi,\phi_0)}{\partial Z_i \partial Y_j} & \frac{\partial^2 V_3(\phi,\phi_0)}{\partial Z_i \partial Z_j} \end{bmatrix} +
\begin{bmatrix} \frac{\partial^2 V_4(r_{ij},r_{0,ij})}{\partial X_i \partial X_j} & \frac{\partial^2 V_4(r_{ij},r_{0,ij})}{\partial X_i \partial Y_j} & \frac{\partial^2 V_4(r_{ij},r_{0,ij})}{\partial X_i \partial Z_j} \\ \frac{\partial^2 V_4(r_{ij},r_{0,ij})}{\partial Y_i \partial X_j} & \frac{\partial^2 V_4(r_{ij},r_{0,ij})}{\partial Y_i \partial Y_j} & \frac{\partial^2 V_4(r_{ij},r_{0,ij})}{\partial Y_i \partial Z_j} \\ \frac{\partial^2 V_4(r_{ij},r_{0,ij})}{\partial Z_i \partial X_j} & \frac{\partial^2 V_4(r_{ij},r_{0,ij})}{\partial Z_i \partial Y_j} & \frac{\partial^2 V_4(r_{ij},r_{0,ij})}{\partial Z_i \partial Z_j} \end{bmatrix} \quad (5.12)
$$

The Hessian matrix is the second derivative of the overall potential (Eq. (5.11)). Let us first consider the first term of the Gō-like potential and let (X_i, Y_i, Z_i) and (X_j, Y_j, Z_j) be the Cartesian coordinates of two consecutive residues i and j.

$$
\begin{aligned} V_1(r, r_0) &= K_r (r - r_0)^2 \\ &= K_r \left\{ \left[(X_j - X_i)^2 + (Y_j - Y_i)^2 + (Z_j - Z_i)^2 \right]^{1/2} - r_0 \right\}^2 \end{aligned} \quad (5.13)
$$

The first and second partial derivatives of V_1 with respect to the X-direction of residue i are

$$
\frac{\partial V_1}{\partial X_i} = -2K_r \left(X_j - X_i \right) \left(1 - r^0 / r \right) \quad (5.14)
$$

$$
\frac{\partial^2 V_1}{\partial X_i^2} = 2K_r \left(1 + r^0 (X_j - X_i)^2 / r^3 - r^0 / r \right) \quad (5.15)
$$

We will get similar results for the $Y-$ and Z-directions of residue i. Since we focus only on the equilibrium fluctuations, we can have $r \cong r^0$ at equilibrium and the first and second partial derivatives of V_1 can be further simplified to the following expressions.

$$\frac{\partial V_1}{\partial X_i} = 0 \tag{5.16}$$

$$\frac{\partial^2 V_1}{\partial X_i^2} = 2K_r\left(X_j - X_i\right)^2/r^2 \tag{5.17}$$

In a similar way, the second cross-derivatives have the following form:

$$\frac{\partial^2 V_1}{\partial X_i \partial Y_j} = -2K_r\left(X_j - X_i\right)\left(Y_j - Y_i\right)/r^2 \tag{5.18}$$

Equations (5.17) and (5.18) give the elements of the first 3×3 matrix of the super element $\boldsymbol{H}_{ij}$ in Eq. (5.6). For the diagonal super elements $\boldsymbol{H}_{ii}$, Eqs. (5.17) and (5.18) are substituted by the following:

$$\frac{\partial^2 V_1}{\partial X_i^2} = -\sum_j 2K_r\left(X_j - X_i\right)^2/r^2 \tag{5.19}$$

$$\frac{\partial^2 V_1}{\partial X_i \partial Y_i} = \sum_j 2K_r\left(X_j - X_i\right)\left(Y_j - Y_i\right)/r^2 \tag{5.20}$$

Now let us consider the second term of the potential in Eq. (5.11) and let (X_i,Y_i,Z_i), (X_j,Y_j,Z_j) and (X_k,Y_k,Z_k) be the Cartesian coordinates of three consecutive residues i, j and k. Suppose θ is the virtual bond angle formed by these three consecutive residues. Since the second term of the potential is $V_2 = K_\theta(\theta - \theta_0)^2$, the first and second partial derivatives of V_2 are

$$\frac{\partial V_2}{\partial X_i} = 2K_\theta\left(\theta - \theta_0\right)\frac{\partial \theta}{\partial X_i} \tag{5.21}$$

$$\frac{\partial^2 V_2}{\partial X_i^2} = 2K_\theta\left(\frac{\partial \theta}{\partial X_i}\right)^2 + 2K_\theta\left(\theta - \theta_0\right)\frac{\partial^2 \theta}{\partial X_i^2} \tag{5.22}$$

Since θ equals θ_0 at equilibrium, $\frac{\partial^2 V_2}{\partial X_i^2}$ can be further simplified as

$$\frac{\partial^2 V_2}{\partial X_i^2} = 2K_\theta\left(\frac{\partial \theta}{\partial X_i}\right)^2 \tag{5.23}$$

Likewise, $\frac{\partial^2 V_2}{\partial X_i \partial X_j}$ becomes

$$\frac{\partial^2 V_2}{\partial X_i \partial X_j} = 2K_\theta \left(\frac{\partial \theta}{\partial X_i}\right)\left(\frac{\partial \theta}{\partial X_j}\right) \tag{5.24}$$

Let $\boldsymbol{p} = (X_i - X_j, Y_i - Y_j, Z_i - Z_j)$ and $\boldsymbol{q} = (X_k - X_j, Y_k - Y_j, Z_k - Z_j)$ and define G as the following.

$$G = \frac{(\boldsymbol{p} \cdot \boldsymbol{q})}{|\boldsymbol{p}|\,|\boldsymbol{q}|} \tag{5.25}$$

The θ can be expressed as

$$\theta = \cos^{-1}\left(\frac{(\boldsymbol{p} \cdot \boldsymbol{q})}{|\boldsymbol{p}|\,|\boldsymbol{q}|}\right) = \cos^{-1}(G) \tag{5.26}$$

The partial derivatives of θ are

$$\frac{\partial \theta}{\partial X_i} = \frac{-1}{\sqrt{1-G^2}}\frac{\partial G}{\partial X_i} \tag{5.27}$$

$$\frac{\partial \theta}{\partial X_j} = \frac{-1}{\sqrt{1-G^2}}\frac{\partial G}{\partial X_j} \tag{5.28}$$

$$\frac{\partial \theta}{\partial X_k} = \frac{-1}{\sqrt{1-G^2}}\frac{\partial G}{\partial X_k} \tag{5.29}$$

The derivative of G is

$$\frac{\partial G}{\partial X_i} = \frac{\partial}{\partial X_i}\left(\frac{(\boldsymbol{p} \cdot \boldsymbol{q})}{|\boldsymbol{p}|\,|\boldsymbol{q}|}\right) = \frac{\left(X_k - X_j\right)|p|\,|q| - (\boldsymbol{p} \cdot \boldsymbol{q})\frac{|q|}{|p|}\left(X_i - X_j\right)}{(|\boldsymbol{p}|\,|\boldsymbol{q}|)^2} \tag{5.30}$$

We can also get $\frac{\partial G}{\partial X_j}$ and $\frac{\partial G}{\partial X_k}$.

$$\frac{\partial G}{\partial X_j} = \frac{\left(2X_j - X_i - X_k\right)|\boldsymbol{p}|\,|\boldsymbol{q}| - (\boldsymbol{p} \cdot \boldsymbol{q})\frac{|q|}{|p|}\left(X_j - X_i\right)}{(|\boldsymbol{p}|\,|\boldsymbol{q}|)^2} - \frac{(\boldsymbol{p} \cdot \boldsymbol{q})\frac{|p|}{|q|}\left(X_j - X_k\right)}{(|\boldsymbol{p}|\,|\boldsymbol{q}|)^2} \tag{5.31}$$

$$\frac{\partial G}{\partial X_k} = \frac{\left(X_i - X_j\right)|\boldsymbol{p}|\,|\boldsymbol{q}| - (\boldsymbol{p} \cdot \boldsymbol{q})\frac{|p|}{|q|}\left(X_k - X_j\right)}{(|\boldsymbol{p}|\,|\boldsymbol{q}|)^2} \tag{5.32}$$

Combined Eqs. (5.23), (5.27) and (5.30), we can get the following formula.

$$\frac{\partial^2 V_2}{\partial X_i^2} = \frac{2K_\theta}{1-G^2}\left(\frac{\left(X_k - X_j\right)|\boldsymbol{p}|\,|\boldsymbol{q}| - (\boldsymbol{p}\cdot\boldsymbol{q})\frac{|\boldsymbol{q}|}{|\boldsymbol{p}|}\left(X_i - X_j\right)}{\left(|\boldsymbol{p}|\,|\boldsymbol{q}|\right)^2}\right)^2 \tag{5.33}$$

Similarly, Combined Eqs. (5.24), (5.27), (5.28), (5.30), and (5.31), the second cross-derivative $\frac{\partial^2 V_2}{\partial X_i X_j}$ becomes

$$\frac{\partial^2 V_2}{\partial X_i X_j} = \frac{2K_\theta}{1-G^2}\frac{\left(X_k - X_j\right)|\boldsymbol{p}|\,|\boldsymbol{q}| - (\boldsymbol{p}\cdot\boldsymbol{q})\frac{|\boldsymbol{q}|}{|\boldsymbol{p}|}\left(X_i - X_j\right)}{\left(|\boldsymbol{p}|\,|\boldsymbol{q}|\right)^2}$$
$$\left(\frac{\left(2X_j - X_i - X_k\right)|\boldsymbol{p}|\,|\boldsymbol{q}| - (\boldsymbol{p}\cdot\boldsymbol{q})\frac{|\boldsymbol{q}|}{|\boldsymbol{p}|}\left(X_j - X_i\right)}{\left(|\boldsymbol{p}|\,|\boldsymbol{q}|\right)^2} - \frac{(\boldsymbol{p}\cdot\boldsymbol{q})\frac{|\boldsymbol{p}|}{|\boldsymbol{q}|}\left(X_j - X_k\right)}{\left(|\boldsymbol{p}|\,|\boldsymbol{q}|\right)^2}\right) \tag{5.34}$$

Following a similar approach, we can get $\frac{\partial^2 V_2}{\partial X_j X_k}$ and $\frac{\partial^2 V_2}{\partial X_k X_i}$ and these second cross-derivatives form the elements of the second 3×3 matrix of the super element $\boldsymbol{H}_{ij}$ in Eq. (5.6).

Due to the complexity of the derivation process of the Hessian matrix for the third (dihedral angle) term of the potential, we omit the derivation process here. The complete derivation can be found in [28].

Finally, let's consider the final (non-local contact) term.

$$V_4 = \varepsilon\left[5\left(\frac{r_{0,ij}}{r_{ij}}\right)^{12} - 6\left(\frac{r_{0,ij}}{r_{ij}}\right)^{10}\right] \tag{5.35}$$

A Taylor expansion will give us the following form.

$$V_4 = -\varepsilon + \frac{120\varepsilon}{r_{0,ij}^2}\left(r_{ij} - r_{0,ij}\right)^2 \tag{5.36}$$

Equation (5.36) has the same harmonic form as the first term but with a different force constant, so the derivation process is the same as the first term. Therefore, we give only the derivation result here.

$$\frac{\partial^2 V_4}{\partial X_i \partial Y_j} = -\frac{240\varepsilon}{r_{0,ij}^2}\left(X_j - X_i\right)\left(Y_j - Y_i\right)/r_{ij}^2 \tag{5.37}$$

After combining the Hessian matrices from all four terms, we can calculate the pseudo inverse of the final Hessian matrix $\boldsymbol{H}$. The mean square displacement $\langle \Delta \mathbf{r}_i^2 \rangle$ and inter residue correlation $\langle \Delta \boldsymbol{r}_i \cdot \Delta \boldsymbol{r}_j \rangle$ can be calculated by summing the elements over the X, Y and Z directions.

$$\langle \Delta \boldsymbol{r}_i^2 \rangle = \frac{k_B T}{\gamma} \left(\boldsymbol{H}^+_{3i-2,3i-2} + \boldsymbol{H}^+_{3i-1,3i-1} + \boldsymbol{H}^+_{3i,3i} \right) \tag{5.38}$$

$$\langle \Delta \boldsymbol{r}_i \cdot \Delta \boldsymbol{r}_j \rangle = \frac{k_B T}{\gamma} \left(\boldsymbol{H}^+_{3i-2,3j-2} + \boldsymbol{H}^+_{3i-1,3j-1} + \boldsymbol{H}^+_{3i,3j} \right) \tag{5.39}$$

5.4.3 Extending STeM to an All-Atom Model

STeM [28] was originally based on the Gō potential [29, 30] and was applied to coarse-grained systems. Here we show how STeM can be extended to all-atom models. Consequently, the force field parameters used in STeM for the interactions among the atoms are adopted from an all-atom force field, for example, the Charmm22 force field.

All-atom STeM is different from NMA. Though all-atom STeM share some similarity with NMA, such as both are all-atom models and can be applied to an equilibrated structure to compute normal modes, STeM is different from NMA in the sense that it is fully spring-based models and does not consider the effect of inter-atomic forces. Indeed, as in Gō model, STeM assumes the input structure is at equilibrium, and in addition, the inter-atomic forces are all zero. NMA, however, does not make the second assumption. NMA has been often applied to locally energetically-minimized structures, where the systems are at equilibrium, but the inter-atomic forces are clearly not zero. Thus, the difference between NMA and STeM mostly represents the effect of inter-atomic forces on a system.

All-atom STeM is also different from all-atom ANM (anisotropic network model [4]). In ANM, atoms interact through two-body Hookean springs only. In STeM, atoms interact via generalized spring tensors (thus the name STeM – spring tensor model) and include three-body and four-body interactions. STeM and ANM do share some similarity. Both models are purely spring-based models and do not take into account the effect of inter-atomic forces when studying protein fluctuations and conformation changes. STeM is especially similar to a particular variant of ANM, the ANMr2, or ANM using $\frac{1}{r^2}$ as spring constants, as was thoroughly investigated in [10]. This is because, the effect of non-bonded terms in STeM, especially the van der Waal interactions, is similar to $\frac{1}{r^2}$ springs [28].

In the following, we will show how STeM is a close approximation of NMA, and how ANM is a further approximation of STeM.

5.4.3.1 The Close Relationship Between NMA to All-Atom STeM

The close relationship between STeM and NMA is illuminated in the following derivation of STeM Hessian matrix.

First, let us consider the three-body interactions, specifically the bond angle interactions. Let $\theta = \angle ijk$ be the instantaneous angle formed by three sequential atoms i, j, and k. The bond angle potential of atoms i, j, and k is defined as $V_\theta = \frac{1}{2}k_\theta(\theta - \theta_0)^2$, where k_θ is the bond angle spring constant, θ_0 is the equilibrium angle. The block Hessian matrix $\boldsymbol{H}_\theta$ for the angle interaction is a 9×9 second derivative matrix of V_θ with respect to x, y and z coordinates of atoms i, j, and k. Write out one component $\frac{\partial V_\theta}{\partial X_i \partial Y_k}$ of $\boldsymbol{H}_\theta$ as follows:

$$\begin{aligned}\frac{\partial^2 V_\theta}{\partial X_i \partial Y_k} &= \frac{\partial}{\partial X_i}\left(\frac{\partial V_\theta}{\partial \theta}\frac{\partial \theta}{\partial Y_k}\right) \\ &= \frac{\partial^2 V_\theta}{\partial \theta^2}\frac{\partial \theta}{\partial X_i}\frac{\partial \theta}{\partial Y_k} + \frac{\partial V_\theta}{\partial \theta}\frac{\partial^2 \theta}{\partial X_i \partial Y_k} \\ &= k_\theta \cdot \frac{\partial \theta}{\partial X_i}\frac{\partial \theta}{\partial Y_k} + f_\theta \cdot \frac{\partial^2 \theta}{\partial X_i \partial Y_k}\end{aligned} \tag{5.40}$$

where $f_\theta = \frac{\partial V_\theta}{\partial \theta}$ is the bending force (which actually is a torque). Notice that Eq. (5.40) is a combination of the physical terms (k_θ and f_θ) and geometric terms (the rest of the terms), which represent the projection of physical interactions into a particular coordinate system. In a similar fashion, the rest of the elements of the block hessian matrix $\boldsymbol{H}_\theta$ can be written out using k_θ and f_θ. Finally, the block Hessian matrix H_θ can be rewritten as a summation of two terms:

$$\boldsymbol{H}_\theta = \boldsymbol{H}_\theta^{\text{NMA}} = k_\theta \cdot \boldsymbol{H}_{\theta|K_\theta} + f_\theta \cdot \boldsymbol{H}_{\theta|f_\theta} \tag{5.41}$$

where $\boldsymbol{H}_{\theta|k_\theta}$ and $\boldsymbol{H}_{\theta|f_\theta}$ are 9×9 matrice that are fully determined by protein geometry and atom coordinates, where k_θ is a force field parameter and $f_\theta = k_\theta(\theta - \theta_0)$ is the torque acting on the bond angle. In STeM, the bending torque f_θ is assumed to be 0, i.e., $f_\theta \to 0$. This simplifies the $\boldsymbol{H}_\theta$ in Eq. (5.41) and it becomes:

$$\boldsymbol{H}_\theta^{\text{STeM}} = k_\theta \cdot \boldsymbol{H}_{\theta|k_\theta}. \tag{5.42}$$

Now for the four-body interactions. Let $\boldsymbol{H}_\phi$ be the 12×12 block Hessian matrix for the dihedral interaction among four atoms i, j, k, and l. Let $k_\phi = \frac{\partial^2 V}{\partial \phi^2}$ and $f_\phi = \frac{\partial V}{\partial \phi}$ be the dihedral spring constant and bending force (torque), respectively. Similar to Eq. (5.41), the Hessian matrix $\boldsymbol{H}_\phi$ can be written as a function of k_ϕ and f_ϕ:

$$\boldsymbol{H}_\phi = \boldsymbol{H}_\phi^{\text{NMA}} = k_\phi \cdot \boldsymbol{H}_{\phi|K_\phi} + f_\phi \cdot \boldsymbol{H}_{\phi|f_\phi}. \tag{5.43}$$

Since $V(\phi) = K_\phi(1 - \cos(n(\phi - \phi_0)))$ in most force fields, where K_ϕ and ϕ_0 are force field parameters and n is the multiplicity, $k_\phi = \frac{\partial^2 V}{\partial \phi^2} = n^2 K_\phi \cos(n(\phi - \phi_0))$. In STeM, the torque f_ϕ is assumed to be zero. In addition, STeM assumes that the input structure has the equilibrium values for all the dihedral angles, i.e., $\phi = \phi_0$. Therefore,

$$\boldsymbol{H}_{\phi}^{\text{STeM}} = k_\phi \cdot \boldsymbol{H}_{\phi\big|K_\phi} = n^2 K_\phi \cdot \boldsymbol{H}_{\phi\big|K_\phi}. \tag{5.44}$$

Improper is a special type of dihedral interactions. Improper potential takes the form of $V(\psi) = K_\psi(\psi - \psi_0)^2$, where K_ψ and ψ_0 are force field parameters. To simplify notations for improper interaction, we define $\boldsymbol{H}_{\psi\big|H_\psi}$ in the same way as $\boldsymbol{H}_{\phi\big|H_\phi}$, and its spring constant $k_\psi = \frac{\partial^2 V(\psi)}{\partial \psi^2} = 2K_\psi$. Therefore,

$$\boldsymbol{H}_{\psi}^{\text{STeM}} = k_\psi \cdot \boldsymbol{H}_{\psi\big|K_\psi} = 2K_\psi \cdot \boldsymbol{H}_{\psi\big|K_\psi}. \tag{5.45}$$

Likewise, the Hessian matrix $\boldsymbol{H}_l$ for two-body interaction between a pair of atoms i and j can be determined: $\boldsymbol{H}_l = k_l \cdot \boldsymbol{H}_{l\big|k_l} + f_l \cdot \boldsymbol{H}_{l\big|f_l}$. In STeM, the force term is again assumed to be zero. As for the first term, there are three types of two-body interactions in an all-atom potential, i.e., bond stretching, van der Waals interactions, and electrostatic interactions, and thus different k_l. For the bond stretching potential, or V_{bond}, which is usually expressed as $V_{\text{bond}} = K_{\text{bond}}(r - r_0)^2$, we have

$$k_{\text{bond}} = \frac{\partial^2 V_{\text{bond}}}{\partial r^2} = 2K_{\text{bond}}. \tag{5.46}$$

For van der Waal term, whose potential is $V_{\text{vdW}} = \epsilon\left(\left(\frac{r_0}{r}\right)^{12} - 2\left(\frac{r_0}{r}\right)^6\right)$, where ϵ and r_0 are force field constants. We have

$$k_{\text{vdW}} = \frac{\partial^2 V_{\text{vdW}}}{\partial r^2} = \frac{12\epsilon}{r^2}\left(13\left(\frac{r_0}{r}\right)^{12} - 7\left(\frac{r_0}{r}\right)^6\right). \tag{5.47}$$

Lastly, for the electrostatic term, since $V_{\text{elec}} = \frac{332 q_i \cdot q_j}{rD}$, where q_i is the partial charge of atom i, and D is the dielectric constant and is set to be 80, k_l is thus:

$$k_{\text{elec}} = \frac{\partial^2 V_{\text{elec}}}{\partial r^2} = \frac{2 \cdot 332 q_i \cdot q_j}{80 r^3} = \frac{8.3 q_i \cdot q_j}{r^3}. \tag{5.48}$$

Finally, the spring constant k_l for two-body interaction is

$$k_l = k_{\text{bond}} + k_{\text{vdW}} + k_{\text{elec}}. \tag{5.49}$$

This spring constant may become negative. In that case, we set k_l to be zero to avoid producing negative eigenvalues from the STeM Hessian matrix.

Finally, let N be the number of atoms, the $3N \times 3N$ full Hessian matrix $\boldsymbol{H}^{\text{NMA}}$ for the whole system can be written as a summation of a spring constant related term $\boldsymbol{H}_{spr}^{\text{NMA}}$ and a force/torque related term $\boldsymbol{H}_{\text{frc}}^{\text{NMA}}$:

$$\boldsymbol{H}^{\text{NMA}} = \boldsymbol{H}^{\text{NMA}}_{\text{spr}} + \boldsymbol{H}^{\text{NMA}}_{\text{frc}}, \tag{5.50}$$

where $\boldsymbol{H}^{\text{NMA}}_{\text{spr}}$ and $\boldsymbol{H}^{\text{NMA}}_{\text{frc}}$ are

$$\boldsymbol{H}^{\text{NMA}}_{\text{spr}} = \sum_{\theta\in\Theta} k_\theta \boldsymbol{H}_{\theta|k_\theta} + \sum_{\phi\in\Phi} k_\phi \boldsymbol{H}_{\phi|k_\phi} + \sum_{\psi\in\Psi} k_\psi \boldsymbol{H}_{\psi|k_\psi} + \sum_{l\in L} k_l \boldsymbol{H}_{l|k_l}, \tag{5.51}$$

$$\boldsymbol{H}^{\text{NMA}}_{\text{frc}} = \sum_{\theta\in\Theta} f_\theta \boldsymbol{H}_{\theta|k_\theta} + \sum_{\phi\in\Phi} f_\phi \boldsymbol{H}_{\phi|k_\phi} + \sum_{\psi\in\Psi} f_\psi \boldsymbol{H}_{\psi|k_\psi} + \sum_{l\in L} f_l \boldsymbol{H}_{l|k_l}, \tag{5.52}$$

where Θ, Φ, Ψ, and L are the sets of angular, dihedral, improper and pairwise interactions.

STeM assumes that all forces and torques are zero. Therefore,

$$\boldsymbol{H}^{\text{STeM}} \approx \boldsymbol{H}^{\text{NMA}}_{\text{spr}}. \tag{5.53}$$

It is approximately equal since STeM additionally assumes that the input structure has the equilibrium values for the dihedral potentials, while NMA does not. Specifically, $\boldsymbol{H}^{\text{STeM}}$ is,

$$\boldsymbol{H}^{\text{STeM}} = \sum_{\theta\in\Theta} k_\theta \boldsymbol{H}_{\theta|K_\theta} + \sum_{\phi\in\Phi} k_\phi \boldsymbol{H}_{\phi|K_\phi} + \sum_{\psi\in\Psi} k_\psi \boldsymbol{H}_{\psi|K_\psi} + \sum_{l\in L} k_l \boldsymbol{H}_{l|K_l} \tag{5.54}$$

Our original work on STeM [28] details how $\boldsymbol{H}^{\text{STeM}}$ can be computed. To compute $\boldsymbol{H}^{\text{NMA}}$, one may use software packages such as gromacs [31] or tinker [27].

5.4.3.2 The Relationship Between STeM and ANM, the Role of Multi-body Interactions

ANM [4] is a widely-used coarse-grained model for proteins. A particular variant of ANM, ANMr2, which uses $\frac{1}{r^2}$ as spring constants, was thoroughly investigated in [10] and was shown to have better performance than the regular ANM.

ANM, particularly ANMr2, is closely related to STeM in that the former is a simplification of the latter [28]. STeM is a close approximation of the NMA, and ANM/ANMr2 is a further approximation of STeM. STeM ignores the contributions of inter-atomic forces that are considered in NMA (Eq. (5.50)), while ANMr2/ANM takes into account only the two-body interactions and ignores the contributions of multi-body interactions (bond angle and torsional angle interactions) that are considered in STeM.

5.4.4 The Protein Sets Studied

To evaluate the STeM model, we apply it to compute B-factors and to study protein conformation changes and compare the results with those computed from ANM and GNM. For B-factors computations, the protein dataset is from [32] and contains 111 proteins. Two proteins, 1CYO and 5PTP, are removed from the dataset because they no longer exist in the current Protein Data Bank [33]. The proteins in the first dataset all have a resolution that is better than or equal to 2.0 Å. For conformation change studies, the dataset is from [20], which contains 20 pairs of protein structures. Each pair of protein structures has significantly large structure difference from each other.

5.4.5 Evaluation Techniques

We used the same evaluation techniques as have been applied before [20, 32]. Specifically, the following three numerical measures are used.

5.4.6 The Correlation Between the Experimental and Calculated B-Factors

The linear correlation coefficient between the experimental and calculated B-factors is calculated using the following formula.

$$\rho = \frac{\sum_i^N (x_i - \overline{x})(y_i - \overline{y})}{\left[\sum_i^N (x_i - \overline{x})^2 \sum_i^N (y_i - \overline{y})^2\right]^{1/2}} \tag{5.55}$$

where x_i and y_i are respectively the experimental and calculated B-factors of the C_α atom of residue i and $\overline{x}$ and $\overline{y}$ are the mean values. N is the number of residues.

5.4.7 The Overlap Between the Experimental Observed Conformation Changes and the Calculated Modes

The overlap measures the directional similarity between a conformation change and a calculated mode. The formula for calculating the overlap is

$$I = \frac{\left|\sum_i^{3N} e_i r_i\right|}{\left[\sum_i^{3N} e_i^2 \sum_i^{3N} r_i^2\right]^{1/2}} \tag{5.56}$$

where e_i is the relative displacement of residue i in a selected mode e and r_i is the conformation displacement of residue i.

5.4.8 The Correlation Between the Experimental Observed Conformation Changes and the Calculated Modes

The correlation measures the magnitude similarity between a conformation change and a calculated mode. The formula used for calculating the correlation is the same as Eq. (5.55), with different meaning for x_i and y_i.

$$\rho = \frac{\sum_i^N (x_i - \overline{x})(y_i - \overline{y})}{\left[\sum_i^N (x_i - \overline{x})^2 \sum_i^N (y_i - \overline{y})^2\right]^{1/2}} \tag{5.57}$$

where x_i is the magnitude of the displacement of residue i in the conformation change and y_i is the magnitude of the displacement of residue i in the selected mode. $\overline{x}$ and $\overline{y}$ are the corresponding mean values.

Acknowledgments Funding from National Science Foundation (CAREER award, CCF-0953517) is gratefully acknowledged.

Competing Interests The authors declare that they have no competing interests.

References

1. Voth GA (2009) Coarse-graining of condensed phase and biomolecular systems. CRC Press, Boca Raton. xviii, 455 p., 16 p. of plates
2. Ma J (2005) Usefulness and limitations of normal mode analysis in modeling dynamics of biomolecular complexes. Structure 13(3):373–380
3. Bahar I, Atilgan AR, Erman B (1997) Direct evaluation of thermal fluctuations in proteins using a single-parameter harmonic potential. Fold Des 2(3):173–181
4. Atilgan AR et al (2001) Anisotropy of fluctuation dynamics of proteins with an elastic network model. Biophys J 80(1):505–515
5. Bahar I, Rader AJ (2005) Coarse-grained normal mode analysis in structural biology. Curr Opin Struct Biol 15(5):586–592

6. Ming D, Bruschweiler R (2006) Reorientational contact-weighted elastic network model for the prediction of protein dynamics: comparison with NMR relaxation. Biophys J 90(10): 3382–3388
7. Song G, Jernigan RL (2006) An enhanced elastic network model to represent the motions of domain-swapped proteins. Proteins 63(1):197–209
8. Song G, Jernigan RL (2007) VGNM: a better model for understanding the dynamics of proteins in crystals. J Mol Biol 369(3):880–893
9. Lu M, Poon B, Ma J (2006) A new method for coarse-grained elastic normal-mode analysis. J Chem Theory Comput 2(3):464–471
10. Yang L, Song G, Jernigan RL (2009) Protein elastic network models and the ranges of cooperativity. Proc Natl Acad Sci USA 106(30):12347–12352
11. Zheng W (2008) A unification of the elastic network model and the Gaussian network model for optimal description of protein conformational motions and fluctuations. Biophys J 94(10):3853–3857
12. Stember JN, Wriggers W (2009) Bend-twist-stretch model for coarse elastic network simulation of biomolecular motion. J Chem Phys 131(7):074112
13. Kundu S, Sorensen DC, Phillips GN Jr (2004) Automatic domain decomposition of proteins by a Gaussian Network Model. Proteins 57(4):725–733
14. Bahar I, Chennubhotla C, Tobi D (2007) Intrinsic dynamics of enzymes in the unbound state and relation to allosteric regulation. Curr Opin Struct Biol 17(6):633–640
15. Zheng WJ, Brooks B (2005) Identification of dynamical correlations within the myosin motor domain by the normal mode analysis of an elastic network model. J Mol Biol 346(3):745–759
16. Yang Z, Majek P, Bahar I (2009) Allosteric transitions of supramolecular systems explored by network models: application to chaperonin GroEL. PLoS Comput Biol 5(4):e1000360
17. Lin TL, Song G (2010) Generalized spring tensor models for protein fluctuation dynamics and conformation changes. BMC Struct Biol 10(Suppl 1):S3
18. Tirion MM (1996) Large amplitude elastic motions in proteins from a single-parameter, atomic analysis. Phys Rev Lett 77(9):1905–1908
19. Hinsen K (1998) Analysis of domain motions by approximate normal mode calculations. Proteins 33(3):417–429
20. Tama F, Sanejouand YH (2001) Conformational change of proteins arising from normal mode calculations. Protein Eng 14(1):1–6
21. Thorpe MF (2007) Comment on elastic network models and proteins. Phys Biol 4(1):60–63; discussion 64–55
22. Koga N, Takada S (2006) Folding-based molecular simulations reveal mechanisms of the rotary motor F1-ATPase. Proc Natl Acad Sci USA 103(14):5367–5372
23. Clementi C, Nymeyer H, Onuchic JN (2000) Topological and energetic factors: what determines the structural details of the transition state ensemble and "en-route" intermediates for protein folding? An investigation for small globular proteins. J Mol Biol 298(5):937–953
24. Koga N, Takada S (2001) Roles of native topology and chain-length scaling in protein folding: a simulation study with a Go-like model. J Mol Biol 313(1):171–180
25. Riccardi D, Cui Q, Phillips GN Jr (2009) Application of elastic network models to proteins in the crystalline state. Biophys J 96(2):464–475
26. Hinsen K et al (2000) Harmonicity in slow protein dynamics. Chem Phys 261(1–2):25–37
27. Ponder JW, Richards FM (1987) An efficient Newton-like method for molecular mechanics energy minimization of large molecules. J Comput Chem 8(7):1016–1024
28. Lin TL, Song G (2010) Generalized spring tensor models for protein fluctuation dynamics and conformation changes. BMC Struct Biol 10(Suppl 1):S3
29. Go N (1983) Protein folding as a stochastic-process. J Stat Phys 30(2):413–423
30. Taketomi H, Ueda Y, Go N (1975) Studies on protein folding, unfolding and fluctuations by computer simulation. I. The effect of specific amino acid sequence represented by specific inter-unit interactions. Int J Pept Protein Res 7(6):445–459

31. Pronk S, Páll S et al (2013) GROMACS 4.5: a high-throughput and highly parallel open source molecular simulation toolkit. Bioinformatics 29(7):845–854
32. Kundu S et al (2002) Dynamics of proteins in crystals: comparison of experiment with simple models. Biophys J 83(2):723–732
33. Berman HM et al (2000) The protein data bank. Nucleic Acids Res 28(1):235–242

Chapter 6
The Joys and Perils of Flexible Fitting

Niels Volkmann

Abstract While performing their functions, biological macromolecules often form large, dynamically changing macromolecular assemblies. Only a relatively small number of such assemblies have been accessible to the atomic-resolution techniques X-ray crystallography and NMR. Electron microscopy in conjunction with image reconstruction has become the preferred alternative for revealing the structures of such macromolecular complexes. However, for most assemblies the achievable resolution is too low to allow accurate atomic modeling directly from the data. Yet, useful models often can be obtained by fitting atomic models of individual components into a low-resolution reconstruction of the entire assembly. Several algorithms for achieving optimal fits in this context were developed recently, many allowing considerable degrees of flexibility to account for binding-induced conformational changes of the assembly components. This chapter describes the advantages and potential pitfalls of these methods and puts them into perspective with alternative approaches such as iterative modular fitting of rigid-body domains.

Keywords Electron microscopy • Fitting • Validation • Statistical methods • Modeling

6.1 Introduction

Cooperative interaction among molecules in large assemblies are fundamental for dynamic processes in living cells. A detailed understanding of how these assemblies work requires structural information at the atomic level. Nuclear magnetic resonance (NMR) spectroscopy and X-ray crystallography are well-established

N. Volkmann (✉)
Bioinformatics and Systems Biology Program, Sanford-Burnham Medical Research Institute, 10901 N Torrey Pines Rd, La Jolla, CA 92037, USA
e-mail: niels@sanfordburnham.org

K. Han et al. (eds.), *Protein Conformational Dynamics*, Advances in Experimental Medicine and Biology 805, DOI 10.1007/978-3-319-02970-2_6,

approaches for obtaining atomic structures of individual molecules and domains. However, atomic structures of large macromolecular assemblies remain more difficult to obtain with these methods. These assemblies can be too large for NMR and often exhibit a large degree of dynamic behavior that hampers crystallization.

Electron microscopy has been a powerful tool for the investigation of biological structures for several decades now. However, only recently steps towards achieving its full potential have begun to come to fruition. Technical advances in electron microscopy equipment, in computational image reconstruction methods, and in methods of specimen preparation have all been essential components in enabling the extraordinary progress in the last few years. Resolutions of 0.5 nm or better can now be achieved not only from two-dimensional crystals [23] or helically symmetrical objects [86], but also from icosahedral virus particles [28] and even from smaller, less symmetric particles [15, 42]. The recent introduction of direct electron detection devices promises to further improve the overall resolution and quality attainable by electron microscopy [5, 9, 27, 44]. Electron tomography, the most widely applicable method for obtaining three-dimensional information by electron microscopy, can now be combined efficiently with localization and dynamics data from light microscopy [29, 52] and potentially allows investigation of entire mammalian cells at molecular resolution [48], paving the way for structure-based systems biology [78].

Since the potential of electron microscopy has been realized, more and more methods are emerging that target incorporation of atomic-level information into reconstructions derived by electron microscopy. Many of these approaches focus on finding new and improved ways to 'mold' the atomic structures into the shape of the electron microscopy reconstructions. The rationale is that binding can lead to significant conformational changes and flexibility has to be introduced into the fitting process to allow for these changes. While this development has lead to several interesting and promising approaches, there are no methods available yet that allow evaluation of the model quality and accuracy. In the remainder of the chapter, we will introduce the general principles behind the various fitting paradigms that are available and compare their relative merits and pitfalls.

6.2 Methods for Fitting High-Resolution Models into Electron Microscopy Densities

Sub-nanometer resolution reconstructions have become more and more common recently, especially for assemblies that are symmetric. However, the most generally achievable resolution for large dynamic assemblies is in the 1–3 nm resolution range. In this range it is possible to directly detect and map individual subunits to understand the general architecture of the assemblies [75]. In addition, this intermediate resolution already gives a solid basis for fitting high-resolution structures of smaller entities into the densities. Generally, if the resolution is higher, the accuracy of the fitting results can be expected to improve. However, there is also a dependency on the shape of the structural element to be fitted. For example, a barrel-shaped structure will exhibit a high degree of ambiguity in the fit at relatively high resolution while a distinct, asymmetrically shaped structure can be fit without ambiguity at rather low resolution.

The models resulting from fitting of high-resolution components into densities of larger, lower-resolution assemblies are often referred to as 'pseudo-atomic' to emphasize the fact that the accuracy of the model is of limited resolution. However, these models are built out of actual atoms and the term 'pseudo atom' is often used to denote atom-like representations of entire residues or other groups of atoms in nuclear magnetic resonance calculations [83], direct phasing approaches [19, 43], and coarse-grained molecular modeling [84]. To avoid confusion, we will use the term 'near-atomic resolution model' instead.

6.2.1 Fitting Paradigms

Fitting methods for combining the information from atomic models with reconstructions from electron microscopy were used as soon as intermediate-resolution electron microscopy densities became available. Since then the field has evolved significantly and a fair number of different methods for fitting are now available. Because of the increased availability in recent years of sub-nanometer resolution reconstructions where secondary structural elements are often visible as rods (α-helices) and sheets (β-sheets), most efforts in the field have been directed towards the development of flexible fitting methods that allow flexibility of some sort in the high-resolution structures in order to accommodate conformational changes that occur upon the interaction of assembly components.

6.2.1.1 Manual Fitting

The first combinations of high-resolution structures with electron microscopy reconstructions were achieved by 'manual fitting'. In this method, the fit of the model is judged by eye and corrected manually using a molecular viewer program such as, initially, O [37] or, at a later stage, Chimera [49] until the fit 'looks good'. Sometimes, this subjective fit is refined locally using, for example, the real-time fitting routines implemented in Chimera. The approach initially gained popularity in the early 1990s and was the prime choice for combining such data for that decade until more sophisticated and more global algorithms were introduced [76, 82]. If the components of the assembly under study are large molecules with distinctive shapes at the resolution of the reconstruction, manual fitting can often be performed with relatively little ambiguity (see for example [51, 64]). However, divergent models of the same complex fitted manually by different investigator teams have also been reported [35, 39].

Surprisingly, manual fitting is still a mainstay in the electron microscopy field despite its obvious shortcomings including user bias and potential for inaccuracies. The argument often invoked for using manual fitting instead of more objective computational procedures involves the perceived complexity of running automated fitting programs [4]. While some of the computational approaches certainly require a rather involved work flow, others are much simpler to use. For example the fitting program CoAn [77] can be run by simply telling the program which density map and which coordinate file to use and – depending on the size of the density map – often provides the final results in shorter time than required to load and orient the model in a molecular viewer.

6.2.1.2 Rigid-Body Fitting

At a later stage, various flavors of automated, global rigid-body searches using density-correlation measures as fitting criteria were developed [11, 54, 55, 76]. These flavors vary in the exact mathematical form of the density correlation, the use of various preprocessing steps including masking and filtering, and in implementation details. Masking operations using the calculated envelope [54] or using the atomic model directly [55] both enhance high-resolution features and suppress low-resolution information, making it somewhat equivalent to high-pass filtering in Fourier space. The amplification of background noise is a common side effect of high-pass filtering [60]. Thus, the success of this type of masking will depend strongly on the noise level in the reconstruction. Convolution with a Laplacian operator [11] is also known to be very sensitive to, and tends to amplify noise [60]. While these filters have the potential to boost the signal in the absence of noise they should be used with care if a significant amount of noise is present in the reconstruction.

6.2.1.3 Flexible Fitting

The interactions between components or interactions with other co-factors often result in dramatic conformational changes within the assembly. The need to accommodate such changes in combination with the increased availability of subnanometer resolution reconstructions where secondary structural elements are often visible as rods (α-helices) and sheets (β-sheets), have led to significant efforts directed towards developing flexible fitting methods that allow the high-resolution structures to be distorted in some way, subject to different types of constraints, in order to improve the fit with the density [17, 24, 33, 36, 38, 58, 59, 68–72, 74, 87, 88].

These methods are designed to refine a predefined starting model and sacrifice the global character of the rigid-body searches. In essence, all available flexible fitting methods try to mold a starting model into the density by balancing force fields that optimize the density fit with force fields that ensure proper stereochemistry in one way or another. Examples include the use of normal modes [33, 69], full fledged molecular dynamics [72], and the use of elastic networks [59]. While these types of methods are capable of significant improvements in the fit, in most cases the number of parameters to be refined (i.e., the number of coordinates) is much larger than the number of experimental observables making these methods potentially susceptible to overfitting and misinterpretation of noisy density features.

6.2.1.4 Iterative Modular Fitting

The first attempt to address conformational changes in the fitted assemblies was to break up the components into smaller domains or 'modules' and to fit these independently as rigid bodies into the density [76, 80]. This approach has later been refined to incorporate iterative refinement of domain boundaries and orientation parameters [79]. As a first step of the method, the target reconstruction is dissected into density modules using density segmentation approaches such as the watershed transform [75].

Next, the structures to be fitted need to be divided into domain modules. If high resolution structures for more than one conformation are available, independently moving domains and hinges can be defined by comparing the two alternative conformations [32]. If only one single conformation is available, normal modes analysis can be used to make this division [34] or the watershed transform can be applied to a low-resolution density calculated from the high-resolution structure to be fitted in order to define the module boundaries. One important step is to identify the correct correspondence between the target density modules and the domain modules. In practice, the correspondence between volume and radius of gyration in combination with connectivity considerations is usually sufficient. Each domain is then fitted into the corresponding density segment using a global rigid-body fitting protocol.

After this initial round, an iterative refinement procedure is applied where for each domain in turn a discrepancy map [80] is generated by removing the contribution of all other fitted domains from the unsegmented target density. Then, the orientation and position of the remaining domain is refined using this discrepancy map. Once done for all domains, the discrepancy-mapping-refinement cycle is repeated until no further changes in orientation and position occurred. The purpose of this refinement step is two-fold. First, it ensures the removal of bias from sharp edges and inaccuracies introduced by the initial watershed segmentation. Second, it removes bias that might be introduced by erroneous modularization of the high-resolution structure.

A natural extension of the concept is the use of real-space refinement techniques [13, 14, 25], where the structure is broken up into ever smaller rigid segments that are then jointly refined into the target density subject to connectivity and energetic restraints [12]. This methodology can accommodate more subtle conformational changes than the domain module approach but needs to be applied with caution for the same reasons as for the flexible fitting approaches.

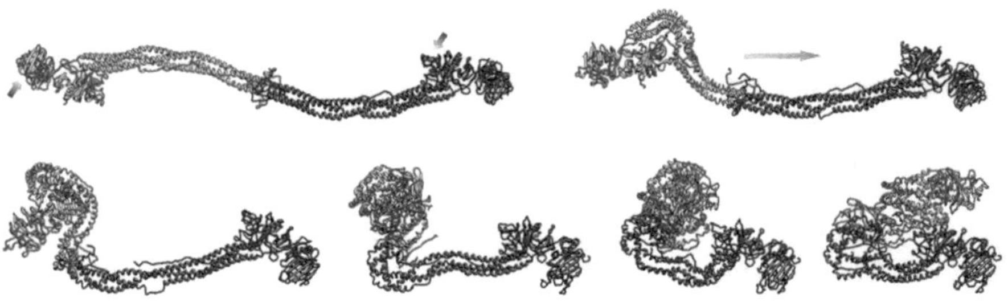

Fig. 6.1 Flexible fitting with sparse constraints. The *upper left* shows the crystal structure of fibrinogen [7]. The constraint used for modeling was that the two residues indicated by *arrows* in the crystal structure are within 0.5 nm under certain conditions. The *arrow* in the *upper right panel* indicates the general direction of the structure deformation. A number of snapshots along the trajectory are shown with the end result at the *lower right corner*

6.2.2 Potential Pitfalls

There are several potential systematic errors originating in the electron microscopy data-collection and reconstruction procedures that can affect the performance of fitting procedures. These include misestimations of the magnification, incomplete corrections of the microscope's contrast transfer function, uneven distribution of projection images in Euler space, and misestimation of the resolution. Other factors that can bias fitting results in significant ways include incompleteness of the structure to be fitted in relation to the target reconstruction and the possible presence of conformational mixtures during the reconstruction process. Generally, rigid-body techniques will be less susceptible to generating artifacts in the presence of such errors simply because the underlying structure is kept intact and only six parameters per module, three for the center-of-mass position and three for the orientation, need to be determined. Flexible fitting methods will tend to allow distortions that accommodate those errors without any straightforward way to detect what is happening.

Unfortunately, the quality of the model in terms of stereochemistry and other physical parameters is not a good indicator for its correctness. To exemplify this issue we run a simulation using the crystal structure of fibrinogen [7] and one single distance constraint (Fig. 6.1). To arrive at a configuration that satisfies this constraint, the structure was systematically and iteratively deformed to minimize the distance between the two residues using the first 150 normal modes. After each iteration a geometric regularization was performed to improve convergence of the procedure. A number of snapshots along the trajectory are shown in Fig. 6.1 with the end result at the lower right corner. The final model does not only fulfill the residue-distance constraint perfectly, it also has perfectly reasonable geometry and contacts, which is also true for all intermediate states along the trajectory. However, the final structure is completely artifactual and is a product of overfitting the single distance constraint by allowing an excessive amount of degrees of freedom to contribute to the deformation. In reality, the constraints are satisfied by interactions between different fibrinogen molecules stacking end-to-end, forming long fibers [7]. Clearly, the fact that a fitted structure makes physical sense is insufficient as a criterion for its correctness.

6.3 Validation

At resolution lower than 0.3 nm the amount of structural information that can be obtained is limited. Care must be taken that the number of the degrees of freedom used during the fitting or modeling procedure does not exceed the number of independent observations for the density maps. Otherwise, overfitting will inevitably ensue. Even fitting of a single rigid body using only the six rotational and translational degrees of freedom can lead to ambiguities in the resulting models at intermediate resolution [77]. Recently developed methods for incorporating data from other data sources such as proteomics [3], Förster resonance energy transfer [85], or sparse distance restraints [10] into the fitting process are helping to resolve some of these ambiguities. There have been several attempts to develop validation methods for this type of situation.

6.3.1 Crossvalidation

While most fitting approaches generally perform quite well with test data where the correct answer is known, there is currently no method available that allows judging fitting quality with experimental data when the answer is unknown. This is especially critical in regards to overfitting because, if too much flexibility is introduced into the fitting process, eventually noise will be fitted. In X-ray crystallography, this problem is solved by using a cross-validated measure of fit in Fourier space, the 'free R-factor' [8]. The free R-factor relies on the fact that Fourier terms are independent in crystallography. In electron microscopy the Fourier terms are strongly correlated so the free R-factor is not applicable in a direct analogy to crystallography. However, it may be possible to remove some of the Fourier-term correlation in an analogous way to treating non-crystallographic symmetry in crystallography [20] and to derive a modified crossvalidation measure. One modification that has been proposed is the use of resolution shells instead of random sets of Fourier terms as is usually done in X-crystallography [21, 62].

A promising idea in the context of crossvalidation is the use of independent reconstructions [18, 79]. The general idea is that a set of particle images is split into two independent sets and reconstructions are built from each set. One of these reconstructions is used for fitting, model building and refinement, and the other is used for crossvalidation. A recent study showed that this type of methodology can be used to select a sensible weighting term between the density and all-atom energy contributions using the program Rosetta [17], provided significant data at resolution higher than 1.2 nm is available [18]. Another recent trend is the use of consensus between multiple flexible fitting approaches [1, 2]. Here, the idea is that models derived with different flexible fitting methods for the same reconstruction that are similar are more likely to be correct than models that are "method dependent". Despite the intuitive appeal of this idea, no formal proof exists that this is indeed the case, especially in the presence of systematic errors where it is likely that all methods that allow flexibility are affected in similar ways.

6.3.2 Statistical Methods

A radically different approach towards the development of validation tools is the use of statistical methods to obtain confidence intervals for the orientation parameters in modular fitting of rigid body domains [77, 79]. In this approach, a global search is followed by a global statistical analysis of the distribution of the fitting criterion. The analysis results in the definition of confidence intervals that lead to the definition solution sets. These sets contain all fits that satisfy the data within the error margin defined by the errors in the data and the chosen confidence level. Structural parameters of interest can then be evaluated as properties of these sets. For example, the uncertainty of each atom position of the fitted structure can be approximated by calculating the root-mean-square deviation for each atom using all members of the solution set. Ambiguities in the fitting are clearly reflected in the shape of the solution set [77]. The size of the solution set can serve as a normalized goodness-of-fit criterion. The smaller the set, the better the data determines the position of the fitted atomic structure. An example for the utility of the method is shown in Fig. 6.2 and Table 6.1.

The statistical nature of the approach allows the use of standard statistical tests, such as Student's t-test, to evaluate the significance of differences between models in different functional states and to help model the corresponding conformational changes in a robust and reliable way. An example is shown in Fig. 6.3. The methodology also allows estimating the probability that a certain residue is involved in the interaction between two components (see for example [30, 80]). This probabilistic ranking of residues in terms of their involvement in binding gives a better starting point for the design of mutagenesis experiments.

Unfortunately, statistical methods are sometimes applied in a very casual manner to density fitting procedures so that the conclusions deduced are not always reliable. A recent example is the comparison of different scoring functions for the quality of fit [73]. The authors calculate and compare confidence intervals by assuming that all scoring functions follow Gaussian, normal distributions. However, it is well known that, for example, the correlation coefficient, one of the scoring functions analyzed, does not follow a Gaussian distribution at all and needs to be subjected to a variance-stabilizing variable transformation [22] before reliable confidence intervals can be calculated. Since no normality tests were performed for the other scoring functions

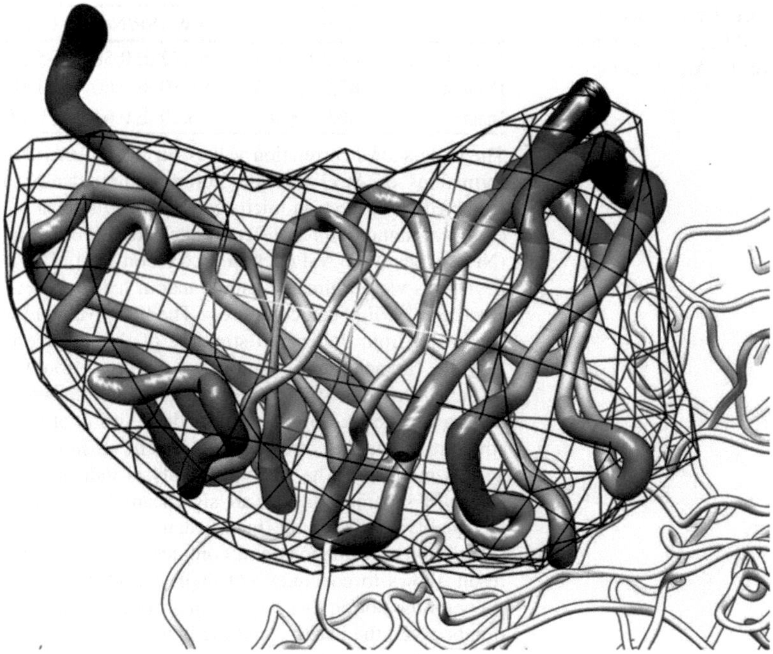

Fig. 6.2 Ensemble of statistically equivalent solutions to a fitting problem. A 2.8-nm resolution map of human rhinovirus complexed with Fab fragments [65] was used. The density corresponding to the Fab fragment (wire frame in the Figure) was derived by difference mapping with the structure of the uncomplexed virion [56]. The structure of the unbound Fab fragment [41] was fitted to the isolated Fab-fragment density using a global rigid-body docking protocol [76] followed by statistical analysis of the results [79]. All fits that satisfy the data at a confidence level of 0.995 are represented by a worm model where the thickness and darkness are proportional to the variability of the positioning of the corresponding structural elements, showing distinct local variations in precision. The structure of the virion is depicted as a *white tube*. The crystal structure of exactly the same construct as investigated by electron microscopy [65] was solved to atomic resolution a few years later by X-ray crystallography [63] giving a one-to-one correspondence between the low-resolution density and the corresponding atomic structure. There are several binding-induced conformational changes in the Fab-fragment, but the uncertainty estimates extracted from the confidence interval capture the correct structure quite well

and a visual inspection of the distributions does not convey a particularly Gaussian shape for any of those, the confidence intervals calculated under the normality assumption can not be considered to be supported by the data in that study.

6.4 Comparison of Flexible and Modular Fitting

Most observed protein conformational changes involve movements of rigid domains that have their internal structure preserved [26, 31, 40]. Iterative modular fitting of rigid-body domains should be adequate to accurately model those types of changes.

Table 6.1 Conformations of Arp3 in different functional states of the Arp2/3 complex

	Scar	NWASP/Nck	Cortactin
Intermediate	88.29 ± 0.49	88.72 ± 0.58	85.55 ± 0.69
Branch	87.27 ± 0.53	88.10 ± 0.60	84.18 ± 0.63
Inactive	86.93 ± 0.58	86.20 ± 0.67	78.78 ± 0.56

The table shows correlation values (in %) for alternative conformations of Arp3 with the corresponding density module extracted from electron microscopy reconstructions of Arp2/3 complex bound to three different nucleation promoting factors (NPFs), Scar, NWASP/Nck and cortactin [85]. The standard deviations were estimated using random half data sets for the respective NPF-bound complexes. The 'inactive' conformation was taken from the crystal structure of the inactive complex [53], the 'branch' conformation was extracted from the model of the entire branch junction [57]. The 'intermediate' conformation was derived by iterative modular fitting of the inactive structure into to the corresponding densities (using subdomains 1–4 as modules). The statistical analysis indicates that – for each NPF separately – there are significant differences between the quality of fit for the intermediate conformation and the other two conformations at a confidence level of 0.995. This result allows to conclude with high confidence that the data supports the notion that Arp3 is in an intermediate conformation between the inactive and the branch conformations when NPFs are bound to the complex

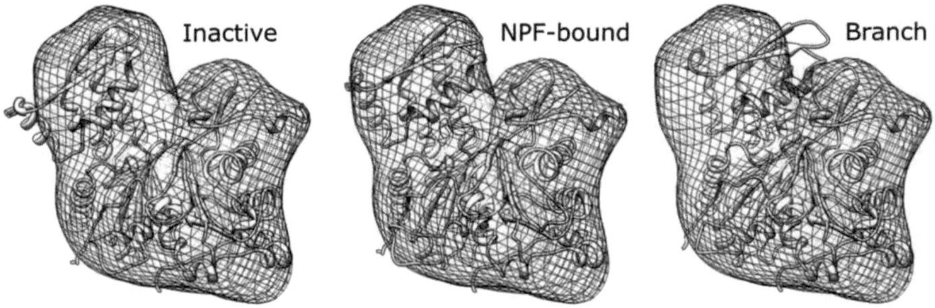

Fig. 6.3 Fit of alternative conformations of Arp3 to the corresponding density module extracted from an electron microscopy reconstruction of Arp2/3 complex with bound nucleation promoting factors (NPFs) [85]. The fitting results and the subsequent statistical analysis (see Table 6.1) clearly indicate that the data support the notion that Arp3 is in a conformation intermediate between that in the inactive complex [53] and that in the branch junction [57]

We compared the performance of the iterative modular fitting approach with the performance of four published flexible fitting methods (Table 6.2). It should be noted that these structures were previously selected as adequate test cases for flexible fitting [36, 71, 72, 81].

Table 6.2 Comparative performance test of flexible and modular fitting strategies

Target	Search	Resolution (nm)	Modules	Residues	Identity (%)	$RMSD_{lsq}$	$RMSD_{mod}$	$RMSD_{flex}$
1oaoC	1oaoD	1.5	3	729	100	0.092	0.111	0.201
1lfh	1lfg	1.5	3	691	100	0.094	0.098	0.189
1hwz	1hrd	1.5	2	491	28	0.258	0.298	0.490
1blb	1a45	1.0	2	172	35	0.157	0.203	1.150

The first two columns are the Protein Data Bank identifiers of the target and search structures. The following columns are the resolution of the target map, the number of domains used as modules, the number of residues, the sequence identity between target and search structures, the root-mean-square deviation between α-carbons of the search and target structures after least-squares fitting of the modules ($RMSD_{lsq}$ in nm), after iterative modular rigid-body docking of the same modules ($RMSD_{mod}$ in nm), and after using flexible fitting protocols ($RMSD_{flex}$ in nm). The values for the last column were taken from [36, 71, 72] (2x)

6.4.1 Alpha Subunit of Acetyl-Coenzyme A Synthase/Carbon Monoxide Dehydrogenase

The crystal structure of the acetyl-coenzyme A synthase/carbon monoxide dehydrogenase assembly [16] revealed two significantly different conformations of the alpha subunit (PDB identifier 1oao, chains C and D). A comparison of the two conformations indicates that this change can be approximated by hinged movements of three rigid domains. Iterative modular fitting was performed using a 1.5-nm resolution calculated density map from chain C and the atomic structure of chain D, broken up into these three domains, as modules to be fitted. The fit resulting from the modular fitting for this test is with root-mean-square deviation (RMSD) of 0.111 nm very close to the 0.092 nm achievable by least-squares fitting of the domains α-carbon atoms to those of the target conformation.

The same fitting problem was tackled as a test case for molecular-dynamics based flexible fitting [72]. The resulting RMSD at 1.5 nm resolution using this approach is with 0.201 nm significantly worse than that from modular fitting and, with 0.125 nm RMSD, is still slightly worse if a target map at 1.0 nm resolution is used. Only if data at 0.5 nm is available, the flexible fitting approach surpasses iterative modular rigid-body fitting with an RMSD of 0.075 nm. This also improves upon the least-squares RMSD, indicating that, at this resolution, non-rigid conformational changes can be picked up correctly by this flexible fitting approach. It is worth noting that, in three dimensions, the amount of information increases by a factor of 3.375 if going from 1.5 to 1.0 nm resolution and by a factor of 27 if going from 1.5 to 0.5 nm. In addition, it should be kept in mind that the tests were performed in the absence of systematic and random errors, which would likely degrade the accuracy of the flexible fitting more than that of the more robust modular fitting approach.

6.4.2 Lactoferrin

The iron-binding protein lactoferrin goes through a large conformational change when iron binds [47]. Comparison of the conformations suggests three hinged rigid-body domain movements to explain the change. We Modular fitting at 1.5 nm resolution using apolactoferrin (PDB identifier 1lfh) as a target and iron-bound lactoferrin (PDB identifier 1lfg), broken up into three domains, as modules. The RMSD after iterative modular fitting was with 0.098 nm almost indistinguishable from the 0.094 nm RMSD achievable from least-squares fitting of the α-carbon atoms.

The same fitting problem was addressed by two flexible docking approaches. One was based on vector quantization and molecular mechanics force fields [81]. That study was also performed at 1.5 nm resolution and the best RMSD achieved with this approach was 0.272 nm, exhibiting local deviations of up to 0.9 nm [77]. The second approach was based on constraint geometric simulations [36]. That study evaluated the RMSD at various resolutions, the lowest of which was 14 Å. At this resolution the best RMSD was 0.189 nm. The best overall RMSD of 0.127 nm was achieved at 0.33 nm target map resolution. Even at near-atomic resolution, this flexible fitting approach does not provide any advantage over modular rigid body fitting at 1.5 nm resolution for this test case.

6.4.3 Glutamate Dehydrogenase

This test involved fitting the structure of the Pyrococcus furiosus glutamate dehydrogenase [6], split into two domains as indicated by the comparison of the conformations, into a 1.5-nm resolution target map calculated from the bovine homologue [66]. The sequence identity between the two is only 28 % and there are several inserts present in the bovine form (PDB identifier 1hwz) that are not present in the Pyrococcus furiosus form (PDB identifier 1hrd). This differences include an extended, finger-like helix-turn-helix motif of 46 residues. This region was easily identified by watershed segmentation and was deleted from the target map after completing the initial step of the modular fitting procedure but before invoking the iterative refinement. Removal of extra density during refinement is not strictly necessary but does tend to increase the accuracy of the fitted structure. In this case, a 0.009 nm improvement in RMSD can be achieved by deleting the density of the helix-turn-helix motif prior to the iterative refinement. The RMSD of the final structure is with 0.298 nm again very close to the least-squares based RMSD of 0.258 nm.

The same fitting task was addressed using a hierarchical flexible fitting procedure involving Monte-Carlo based refinement of successively smaller structure fragments [71]. That study was performed with target maps calculated at 1.0 nm resolution. Despite the significant increase in information corresponding to the use of higher resolution data, the best RMSD achieved with this method was with 0.49 nm significantly higher than the RMSD achieved by iterative modular fitting.

6.4.4 *Eye Lens Crystallin*

This test case also involves fitting of homologous structures, in this case the structure of β-crystallin [45] was the target and γ-crystallin [46], divided into two domains, provided the modules for the fitting. The sequence identity between the two is 35 %. This case was difficult in several ways. (i) The structure is fairly small, the single domain modules are only ~80 residues. This fact implies that less information than in the other test cases is available for fitting. (ii) Both modules are highly similar in structure and shape, consisting of relatively symmetrical β-barrels. As a consequence, the assignment of each domain to its corresponding segment is not trivial and, after fitting, the barrel alignment might be out of register. (iii) The conformational change is large. The centers of masses of the barrels in the extended γ-crystallin (PDB identifier 1blb) are 4.3 nm apart whereas this distance is only 2.4 nm in the more compact β-crystallin structure (PDB identifier 1a45). This change is primarily achieved by stretching the linker between the two barrels.

The iterative modular fitting approach with a target map at 1.5 nm resolution, yielded an alignment of one of the barrels aligned within 0.25 nm RMSD whereas the second barrel was out of register by one β-strand. It appears that, at this resolution, this configuration is the true correlation maximum because, even using only local refinement and the correct least-squares based alignment as a starting point, the structures would align out of register. This result has far-reaching consequences for fitting strategies. The hierarchical multi-resolution strategy often employed in registration of volumes derived by MRI or other clinical imaging techniques [67], needs to be applied with caution in the case of fitting atomic structures into low-resolution density maps. This test case shows that there is a real danger of getting stuck in the wrong local maximum of the score function. It appears that the safer option is to perform global searches at the highest available resolution.

In order to obtain the correct registration of the β-strand, data up to a minimum of 1.0 nm needs to be included for the iterative modular fitting. The RMSD of the resulting model with the target structure is 0.203 nm and compares quite favorably with the least-squares based RMSD of 0.157 nm. The crystallin fitting was also chosen as a test case for the hierarchical approach mentioned in the last paragraph [71]. However, this method fails to align the β-barrels correctly even at 1.0 nm resolution.

6.4.5 *Summary*

In all four test cases, the iterative modular fitting approach yields RMSDs within 0.05 nm of what is achievable by least squares fitting of the α-carbon coordinates. While quite remarkable, this is not necessarily a surprising result. For each domain or module, only six parameters (three translational and three rotational) need to be determined. This problem is highly over-determined for most practical cases. This also makes the method very resilient against random and systematic errors. The question is not if there are enough bits of independent information in the data to support the degrees of freedom used for fitting (as one would need to ask for flexible fitting approaches), the question is whether the level of detail in the density is fine enough to nail down the six parameters for each module accurately and uniquely. With the availability of confidence intervals, these questions become testable hypotheses.

In all tests presented here, flexible fitting protocols appear to significantly deteriorate the RMSD (Table 6.2), most likely due to overfitting and inadequate distortions of the fitted structures. Only at resolution better than about 0.5 nm, a 27-fold increase in information content over 1.5 nm resolution, one of the flexible fitting methods [72] appears to pick up conformational changes that can not be adequately modeled as movements of rigid domains and improves upon the results obtained by modular rigid-body fitting done at 1.5 nm resolution.

6.5 Future Directions

It is clear that rigorous and objective evaluation criteria are still needed to corroborate conclusions drawn from models derived from fitting of high-resolution structures into lower-resolution densities from electron microscopy or alternative imaging approaches such as Small Angle X-ray Scattering. Especially in the case of flexible fitting methods, it is essential to find ways to validate results in order to avoid over fitting and to increase robustness in the presence of systematic and random errors. Crossvalidation procedure analogous to those used in X-ray crystallography and NMR are difficult to implement owing to the strong correlation of Fourier terms in electron microscopy reconstruction but are actively pursued by several research groups. The use of statistical methods to evaluate the quality of rigid body fits either with complete assembly models or using iterative modularization are well under way. Extension of the concept to provide confidence intervals of fits derived by flexible fitting methods would be very valuable in the context of validation as well.

Another, recently proposed, promising direction is the uncoupling of the modeling step from the fitting. Instead of steering the modeling procedure using low-resolution density constraints, simulations or modeling attempts are performed independently, without introducing knowledge about the density at the modeling stage. This ensures that the resulting models and/or trajectories are completely free of bias from the low-resolution density. A-posteriori comparison with the low-resolution densities will then give an unbiased idea how well the models [61] or trajectories [50] match the data, allowing robust statistical evaluation of the results.

6.6 Conclusions

Flexible fitting methods that allow distortions of the initial component structures have become a popular approach for providing near-atomic resolution models of dynamic assemblies. However, there has been little effort devoted to the development of validation methods that allow to test whether these models are free of overfitting artifacts. This problem tends to be particularly severe if systematic errors are present in the model or in the density map. Thus, open issues in this area include estimation of fitting quality, validation of results, estimation of fitting errors, and detection of ambiguities. As a consequence, extreme care needs to be exercised in the interpretation of models from flexible fitting methods.

For resolutions lower than 0.5 nm, the alternative method of iterative modular fitting using rigid-body modules currently appears to be the better choice. Modular fitting is not only less sensitive to systematic errors, it also is amenable to statistical analysis, which provides objective criteria for the quality of prospective fits and allows the derivation of significance levels for differences in conformations as well as the detection of ambiguities in the fit.

Acknowledgements This work was supported by National Institutes of Health grant P01 GM066311.

References

1. Ahmed A, Tama F (2013) Consensus among multiple approaches as a reliability measure for flexible fitting into cryo-EM data. J Struct Biol 182:67–77
2. Ahmed A, Whitford PC, Sanbonmatsu KY, Tama F (2012) Consensus among flexible fitting approaches improves the interpretation of cryo-EM data. J Struct Biol 177:561–570
3. Alber F, Dokudovskaya S, Veenhoff LM, Zhang W, Kipper J, Devos D, Suprapto A, Karni-Schmidt O, Williams R et al (2007) Determining the architectures of macromolecular assemblies. Nature 450:683–694
4. Allen GS, Stokes DL (2013) Modeling, docking, and fitting of atomic structures to 3D maps from cryo-electron microscopy. Methods Mol Biol 955:229–241
5. Bammes BE, Rochat RH, Jakana J, Chen DH, Chiu W (2012) Direct electron detection yields cryo-EM reconstructions at resolutions beyond 3/4 Nyquist frequency. J Struct Biol 177:589–601
6. Britton KL, Baker PJ, Rice DW, Stillman TJ (1992) Structural relationship between the hexameric and tetrameric family of glutamate dehydrogenases. Eur J Biochem 209:851–859
7. Brown JH, Volkmann N, Jun G, Henschen-Edman AH, Cohen C (2000) The crystal structure of modified bovine fibrinogen. Proc Natl Acad Sci USA 97:85–90
8. Brünger AT (1992) Free R-value – a novel statistical quantity for assessing the accuracy of crystal-structures. Nature 355:472–475
9. Campbell MG, Cheng A, Brilot AF, Moeller A, Lyumkis D, Veesler D, Pan J, Harrison SC, Potter CS et al (2012) Movies of ice-embedded particles enhance resolution in electron cryo-microscopy. Structure 20:1823–1828
10. Campos M, Francetic O, Nilges M (2011) Modeling pilus structures from sparse data. J Struct Biol 173:436–444

11. Chacon P, Wriggers W (2002) Multi-resolution contour-based fitting of macromolecular structures. J Mol Biol 317:375–384
12. Chapman MS (1995) Restrained real-space macromolecular atomic refinement using a new resolution-dependent electron density function. Acta Crystallogr A51:69–80
13. Chapman MS, Trzynka A, Chapman BK (2013) Atomic modeling of cryo-electron microscopy reconstructions – joint refinement of model and imaging parameters. J Struct Biol 182(1):10–21
14. Chen JZ, Furst J, Chapman MS, Grigorieff N (2003) Low-resolution structure refinement in electron microscopy. J Struct Biol 144:144–151
15. Cong Y, Baker ML, Jakana J, Woolford D, Miller EJ, Reissmann S, Kumar RN, Redding-Johanson AM, Batth TS et al (2010) 4.0-Å resolution cryo-EM structure of the mammalian chaperonin TRiC/CCT reveals its unique subunit arrangement. Proc Natl Acad Sci USA 107:4967–4972
16. Darnault C, Volbeda A, Kim EJ, Legrand P, Vernede X, Lindahl PA, Fontecilla-Camps JC (2003) Ni-Zn-[Fe4-S4] and Ni-Ni-[Fe4-S4] clusters in closed and open subunits of acetyl-CoA synthase/carbon monoxide dehydrogenase. Nat Struct Biol 10:271–279
17. DiMaio F, Tyka MD, Baker ML, Chiu W, Baker D (2009) Refinement of protein structures into low-resolution density maps using Rosetta. J Mol Biol 392:181–190
18. DiMaio F, Zhang J, Chiu W, Baker D (2013) Cryo-EM model validation using independent map reconstructions. Protein Sci 22:865–868
19. Dorset DL (1997) Direct phase determination in protein electron crystallography: the pseudo-atom approximation. Proc Natl Acad Sci USA 94:1791–1794
20. Fabiola F, Korostelev A, Chapman MS (2006) Bias in cross-validated free R factors: mitigation of the effects of non-crystallographic symmetry. Acta Crystallogr D62:227–238
21. Falkner B, Schröder GF (2013) Cross-validation in cryo-EM-based structural modeling. Proc Natl Acad Sci USA 110:8930–8935
22. Fisher RA (1921) On the 'probable error' of a coefficient of correlation deduced from a small sample. Metron 1:1–32
23. Fujiyoshi Y, Unwin N (2008) Electron crystallography of proteins in membranes. Curr Opin Struct Biol 18:587–592
24. Gao H, Frank J (2005) Molding atomic structures into intermediate-resolution cryo-EM density maps of ribosomal complexes using real-space refinement. Structure 13:401–406
25. Gao H, Sengupta J, Valle M, Korostelev A, Eswar N, Stagg SM, Van Roey P, Agrawal RK, Harvey SC et al (2003) Study of the structural dynamics of the E coli 70S ribosome using real-space refinement. Cell 113:789–801
26. Gerstein M, Krebs W (1998) A database of macromolecular motions. Nucleic Acids Res 26:4280–4290
27. Grigorieff N (2013) Direct detection pays off for electron cryo-microscopy. Elife 2:e00573
28. Grigorieff N, Harrison SC (2011) Near-atomic resolution reconstructions of icosahedral viruses from electron cryo-microscopy. Curr Opin Struct Biol 21:265–273
29. Hanein D, Volkmann N (2011) Correlative light-electron microscopy. Adv Protein Chem Struct Biol 82:91–99
30. Hanein D, Volkmann N, Goldsmith S, Michon AM, Lehman W, Craig R, DeRosier D, Almo S, Matsudaira P (1998) An atomic model of fimbrin binding to F-actin and its implications for filament crosslinking and regulation. Nat Struct Biol 5:787–792
31. Hayward S (1999) Structural principles governing domain motions in proteins. Proteins 36:425–435
32. Hayward S, Berendsen HJ (1998) Systematic analysis of domain motions in proteins from conformational change: new results on citrate synthase and T4 lysozyme. Proteins 30:144–154
33. Hinsen K, Reuter N, Navaza J, Stokes DL, Lacapère JJ (2005) Normal mode-based fitting of atomic structure into electron density maps: application to sarcoplasmic reticulum Ca-ATPase. Biophys J 88:818–827
34. Hinsen K, Thomas A, Field MJ (1999) Analysis of domain motions in large proteins. Proteins 34:369–382

35. Hoenger A, Sack S, Thormahlen M, Marx A, Muller J, Gross H, Mandelkow E (1998) Image reconstructions of microtubules decorated with monomeric and dimeric kinesins: comparison with x-ray structure and implications for motility. J Cell Biol 141:419–430
36. Jolley CC, Wells SA, Fromme P, Thorpe MF (2008) Fitting low-resolution cryo-EM maps of proteins using constrained geometric simulations. Biophys J 94:1613–1621
37. Jones TA, Zou J-Y, Cowan SW, Kjeldgaard M (1991) Improved methods for the building of protein models in electron density maps and the location of errors in these models. Acta Crystallogr A 47:110–119
38. Kovacs JA, Yeager M, Abagyan R (2008) Damped-dynamics flexible fitting. Biophys J 95:3192–3207
39. Kozielski F, Arnal I, Wade R (1998) A model of the microtuble-kinesin complex based on electron cryomicroscopy and X-ray crystallography. Curr Biol 8:191–198
40. Krebs WG, Gerstein M (2000) SURVEY AND SUMMARY: the morph server: a standardized system for analyzing and visualizing macromolecular motions in a database framework. Nucleic Acids Res 28:1665–1675
41. Liu H, Smith TJ, Lee WM, Mosser AG, Rueckert RR, Olson NH, Cheng RH, Baker TS (1994) Structure determination of an Fab fragment that neutralizes human rhinovirus 14 and analysis of the Fab-virus complex. J Mol Biol 240:127–137
42. Ludtke SJ, Baker ML, Chen DH, Song JL, Chuang DT, Chiu W (2008) De novo backbone trace of GroEL from single particle electron cryomicroscopy. Structure 16:441–448
43. Lunin VY, Lunina NL, Petrova TE, Vernoslova EA, Urzhumtsev AG, Podjarny AD (1995) On the ab initio solution of the phase problem for macromolecules at very low resolution: the few atoms model method. Acta Crystallogr D Biol Crystallogr 51:896–903
44. Milazzo AC, Leblanc P, Duttweiler F, Jin L, Bouwer JC, Peltier S, Ellisman M, Bieser F, Matis HS et al (2005) Active pixel sensor array as a detector for electron microscopy. Ultramicroscopy 104:152–159
45. Nalini V, Bax B, Driessen H, Moss DS, Lindley PF, Slingsby C (1994) Close packing of an oligomeric eye lens beta-crystallin induces loss of symmetry and ordering of sequence extensions. J Mol Biol 236:1250–1258
46. Norledge BV, Hay RE, Bateman OA, Slingsby C, Driessen HP (1997) Towards a molecular understanding of phase separation in the lens: a comparison of the X-ray structures of two high Tc gamma-crystallins, gammaE and gammaF, with two low Tc gamma-crystallins, gammaB and gammaD. Exp Eye Res 65:609–630
47. Norris GE, Anderson BF, Baker EN (1991) Molecular replacement solution of the structure of apolactoferrin, a protein displaying large-scale conformational change. Acta Crystallogr B47:998–1004
48. Noske AB, Costin AJ, Morgan GP, Marsh BJ (2008) Expedited approaches to whole cell electron tomography and organelle mark-up in situ in high-pressure frozen pancreatic islets. J Struct Biol 161:298–313
49. Pettersen EF, Goddard TD, Huang CC, Couch GS, Greenblatt DM, Meng EC, Ferrin TE (2004) UCSF Chimera–a visualization system for exploratory research and analysis. J Comput Chem 25:1605–1612
50. Pfaendtner J, Volkmann N, Hanein D, Dalhaimer P, Pollard TD, Voth GA (2012) Key structural features of the actin filament Arp2/3 complex branch junction revealed by molecular simulation. J Mol Biol 416:148–161
51. Rayment I, Holden HM, Whittaker M, Yohn CB, Lorenz M, Holmes KC, Milligan RA (1993) Structure of the actin-myosin complex and its implications for muscle contraction. Science 261:58–65
52. Rigort A, Bäuerlein FJ, Leis A, Gruska M, Hoffmann C, Laugks T, Böhm U, Eibauer M, Gnaegi H et al (2010) Micromachining tools and correlative approaches for cellular cryo-electron tomography. J Struct Biol 172:169–179
53. Robinson RC, Turbedsky K, Kaiser DA, Marchand JB, Higgs HN, Choe S, Pollard TD (2001) Crystal structure of Arp2/3 complex. Science 294:1679–1684

54. Roseman AM (2000) Docking structures of domains into maps from cryo-electron microscopy using local correlation. Acta Crystallogr D Biol Crystallogr 56:1332–1340
55. Rossmann MG (2000) Fitting atomic models into electron-microscopy maps. Acta Crystallogr D Biol Crystallogr 56:1341–1349
56. Rossmann MG, Arnold E, Erickson JW, Frankenberger EA, Griffith JP, Hecht HJ, Johnson JE, Kamer G, Luo M et al (1985) Structure of a human common cold virus and functional relationship to other picornaviruses. Nature 317:145–153
57. Rouiller I, Xu XP, Amann KJ, Egile C, Nickell S, Nicastro D, Li R, Pollard TD, Volkmann N, Hanein D (2008) The structural basis of actin filament branching by Arp2/3 complex. J Cell Biol 180:887–895
58. Rusu M, Birmanns S, Wriggers W (2008) Biomolecular pleiomorphism probed by spatial interpolation of coarse models. Bioinformatics 24:2460–2466
59. Schröder GF, Brünger AT, Levitt M (2007) Combining efficient conformational sampling with a deformable elastic network model facilitates structure refinement at low resolution. Structure 15:1630–1641
60. Seul M, O'Gorman L, Sammon MJ (2000) Practical algorithms for image analysis: descriptions, examples, and code. Cambridge University Press, Cambridge
61. Shacham E, Sheehan B, Volkmann N (2007) Density-based score for selecting near-native atomic models of unknown structures. J Struct Biol 158:188–195
62. Shaikh TR, Hegerl R, Frank J (2003) An approach to examining model dependence in EM reconstructions using cross-validation. J Struct Biol 142:301–310
63. Smith CA, Rayment I (1996) X-ray structure of the magnesium(II).ADP.vanadate complex of the *Dictyostelium discoideum* myosin motor domain to 1.9 Å resolution. Biochemistry 35:5404–5417
64. Smith TJ, Olson NH, Cheng RH, Chase ES, Baker TS (1993) Structure of a human rhinovirus-bivalently bound antibody complex: implications for viral neutralization and antibody flexibility. Proc Natl Acad Sci USA 90:7015–7018
65. Smith TJ, Olson NH, Cheng RH, Liu H, Chase ES, Lee WM, Leippe DM, Mosser AG, Rueckert RR, Baker TS (1993) Structure of human rhinovirus complexed with Fab fragments from a neutralizing antibody. J Virol 67:1148–1158
66. Smith TJ, Peterson PE, Schmidt T, Fang J, Stanley CA (2001) Structures of bovine glutamate dehydrogenase complexes elucidate the mechanism of purine regulation. J Mol Biol 307:707–720
67. Studholme C, Hill DL, Hawkes DJ (1996) Automated 3-D registration of MR and CT images of the head. Med Image Anal 1:163–175
68. Suhre K, Navaza J, Sanejouand YH (2006) NORMA: a tool for flexible fitting of high-resolution protein structures into low-resolution electron-microscopy-derived density maps. Acta Crystallogr D Biol Crystallogr 62:1098–1100
69. Tama F, Miyashita O, Brooks CL (2004) Flexible multi-scale fitting of atomic structures into low-resolution electron density maps with elastic network normal mode analysis. J Mol Biol 337:985–999
70. Tan RKZ, Devkota B, Harvey SC (2008) YUP.SCX: coaxing atomic models into medium resolution electron density maps. J Struct Biol 163:163–174
71. Topf M, Lasker K, Webb B, Wolfson H, Chiu W, Sali A (2008) Protein structure fitting and refinement guided by cryo-EM density. Structure 16:295–307
72. Trabuco LG, Villa E, Mitra K, Frank J, Schulten K (2008) Flexible fitting of atomic structures into electron microscopy maps using molecular dynamics. Structure 16:673–683
73. Vasishtan D, Topf M (2011) Scoring functions for cryoEM density fitting. J Struct Biol 174:333–343
74. Velazquez-Muriel JA, Valle M, Santamaria-Pang A, Kakadiaris IA, Carazo JM (2006) Flexible fitting in 3D-EM guided by the structural variability of protein superfamilies. Structure 14:1115–1126
75. Volkmann N (2002) A novel three-dimensional variant of the watershed transform for segmentation of electron density maps. J Struct Biol 138:123–129

76. Volkmann N, Hanein D (1999) Quantitative fitting of atomic models into observed densities derived by electron microscopy. J Struct Biol 125:176–184
77. Volkmann N, Hanein D (2003) Docking of atomic models into reconstructions from electron microscopy. Methods Enzymol 374:204–225
78. Volkmann N, Hanein D (2009) Electron microscopy in the context of systems biology. In: Gu J, Bourne PE (eds) Structural bioinformatics. Wiley-Blackwell, New York, pp 143–170
79. Volkmann N (2009) Confidence intervals for fitting of atomic models into low-resolution densities. Acta Crystallogr D Biol Crystallogr 65:679–689
80. Volkmann N, Hanein D, Ouyang G, Trybus KM, DeRosier DJ, Lowey S (2000) Evidence for cleft closure in actomyosin upon ADP release. Nat Struct Biol 7:1147–1155
81. Wriggers W, Birmanns S (2001) Using situs for flexible and rigid-body fitting of multiresolution single-molecule data. J Struct Biol 133:193–202
82. Wriggers W, Milligan RA, McCammon JA (1999) Situs: a package for docking crystal structures into low-resolution maps from electron microscopy. J Struct Biol 125:185–195
83. Wüthrich K, Billeter M, Braun W (1983) Pseudo-structures for the 20 common amino acids for use in studies of protein conformations by measurements of intramolecular proton-proton distance constraints with nuclear magnetic resonance. J Mol Biol 169:949–961
84. Xia Z, Gardner DP, Gutell RR, Ren P (2010) Coarse-grained model for simulation of RNA three-dimensional structures. J Phys Chem B 114:13497–13506
85. Xu XP, Rouiller I, Slaughter BD, Egile C, Kim E, Unruh JR, Fan X, Pollard TD, Li R et al (2011) Three-dimensional reconstructions of Arp2/3 complex with bound nucleation promoting factors. EMBO J 31:236–247
86. Yonekura K, Maki-Yonekura S, Namba K (2003) Complete atomic model of the bacterial flagellar filament by electron cryomicroscopy. Nature 424:643–650
87. Zheng W (2011) Accurate flexible fitting of high-resolution protein structures into cryo-electron microscopy maps using coarse-grained pseudo-energy minimization. Biophys J 100:478–488
88. Zhu J, Cheng L, Fang Q, Zhou ZH, Honig B (2010) Building and refining protein models within cryo-electron microscopy density maps based on homology modeling and multiscale structure refinement. J Mol Biol 397:835–851

Chapter 7
Coarse-Grained Models of the Proteins Backbone Conformational Dynamics

Tap Ha-Duong

Abstract Coarse-grained models are more and more frequently used in the studies of the proteins structural and dynamic properties, since the reduced number of degrees of freedom allows to enhance the conformational space exploration. This chapter attemps to provide an overview of the various coarse-grained models that were applied to study the functional conformational changes of the polypeptides main chain around their native state. It will more specifically discuss the methods used to represent the protein backbone flexibility and to account for the physico-chemical interactions that stabilize the secondary structure elements.

Keywords Protein coarse-grained models • Molecular dynamics simulation • Backbone flexibility • Functional conformational changes • Effective physico-chemical potentials

7.1 Introduction

The biological function of proteins is not only related to their tridimensional structure, but also to their conformational dynamics, particularly their backbone motions [26]. Among many examples, the opening and reclosing motion of the two flaps that protect the active site of the HIV-1 protease, is one of the key steps of its enzymatic mechanism [33, 73]. The allosteric effects in proteins which regulate their biological activities, are also well known to involve conformational rearrangements of their backbone [22] and/or alterations of their dynamic properties [53]. Conformational changes of proteins also play an important role in their association and self-assemblies, such as for the peptide Aβ associated to the Alzheimer disease,

T. Ha-Duong (✉)
BIOCIS – UMR CNRS 8076, Faculté de Pharmacie – Université Paris Sud, 5 rue Jean-Baptiste Clément, 92296 Châtenay-Malabry, France
e-mail: tap.ha-duong@u-psud.fr

K. Han et al. (eds.), *Protein Conformational Dynamics*, Advances in Experimental Medicine and Biology 805, DOI 10.1007/978-3-319-02970-2_7,

which undergoes a structural transition in its aggregation pathway from oligomers to fibrils [1].

As for the determination of their tridimensional structure, the proteins dynamics is mostly experimentally probed at the atomic scale by X-ray crystallography and NMR spectroscopy. Information about proteins fluctuations is provided by X-ray diffraction through the Debye-Waller or temperature B factors. These latter are related to the vibrational movements of the proteins atoms around their average position in the crystal. With the help of molecular mechanics calculations, such as the Translation/Libration/Screw model [58] or Normal Modes analysis [36], these experimental data can be interpreted in terms of correlated atomic displacements of the proteins in a crystal environment. In solution state, the fast and slow proteins dynamics can be probed by the acquisition and interpretation of NMR dipolar couplings [68]. Notably, their backbone movements can be characterized by the measurement of the N–H order parameters S^2 which reflect the angular fluctuation of the N–H bonds and thus the flexibility of the polypeptide chains [32].

Using theoretical approaches, the proteins conformational changes are mostly examined at the atomic scale by Normal Modes (NM) calculations and Molecular Dynamics (MD) simulations. NM calculations determine the frequencies and directions of the proteins collective vibrations around a stable structure. This technique can be applied to large biomolecules but because of the quadratic approximation of their energy function, it can hardly study anharmonic conformational transitions [43]. All-atom MD allows to simulate the proteins dynamics in solution by numerically integrating the Newton laws of movement for each particle [35]. Nevertheless, because proteins are generally solvated with a huge number of explicit water molecules, this computational approach generally meets difficulties to study long timescale movements of large proteins, despite the availability of high performance parallel computing platforms. Actually, when using proteins models at the atomic level, both NM and MD approaches struggle to provide relevant information on large biomolecular systems dynamics because of their too large number of degrees of freedom and their very rough potential energy surface. Hence the development of novel methodologies to investigate the functional internal motions of large biomolecular systems is still a very active research field.

In this perspective, Coarse-Grained (CG) models of polymers [50], particularly proteins, are becoming very popular, since the reduction in the number of particles smooths their potential energy surface, enhances the phase space exploration [79], speeds up the computer calculations and allows to gain insight into up to microsecond timescale biological processes [19, 37]. Among them, simplified models at the residue scale, which describe each amino-acid with one or few beads, succeed in combining computational efficiency and realistic description of structural protein details. Thus, since the pioneer work of Levitt in 1976 [42], a large number of CG proteins force fields were developed, mainly in order to tackle the protein folding issue, but also to simulate the conformational dynamics of large proteins [41, 69]. CG proteins models were also applied to the protein-protein recognition problem, since the bead softness can implicitly account for the side-chain local flexibility and improve the predictions of matching interfaces between proteins [8].

The internal movements of CG proteins can be efficiently studied using "structure-based" models which explicitly need the native structure as input. Among them, the "Elastic Network" models replace all the interactions between all pairs of beads that are spatially close with quadratic potentials [67]. In a similar spirit but in order to allow protein unfolding, the Go models distinguish the native local contacts, which are still modelled by harmonic springs, from the non-local contacts, whose energy potential is generally a Lennard-Jones dissociative functions [13,74]. Despite their simplicity, these models can capture the essential features of the functional large deformations of proteins around their native state [3, 29, 31, 46, 48, 65]. An extension of these models even introduces effective potentials with two energy minima, either for local interactions or for the protein global energy, in order to study conformational transitions between two stable states [12,40,51,64]. However, the main drawback of the "structure-based" models is the lack of physico-chemical interactions, such as the hydrophobic ones and the hydrogen bonds, that can be formed or broken during the dynamic course of the proteins. An alternative to these models is the "molecular-mechanics-based" CG models which were initially developed to study the protein folding. These models are now more and more reliable to simulate their conformational dynamics.

Many excellent review papers already report overviews of the progress and applications of the proteins CG models in the structural biology field [11, 27, 41, 62, 69, 70]. In this chapter, we aim at reviewing the "molecular-mechanics-based" CG models of proteins that were applied to study their conformational dynamics. In particular, we will discuss how the models introduce the polypeptide main chain flexibility and how they account for the physico-chemical forces that stabilize the secondary structure elements. Indeed, the reliability of the reduced proteins models depends on the fine balance between the different terms of the force fields. As in classical all-atoms models, "molecular-mechanics-based" CG force fields generally have a nonbonded (or long-range) contribution, which includes van-der-Waals and electrostatic interactions, and a bonded (or short-range) one, that determines the local geometry and flexibility of polypeptide chains [4, 15, 17, 24, 42, 44, 49, 56, 75, 77]. Whereas physical basis can guide the building of nonbonded potentials between coarse grains [5, 6, 20, 34, 54, 55], the empirical parameterization of the bonded terms is not straightforward, since their ability to reproduce proteins secondary structures depends on the details of the nonbonded interactions, particularly the hydrogen bonds.

This discussion will be organised according to the different proteins coarse-grained levels and will be divided into three sections: the first one will be focused on the "high resolution" CG models, the second one on "intermediate resolution" proteins description, and the last one on the so-called "one bead" models.

7.2 High Resolution CG Models

This "backbone centric" category includes the CG models that keep a detailed representation of the polypeptides main chain and a more coarse-grained description of the side chains. The less simplified models of the protein backbone use four united atoms to represent its geometry, one for the nitrogen and its hydrogen, another for the α-carbon and its hydrogen, a third one for the carbonyl carbon and a last one for its oxygen (Fig. 7.1) [15, 18, 25, 30, 60, 61]. Slightly more coarse-grained models group together the carbonyl carbon and its oxygen into one particle, reducing the number of backbone grains to three [63, 75, 76, 79]. In most of these models, the side chains atoms are grouped together into one single bead, except in the models by Hoang et al. and Ding et al. which describe the side chains with one to four united atoms [18, 30].

Using these descriptions, the two backbone torsional degrees of freedoms $\Phi_k = C_{k-1} - N_k - C_k^{\alpha} - C_k$ and $\Psi_k = N_k - C_k^{\alpha} - C_k - N_{k+1}$ are naturally defined as those in all-atom models. The profile of their energy functions can be similar to the atomic force fields one, and their parameters can be extracted from a dataset of known protein structures [61] or from atomic MD simulations [79]. They can also be empirically calibrated in order to reproduce the Ramachandran energy landscapes [25, 63, 75]. The other advantage of these models is that the two backbone united atoms NH and CO allow to naturally introduce the hydrogen bonds that stabilize the protein secondary structures. Despite the electrostatic nature of these interactions, the hydrogen bonds between the backbone grains are generally modelled with effective attractive potentials, such as square-well functions [18, 76] or Lennard-Jones like potentials [25, 63, 75].

Most of these "backbone centric" CG models were developed to simulate the proteins folding process, using Discontinuous MD [18, 76], classical MD [25, 63, 79], or Langevin dynamics [75] methods. In theory, these models can also be used to study the conformational dynamics of proteins around their folded native structure. But curiously, this kind of CG models was seldom used to simulate the

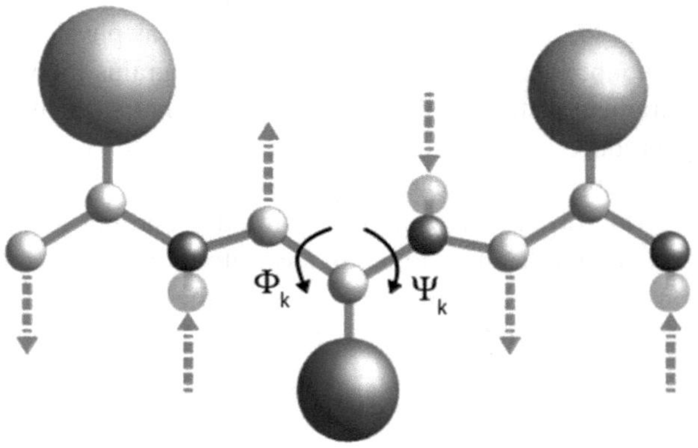

Fig. 7.1 High resolution CG models. The *dashed arrows* indicate the hydrogen bonds that can be formed between the backbone beads

polypeptidic chains dynamics over long trajectories. Among them, the OPEP model developed by Derreumaux et al. was proved to be able to reproduce the structural and thermodynamics properties of several medium-sized proteins with stable MD trajectories over several hundreds nanoseconds [10,16]. OPEP was also successfully used to sample the transient metastable conformations of the peptide Aβ [9]. Recently, PaLaCe, another "high resolution" CG models developed by Pasi et al. was able to generate long stable dynamics trajectories of a large set of 98 proteins very close to their experimental conformations. The dynamics fluctuations observed in these CG simulations also well reproduced the crystallographic B factors profiles and those from atomistic MD simulations in explicit solvent [52].

7.3 Intermediate CG Models

Many proteins CG models attempt to reduce the representation of the backbone atoms of each amino-acid to one grain, generally located at the C$^\alpha$. In these representations, the proteins main chain geometry is principally described by the pseudo-valence $\theta_k = C^\alpha_{k-1} - C^\alpha_k - C^\alpha_{k+1}$ and the pseudo-dihedral $\tau_k = C^\alpha_{k-1} - C^\alpha_k - C^\alpha_{k+1} - C^\alpha_{k+2}$ angles (Fig. 7.2). In most cases, the energy functions for these degrees of freedom are derived from a statistical analysis of known proteins structures, using a Boltzmann inversion procedure (Fig. 7.3) [3, 17, 23, 42, 47]. However these bendings and torsions energy functions alone seem insufficient to stabilize the proteins backbone into their preferential secondary structures, during MD simulations.

In order to account for the hydrogen bonds that stabilize the α-helices and the β-sheets conformations, several of these models introduced one or two additional interacting virtual atoms to the polypeptides backbone representation [2, 23, 42, 44, 47]. These "intermediate CG" models do not artificially restrain the protein backbone in any secondary structure with biasing potentials, but generally model the hydrogen bonding as physical dipole-dipole interactions (Fig. 7.2). In his pioneer work, Levitt introduced two effective atoms, one located in the middle of each C$^\alpha_{k-1}$$-$$C^\alpha_k$ bond (N$'_k$) and the other (O$'_k$) displaced of 1 Å from N$'_k$ perpendicularly to the plane C$^\alpha_{k-1}$$-C^\alpha_k-C^\alpha_{k+1}$. All these virtual atoms were assigned a partial charge,

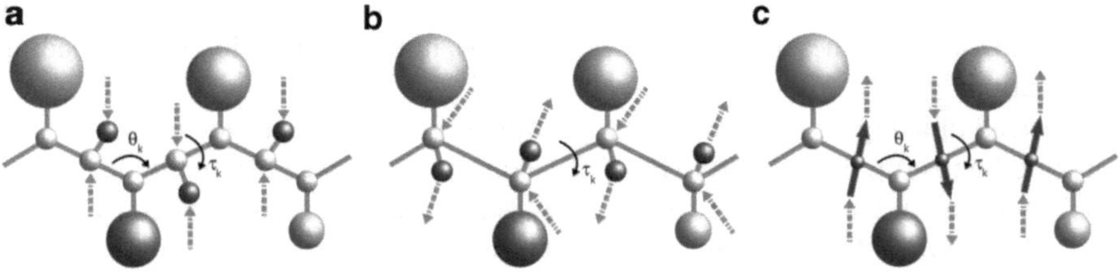

Fig. 7.2 Intermediate CG models. *Left*: Levitt [42]. *Middle*: Ha-Duong [23]. *Right*: Liwo et al. [44], Majek and Elber [47] or Alemani et al. [2]

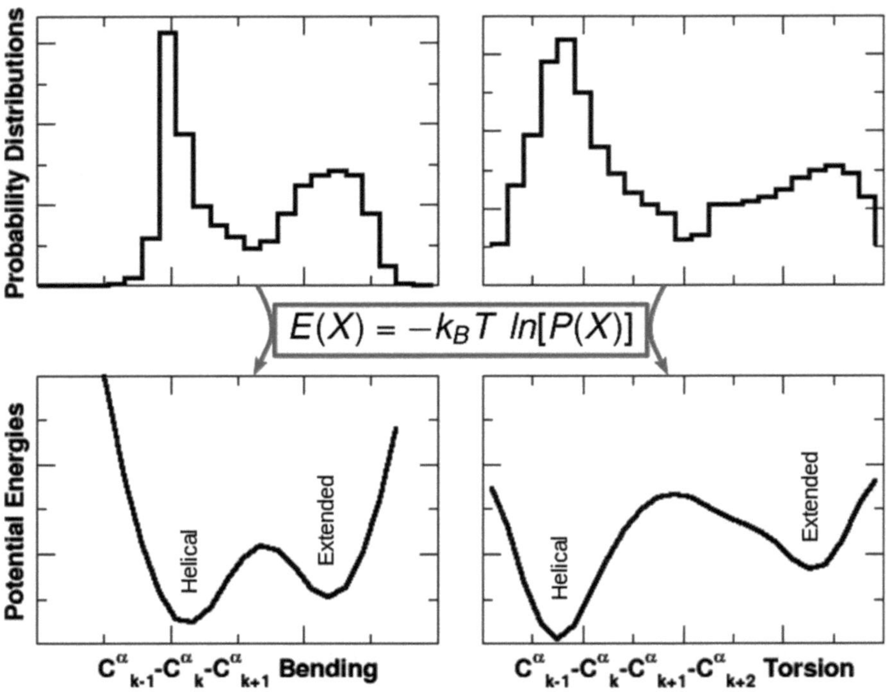

Fig. 7.3 *Left*: Backbone bending angle probability and energy profile. *Right*: Backbone torsion angle probability and energy profile. The "helical" and "extended" legends indicate the two preferential backbone conformations

respectively $q_{N'} = 0.74e$ and $q_{O'} = -0.74e$, and interact with each other through the Coulomb law [42]. A recent variant of the Levitt model, which was developed by Ha-Duong, locates the two backbone pseudo-atoms in a different manner: The first bead B_k groups the four atoms N, C^α, C and O, and is positionned at their geometric center. It carries a negative charge $q = -0.5e$. The second one (H_k) is introduced to account for the dipolar property of the backbone bead: It carries the opposite charge and the B_k–H_k bond length is such as its product by $|q|$ is equal to the average peptide dipole. The average orientation of the B_k–H_k bond relative to the B_{k-1}–B_k–B_{k+1} plane is extracted from a statistical analysis of known structures [23]. Then, like in the Levitt model, all the protein backbone charges interact with each other through the Coulomb law, mimicking the hydrogen bonding interactions.

Instead of using two separated charged atoms, one can use a vector to model the peptide bonds dipolar property (Fig. 7.2). This approach was originally developed by Liwo et al. in their UNRES force field [44] and more recently adopted by Majek and Elber in their FREADY model [47]. These authors introduced a dipolar vector $\mathbf{P}_k$ at the peptide center which is assumed to be in the middle of the $C_k^\alpha - C_{k+1}^\alpha$ bond.

Then the dipolar vectors interact with each other through a mean-field effective potential depending on the relative orientation of the two virtual peptide bonds in interaction. Interestingly, a variant of this vectorial approach was published by Alemani et al. In their model, the orientation of the dipoles $\mathbf{P}_k$ are related to the pseudo-valence angles $C^{\alpha}_{k-1} - C^{\alpha}_{k} - C^{\alpha}_{k+1}$ and interact with each other through the exact dipole-dipole energy function [2].

The presence of a physical and non-biased representation of the hydrogen bonds in these CG models allows to simulate various aspects of the proteins dynamics, such as the folding processes, transitions between different conformations, and the large amplitude conformational fluctuations around stable structures. So, the model by Alemani et al. is able to generate not only stable MD trajectories of α-helices and β-sheets, but also transitions from helices to helix-coil-helix or β-hairpin motifs [2]. The CG force field by Ha-Duong can also generate stable trajectories for various proteins in the neighborhood of their experimental conformations. In addition, the simulated dynamic conformations are in overall good agreement with the experimental probes, particularly the NMR measurements of the N–H parameters S^2 which can be directly compared to the B_k–H_k virtual bonds order parameters [23]. The UNRES model by Liwo et al. and the FREADY one by Majek and Elber are both able to fold proteins in a satisfactory way [45], as well as to simulate their conformational dynamics around their native state in an acceptable agreement with the crystallographic B factors [47]. Recently, the CG model UNRES was used to simulate by REMD techniques the conformational transitions of Hsp70 chaperones, as a function of their nucleotide-binding states, providing a detailed description of the action mechanism of these proteins [21].

7.4 One-Bead Models

This category of low resolution models includes those which represent the backbone atoms with a single coarse grain (and possibly other beads for the side chains). In these models, the preferential conformations of the proteins main chain are captured with effective potentials functions for the pseudo-bendings $\theta_k = C^{\alpha}_{k-1} - C^{\alpha}_{k} - C^{\alpha}_{k+1}$ and pseudo-torsions $\tau_k = C^{\alpha}_{k-1} - C^{\alpha}_{k} - C^{\alpha}_{k+1} - C^{\alpha}_{k+2}$, which are generally extracted from statistical analysis of known proteins structures, using a Boltzmann inversion procedure (Fig. 7.3) [3, 17, 56]. However, as previously mentioned, in the absence of hydrogen bonding interactions, these models seem to be unable to yield stable trajectories of CG proteins using MD simulations. For this reason, the study of the proteins conformational dynamics with these models requires more or less some bias potentials which maintain their secondary structural elements in their initial state.

For instance, in the model by Klimov et al., two virtual atoms representing the carbonyl CO group and the amide NH one, are placed on the $C^{\alpha}_{k} - C^{\alpha}_{k+1}$ bonds, and an angular gaussian potential mimick the hydrogen bonds between these two grains.

However, in contrast to the previous "intermediate CG" models, these empirical interactions can only occur between pairs of atoms predefined by the authors, such as between the groups CO_i and NH_{i+4} in α-helices or between CO_i and NH_{17-i} in an anti-parallel β-sheet [38, 39]. In a similar spirit but less restrictive, the group of Head-Gordon introduced in their one-bead model a potential of mean force accounting for the backbone hydrogen bonds. However, according to their propensity to form a secondary structure, each C_k^α bead is initially pre-assigned one hydrogen bond forming capability among three possibilities: Either it interacts with the C_{k+3}^α if this latter is similarly assigned (helical type), either it interacts with other C^α similarly assigned and within a certain cutoff distance (sheet type), or it cannot make such interactions [78]. Using Langevin dynamics, the Klimov et al. and Head-Gordon CG models were applied to provide a description of the thermodynamic stability of small proteins.

If we are interested in the dynamic properties of proteins around their folded conformation, it is even possible to introduce more drastic bias potentials in their CG models. For instance, in the one-bead model by Tozzini et al., the backbone pseudo-bendings $\theta_k = C_{k-1}^\alpha - C_k^\alpha - C_{k+1}^\alpha$ potential has two minima corresponding to the two preferential conformations observed in the compact helices and extended β-strands (Fig. 7.3). However, the pseudo-torsions $\tau_k = C_{k-1}^\alpha - C_k^\alpha - C_{k+1}^\alpha - C_{k+2}^\alpha$ are prevented to undergo conformational transitions by using harmonic potentials. Because the backbone torsions were restrained, Tozzini et al. were able to generate, using MD or Langevin simulations, long stable trajectories of the native structure of the HIV-1 protease. The simulated conformational fluctuations of the protein core were in good agreement with the experimental B factors of the crystallographic structure 1HHP. Their simulations reveal in addition several events of opening and closing of the flaps that control the access to the binding site [71, 73].

In the popular MARTINI model of proteins, the backbone flexibility is also restrained by harmonic potentials whose parameters depend on the residues helical or extended conformation. This allows to conserve the proteins secondary structures along MD simulations and so to study only movements of secondary structure elements relative to each other [49]. For instance, the model was used to follow the opening and closing mechanism of several protein channels in their membrane environment, such as voltage-gated potassium or mechanosensitive channels [57, 72]. In a variant of the MARTINI force field, developed in the group of Sansom, the bias potentials for the backbone pseudo-torsions are replaced with harmonic distance restraints between backbone beads mimicking the hydrogen bonds in the secondary structures (Fig. 7.4) [7]. This model was used to study the conformational dynamics of an R-SNARE peptide inserted in a lipid bilayer. The microsecond-scale MD simulations reveal long-lived conformational sub-states in agreement with most experimental data on this system and which give insights into its role for the membrane fusion mechanism [19]. The Sansom's model was also applied to simulate the conformational dynamics of the membrane-bound CYP2C9 enzyme, in order to study how the lipid bilayer influences the opening and closing of different tunnels to access its catalytic site [14].

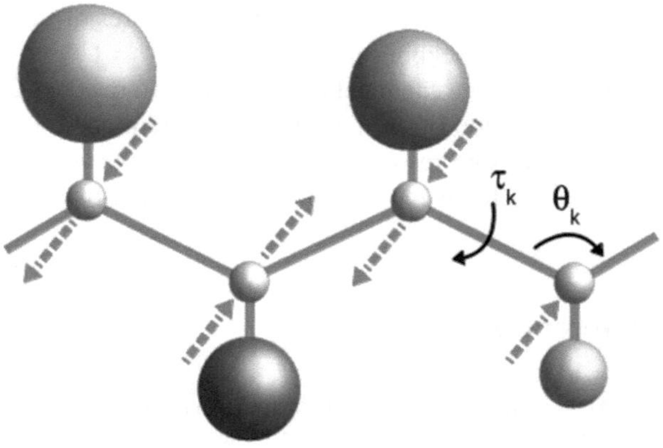

Fig. 7.4 One-Bead CG models. The *dashed arrows* indicate the bias effective potentials introduced to maintain the secondary structures

7.5 Perspectives

In constrat to the "structure-based" models, the "molecular-mechanics-based" CG models are quite delicate to be parameterized in the view of correctly simulate the proteins conformational fluctuations and transitions, because these properties subtly depend on the balance between the different physico-chemical energy contributions. The more the model is coarse-grained, the less the effective interactions description is straightforward. Regarding the "one-bead" CG approaches, most of the applications using these models need some bias potentials to maintain the proteins secondary structures during the MD simulations.

However, recently, new progress seem to be made by several groups in the development of physico-chemical one-bead models: In constrast to knowledge-based potentials derived from experimental structures, new effective potentials for the pseudo-torsions $\tau_k = C^{\alpha}_{k-1} - C^{\alpha}_{k} - C^{\alpha}_{k+1} - C^{\alpha}_{k+2}$ were developed on the basis of all-atom simulations of peptides and proteins. The group of Voth developed such a model which was used to successfully simulate the folding of the Trpzip, Trp-cage and Adenylate Kinase proteins. In addition, their model generated stable dynamic conformations of these proteins around their native state, and can be used to monitor the conformational transition between open and close states [28]. Using a similar procedure to parameterize the bonded potentials, the one-bead protein model developed in the group of Takada was applied to the disordered N-terminal domain of p53. Their MD simulations generated an ensemble of conformations that reproduced NMR residual dipolar coupling and SAXS profiles very accurately [66]. Another "atomistic-MD-based" backbone potentials was also integrated into the MARTINI model in order to better describe the polypeptides main chain flexibility. Without constrains to impose secondary structures and without a specific model of hydrogen bonds, this extension of MARTINI is able to reproduce quite well the dynamic conformations of Amyloid and Elastin-like peptides calculated by all-atom trajectories [59].

It seems that the two common key features of these three recent developments are (i) that the backbone pseudo-torsions $\tau_k = C^{\alpha}_{k-1} - C^{\alpha}_{k} - C^{\alpha}_{k+1} - C^{\alpha}_{k+2}$ potentials are specific to the chemical nature of the two residues C^{α}_{k} and C^{α}_{k+1} and (ii) that their parameterization is based on all-atom MD simulations of proteins in explicit solvent. These promising results open the route to physics-based one-bead CG models as robust and effective, if not more, as "structure-based" approaches.

References

1. Ahmed M, Davis J, Aucoin D, Sato T, Ahuja S, Aimoto S, Elliott JI, Van Nostrand WE, Smith SO (2010) Structural conversion of neurotoxic amyloid-beta(1–42) oligomers to fibrils. Nat Struct Mol Biol 17:561
2. Alemani D, Collu F, Cascella M, Dal Peraro M (2010) A nonradial coarse-grained potential for proteins produces naturally stable secondary structure elements. J Chem Theory Comput 6:315
3. Bahar I, Atilgan A, Erman B (1997) Direct evaluation of thermal fluctuations in proteins using a single-parameter harmonic potential. Fold Des 2:173
4. Bahar I, Kaplan M, Jernigan R (1997) Short-range conformational energies, secondary structure propensities, and recognition of correct sequence-structure matches. Proteins 29:292
5. Basdevant N, Borgis D, Ha-Duong T (2007) A coarse-grained protein-protein potential derived from an all-atom force field. J Phys Chem B 111:9390
6. Basdevant N, Borgis D, Ha-Duong T (2013) Modeling protein-protein recognition in solution using the coarse-grained force field SCORPION. J Chem Theory Comput 9:803
7. Bond PJ, Holyoake J, Ivetac A, Khalid S, Sansom MSP (2007) Coarse-grained molecular dynamics simulations of membrane proteins and peptides. J Struct Biol 157:593
8. Bonvin A (2006) Flexible protein-protein docking. Curr Opin Struct Biol 16:194
9. Chebaro Y, Mousseau N, Derreumaux P (2009) Structures and thermodynamics of alzheimer's amyloid-β aβ(16–35) monomer and dimer by replica exchange molecular dynamics simulations: implication for full-length aβ fibrillation. J Phys Chem B 113:7668
10. Chebaro Y, Pasquali S, Derreumaux P (2012) The coarse-grained opep force field for non-amyloid and amyloid proteins. J Phys Chem B 116:8741
11. Chng C-P, Yang L-W (2008) Coarse-grained models reveal functional dynamics–II. Molecular dynamics simulation at the coarse-grained level – theories and biological applications. Bioinform Biol Insights 2:171
12. Chu J-W, Voth GA (2007) Coarse-grained free energy functions for studying protein conformational changes: a double-well network model. Biophys J 93:3860
13. Clementi C, Nymeyer H, Onuchic JN (2000) Topological and energetic factors: what determines the structural details of the transition state ensemble and "en-route" intermediates for protein folding? An investigation for small globular proteins. J Mol Biol 298:937
14. Cojocaru V, Balali-Mood K, Sansom MSP, Wade RC (2011) Structure and dynamics of the membrane-bound cytochrome P450 2C9. PLoS Comput Biol 7:e1002152
15. Derreumaux P (1999) From polypeptide sequences to structures using Monte Carlo simulations and an optimized potential. J Chem Phys 111:2301
16. Derreumaux P, Mousseau N (2007) Coarse-grained protein molecular dynamics simulations. J Chem Phys 126:025101
17. DeWitte R, Shakhnovich E (1994) Pseudodihedrals: simplified protein backbone representation with knowledge-based energy. Protein Sci 3:1570
18. Ding F, Buldyrev SV, Dokholyan NV (2005) Folding Trp-cage to NMR resolution native structure using a coarse-grained protein model. Biophys J 88:147

19. Durrieu M, Bond P, Sansom M, Lavery R, Baaden M (2009) Coarse-grain simulations of the R-SNARE fusion protein in its membrane environment detect long-lived conformational substates. Chem Phys Chem 10:1548
20. Gabdoulline R, Wade R (1996) Effective charges for macromolecules in solvent. J Phys Chem 100:3868
21. Goaś E, Maisuradze GG, Senet P, Ołdziej S, Czaplewski C, Scheraga HA, Liwo A (2012) Simulation of the opening and closing of Hsp70 chaperones by coarse-grained molecular dynamics. J Chem Theory Comput 8:1750
22. Gunasekaran K, Ma B, Nussinov R (2004) Is allostery an intrinsic property of all dynamics proteins? Proteins 57:433
23. Ha-Duong T (2010) Protein backbone dynamics simulations using coarse-grained bonded potentials and simplified hydrogen bonds. J Chem Theory Comput 6:761
24. Haliloglu T, Bahar I (1998) Coarse-grained simulations of conformational dynamics of proteins: application to apomyoglobin. Proteins 31:271
25. Han W, Wu Y-D (2007) Coarse-grained protein model coupled with a coarse-grained water model: molecular dynamics study of polyalanine-based peptides. J Chem Theory Comput 3:2146
26. Henzler-Wildman K, Kern D (2007) Dynamic personalities of proteins. Nature 450:964
27. Hills RD, Brooks CL (2009) Insights from coarse-grained Go models for protein folding and dynamics. Int J Mol Sci 10:889
28. Hills RD, Lu L, Voth GA (2010) Multiscale coarse-graining of the protein energy landscape. PLoS Comput Biol 6:e1000827
29. Hinsen K (1998) Analysis of domain motions by approximate normal mode calculations. Proteins 33:417
30. Hoang TX, Seno F, Banavar JR, Cieplak M, Maritan A (2003) Assembly of protein tertiary structures from secondary structures using optimized potentials. Proteins 52:155–165
31. Hyeon C, Onuchic JN (2007) Internal strain regulates the nucleotide binding site of the kinesin leading head. Proc Natl Acad Sci USA 104:2175
32. Ishima R, Torchia D (2000) Protein dynamics from NMR. Nat Struct Biol 7:740
33. Ishima R, Freedberg D, Wang Y, Louis J, Torchia D (1999) Flap opening and dimer-interface flexibility in the free and inhibitor-bound HIV protease and their implications for function. Structure 7:1047
34. Izvekov S, Voth G (2005) Multiscale coarse graining of liquid-state systems. J Chem Phys 123:134105
35. Karplus M, McCammon J (2002) Molecular dynamics simulations of biomolecules. Nat Struct Biol 9:646
36. Kidera A, Go N (1990) Refinement of protein dynamic structure: normal mode refinement. Proc Natl Acad Sci USA 87:3718
37. Klein M, Shinoda W (2008) Large-scale molecular dynamics simulations of self-assembling systems. Science 321:798
38. Klimov D, Thirumalai D (2000) Mechanisms and kinetics of beta-hairpin formation. Proc Natl Acad Sci USA 97:2544
39. Klimov D, Betancourt M, Thirumalai D (1998) Virtual atom representation of hydrogen bonds in minimal off-lattice models of alpha-helices: effects on stability, cooperativity and kinetics. Fold Des 3:481
40. Koga N, Kameda T, Okazaki K-i, Takada S (2009) Paddling mechanism for the substrate translocation by AAA+ motor revealed by multiscale molecular simulations. Proc Natl Acad Sci USA 106:18237–18242
41. Kolinski M, Skolnick J (2004) Reduced models of proteins and their applications. Polymer 45:511
42. Levitt M (1976) A simplified representation of protein conformations for rapid simulation of protein folding. J Mol Biol 104:59

43. Levy R, Perahia D, Karplus M (1982) Molecular dynamics of an α-helical polypeptide: temperature dependence and deviation from harmonic behavior. Proc Natl Acad Sci USA 79:1346
44. Liwo A, Pincus M, Wawak R, Rackovsky S, Scheraga H (1993) Prediction of protein conformation on the basis of a search for compact structures: test on avian pancreatic polypeptide. Protein Sci 2:1715
45. Liwo A, Khalili M, Scheraga H (2005) Ab initio simulations of protein-folding pathways by molecular dynamics with the united-residue model of polypeptide chains. Proc Natl Acad Sci USA 102(7):2362
46. Lu Q, Lu HP, Wang J (2007) Exploring the mechanism of flexible biomolecular recognition with single molecule dynamics. Phys Rev Lett 98:128105
47. Majek P, Elber R (2009) A coarse-grained potential for fold recognition and molecular dynamics simulations of proteins. Proteins 76:822
48. Miyashita O, Onuchic JN, Wolynes PG (2003) Nonlinear elasticity, proteinquakes, and the energy landscapes of functional transitions in proteins. Proc Natl Acad Sci USA 100:12570
49. Monticelli L, Kandasamy S, Periole X, Larson R, Tieleman D, Marrink S (2008) The MARTINI coarse-grained force-field: extension to proteins. J Chem Theory Comput 4:819
50. Muller-Plathe F (2002) Coarse-graining in polymer simulation: from the atomistic to the mesoscopic scale and back. Chem Phys Chem 3:755
51. Okazaki K-i, Koga N, Takada S, Onuchic JN, Wolynes PG (2006) Multiple-basin energy landscapes for large-amplitude conformational motions of proteins: structure-based molecular dynamics simulations. Proc Natl Acad Sci USA 103:11844
52. Pasi M, Lavery R, Ceres N (2013) PaLaCe: a coarse-grain protein model for studying mechanical properties. J Chem Theory Comput 9:785
53. Popovych N, Sun S, Ebright R, Kalodimos C (2006) Dynamically driven protein allostery. Nat Struct Mol Biol 13:831
54. Prampolini G (2006) Parametrization and validation of coarse grained force-fields derived from ab initio calculations. J Chem Theory Comput 2:556
55. Reith D, Putz M, Muller-Plathe F (2003) Deriving effective mesoscale potentials from atomistic simulations. J Comput Chem 24:1624
56. Reva B, Finkelstein A, Sanner M, Olson A (1997) Residue-residue mean-force potentials for protein structure recognition. Protein Eng 10:865
57. Samuli Ollila OH, Louhivuori M, Marrink SJ, Vattulainen I (2011) Protein shape change has a major effect on the gating energy of a mechanosensitive channel. Biophys J 100:1651
58. Schomaker V, Trueblood K (1968) On the rigid-body motion of molecules in crystals. Acta Crystallogr B 24:63
59. Seo M, Rauscher S, Pomès R, Tieleman DP (2012) Improving internal peptide dynamics in the coarse-grained MARTINI model: toward large-scale simulations of amyloid- and elastin-like peptides. J Chem Theory Comput 8:1774
60. Srinivasan R, Rose GD (1999) A physical basis for protein secondary structure. Proc Natl Acad Sci USA 96:14258
61. Sun S (1993) Reduced representation model of protein structure prediction: statistical potential and genetic algorithms. Protein Sci 2:762
62. Takada S (2012) Coarse-grained molecular simulations of large biomolecules. Curr Opin Struct Biol 22:130
63. Takada S, Luthey-Schulten Z, Wolynes P (1999) Folding dynamics with nonadditive forces: a simulation study of a designed helical protein and a random heteropolymer. J Chem Phys 110:11616
64. Takagi F, Kikuchi M (2007) Structural change and nucleotide dissociation of Myosin motor domain: dual Go model simulation. Biophys J 93:3820
65. Tama F, Sanejouand Y (2001) Conformational change of proteins arising from normal mode calculations. Protein Eng 14:1
66. Terakawa T, Takada S (2011) Multiscale ensemble modeling of intrinsically disordered proteins: p53 N-terminal domain. Biophys J 101:1450

67. Tirion M (1996) Large amplitude elastic motions in proteins from a single-parameter atomic analysis. Phys Rev Lett 77:1905
68. Tolman JR, Ruan K (2006) NMR residual dipolar couplings as probes of biomolecular dynamics. Chem Rev 106:1720
69. Tozzini V (2005) Coarse-grained models for proteins. Curr Opin Struct Biol 15:144
70. Tozzini V (2010) Minimalist models for proteins: a comparative analysis. Q Rev Biophys 43:333
71. Tozzini V, Trylska J, Chang C-e, McCammon JA (2007) Flap opening dynamics in HIV-1 protease explored with a coarse-grained model. J Struct Biol 157:606
72. Treptow W, Marrink S, Tarek M (2008) Gating motions in voltage-gated potassium channels revealed by coarse-grained molecular dynamics simulations. J Phys Chem B 112:3277
73. Trylska J, Tozzini V, Chang C, McCammon J (2007) HIV-1 protease substrate binding and product release pathways explored with coarse-grained molecular dynamics. Biophys J 92:4179
74. Ueda Y, Taketomi H, Go N (1978) Studies on protein folding, unfolding, and fluctuations by computer simulation. II. A. three-dimensional lattice model of lysozyme. Biopolymers 17:1531
75. Van Giessen A, Straub J (2006) Coarse-grained model of coil-to-helix kinetics demonstrates the importance of multiple nucleation sites in helix folding. J Chem Theory Comput 2:674
76. Voegler Smith A, Hall C (2001) Alpha-helix formation: discontinuous molecular dynamics on an intermediate-resolution protein model. Proteins 44:344
77. Wallqvist A, Ullner M (1994) A simplified amino acid potential for use in structure predictions of proteins. Proteins 18:267
78. Yap E, Fawzi N, Head-Gordon T (2008) A coarse-grained alpha-carbon protein model with anisotropic hydrogen-bonding. Proteins 70:626
79. Zhou J, Thorpe I, Izvekov S, Voth G (2007) Coarse-grained peptide modeling using a systematic multiscale approach. Biophys J 92:4289

Chapter 8
Simulating Protein Folding in Different Environmental Conditions

Dirar Homouz

Abstract Molecular dynamics simulations have become an invaluable tool in investigating the dynamics of protein folding. However, most computational studies of protein folding assume dilute aqueous simulation conditions in order to reduce the complexity of the system under study and enhance the efficiency. Nowadays, it is evident that environmental conditions encountered *in vivo* (or even *in vitro*) play a major role in regulating the dynamics of protein folding especially when one considers the highly condensed environment in the cellular cytoplasm. In order to factor in these conditions, we can utilize the high efficiency of well-designed low resolution (coarse-grained) simulation models to reduce the complexity of these added protein-milieu interactions involving different time and length scales. The goal of this chapter is to describe some recently developed coarse-grained simulation techniques that are specifically designed to go beyond traditional aqueous solvent conditions. The chapter also gives the reader a flavor of the things that we can study using such "smart" low resolution models.

Keywords Molecular dynamics • All-atom models • Coarse-grained models • Multi-scale methods • Proteins • Folding • Crowding • Urea • HP model • Gō model • Statistical potential • C_α models • Side-chain-C_α Model (SCM) • Boltzmann inversion • SCAAL • MultiSCAAL

D. Homouz (✉)
AMS Department, Khalifa University, P.O. Box 127788, Abu Dhabi, UAE
e-mail: dirar.homouz@kustar.ac.ae

K. Han et al. (eds.), *Protein Conformational Dynamics*, Advances in Experimental Medicine and Biology 805, DOI 10.1007/978-3-319-02970-2_8,

8.1 Introduction

Molecular Dynamics (MD) simulation is a computer computational method that utilizes the laws of classical statistical physics in order to predict the behavior of many particle systems. The history of MD is tied to the history of the development of computer technology. The first real system to be studied using MD simulations was in 1964 by Rahman [1] who simulated liquid argon at 94.4 K. The system simulated by Rahman was limited to only 864 particles. Studying bigger systems with more particles became increasingly more feasible with the continual growth in computational power and speed. The pioneering work of McCammon et al. [2] marked the beginning of new era in using MD simulations in the very important biological problem of protein folding.

Most of the functions performed in a living cell are carried out by different proteins. In order for these proteins to function properly they have to be in their functional shape or fold. Proteins are large biomolecules that consist of one or more chains of amino acids. Thus, understanding the dynamics of how a protein can go from unfolded sequence of amino acids into its functional three dimensional fold is one of the fundamental problems in biology. MD simulations became an invaluable tool for studying protein folding and unfolding dynamics. It is used in conjunction with several experimental techniques in order to understand and interpret the experimental results at the atomic level. For more details on the MD history and techniques in protein folding studies we refer the reader to the following review articles [3–5].

In recent years, it became very obvious that the folding of proteins is highly dependent on their environmental conditions. Thus, the native protein folds are likely to be different from the ones usually determined by experimental techniques such as x-ray crystallography and NMR as these methods don't account for the densely crowded cellular environment. Several experimental studies have recently started factoring in these crowding effects in their experimental design by adding synthetic chowders to mimic the macromolecular crowding in the cell [6–14]. In addition to crowding, other cellular conditions can affect protein folding and stability such as the concentration of different ions. Well-designed computer simulation schemes are needed in order to better understand the role that all these environmental factors play in determining protein structure. In order to efficiently simulate protein interactions *in vivo*, one has to account for different sizes of interacting particles and different time scales.

In this chapter we present a multi-scale molecular dynamics scheme that can be used to simulate protein interactions in different crowding and solvent conditions. This scheme is based on a low resolution simulation model Side-chain C_α Model (SCM) [15] that was previously implemented in studying the protein folding dynamics in crowded environment. However, this model can't handle other environmental factors with small length scales besides the large crowders. Thus, SCM is integrated into a multi-scale algorithm (MultiSCAAL) [16] that deals with both large macromolecular crowders and small interfering chemicals. This scheme enables us to simulate proteins in many cellular as well as experimental conditions.

The material in this chapter is organized as follows: Sect. 8.2 gives a short overview of molecular dynamics simulations in the context of protein folding applications. In Sect. 8.3 we describe SCM and how it is integrated into MultiSCAAL scheme. In Sect. 8.4 we discuss some of the applications of these various techniques. Finally, we close this chapter with conclusions.

8.2 Molecular Dynamics and Protein Folding

8.2.1 All-Atom Versus Coarse-Grained

Different Molecular Dynamics simulation schemes are distinguished by the models they use to represent proteins and their interactions. These models differ in the level of detail, or resolution, that they reflect. Traditionally, these models are classified into two classes; All-Atom (AA) and Coarse-Grained (CG) models. AA models, with their explicit solvent representation, provide a great deal of detail at very short time scales (picoseconds). However, the inverse relationship between the resolution and computational cost usually limits the applicability of AA models when it comes to simulating protein folding trajectories with long timescales (microseconds). In addition, the computational cost grows exponentially when one considers environmental interactions with solvent, crowders, and other ions.

On the other hand, CG models with implicit solvents average out all amino acid atomic sites and replace them with a smaller number of beads, typically one or two. Thus, with these CG models, the accuracy of atomistic details and the reliability of energy functions are reduced. However, this is the price that one has to pay in order to capture the main features of protein folding over reasonable biological times. CG models are capable of increasing the timescale of molecular simulations due to the huge reduction in the number of degrees of freedom in the systems simulated mainly due to replacing all the degrees of freedom of the solvent with a mean field implicit solvent representation with zero degrees of freedom. Thus, with existing computer technology, CG simulations seem to be the only viable solution in order to study protein folding especially when the right environmental conditions are considered.

8.2.2 Coarse-Grained Models for Protein Folding

The famous experiments of Anfinsen et al. [17] in the early 1960s have instigated a large interest in the problem of protein folding. These experiments show that proteins can fold and refold reversibly to the same native state (functional state) which means that this state is thermodynamically stable and forms a global minimum. This conclusion raised the question of how can proteins reach this minimum starting from an unfolded state in a relatively short time ($\sim$ms) given the large number of possible conformations of any given protein. Levinthal [18] tried to resolve this paradox by suggesting that proteins follow a specified (encoded) kinetic "folding pathway" to reach its global minimum.

Several objections were raised against the idea of folding pathways and alternative views were proposed [19]. Among these alternative views, the Energy Landscape Theory was the most acceptable one. According to the Energy Landscape Theory, proteins don't follow a single pathway to reach the native state. Rather, they can follow multiple routes down a biased energy landscape towards the global minimum [20–22]. In other words, the energy landscape of protein folding process has a funnel-like shape and the folding is viewed as a flow process of an ensemble of routes down this funnel. The energy funnel is controlled by both its bias towards the native state and its roughness. In order for the protein to have fast folding, the roughness has to be small compared to the bias. This concept gave rise to the Principle of Minimal Frustration [23, 24] which can be justified by the fact that folding processes have evolved to make the native state more stable, favor stabilizing interactions, and make folding processes fast [25].

Coarse-grained computer models of proteins tried to conform to these competing views of protein folding processes. Early models used simplified geometries as well as energy functions. Lattice models achieved an early success due to the great simplification in the simulation geometry [24, 26–28]. In these models, proteins were modeled as self-avoiding polymer chains of one-bead amino acids where the beads on the chain are confined to move on a fixed three dimensional cubic lattice. These simplified models used fictitious energy functions such as HP [28] and Gō [29] energy functions. The HP model distinguishes between two types of monomers, H (Hydrophobic) and P (Polar), and assumes an attractive interaction between HH pairs and none between all other pairs. Gō model on the other hand tries to bias the energy function towards the native state by assuming attractive interactions for native contacts and repulsive interactions for none-native contacts. The Gō model gained more recognition later since it conforms to the Energy Landscape Theory and the principle of minimal frustration. Several Gō-like energy functions were developed later to be used with more advanced CG models [30].

The lattice models gave way to off-lattice models as computer power improved. This development allowed for more realistic representation of protein's geometry. Most of the early off-lattice models relied on simplified energy functions and one-bead amino acid representation [31–33]. These models are typically called C_α models since each amino acid is represented by one site located at the C_α carbon position. These C_α models started to take shape and give more faithful representation of protein by adopting more sophisticated energy functions (force fields) that included different type of structural as well as non-bonded interactions.

The difficulty in designing these dimensionally reduced C_α models lies in choosing the proper force field. There were different strategies for choosing the interaction energies between the 20 different types of beads (20 different amino acids). The structural energy terms (bond, angle, dihedral) were typically chosen such that they produce a thermodynamically stable structure. The non-bonded interactions could be still borrowed from earlier fictitious energy functions such as Gō model. However, more improved models tried to base these interaction energies on measured experimental values of amino acid pair potentials. Examples of such interaction maps are the Betancourt-Thirumalai (BT) statistical potential [34] and the Miyazawa-Jernigan (MJ) potential [35].

The C_α models gave way to more advanced models that incorporate more structural details of proteins. Cheung et al. [15] introduced one such a model in which each amino acid is represented by two beads; one at the C_α position and the second one at the center of mass of the side chain. This model called Side-chain C_α Model (SCM) falls between C_α and AA models and is capable of accounting for side-chain packing while keeping the computational cost low. This model was very successful in addressing protein folding interactions in crowded medium and confined geometries [6, 7, 36, 37]. With such improvements, the CG models start to look more like AA models and include more interactions which enable them to simulate different biological and experimental conditions. More information about CG models of protein folding can be found in these reviews [38, 39].

8.3 Flexible Low Resolution Simulation Techniques

The success of CG molecular dynamics stems from their ability to simulate protein folding and refolding events over large time scales. They do so by capturing the main features of the protein, stripping away complex details, and using implicit solvent models. In fact the greatest reduction in computation cost and time comes from replacing the atomic details of water with implicit solvent model. Thus, this approach works well for studying folding dynamics of isolated proteins or protein-protein interactions. In addition, the same CG models can be easily extended to studying protein folding in crowded medium where the dominant crowding agents are large macromolecules that can be themselves coarse-grained. However, this approach will be useless if one has to deal with environmental conditions that are controlled by small particles ($\sim$water molecule size) like urea. The reason being that the simplification and reduction in computational time achieved by removing water molecules will be undone by including a large number of these additional small molecules.

Taking these points into consideration, CG models have to be modified and a multi-scale approach is needed in order to capture both protein and environment details without sacrificing the computational efficiency. Here we present the details of the modifications that can be done to a simple two-bead model in order to develop it into a multi-scale algorithm. This is done by using SCM at the core to model proteins and large crowders, Langevin Dynamics to represent water solvent conditions, and adjusting force field parameters for different solvent conditions in order to account for chemical interference effects. The main elements of the final

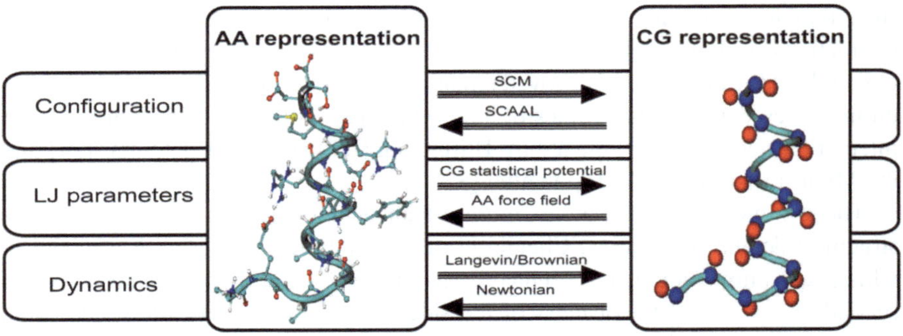

Fig. 8.1 A schematic diagram in a multi-scale algorithm where a protein configuration switches from all-atomistic (AA) to coarse-grained (CG) representation and vice versa. A side-chain-C_α model (SCM) is used as a coarse-grained model. The reconstruction of a protein in an AA representation from CG representation is achieved by SCAAL. The Lennard-Jones (LJ) parameters for an AA representation follow atomistic force field, while for a CG representation they follow a statistical potential based on bioinformatics and the potential of mean force from the AA molecular dynamic simulations via Boltzmann inversion method. The dynamics of an AA protein is governed by the Newtonian equations of motion. The dynamics of a CG protein is governed by the Langevin/Brownian equations of motion

multi-scale scheme, MultiSCAAL, are shown in Fig. 8.1 where we can see that SCM model is used to build the coarse-grained model starting from the corresponding all-atom representation. The scheme also includes the algorithm, Side-chain C Alpha to All-atom (SCAAL), which enables us to construct the all-atom representation of a protein starting from its course grained model. The Lennard-Jones (LJ) parameters for nonbonded interactions are based on a CG statistical potential. The dynamics that we use to sample the phase space of the protein is the Langevin Dynamics in order to account for the water solvent conditions implicitly. The details of these different elements and the implementation of the MultiSCAAL algorithm are given in the subsections below.

8.3.1 SCM Model (Representation & Hamiltonian)

A Sidechain-C_α (SCM) [15] coarse-grained model is used to represent proteins where each amino acid (except glycine) is modeled by two beads: a C_α bead and a side-chain bead located at the center of mass of the side-chain. The potential energy of a protein, E_p is the sum of three terms; the structural energy (E_{Struc}), the nonbonded energy (E_{NB}), and the Hydrogen bond energy (E_{HB})

$$E_p = E_{Struc} + E_{NB} + E_{HB} \tag{8.1}$$

8.3.1.1 Structural Energy

The structural energy, E_{Struc}, consists of the terms that account for all of the topological constraints of our structure. It is the sum of bond-length potential (E_{bond}), bond-angle potential (E_{angle}), dihedral potential (E_{dih}), and chiral interactions (E_{chi}).

$$E_{Struc} = E_{bond} + E_{angle} + E_{dih} + E_{chi} \tag{8.2}$$

The bond-length potential (E_{bond}) and the bond-angle potential (E_{angle}) are represented by harmonic springs as follows:

$$E_{bond} = \sum_{bonds} k_b (r - r_0)^2 \tag{8.3}$$

$$E_{angle} = \sum_{angles} k_\theta (\theta - \theta_0)^2 \tag{8.4}$$

Dihedral potential (E_{dih}) for every four consecutive C_α beads is represented by:

$$E_{dih} = \sum_{dihedrals}^{C_\alpha - C_\alpha - C_\alpha - C_\alpha} k_\phi^{(n)} \left[1 - \cos\left(n\left(\phi - \phi_0\right)\right)\right] \tag{8.5}$$

where ϕ is the dihedral angle, r is the distance between two adjacent beads and θ is the angle of three consecutive beads. The equilibrium values of ϕ_0, θ_0, and r_0 are calculated based on the native all-atom structure of a protein. The force constants are given these values $k_b = 100\varepsilon$, $k_\theta = 20\varepsilon$, $k_\phi^{(1)} = \varepsilon$, and $k_\phi^{(3)} = 0.5\varepsilon$, where $\varepsilon = 0.6$ kcal/mol.

The chiral energy (E_{chi}) accounts for an L-isoform preference of side chains. This energy is given by:

$$E_{chi} = \sum_{chiral} k_c (c - c_0)^2 \tag{8.6}$$

where c is the triple scalar product defined as $c = \vec{r}_{C_\alpha^i C_{SC}^i} \cdot \left(\vec{r}_{C_\alpha^i C_\alpha^{i-1}} \times \vec{r}_{C_\alpha^i C_\alpha^{i+1}}\right)$, c_0 is determined based on the native structure of the protein and $k_c = 20\varepsilon$. C_α^i and C_{SC}^i are the C_α bead and side-chain bead of the ith residue of the protein, respectively.

8.3.1.2 Nonbonded Energy

Nonbonded interaction energy E_{NB}^{ij} between a pair of i and j side-chain beads at a distance r has an LJ potential of the form,

$$E^{ij}_{NB} = \varepsilon_{ij}\left[\left(\frac{\sigma_{ij}}{r_{ij}}\right)^{12} - 2\left(\frac{\sigma_{ij}}{r_{ij}}\right)^{6}\right] \tag{8.7}$$

where $\sigma_{ij} = f(\sigma_i + \sigma_j)$, σ_i and σ_j are the Van der Waals (VdW) radii of side-chain beads, $|i - j| > 2$, and f is a control scaling factor that is used to prevent clashes that might destabilize the native state. The values of ε_{ij} are based on the solvent-mediated interaction between pairs of residues. For water solvent conditions we use the Betancourt-Thirumalai statistical potential map [34]. For other solvents this map can be modified according to the recipe give in Sect. 8.3.3.

Repulsive hard-core potential is used to model excluded volume interactions between C_α–Side-chain nonbonded pairs. This potential is given by this form:

$$E^{ij}_{NBrep} = \varepsilon\left(\frac{\sigma_{ij}}{r_{ij}}\right)^{12} \tag{8.8}$$

8.3.1.3 Hydrogen Bond Energy

For backbone hydrogen bonding interactions, an angular-dependent function is used to capture directional properties of backbone hydrogen bonds. For a pair of i and j C_α beads, the hydrogen bond interaction is given by:

$$E^{ij}_{HB} = A(\rho)\, E^{ij}_{NB} \tag{8.9}$$

$$A(\rho) = \frac{1}{\left[1 + (1 - \cos^2\rho)\left(1 - \frac{\cos\rho}{\cos\rho_a}\right)\right]^2} \tag{8.10}$$

where E^{ij}_{NB} has the same form as in Eq. (8.8), except that ε_{ij} for backbone hydrogen bonding is 0.6 kcal/mol and σ_{ij} is the hydrogen bond length, 4.6 Å.

The Lorentzian function $A(\rho)$ in Eq. (8.10) restricts the structural alignment of two interacting strands such that local backbone orientational configurations of parallel β sheets, antiparallel β sheets, or left and right-handed α helices are favored. The parameter ρ is the pseudo-dihedral angle between two interacting strands of the backbone. The function $A(\rho)$ will have its maximum value of 1 when $\rho = 0$ (the alignment that points to β-strands or α-helices) or when $\rho = \rho_a$ (the pseudo-dihedral angle of a canonical helical turn, 0.466 rad). For all other pseudo-dihedral angles (ρ) the value of $A(\rho)$ will be diminished (much smaller than 1).

8.3.1.4 SCM with Gō-Like Hamiltonian

The energy terms presented above are used to model proteins with non-specific nonbonded interactions. However, these terms can be manipulated easily to produce a topologically based Gō-like model that provides a minimally frustrated energy landscape. In such a model, the nonbonded interactions found in the native structure of the protein retain their sticky interaction represented by LJ potential of the form give in Eq. (8.7). All other non-native nonbonded pairs will be assigned repulsive interaction of the form in Eq. (8.8). The same rule can be applied to hydrogen bonding where native interactions will be represented by Eq. (8.9) while the non-native ones are represented by repulsive potential.

This kind of flexibility enables to tailor the SCM model to our computational needs. While the SCM model with non-specific Hamiltonian can explore bigger regions of the energy landscape than a one with Gō-like Hamiltonian, it is more expensive computationally. Thus, when we are interested in protein folding problems where the focus is on transitions out or into the native state we can utilize the Gō-like based SCM model.

8.3.2 Langevin Dynamics (Implicit Solvent)

To account for the effect of the solvent on the protein dynamics the Langevin equation of motion [40] is used to describe the dynamics in SCM coarse-grained molecular simulations. The solvent is treated implicitly in the Langevin equation through a stochastic term. The Langevin equation of motion for a general coordinate x is:

$$m\ddot{x} = -\frac{\partial U}{\partial x} - \zeta\dot{x} + \Gamma, \tag{8.11}$$

where m is the mass and U is the potential energy of the molecule. The drag term, $-\zeta\dot{x}$, or the dissipation term, is caused by friction which is compensated by a random force Γ representing random collisions with solvent molecules. Γ is sampled from a distribution of a white noise (Gaussian noise).

Fast motions of large biomolecules are quickly damped in a viscous solvent such as water. As a result, they follow random trajectories referred to as the Brownian motion. The inertia term is dropped in Eq. (8.11) and we get the first order ordinary differential equation for the Brownian motion given by:

$$\zeta\dot{x} = -\frac{\partial U}{\partial x} + \Gamma. \tag{8.12}$$

8.3.3 Different Solvent Conditions (Modifying LJ Parameters)

The techniques implemented in SCM were designed to simulated protein folding in aqueous medium. However, we are presented with many situations where it is important to study protein folding/refolding in different solvent conditions. One such a situation arises when one wants to simulate the experimental unfolding of proteins in different urea concentrations or the experimental folding of a protein in the presence of small molecules such as salt and alcohol, or small crowders such as glycerol. Extending SCM to cover these situations presents us with great challenge since these small molecules have the same length scale as water. These solvent conditions can be readily handled in AA simulations. Thus, one has to devise a multi-scale approach that can benefit from AA models of these solvents and feeds back into GC simulations. This is the approach used here in order to implicitly account for chemical interference in solvents by adjusting the solvent-mediated amino acid pair interaction energies. The details of the technique used to adjust these parameters are given below.

8.3.3.1 The Choice of Parameters ε_{ij}

In order to design a coarse-grained model that can accommodate the chemical properties of different amino acids we chose our nonbonded LJ interaction parameters in Eq. (8.7) based on knowledge-based potentials. These knowledge-based (or statistical) potentials are matrices (of 210 elements) that give the solvent-mediated interaction energies between all pairs of amino acids. There are several schemes for calculating these potentials such as those of Miyazawa and Jernigan [35], Kolinski and Skolnick [41], or Betancourt-Thirumalai [34]. Our model is based on the Betancourt-Thirumalai statistical potential [34]. This statistical potential addresses sequence variations where the reference interaction, $\varepsilon = 0.6$ kcal/mol, is based on the Thr-Thr pairwise interaction.

8.3.3.2 The Statistical Potential Map in a Different Solvent

All of the available statistical potential maps give the interactions energies between amino acids in water. Using SCM model to simulate proteins in other solvents such as urea requires expanding the idea of statistical potential maps to other solvents. In principle, the statistical potential between two residues should be the same as the potential of mean force (PMF) between these residues. The effect of the solvent is implicitly accounted for in the statistical potential. Calculating the potential of mean force is inherently complex and inefficient. The direct calculation of the residue-residue interaction from the PMF is therefore not attainable. However, creating the statistical potential parameter map (SPPM) is a much simpler problem.

In order to get the statistical potential parameter map (SPPM) for a certain solvent, we compute the PMFs of pairs of amino acids using all-atom simulations of free residues in that solvent. We circumvent the inherent difficulty of calculating this PMF by simulating a large number of copies of each pair at once, instead of one pair. For instance, in order to calculate the parameter ε_{TT} between two Threonine (Thr) residues we run a simulation of a large number of solvated free Thr residues. This method helps enhance the sampling and converge the PMF for this pair of residues. We make two approximations in order to further simplify the calculation of the statistical potential map. First, we approximate the PMF between a pair of amino acids by a two point correlation function of the distance between the two centers of mass of the side chains. Second, we fit the calculated PMF to a Lennard Jones (LJ) potential and set the statistical potential parameter to be equal to the depth of the resulting LJ potential. This calculation is done by using the Boltzmann inversion method discussed below.

8.3.3.3 Boltzmann Inversion

The CG energy function that accommodates chemical interference can be created using Boltzmann inversion [42–44] based on data obtained from all-atomistic molecular dynamics simulations. The pair correlation function between any two amino acid types i and j at a distance r in type α solvent is $g_{ij}^{\alpha}(r)$. This function is related to the potential of mean force, $U_{ij}^{\alpha}(r)$, between the same pair of amino acids through Boltzmann inversion at temperature T by the following formula [45]:

$$U_{ij}^{\alpha}(r) = -k_B T \ln\left[\frac{g_{ij}^{\alpha}(r)}{\rho_o}\right], \tag{8.13}$$

where ρ_o is the average density of the system (amino acid pairs and the solvent) and k_B is the Boltzmann constant. The average density ρ_o is used to normalize the pair correlation function at distances greater than the excluded volume radius. The solvent mediated interactions $\varepsilon'^{\alpha}_{ij}$ for every pair of amino acids i and j is equal to $U^{\alpha}{}_{ij}(r*)$

$$\varepsilon'^{\alpha}_{ij} = U_{ij}^{\alpha}\left(r^*\right), \tag{8.14}$$

where $r*$ denotes the first highest peak of $g_{ij}^{\alpha}(r)$. Next $\varepsilon'^{\alpha}_{ij}$ is shifted by a constant, V_o.,

$$\varepsilon_{ij}^{\alpha} = \varepsilon'^{\alpha}_{ij} + V_o. \tag{8.15}$$

where V_o is obtained from a Threonine pair by setting $\varepsilon'^{\alpha}_{TT}$ (in water) from the simulation equal to $\varepsilon^{\alpha}_{TT}$ from the statistical potential of the same amino acid pair [34].

A Lennard-Jones potential (LJ), $V_{ij}^{\alpha}(r)$, is used to approximate the overall profile of $U_{ij}^{\alpha}(r)$ [46] and it is the energy function for the same type of amino acids in coarse-grained molecular simulation:

$$V_{ij}^{\alpha}(r) = \varepsilon_{ij}^{\alpha}\left[\left(\frac{r_{ij}^{o}}{r}\right)^{12} - 2\left(\frac{r_{ij}^{o}}{r}\right)^{6}\right]. \tag{8.16}$$

$\varepsilon_{ij}^{\alpha}$ is the solvent-mediated interaction of an amino acid pair i and j in solvent type α. r_{ij}^{o} is the bonding distance. Figure 8.2 shows how Boltzmann inversion

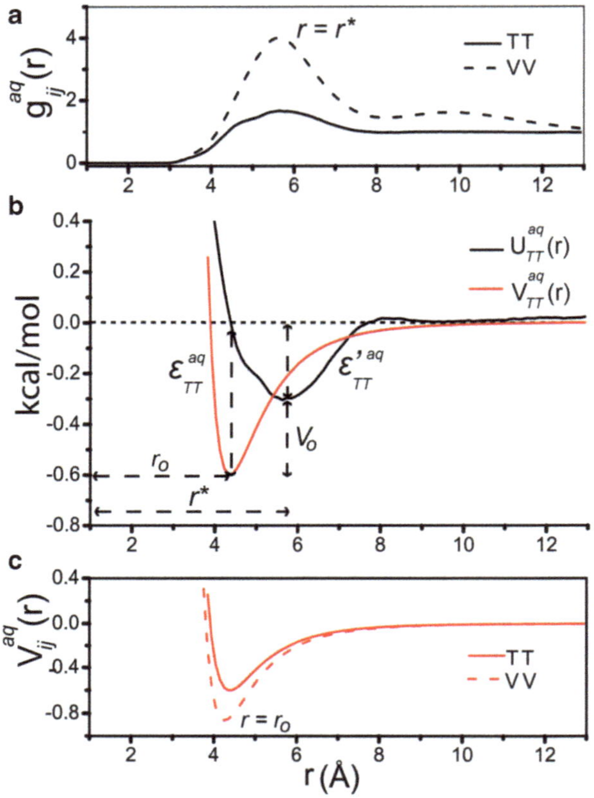

Fig. 8.2 (**a**) Pair correlation function $g_{ij}^{aq}(r)$ for Thr-Thr (*solid line*) and Val-Val pairs (*dotted line*) derived from all–atomistic molecular dynamics simulations in aqueous condition. $r = r*$ is at the maximum $g_{ij}^{aq}(r*)$. (**b**) The potential of mean force $U_{TT}^{aq}(r)$ of Thr-Thr interaction (in *black*) is obtained from all–atomistic molecular simulations under aqueous condition through Boltzmann inversion (Eq. 8.13) as a function of *r*, distance between the chosen atoms (i.e. C_β atom for Thr.) that are in closest proximity to the center of mass of the side chain in threonine. *r** denotes the position of the major peak of the pair correlation function $g^{aq}TT(r)$ in (**a**) and $\varepsilon'^{aq}_{TT} = U_{TT}^{aq}(r*)$. The Betancourt-Thirumalai statistical potential follows a Lennard-Jones interaction $V_{TT}^{aq}(r)$ (Eq. 8.16) for the same pair of amino acid in coarse-grained molecular simulations (in *red*). *r* is the interacting distance between the coarse-grained side-chain beads of the amino acids (i.e. center of mass of side chains). r_o is the bonding distance σ_{TT} in Eq. (8.7). $\varepsilon_{TT}^{aq} = V_{TT}^{aq}(r_o)$ is taken from the Betancourt–Thirumalai statistical potential. The reference potential from Eq. (8.15) is V_o. (**c**) $V_{ij}^{aq}(r)$ for Thr-Thr (*solid line*) and Val-Val pairs (*dotted line*) in aqueous solvent. r_o is the same bonding distance in (**b**)

is applied in practice to generate LJ parameters for amino acid pairs in water. In addition, Fig. 8.3 shows the accuracy of this process by comparing the SPPM for all amino acid pairs in water with the BT map. The end process result of this process is to generate a new SPPM of the parameters $\varepsilon_{ij}^{\alpha}$. Once this map is generated it can be then used for any CG simulation with the corresponding solvent. Important examples of these maps would be the maps of solvent mediated interaction for all 210 amino acid pairs in different concentrations of urea published in [16].

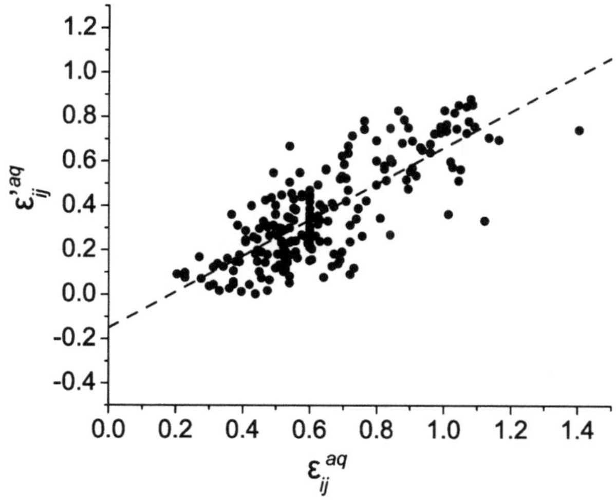

Fig. 8.3 The correlation between the aqueous solvent-mediated interactions between amino acids i and j, ε'^{aq}_{ij}, which are derived from the molecular dynamics simulations and the ones from the Betancourt-Thirumalai statistical potential ε^{aq}_{ij}. The linear correlation coefficient is 0.79

8.3.4 Crowders and Ions

More modifications were devised in order to account for other environmental factors such as large molecular crowders, electrostatic interaction, and ions. A short description of these modifications follows.

8.3.4.1 Macromolecular Crowders

Intracellular crowding can be mimicked experimentally by adding high concentrations of inert synthetic or natural macromolecules, termed crowding agents, to the systems *in vitro*. Inert large synthetic macromolecules such as Ficoll 70 and dextran can be readily included in CG simulations because of their large sizes. The atomic details of these particles will be irrelevant when we investigate their excluded volume effect on protein folding. Thus, they can be represented as hard particles with shapes that capture the geometry of each molecule. For instance Ficoll 70 can be modeled as a hard sphere and dextran as a hard dumbbell (two bonded spheres) of relevant size. In terms of the Hamiltonian, all the interactions that involve crowders (crowder-crowder, crowder-protein) will be repulsive with the same form given in Eq. (8.8). These repulsive interactions model the nonspecific steric space-filling repulsions due to the excluded volume effect of crowding. For other types of crowders such as the macromolecules in the cellular environment, a polydisperse CG model of these particles can be employed in order to mimic their different sizes and shapes.

8.3.4.2 Electrostatics and Ionic Concentration

In order to improve the accuracy and the performance of the coarse-grained (SCM) model, we included electrostatic interactions by adding a Debye-Hückel energy term [47]. This added term is supposed to represent screened Columbic interactions between charged sites. The charges are obtained using quantum chemistry calculations of the electronic structures of the all-atomistic representation of all residues in the protein. However, adding this term means that our Lennard Jones (LJ) potential parameters have to be adjusted. The original LJ parameters in the coarse-grained model were obtained from knowledge-based statistical potential which measures the solvent mediated interaction energies between different amino acid pairs including electrostatic interactions.

In order to adjust the LJ parameters in the coarse-grained simulation we first adjust the LJ parameters for every amino acid pair *(i,j)* as follows:

$$\varepsilon'_{ij} = \varepsilon_{ij} + \frac{e^2}{4\pi\varepsilon}\left(\frac{q_i q_j}{\sigma_{ij}}\right) = \varepsilon_{ij} + \alpha\left(\frac{q_i q_j}{\sigma_{ij}}\right). \tag{8.17}$$

where q_i and q_j are the charges of the two amino acids and σ_{ij} is the position of the minimum in the original LJ potential. Then, we can adjust the nonbonded interactions to have this form:

$$E^{ij}_{NB+Elect} = \varepsilon'_{ij}\left[\left(\frac{\sigma_{ij}}{r_{ij}}\right)^{12} - 2\left(\frac{\sigma_{ij}}{r_{ij}}\right)^{6}\right] + \alpha\left(\frac{q_i q_j}{r_{ij}}\right). \tag{8.18}$$

The effects of ionic concentration in the solvent will be captured through a screening factor that changes Eq. (8.18) to this form

$$E^{ij}_{NB+Elect+I} = \varepsilon'_{ij}\left[\left(\frac{\sigma_{ij}}{r_{ij}}\right)^{12} - 2\left(\frac{\sigma_{ij}}{r_{ij}}\right)^{6}\right] + \alpha\left(\frac{q_i q_j}{r_{ij}}\right)\exp\left(-r_{ij}\sqrt{(4\pi\alpha I/k_B T)}\right), \tag{8.19}$$

where I is the ionic concentration. The LJ parameters in Eq. (8.19) retain the same modified values according to Eq. (8.17).

8.3.5 Reconstructing the AA Coordinates (SCAAL)

Several methods of reconstructing reduced representation into all-atomistic structures have been developed over the last few years [48–51]. These include methods that can either recover the atomistic details of a protein's backbone with the knowledge of C_α beads [48], or reconstructing a full protein with the knowledge of its four heavy backbone atoms [49]. Methods that reconstruct all–atomistic structures from the information provided by a C_α bead and the center of mass of the side chain are also available [50]. However, the main purpose of the methods above is to reconstruct protein conformations that are very close to the crystal structures obtained by X-ray or NMR experiments. The use of rotamer libraries, obtained from PDB structures, in all these algorithms has made possible the development of very fast and accurate reconstruction methods. However, when reconstructing far – from the native state protein structures, which is most often the case in the course of a multi-scale simulation, it is questionable whether the accuracy of such methods can still be achieved. For this reason we have used a very simple approach based on the physics principle of harmonic constraints to reconstruct all-tom structures from coarse-grained ones in multi-scale simulation scheme.

In order to reconstruct the desired all-atom structure from coarse-grained models we use the positions from coarse-grained SCM as a part of harmonic constraints and apply them to an all-atom protein template through a process of energy minimization. For each residue, C_α positions from the SCM will be used as position constraints for C_α in the backbones from the all-atom template. As for the constraint of a side-chain position, it will be imposed on a heavy atom with the closest proximity to the actual center of mass of the side chain, in which the distance between the two is typically less than 1 Å. By doing this, the calculation of the center of mass of the side chain during a reconstruction algorithm is avoided by paying a small price on accuracy as long as we keep the harmonic spring constants at a reasonable range. During the reconstruction procedure that takes in both a SCM protein structure and an all-atom template as an input, the harmonic constraints imposed by a few chosen beads will carry the all-atom template to the desired structure, through driving forces of energy minimization, without the need for building a protein from individual atoms. The use of this "template concept" for protein reconstruction is depicted schematically in Fig. 8.4a and the flowchart of the SCAAL reconstruction algorithm is shown in Fig. 8.4b. The details of this method can be found in previous studies [7, 16].

8.3.6 MultiSCAAL: SCM + SCAAL

The improvements described above have transformed the SCM into a multifaceted algorithm that can be used to simulate protein folding in many different conditions. It can simulate the folding behavior in crowded environment that resembles the cellular conditions or reproduce the effect of synthetic crowding agents used in experimental studies to mimic cellular crowding. The modified SCM is capable of simulating experimental refolding events in the presence of denaturing factors such as urea or in the presence of other ions. Any combination of these different conditions (crowding, urea, ionic concentration) becomes accessible for simulation using low resolution protein representation.

In addition, combining reconstruction algorithm SCAAL with SCM results in a more sophisticated multi-scale scheme that combines AA simulations with CG ones.

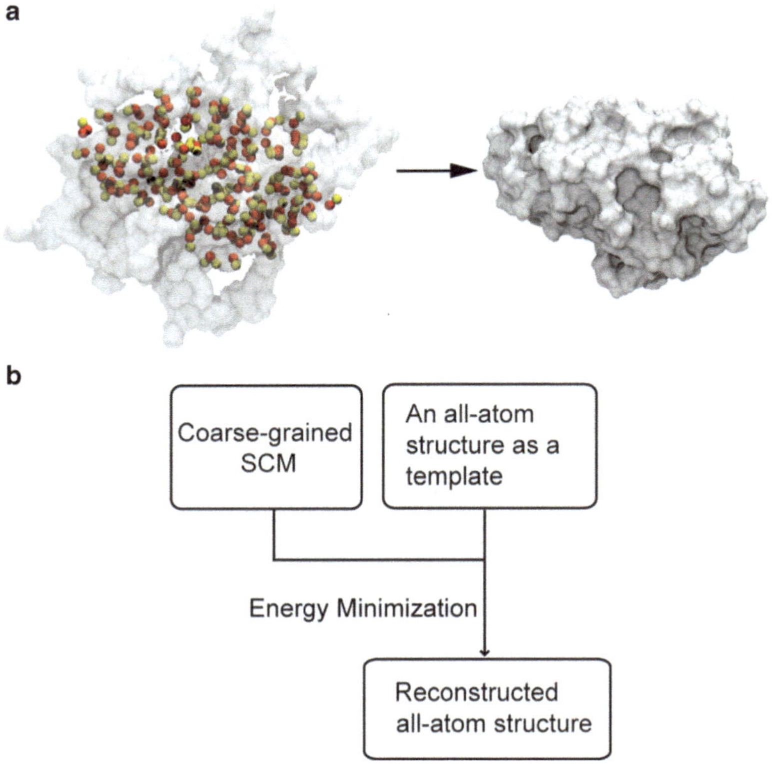

Fig. 8.4 A schematic representation of the SCAAL reconstruction method with the use of an all-atomistic protein structure as a template and the positions of coarse-grained side-chain-C_α model (SCM) as harmonic constraints. (**a**) (*Left*) C_α beads are in *red* and the heavy side-chain beads are in *yellow*. The two beads hold the positions through harmonic constraints for a projected reconstructed all-atomistic protein model. A randomly chosen all-atomistic protein structure that can be far from the crystal structure is introduced as a structural template and shown in a solvent accessible surface area mode. (*Right*) After the structural reconstruction by SCAAL, an all-atomistic representation of a projected protein structure is created (myoglobin, PDBID 1A6M, is used for illustration). (**b**) Flow chart of the SCAAL algorithm

This combined multi-scale scheme "MultiSCAAL" builds on the capabilities of the modified SCM which can handle different solvent and environment condition and on the accurate reconstruction of all-atom protein structures from SCM provided by SCAAL. Both these steps are necessary to incorporate crowding and chemical interferences in a multi-scale molecular simulation.

The MultiSCAAL scheme works on enhancing the sampling of all-atomistic simulations by utilizing a large set of initial conditions sampled from the SCM distributions. These selected initial CG structures are reconstructed into AA ones using SCAAL. Then we let the all-atom simulation visit and refine all the conformations that are predicted by the more efficient SCM model. Our scheme is not based on the concept of Resolution Exchange. Thus, we don't perform any conformation exchanges between the CG and AA simulations. Instead, we concentrate on the proper selection of initial AA conformations based on a knowledge-based CG model that can be adjusted to different environmental conditions.

In summary, the MultiSCAAL scheme follows these steps:

(1) The energy function for SCM molecular dynamics simulations is derived from the potential of mean force (PMF) from the all-atomistic simulations that contain certain chemical interference using Boltzmann inversion method.
(2) SCM protein representations in a thermodynamic ensemble of interest are selected according to a Metropolis criterion [52] and all-atomistic protein conformation are promptly reconstructed using SCAAL.
(3) Folding free energy landscape of a protein is effectively simulated by all-atomistic molecular dynamics that uses reconstructed all-atomistic protein models built from step (2) as initial conformations.

8.4 Protein Folding in Different Conditions: Examples

8.4.1 Crowding and Protein Folding

The living cell is a highly crowded environment due to the presence of large amounts of soluble and insoluble macromolecules, including proteins, nucleic acids, ribosomes, and carbohydrates. This cellular crowding limits the available space for biochemical interactions including protein folding. It is estimated that the concentration of macromolecules in the cytoplasm is in the range of 80–400 mg/ml which amounts to a volume fraction between 10 and 40 % [53–56]. Crowding can be mimicked experimentally by adding high concentrations of inert synthetic crowders. In addition, crowding can be modeled using CG molecular dynamics simulations. There are established effects of crowding on protein folding such that crowding stabilizes the folded protein, compacts denatured states. These effects have been investigated using different theoretical and experimental techniques [6, 8, 36, 53, 57, 58]. Here, we present examples of other interesting effects of macromolecular crowding on protein folding. These studies utilized the power and efficiency of CG simulations based on the SCM model.

8.4.1.1 Crowding Changes Protein Shape

SCM based molecular dynamics [7] simulations were used to investigate the secondary structure changes in protein *Borrelia burgdorferi* VlsE in experimental crowded conditions [59]. VlsE is an aspherical protein with marginal stability: It is best described as having an elongated football shape with a helical core surrounded by floppy loops at each end [60]. Experiments using Ficoll 70 as an inert synthetic crowding agent have shown that VlsE folded state is stabilized in the presence of increasing concentration of crowders. However, when the same crowding experiments were repeated in the presence of urea, crowding seemed to destabilize the folded state.

In order to understand these varying effects of crowding on the folding of VlsE, CG molecular simulations were used to calculate energy landscape of VlsE in different volume fractions of Ficoll 70 and at different temperatures. VlsE was modeled using SCM with nonspecific interactions. Ficoll 70 molecules are modeled as hard spheres that provide nonspecific repulsive interactions in the simulations. The thermodynamic properties of VlsE in aqueous solvent and in crowded environments (volume fractions, ϕ_c, of 0, 15, and 25 %) were studied by molecular simulations with Langevin dynamics. The replica exchange method (REM) [61, 62] was used in order to enhance the efficiency of sampling. The resulting trajectories were analyzed using the weighted histogram analysis method (WHAM) [63, 64].

The resulting energy landscape is shown in Fig. 8.5. This energy landscape shows that the combination of crowding and denaturing agents (temperature in simulations versus urea in experiments) can produce conformational changes in VlsE between three dominant states. These three states are the native structure (football shaped), a bean-like structure, and a collapsed globular structure. The all-atomic structures for these three states were reconstructed using SCAAL as shown in Fig. 8.6. The simulations have also shown that these conformational (shape) changes were accompanied by secondary structure transformations that lead to the exposure of a hidden antigenic region in agreement with experiments.

8.4.1.2 Crowding and Protein Folding Routes

The folding energy landscape of an α/β protein, apoflavodoxin, in the presence of inert macromolecular crowding agents was studied using *in silico* and *in vitro* approaches [65]. The crowding conditions were created using two crowding agents with different shapes, the spherical Ficoll 70 and the rod-like dextran. Parallel kinetic folding experiments were performed on purified apoflavodoxin in the presence of Ficoll 70 and dextran. These experiments have shown that time-resolved folding pathway of apoflavodoxin is modulated by crowding agent geometry.

In the CG molecular simulations, apoflavodoxin was constructed using the SCM model with a Gō-like Hamiltonian. Ficoll 70 was modeled as hard sphere. The rod-like dextran was modeled as dumbbell consisting of two bonded hard spheres (Ficoll 70). As with VlsE above, Langevin dynamics, REM, and WHAM were used. The results of the simulations showed that these different types of crowders stabilize the native state of apoflavodoxin (Fig. 8.7). In addition, the geometry of the crowder tends to play an important role in manipulating the folding route. The simulations show that the early formation of contacts around the β_1 sheet of apoflavodoxin creates a topological frustrated structure. In order for the protein to proceed in its folding, it has to unfold and undo these early formed contacts. This topological frustration is affected by the crowded environment. More specifically, the shape of the crowder can worsen or remedy the early topological frustration as can be seen in Fig. 8.8.

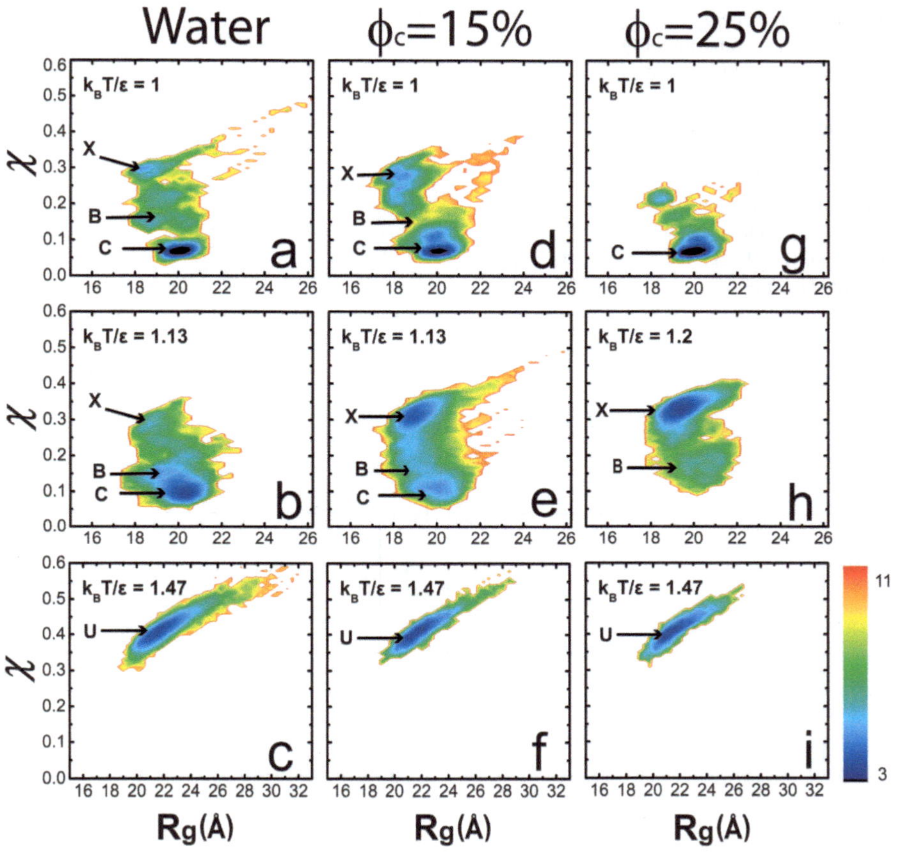

Fig. 8.5 Free-energy diagram as a function of radius of gyration R_g and the overlap function (χ) for $\varphi_c = 0$ % (water) (**a**, **b**, and **c**), 15 % (**d**, **e**, and **f**), 25 % (**g**, **h**, and **i**) at various temperatures expressed in $k_B T/\varepsilon$. χ measures the deviation from crystal structure (χ=0). The *color* is scaled by $k_B T$. The native *football-shaped* species is labeled *C*, the *bean structure* is labeled *B*, the *spherical state* is named *X*, and the unfolded state is indicated by *U*

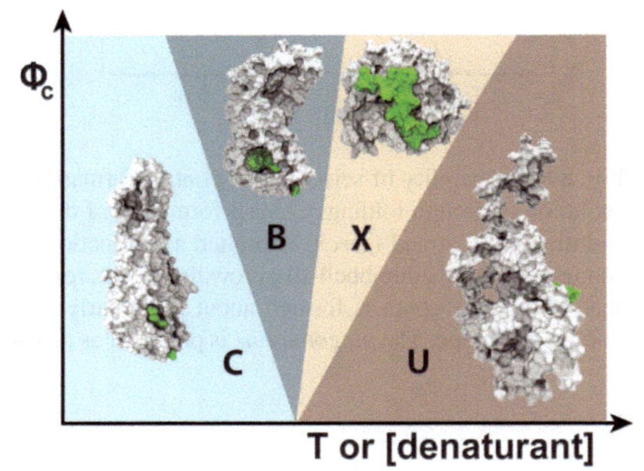

Fig. 8.6 A schematic phase diagram of VlsE conformations in the φ_c–T (or urea) plane. The antigenic IR6 sequence is shown in *green* for all representative states *C*, *B*, *X*, and *U*

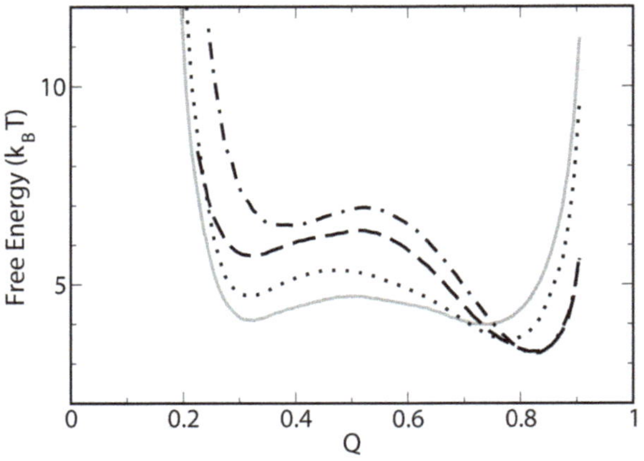

Fig. 8.7 Free energy profiles are plotted as a function of Q (the fraction of native contact formation) at different crowding conditions at 360 K. φ_c (water) = 0, *solid line*; φ_c (Ficoll70) = 25 %, *dotted line*; φ_c (Ficoll70) = 40 %, *dashed line*; and φ_c (dumbbell) = 40 %, *dot-dashed line*

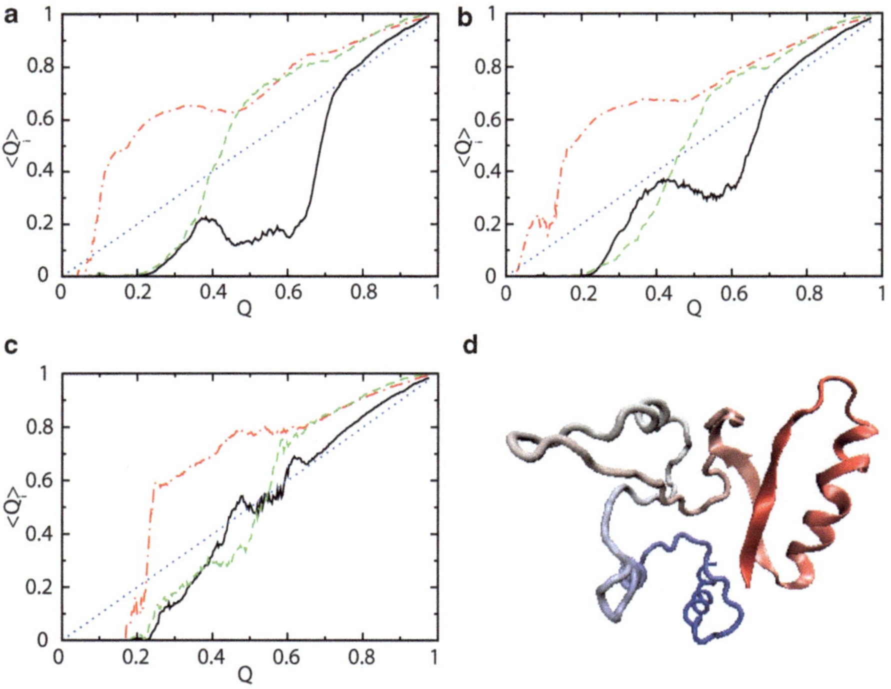

Fig. 8.8 Probability of select native contact formation $<Q>_i$ at the *i*th region of a protein in the evolution of protein folding. Contact formation of the first β-strand (*black*), the first α-helix (*red*), and the third β-strand (*green*) is plotted as a function of Q in (**a**) water, (**b**) φ_c =40 %, Ficoll70, and (**c**) φ_c =40 %, dumbbell-like crowding agent, respectively. (**d**) A conformation in the unfolded state with some contacts formed about β_1 in early Q that causes topological frustrations in the folding landscape. The *diagonal line* is provided as a visual guidance for a mean-field like behavior

8.4.2 Multi-scale Simulation of Protein Folding with Chemical Interference

8.4.2.1 Protein Folding in Urea

The mutliscale simulations using MultiSCAAL were used in order to investigate the effect of urea on the folding landscape of Trp-cage [16]. In this approach, CG simulations of Trp-cage in urea were performed first and then structures fished from these simulations are fed into AA simulations in order to zoom in on important details. In order to perform the CG simulation, statistical potential maps of amino acid LJ parameters were created for different concentrations of urea. These maps were created using the Boltzmann Inversion technique presented above.

The results obtained from the MultiSCAAL simulations were compared with those of AA simulations performed in the same study. The AA atom simulation of Trp-cage utilized the enhanced sampling technique of Replica Exchange Method (REM). AA-REM and MultiSCAAL simulations were performed in aqueous and 8 M urea solvent conditions. MultiSCAAL were shown to be more accurate and more efficient that AA-REM.

In terms of accuracy, MultiSCAAL samples a broader energy landscape, with a wide distribution of ensemble structures as can be seen in Fig. 8.9. Interestingly, in the case of 8 M urea the dominant structure sampled by MultiSCAAL matches better with interatomic distances obtained by NMR experiments [66]. By using a reduced representation in side-chain beads in the CG model, without explicit solvent molecules, the protein can explore different side-chain orientations faster. This allows the indole group of Trp 6 to exit the hydrophobic core of the protein and this structural feature can account for the shorter distances between Trp 6 and other amino acids.

In terms of efficiency, MultiSCAAL simulation was shown to provide a considerably enhanced sampling efficiency and lower computational cost than the standard AA-REMD simulations with the total simulation length being $\sim$25 times greater in less computational hours ($<1/2$).

8.4.2.2 Protein Folding and Ionic Concentration

Calmodulin (CaM) is the smallest known functional protein and plays an important role in regulating intercellular signaling. CaM possesses a great conformational flexibility as it can bind over 300 targets when fully saturated with calcium [67]. SCM based coarse-grained simulations were used to study the crowding effects on the conformational states of apoCaM [68]. In addition, these calculations were extended using a multi-scale approach to include electrostatics in studying the conformational states of both apoCaM and holoCaM at different salt concentrations

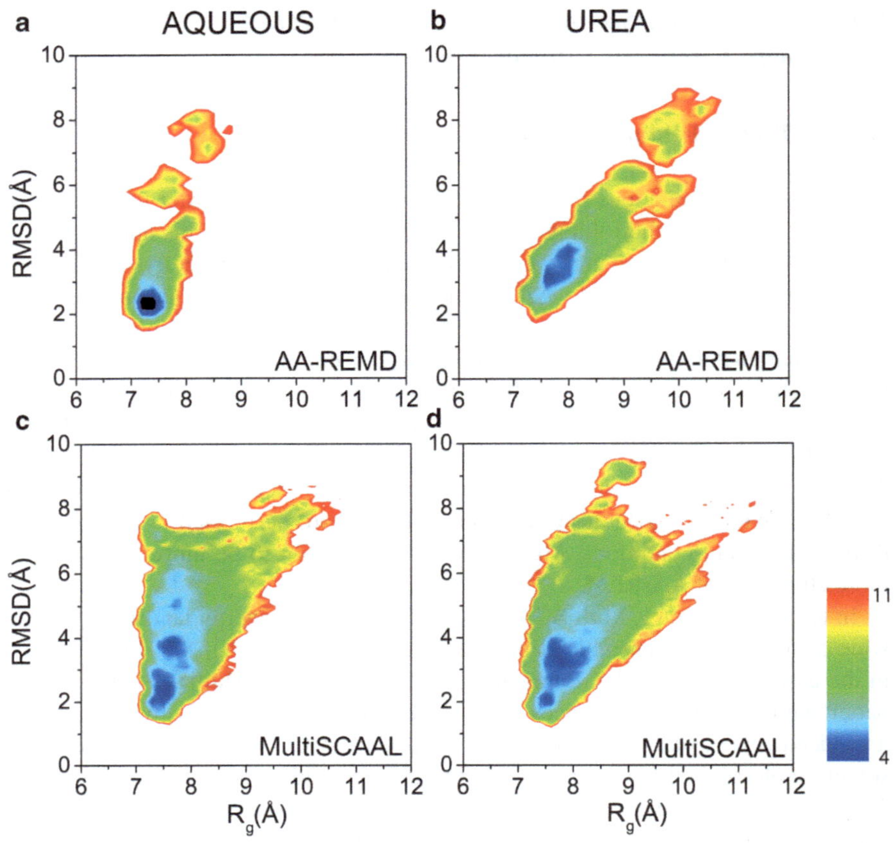

Fig. 8.9 Two–dimensional free energy landscape for Trp-cage as a function of the radius of gyration (R_g) and the root-mean-square-deviation (RMSD) under (**a**, **c**) aqueous and (**b**, **d**) urea conditions based on two different simulation schemes at 300 K: (**a**, **b**) simulations using AA-REMD; (**c**, **d**) simulations using MultiSCAAL. The free energy is *colored* by k_BT

in crowded environment [69]. This was done by developing a unique multi-scale solution of charges computed from quantum chemistry, together with SCAAL protein reconstruction, SCM coarse-grained molecular simulations, and statistical physics, to represent the charge distribution in the transition from apoCaM to holoCaM upon calcium binding. The simulations were performed at different salt concentrations, different volume fraction of crowding agents, and a combination of both. These simulations showed that increased levels of macromolecular crowding, in addition to calcium binding and ionic strength typical of that found inside cells, can impact the conformation, secondary structure and the EF hand orientation of CaM [69].

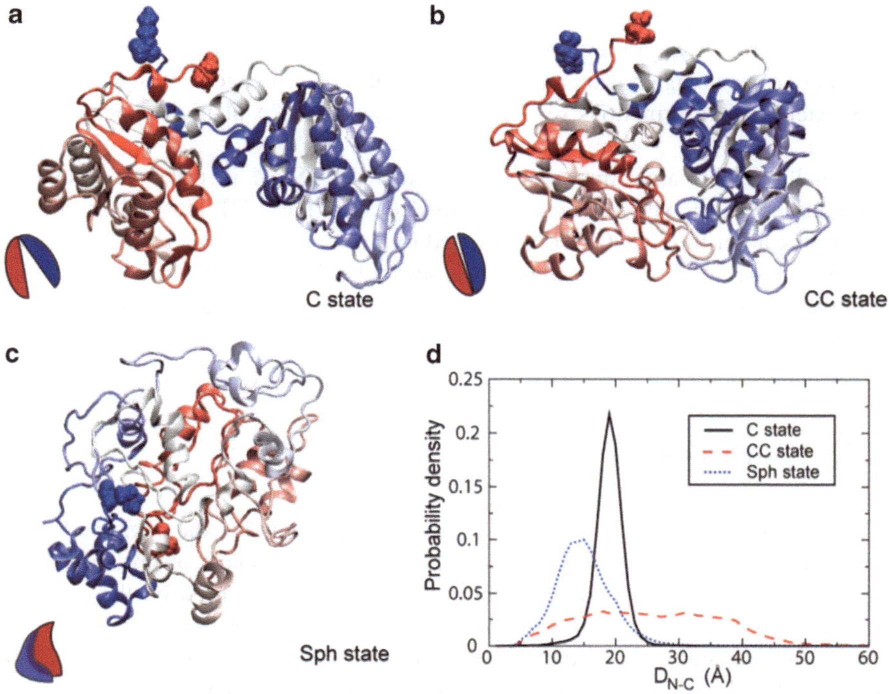

Fig. 8.10 Structural characteristics of the dominant compact ensemble structures of PGK in cartoon representation. (**a**) Crystal state C, (**b**) Collapsed crystal state CC and (**c**) Spherical state Sph. The coloring of each protein model ranges from N-terminus (*red*) to C-terminus (*blue*). The -N and -C termini are represented with Van der Waals spheres. The schematic representation at the *bottom left* of each panel is to address a simplistic view of the arrangement of the N- and C-lobes in each conformation. (**d**) The probability distribution of the distance between N- and C- termini of the three dominant structures of PGK under the condition when each prevails in the simulations. C state (*solid black*), Collapsed Crystal CC (*dashed red*) and Spherical Sph (*dotted blue*)

8.4.3 Other Applications

SCM coarse-grained molecular dynamics simulations were used to investigate the effect of macromolecular crowding on the folding and enzymatic activity of phosphoglycerate kinase (PGK) [70]. Experiments suggested that PGK in a crowded medium adopts conformations that are not seen in dilute conditions. In addition, crowding was shown to enhance the enzymatic activities of PGK by more than 15 times. In the SCM coarse-grained molecular simulations, three possible compact ensembles of PGK were identified as shown in Fig. 8.10. These results suggest that rather than undergoing a hinge motion, the ADP and substrate sites at the inner parts of two domains of PGK are already located in proximity in compact form under crowded or even *in vivo*.

SCM coarse-grained simulations were also used to investigate the competing effects of crowding and urea on the folding of protein Trp-cage [71]. This study shows that crowding enhancement of folding rates of Trp-cage is most pronounced for extended conformations of Trp-cage in the presence of high concentrations of urea.

Finally, a new algorithm was recently added to SCM in order to extend its capabilities to deal with more realistic crowded conditions [72]. This self-assembled clustering algorithm (CGCYTO) was used to produce a polydisperse (PD) coarse-grained model for *E. coli* cytoplasm. It is shown by SCM coarse-grained molecular simulations that the folding temperature of a test protein apoazurin in a PD cytoplasm model is $\sim 5°$ greater than that in a Ficoll 70 model [72].

8.5 Conclusion

This chapter presented some of the recent developments in coarse-grained (CG) molecular dynamics techniques when it applies to the problem of protein folding in varying crowding and solvent conditions. We mainly focused on the evolving (Side-chain-C_α Model, [15]) SCM-based techniques. SCM molecular simulations were used to study the protein folding dynamics in crowded conditions that mimic the highly condensed cellular cytoplasm. In these studies, the computational efficiency of simulations based on a minimalist model is utilized to incorporate the additional crowding particles. Several studies have used the SCM simulations to model different types, shapes, and concentration of crowders. SCM simulations achieved a great success in explaining and predicting the behavior of protein folding dynamics in crowded medium as can be seen in the example studies discussed in this chapter.

Additional techniques can extend the capabilities of a CG model to address different types of environmental conditions such as solvent, denaturants, and ions. Several examples of these techniques were presented in this chapter in addition to some applications of SCM-based simulations. A growing trend now in computational studies is to design a multi-scale approach to simulate biophysical systems. This approach tries to combine the advantages of both the more detailed atomic simulations with the efficiency of coarse-grained ones. The chapter presented an example of these multi-scale approaches, MultiSCAAL. MultiSCAAL uses CG simulations in order to speed up and expand the sampling of the all-atom protein folding landscape. All the techniques and the examples discussed here show that well-designed coarse-grained molecular simulations can be a great tool in addressing complicated problems such as protein folding. With the new emerging techniques and with the help of coarse-grained models we can achieve significant progress in understanding complicated systems, especially when they are coupled with experimental methods or with higher resolution (All-atom or Quantum) simulations.

References

1. Rahman A (1964) Correlations in the motion of atoms in liquid argon. Phys Rev 136:405–411
2. McCammon A, Gelin B, Karplus M (1977) Dynamics of folded proteins. Nature 267:585–590
3. Buchner GS, Murphy RD, Buchete NV, Kubelka J (2011) Dynamics of protein folding: probing the kinetic network of folding-unfolding transitions with experiment and theory. Biochim Biophys Acta-Proteins Proteomics 1814:1001–1020
4. Scheraga HA, Khalili M, Liwo A (2007) Protein-folding dynamics: overview of molecular simulation techniques. Annu Rev Phys Chem 58:57–83
5. Dror RO, Dirks RM, Grossman JP, Xu HF, Shaw DE (2012) Biomolecular simulation: a computational microscope for molecular biology. Annu Rev Biophys 41, D. C. Rees, Ed., ed Palo Alto: Annual Reviews, pp 429–452
6. Stagg L, Zhang SQ, Cheung MS, Wittung-Stafshede P (2007) Molecular crowding enhances native structure and stability of alpha/beta protein flavodoxin. Proc Natl Acad Sci USA 104: 18976–18981
7. Homouz D, Perham M, Samiotakis A, Cheung MS, Wittung-Stafshede P (2008) Crowded, cell-like environment induces shape changes in aspherical protein. Proc Natl Acad Sci USA 105: 11754–11759
8. van den Berg B, Wain R, Dobson CM, Ellis RJ (2000) Macromolecular crowding perturbs protein refolding kinetics: implications for folding inside the cell. EMBO J 19:3870–3875
9. Ai X, Zhou Z, Bai Y, Choy W-Y (2006) 15N NMR spin relaxation dispersion study of the molecular crowding effects on protein folding under native conditions. J Am Chem Soc 128:3916–3917
10. Charlton LM, Barnes CO, Li C, Orans J, Young GB, Pielak GJ (2008) Residue-level interrogation of macromolecular crowding effects on protein stability. J Am Chem Soc 130:6826–6830
11. Sasahara K, McPhie P, Minton AP (2003) Effect of dextran on protein stability and conformation attributed to macromolecular crowding. J Mol Biol 326:1227–1237
12. Kozer N, Kuttner YY, Haran G, Schreiber G (2007) Protein-protein association in polymer solutions: from dilute to semidilute to concentrated. Biophys J 92:2139
13. Snoussi K, Halle B (2005) Protein self-association induced by macromolecular crowding: a quantitative analysis by magnetic relaxation dispersion. Biophys J 88:2855–2866
14. Rivas G, Fernández JA, Minton AP (2001) Direct observation of the enhancement of noncooperative protein self-assembly by macromolecular crowding: indefinite linear self-association of bacterial cell division protein FtsZ. Proc Natl Acad Sci USA 98:3150–3155
15. Cheung MS, Finke JM, Callahan B, Onuchic JN (2003) Exploring the interplay between topology and secondary structural formation in the protein folding problem. J Phys Chem B 107:11193–11200
16. Samiotakis A, Homouz D, Cheung MS (2010) Multiscale investigation of chemical interference in proteins. J Chem Phys 132:175101
17. Anfinsen CB, Haber E, Sela M, White FH (1961) Kinetics of formation of native ribonuclease during oxidation of reduced polypeptide chain. Proc Natl Acad Sci USA 47:1309–1314
18. Levinthal C (1968) Are there pathways for protein folding? J Chim Phys Phys-Chim Biol 65:44
19. Dill KA, Ozkan SB, Shell MS, Weikl TR (2008) The protein folding problem. Annu Rev Biophys 37:289
20. Onuchic JN, Luthey-Schulten Z, Wolynes PG (1997) Theory of protein folding: the energy landscape perspective. Annu Rev Phys Chem 48:545–600
21. Leopold PE, Onuchic JN, Montal M (1992) Protein folding funnels: a kinetic approach to the sequence-structure relationship. Proc Natl Acad Sci USA 89:8271–8275
22. Dill KA, Chan HS (1997) From Levinthal to pathways to funnels. Nat Struct Biol 4:10–19
23. Bryngelson JD, Wolynes PG (1987) Spin-glasses and statistical mechanics of protein folding. Proc Natl Acad Sci USA 84:7524–7528
24. Go N (1983) Theoretical studies of protein folding. Annu Rev Biophys Bioeng 12:183–210

25. Baker D (2000) A surprising simplicity to protein folding. Nature 405:39–42
26. Socci ND, Onuchic JN (1994) Folding kinetics of protein like heteropolymers. J Chem Phys 101:1519–1528
27. Taketomi H, Ueda Y, Gō N (1975) Studies on protein folding, unfolding and fluctuations by computer simulations. Int J Pept Protein Res 7:445–459
28. Dill KA (1985) Theory for the folding and stability of globular proteins. Biochemistry 24:1501–1509
29. Ueda Y, Taketomi H, Gō N (1978) Studies on protein folding, unfolding, and fluctuations by computer simulation. II. A. Three-dimensional lattice model of lysozyme. Biopolymers 17:1531–1548
30. Clementi C, Nymeyer H, Onuchic JN (2000) Topological and energetic factors: what determines the structural details of the transition state ensemble and "en-route" intermediates for protein folding? An investigation for small globular proteins. J Mol Biol 298:937–953
31. Thirumalai D, Guo Z (1995) Nucleation mechanism for protein folding and theoretical predictions for hydrogen-exchange labeling experiments. Biopolymers 35:137–140
32. Wolynes PG, Onuchic JN, Thirumalai D (1995) Navigating the folding routes. Science 267:1619–1620
33. Shea J-E, Onuchic JN, Brooks CL III (1999) Exploring the origins of topological frustration: design of a minimally frustrated model of fragment B of protein A. Proc Natl Acad Sci USA 96:12512–12517
34. Betancourt MR, Thirumalai D (1999) Pair potentials for protein folding: choice of reference states and sensitivity of predicted native states to variations in the interaction schemes. Protein Sci 8:361–369
35. Miyazawa M, Jernigan RL (1985) Estimation of interresidue contact energies from protein crystal structures: quasi-chemical approximation. Macromolecules 18:534–552
36. Cheung MS, Klimov D, Thirumalai D (2005) Molecular crowding enhances native state stability and refolding rates of globular proteins. Proc Natl Acad Sci USA 102:4753–4758
37. Cheung MS, Thirumalai D (2006) Nanopore-protein interactions dramatically alter stability and yield of the native state in restricted spaces. J Mol Biol 357:632–643
38. Cheung MS, Chavez L, Onuchic JN (2004) The energy landscape for protein folding and possible connections to functions. Polymer 45:547–555
39. Hyeon C, Thirumalai D (2011) Capturing the essence of folding and functions of biomolecules using coarse-grained models. Nat Commun 2:487
40. Veitshans T, Klimov D, Thirumalai D (1997) Protein folding kinetics: timescales, pathways, and energy landscapes in terms of sequence-dependent properties. Fold Des 2:1–22
41. Kolinski A, Skolnick J (1992) Discretized model of proteins. I. Monte Carlo study of cooperativity in homopolypeptides. J Chem Phys 97:9412–9426
42. Sippl MJ (1990) Calculation of conformational ensembles from potentials of mean force an approach to the knowledge-based prediction of local structures in globular proteins. J Mol Biol 213:859–883
43. Sope AK (1996) Empirical potential Monte Carlo simulation of fluid structure. Chem Phys 202:295–306
44. Reith D, Putz M, Muller-Plathe F (2003) Deriving effective mesoscale potentials from atomistic simulations. J Comput Chem 24:1624–1636
45. Betancourt MR, Omovie SJ (2009) Pairwise energies for polypeptide coarse-grained models derived from atomic models. J Chem Phys 130:195103
46. Makowski M, Liwo A, Makowska MKJ, Scheraga HA (2007) Simple physics-based analytical formulas for the potentials of mean force for the interaction of amino acid side chains in water. 2. Tests with simple spherical systems. J Phys Chem B 111:2917–2924
47. Debye P, Hückel E (1923) The theory of electrolytes. I. Lowering of freezing point and related phenomena. Physikalische Zeitschrift 24:185–206
48. Gront D, Kmiecik S, Kolinski A (2007) Backbone building from quadrilaterals: a fast and accurate algorithm for protein backbone reconstruction from alpha carbon coordinates. J Comput Chem 28:1593–1597

49. Canutescu A, Shelenkov A, Dunbrack R (2003) A graph-theory algorithm for rapid protein side-chain prediction. Protein Sci 12:2001–2014
50. Rotkiewicz P, Skolnick J (2008) Fast procedure for reconstruction of full-atom protein models from reduced representations. J Comput Chem 29:1460–1465
51. Heath AP, Kavraki LE, Clementi C (2007) From coarse-grain to all-atom: toward multiscale analysis of protein landscapes. Proteins Struct Funct Bioinform 68:646–661
52. Frenkel D, Smit B (2001) Understanding molecular simulation: from algorithms to applications. Academic Press, San Diego, CA
53. van den Berg B, Ellis RJ, Dobson CM (1999) Effects of macromolecular crowding on protein folding and aggregation. EMBO J 18:6927–6933
54. Rivas G, Ferrone F, Herzfeld J (2004) Life in a crowded world. EMBO Rep 5:23–27
55. Record MT, Courtenay ES, Cayley S, Guttman HJ (1998) Biophysical compensation mechanisms buffering E. coli protein-nucleic acid interactions against changing environments. Trends Biochem Sci 23:190–194
56. Ellis RJ, Minton AP (2003) Cell biology – join the crowd. Nature 425:27–28
57. Minton AP (2005) Models for excluded volume interaction between an unfolded protein and rigid macromolecular cosolutes: macromolecular crowding and protein stability revisited. Biophys J 88:971–985
58. Zhou HX, Dill KA (2001) Stabilization of proteins in confined spaces. Biochemistry 40: 11289–11293
59. Perham M, Stagg L, Wittung-Stafshede P (2007) Macromolecular crowding increases structural content of folded proteins. FEBS Lett 581:5065–5069
60. Eicken C, Sharma V, Klabunde T, Lawrenz MB, Hardham JM, Norris SJ et al (2002) Crystal structure of Lyme disease variable surface antigen VlsE of *Borrelia burgdorferi*. J Biol Chem 277:21691–21696
61. Sanbonmatsu KY, Garcia AE (2002) Structure of Met-enkephalin in explicit aqueous solution using replica exchange molecular dynamics. Proteins 46:225–234
62. Sugita Y, Okamoto Y (1999) Replica-exchange molecular dynamics method for protein folding. Chem Phys Lett 314:141–151
63. Kumar S, Bouzida D, Swendsen RH, Kollman PA, Rosenberg JM (1992) The weighted histogram analysis method for free-energy calculations on biomolecules I. The method. J Comput Chem 13:1011–1021
64. Chodera JD, Swope WC, Pitera JW, Seok C, Dill KA (2007) Use of the weighted histogram analysis method for the analysis of simulated and parallel tempering simulations. J Chem Theory Comput 3:26–41
65. Homouz D, Stagg L, Wittung-Stafshede P, Cheung MS (2009) Macromolecular crowding modulates folding mechanism of alpha/beta protein apoflavodoxin. Biophys J 96:671–680
66. Mok KH, Kuhn LT, Goez M, Day I, Lin J, Andersen NH et al (2007) A pre-existing hydrophobic collapse in the unfolded state of an ultrafast folding protein. Nature 447:106–109
67. Means AR, Dedman JR (1980) Calmodulin – an intracellular calcium receptor. Nature 285:73–77
68. Homouz D, Sanabria H, Waxham MN, Cheung MS (2009) Modulation of calmodulin plasticity by the effect of macromolecular crowding. J Mol Biol 391:933–943
69. Wang Q, Liang K-C, Czader A, Waxham MN, Cheung MS (2011) The effect of macromolecular crowding, ionic strength and calcium binding on calmodulin dynamics. PLoS Comput Biol 7:e1002114
70. Dhar A, Samiotakis A, Ebbinghaus S, Nienhaus L, Homouz D, Gruebele M et al (2010) Structure, function, and folding of phosphoglycerate kinase are strongly perturbed by macromolecular crowding. Proc Natl Acad Sci USA 107:17586–17591
71. Samiotakis A, Cheung MS (2011) Folding dynamics of Trp-cage in the presence of chemical interference and macromolecular crowding. I. J Chem Phys 135:175101-175101-16
72. Wang Q, Cheung M (2012) A physics-based approach of coarse-graining the cytoplasm of E. coli Biophys J 102:2353–2361

Chapter 9
Simulating the Peptide Folding Kinetic Related Spectra Based on the Markov State Model

Jian Song and Wei Zhuang

Abstract Optical spectroscopic tools are used to monitor protein folding/unfolding dynamics after a fast triggering such as the laser induced temperature jump. These techniques provide new opportunities for comparison between theory and simulations and atom-level understanding protein folding mechanism. However, the direct comparison still face two main challenges: a gap between folding relevant timescales (microseconds or above) and length of molecular dynamics simulations (typically tens to hundreds of nanoseconds), and difficulty in directly calculating spectroscopic observables from simulation configurations. Markov State Model (MSM) approach is one of the most powerful means which can increase simulations timescale up to microsecond or even millisecond. We address progress on modeling infrared and fluorescence spectroscopic signals of temperature jump induced unfolding dynamics for a few small proteins. The harmoniousness between experiment and theoretical can both improve our understanding of protein folding mechanisms and provide direct validation of those theoretical models.

Keywords Protein folding/unfolding • Markov state model • Folding intermediates • Two dimensional infrared spectroscopy • Fluorescence spectroscopy

---

J. Song
Department of Physics, HeNan Normal University, XinXiang, 453003, China

State Key Lab of Molecular Reaction Dynamics, Dalian Institute of Chemical Physics, Dalian, 116023, China

W. Zhuang (✉)
State Key Lab of Molecular Reaction Dynamics, Dalian Institute of Chemical Physics, Dalian, 116023, China
e-mail: wzhuang@dicp.ac.cn

K. Han et al. (eds.), *Protein Conformational Dynamics*, Advances in Experimental Medicine and Biology 805, DOI 10.1007/978-3-319-02970-2_9,

Abbreviations

2DIR	Two-dimensional infrared spectroscopy
CD	Circular dichroism
CGF	Gaussian fluctuation
DFT	Density functional theory
FE	Frenkel exciton
FTIR	Fourier transform infrared spectroscopy
GE	Generalized ensemble
GLDP	Glycine dipeptide
IR	Infrared
MD	Molecular dynamics
MSM	Markov state model
MSMs	Markov state models
NEP	Nonlinear exciton propagation
NMA	N-methyl acetamide
NMR	Nuclear magnetic resonance
QM/MM	Quantum mechanics and molecular mechanics
REM	Replica exchange method
RHF	Restricted Hartree-Fock
SHC	Superlevel-set hierarchical clustering
TCC	Transition charge coupling
TC-n	Typical conformation n
TDC	Transition dipole coupling
T-jump	Temperature jump

9.1 Introduction

Understanding the mechanism of protein folding and unfolding is always one of significant tasks in life sciences [1, 9, 12, 14, 15]. Amino acids interact with each other to produce a well-defined three-dimensional structure and carry out unique functions. Incorrectly folded proteins can induce many serious and fatal neurodegenerative diseases, such as mad cow disease, Alzheimer's disease and Creutzfeldt-Jakob disease [10, 31, 39]. In the past decades, tremendous efforts in theoretical and experimental fields have been devoted into understanding the protein folding mechanisms. However, many aspects of this issue remains unclear even for the small, single domain peptides [40].

X-ray crystallography [2, 30] has been used extensively to determine three-dimensional atomic-resolution protein structure. Exploring the folding dynamics usually employs the spectroscopy techniques, such as nuclear magnetic resonance (NMR) [4, 28, 47], circular dichroism (CD), Infrared (IR) spectroscopy, and fluorescence spectroscopy, to monitor conformational change in adequate temporal and spatial resolution [7, 13, 32]. Spectroscopy signals are, however, the eigenspace reflections of the structure and dynamics in the real space. Understanding the physics behind these signals is usually a nontrivial work, especially for complex molecular systems such as the protein aqueous solutions.

Theoretical approaches such as Molecular Dynamics (MD) generate the molecular level pictures of protein folding thermodynamics and kinetics. A proper combination of the spectroscopic experiments and the MD simulation thus has the potential to significantly improve our understanding of the protein folding events. However, direct comparison between theory and experiment for protein folding is still difficult due to two major challenges. First, a timescale gap exists between experimental protein folding times and capability of MD simulations. Even for the small and single domain protein and peptides, their folding time are usually microsecond or above [11, 35]. The capability of MD simulation nowaday is, however, usually submicrosecond. Second, even with sufficiently long simulations, it is still a difficult task to carry out a direct comparison between MD simulations and experimental results those are often optical spectroscopic observables.

In recent years, a number of research groups have been working toward the direction to bridge the gap between MD simulations and experimental spectroscopic investigations in order to achieve a better understanding of protein folding. Coarse-grained simulations [45] with simplified representations of proteins are a natural solution to bridge the timescale gap. But they sacrifice the atomistic details needed for the simulation of the spectroscopic signals comparable with the experiment. Another solution to fulfil the gap is to develop algorithms that can construct models from short simulations to predict long timescale dynamics for protein folding. Markov State Model (MSM) approach [5,6,8,25,38,42] is one of the most powerful means that have recently shown success in investigating protein folding kinetic at microsecond or even millisecond timescales.

More meaningful comparison between the theoretical and experimental results requires the modeling of spectra based on the MD simulation trajectory ensembles, which is in general complicated due to the entangled and congested nature of the optical transitions. The hybrid quantum mechanics and molecular mechanics method is commonly used to model the spectroscopic lineshape in the protein-solvent systems. For multi-chromophoric systems with weak couplings between the units, one can adapt the Frenkel exciton model for the Hamiltonian construction, which is widely used in optical response calculation of biological system.

In the current manuscript, we introduce our effort in modeling the temperature jump (T-jump) triggered peptide long time unfolding related Infrared, Two-dimensional infrared spectroscopy (2DIR) and fluorescence spectra based on the MSM approach. We will first briefly describe the MSM approach for generating the long time peptide unfolding pathways triggered by the T-jump technique, then discuss how to simulate the related Infrared, 2DIR and fluorescence spectra, respectively.

9.2 Long Time-Scale Molecular Dynamic Simulation: Markov State Models

In this section, we only give a brief introduction on the basic idea of MSM. For the complete and thorough discussions on this method, one should read those original manuscripts [6,8,19,25,55]. In the Markov State Models (MSMs), the phase space is partitioned into a group of metastable states, the intra-state transition is designed to be much faster than the inter-state transitions, so that the kinetics can be considered as Markovian, and described using a memoryless master equation (9.1),

$$P(n\Delta t) = [T(\Delta t)]^n P(0) \tag{9.1}$$

in which $P(n\Delta t)$ is the state populations at time $n\Delta t$ and T is the transition probability matrix. Δt, the time interval for transitions, is the lag time. T is calculated by normalizing the number of transitions between each pair of states after a lag time in the simulation database [8].

Building MSMs with good state decomposition is challenging. Typically, a two-step procedure is adopted: First, the massive number of MD conformations is divided into a large set of microstates by geometrical similarity. These microstates have to be fine enough so that they do not combine kinetically separated regions of the phase space. Second, a kinetic clustering is performed to group these microstates into a number of metastable states so that transitions between microstates within the same metastable state are much faster than transitions between different metastable states. MSMs can also be considered as a data-mining tool to generate comprehensive folding models from massive simulation datasets. One of the major challenges for the MSM is to ensure that all the relevant conformation states have been visited. This issue may be alleviated by Generalized Ensemble (GE) algorithms [22, 23], which can enhance conformational sampling by inducing a random walk in Temperature or Hamiltonian space. In recent years, GE algorithms especially Replica Exchange Method (REM) [41] have been widely applied in protein folding studies. Huang et al. [18] have used non-equilibrium GE simulations to explore the phase space, and then seed short simulations at constant temperature from GE conformations to construct MSMs to obtain both equilibrium thermodynamics and kinetics. Adaptive sampling, allowing one to use an initial MSM to decide where to run new simulations, is another solution to alleviate the sampling issue. In the adaptive sampling, new simulations are started from those states that contribute most to the statistical uncertainty in kinetic properties of interest calculated from the initial MSM. It has been shown that performing adaptive sampling and constructing MSMs iteratively can quickly yield a good model.

In the following, we discuss the modeling of the IR and 2DIR spectra related to the model peptide trpzip2 T-jump unfolding based on extended MSMs generated by trajectory ensembles.

9.3 Modeling Infrared (IR) and Two Dimensional Infrared (2DIR) Spectroscopies

IR spectroscopy is widely used to monitor local environments and dynamics in proteins. Vibrational transitions are sensitive to the local structure and bonding environment, thus are ideal for distinguishing between various secondary structural motifs and monitoring the effects of changing environments through hydrogen bonding and electrostatic focus. For example, α-helix regions absorb near 1,650 cm^{-1}, while β-sheets absorb at 1,620 and 1,675 cm^{-1}. IR pulses can provide 50 fs snapshots of dynamical events. The 1,600 − 1,700 cm^{-1} amid I band which originates from the stretching motion of the C = O peptide bond (coupled to in-phase N–H bending and C–H stretching) is particularly useful for structural studies, since it has a strong transition dipole moment and is spectrally well separated from other vibrational modes. The up to 20 cm^{-1} variation of the amid I frequency with secondary structure and conformation, is widely used as a marker in polypeptide and protein structure determination.

Coherent IR techniques, which record the molecular response to sequences of pulses, provide a multidimensional view of protein structure. 2DIR is one of a rapidly expanding class of new ultrafast coherent vibrational spectroscopies that are finding broad use in studies of molecular structure and dynamics that probe peptides, proteins, DNA, chemical exchange kinetics, hydrogen bonding, and rapidly initiated chemical reactions. The first frequency-frequency 2DIR measurement was carried out by Hamm and Hochstrasser, who employed a pump-probe technique with two IR pulses with a narrow (ca. 10 cm^{-1}) pump and a broad (130 cm^{-1}) probe pulse. A heterodyne-detected 2DIR experiment (Fig. 9.1) involves the interaction

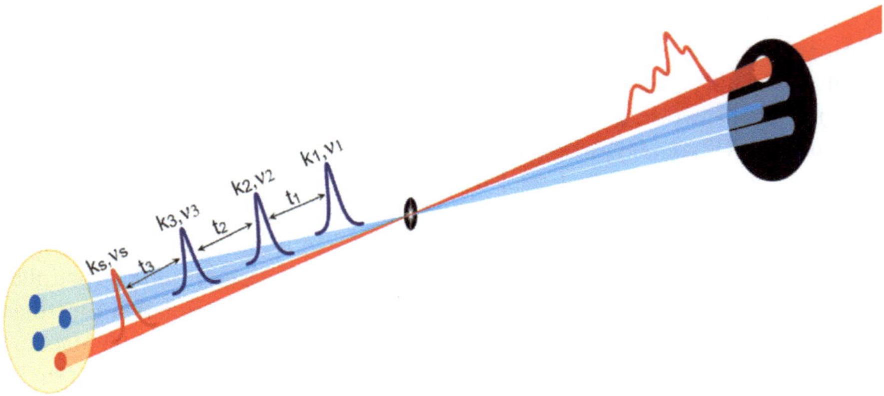

Fig. 9.1 Schematic experimental setup for a heterodyne detected four-wave mixing experiment. Signals are recorded as the function of three time delays and displayed as 2D correlation plots involving the double fourier transform of two time delays, holding the third fixed

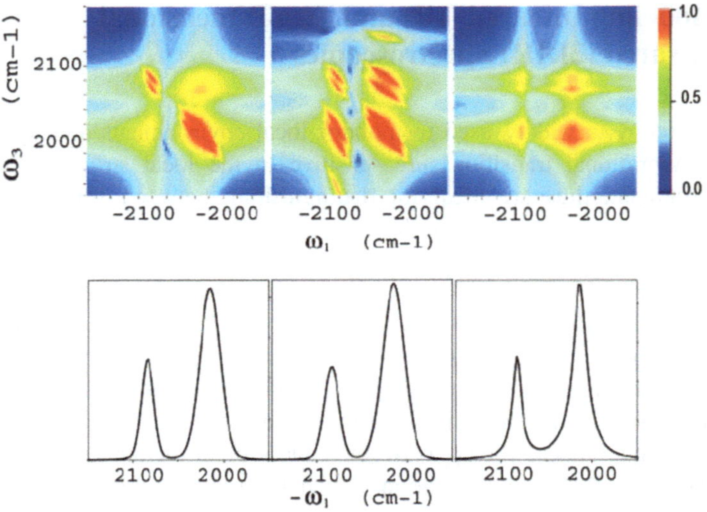

Fig. 9.2 (*Top*) 2D spectra of two coupled vibrations. The frequency fluctuations of the two modes are slow and anticorrelated in the *left*, slow and correlated in the *middle*, fast and anticorrelated in the *right* (Adapted from Ref. [46]). (*Bottom*) Linear absorptions for the three models

of three laser pulses with wave vectors $\mathbf{k}_1$, $\mathbf{k}_2$, $\mathbf{k}_3$, (in chronological order) with the peptide. A signal field is then generated coherently in the directions: $\mathbf{k}_4 = \pm\mathbf{k}_1 \pm \mathbf{k}_2 \pm \mathbf{k}_3$. This signal field is detected by interference with a 4th pulse with the same wavevector $\mathbf{k}_4$. The signal $\mathbb{S}(t_3, t_2, t_1)$ is defined as the intensity difference of the pulse before and after the interaction with the sample, and depends parametrically on the time intervals between pulses t_1, t_2 and t_3. The 2D IR signal is obtained by displaying it as a two-dimensional correlation plot with respect to two of these intervals, say t_1 and t_3, holding the third (t_2) fixed. Such plots are highly oscillatory. For a clearer picture, the signal is double fourier transformed with respect to two time variables to generate a frequency/frequency correlation plot such as $\mathbb{S}(\Omega_1, t_2, \Omega_3)$ where Ω_1 and Ω_3 are the frequency conjugates to t_1 and t_3 (holding t_2 fixed Fig. 9.1). Coupled vibrational modes create new resonances at combinations of single-mode frequencies. The intensities and profiles of these *cross-peaks*, give direct zero-background signatures related to the correlations between transitions. Correlation plots of dynamical events taking place during controlled evolution periods can be interpreted in terms of multipoint correlation functions. These carry considerably more information than the two point correlation functions of linear spectroscopy, and can distinguish between possible models whose 1D responses are virtually identical (see Fig. 9.2).

The most commonly employed method to model IR spectroscopy is the normal mode analysis. However, it is difficult to model IR spectroscopy of large molecular systems such as proteins using the normal mode analysis. Since classical normal mode analysis can not reproduce the high frequency bands accurately while the quantum normal mode analysis is obviously too expensive.

A peptide can be viewed as a chain of beads, connected by amide bonds (O=C–N–H). The amide I vibrations [20] are localized on the backbone peptide bonds, and these excitations have nonoverlaping transition charge densities, well localized in space. These can be described by the Frenkel exciton model [20, 43, 54].

We assume the following form for the vibrational exciton Hamiltonian:

$$\hat{H} = \hat{H}_S + \hat{H}_F, \tag{9.2}$$

where

$$\hat{H}_S = \sum_m \varepsilon_m(Q)\hat{B}_m^\dagger \hat{B}_m + \sum_{mn}^{m\neq n} J_{mn}(Q)\hat{B}_m^\dagger \hat{B}_n - \frac{1}{2}\sum_m \Delta_m(Q)\hat{B}_m^\dagger \hat{B}_m^\dagger \hat{B}_m \hat{B}_m \tag{9.3}$$

is the system Hamiltonian and $\hat{H}_F$ represents the interaction with the optical field, $E(t)$:

$$\hat{H}_F = -\mathbf{E}(t) \cdot \sum_m \mu_m(\hat{B}_m^\dagger + \hat{B}_m), \tag{9.4}$$

$\hat{B}_m^\dagger$ ($\hat{B}_m$) is the creation (annihilation) operator for the m'th amide I mode, localized within the amide unit (O=C–N–H), with frequency ε_m, anharmonicity Δ_m and transition dipole moment $\boldsymbol{\mu}_m$. These operators satisfy the Bose commutation relations $[\hat{B}_m, \hat{B}_n^\dagger] = \delta_{mn}$. J_{mn} are the harmonic inter-mode couplings. Diagonal elements of the Hamiltonian matrix give the zero-order local mode frequencies while off-diagonal elements represent their couplings. All parameters of $\hat{H}_S$ fluctuate due to conformational changes of the backbone, as well as solvent and side-chain dynamics. These other degrees of freedom are represented collectively by Q. Starting with the Hamiltonian in cartesian coordinates we define: $\hat{B}_m = \frac{1}{\sqrt{2}}(q_m + ip_m)$, $\hat{B}_m^+ = \frac{1}{\sqrt{2}}(q_m - ip_m)$. Alternatively we can start with a highly anharmonic local Hamiltonian, calculate the eigenstates and use a bosonization procedure to bring the Hamiltonian to this form. This can account for local anharmonicities to all orders. Only the nonlocal couplings are expanded to quartic order. The Hamiltonian matrix for large globular proteins may now be constructed using parameters obtained from electronic structure calculations performed on small segments, (which constitute the individual chromophores) [20, 43, 54]. It then becomes possible to have a fairly accurate description of high frequency vibrational Hamiltonians for large systems.

A full microscopic simulation of the lineshape will require the construction of a Hamiltonian at each joint along the MD trajectory with around 100,000–500,000 snapshots. The repeated electronic structure calculations, even for a single peptide residue in solution, are very expensive.

We can adopt an alternative strategy for the construction of instant vibrational frequency and transition dipole moments which avoids the repeated calculations along the trajectory. The basic idea is to describe the Hamiltonian and transition dipole elements as a function of some fluctuating parameters of the molecular system which can be easily obtained from the MD simulation trajectories.

Several maps which correlate site frequencies with the electrostatic potential (Cho map), electric field (Skinner map) as well as the electric multipole field up to 2nd derivatives of the electric field (Mukamel map) evaluated at or between the atoms in the amide bond have been proposed. These maps were constructed from electronic structure calculations of N-methyl acetamide (NMA). Cho's electric potential map and Skinner's electric-field map assume a linear correlation between the frequency and the electric fields which projected on the such as C, O, N and D atoms. However, these maps focus on individual amide units, and do not take into account coupling between different amides. The Mukamel 's map was constructed by amide I frequency calculations of a single NMA molecule subject to a set of nonuniform multipole electric fields. The amide I frequency was parametrized as a quadratic function of the electric field, its gradients and the second derivatives. A similar approach was later adopted by Knoester who parametrized the frequency with the electric field and gradients at the C, O, N and H atoms. That calculation was carried out using a NMA molecule embedded in a set of electric charge distributions.

The couplings J between amide I vibrations of different peptide units are usually assumed to depend only on the peptide backbone structure and not the electric field. This coupling was first calculated using the transition dipole coupling (TDC) model.

$$J_{m,n} = \frac{0.1A}{\epsilon} \frac{(\boldsymbol{\mu}_m \cdot \boldsymbol{\mu}_n) - 3[\boldsymbol{\mu}_m \cdot \mathbf{e}_{mn}][\boldsymbol{\mu}_n \cdot \mathbf{e}_{mn}]}{r_{mn}^3} \tag{9.5}$$

where $\boldsymbol{\mu}_m$ is the transition dipole in (D Å^{-1} u$^{-1/2}$) units, r_{mn} is the distance between dipoles (in Å), $\mathbf{e}_{mn}$ is the unit vector connecting m and n and $\epsilon = 1$ is the dielectric constant.

Torii and Tasumi had shown that this model fails to describe the coupling of neighboring peptide units. They constructed an *ab initio* map [44] of the coupling as function of the Ramachandran angles between the neighboring peptide units (the Tasumi map). This was done using restricted Hartree-Fock (RHF) electronic structure calculations on an ensemble of glycine dipeptide (GLDP) configurations with a 30° grid of the Ramachandran angles. This type of map should be transferable between different peptides. Stock and Cho had independently derived similar maps with higher level quantum chemistry protocol and a finer (1°) grid. Both the stock and the Cho maps give similar values; half those of the Tasumi's map. Hamm and Woutersen had suggested a transition charge coupling (TCC) model, which extends the TDC model to include higher-order multipoles. The model agreed reasonably with the coupling constant calculated with density functional theory (DFT) on

GLDP. The remaining discrepancy between the DFT calculation and the TCC model can be mainly attributed to through-bond coupling, which cannot be described with an electrostatic model, such as the TCC.

Infrared spectra of molecular vibrations can be simulated as the function of the time correlation functions. The Cumulant expansion of Gaussian Fluctuation (CGF) fluctuation models which may be solved exactly and provides a compact closed form of expressions for the response functions. This model assume diagonal (energy) fluctuations with Gaussian statistics and the transition dipoles do not fluctuate.

Line broadening functions $g_{mn}(t)$ is given by the double time integral of the time correlation function:

$$g_{mn}(t) = \int_0^t d\tau_1 \int_0^{\tau_1} d\tau_2 C_{mn}(\tau_1) \tag{9.6}$$

here:

$$\begin{aligned} C_{mn}(\tau_1, \tau_2) &= \frac{1}{\hbar^2} \langle \Delta H_{ma}(\tau_1) \Delta H_{na}(\tau_2) \rangle \\ &\equiv C'(\tau_{12}) + iC''(\tau_{12}) \end{aligned} \tag{9.7}$$

$\Delta H_{\alpha\beta}(t) = H_{\alpha\beta}(t) - \bar{H}_{\alpha\dot{\beta}}$ represents the fluctuations of the transition frequencies $g_{mn}(t)$ can be conveniently expressed in terms of using the spectral density:

$$\begin{aligned} g_{mn}(t) = &\int_{-\infty}^{\infty} \frac{d\omega}{2\pi} \frac{1 - \cos(\omega t)}{\omega^2} \coth\left(\frac{\hbar\omega}{2k_B T}\right) C''_{mn}(\omega) \\ &+ i \int_{-\infty}^{\infty} \frac{d\omega}{2\pi} \frac{\sin(\omega t) - \omega t}{\omega^2} C''_{mn}(\omega) \end{aligned} \tag{9.8}$$

with

$$C''_{mn}(\omega) = 2 \int_0^{\infty} dt \sin(\omega t) C''_{mn}(t) \tag{9.9}$$

where $C''_{mn}(t)$ is the imaginary part of the two time correlation function.

We first generate the fluctuating eigenspace Hamiltonian along the MD trajectory using the protocol discussed above. We then define a reference Hamiltonian and calculate the fluctuating energy $U_{mn}(t) = H_{mn}(t) - \bar{H}_{mn}$. The next step is to calculate the Fourier transform of $U_{mn}(t)$, $\tilde{U}_{mn}(\omega)$:

$$\tilde{U}_{mn}(\omega) = \int_{-\infty}^{\infty} dt U_{mn}(t) \exp(i\omega t) \tag{9.10}$$

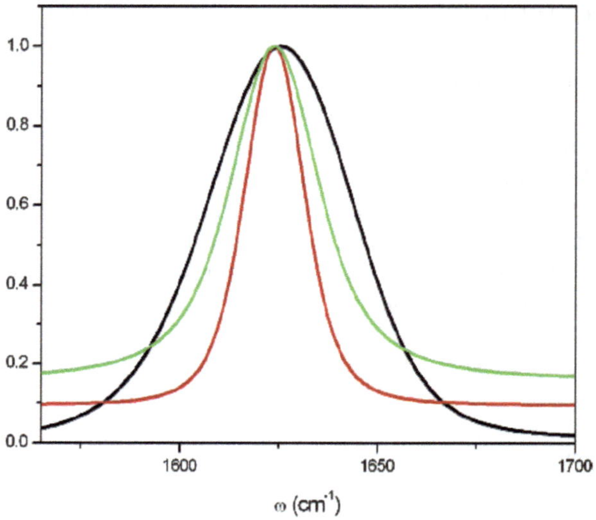

Fig. 9.3 Simulated absorption spectra of the amide I band of NMA in water at room temperature. CGF, excluding lifetime broadening (*green*), FWHM ($19\,\mathrm{cm}^{-1}$), with lifetime broadening (*red*), the FWHM becomes $30\,\mathrm{cm}^{-1}$, which is very close to the experiment ($29\,\mathrm{cm}^{-1}$) [51]. The inhomogeneous SOS simulated signal (*black*) overestimates the linewidth ($52\,\mathrm{cm}^{-1}$)

The correlation function is finally given in the frequency domain:

$$\tilde{C}^{c}_{mn}(\omega) = \frac{1}{2\tau\hbar^2}|\tilde{U}_{mn}(\omega)|^2$$
$$C''_{mn}(\omega) = \frac{\hbar\omega}{2k_bT}\tilde{C}^{c}_{mn}(\omega) \tag{9.11}$$

2τ is the length of the trajectory.

Figure 9.3 shows the simulated CGF linear absorption spectra (give formula) of the amide I band of NMA in water. Neglecting vibrational relaxation is (green line), the Full width at half maximum (FWHM) is $19\,\mathrm{cm}^{-1}$, by adding the experimental lifetime broadening give number (red line), the FWHM becomes $30\,\mathrm{cm}^{-1}$, which is very close to experiment ($29\,\mathrm{cm}^{-1}$).

A commonly used method to trigger the peptide and protein unfolding processes is T-jump, which uses a intense laser pulse to create a significant temperature jump in nanoseconds. This then generates an unstable conformational distribution, the relaxation of this distribution to the equilibrium can be monitored using optical spectroscopic tools such as 2DIR. Even for small peptides with single structural domain, the unfolding time is usually in the microsecond timescale, while the straightforward MD simulations can usually achieve good statistics within several hundred nanoseconds. One possible way to bridge this time gap is to extend MSM method, which is described above and has recently shown success in investigating protein folding kinetic at microsecond or even millisecond timescales, to simulate the T-jump triggered folding events.

We've developed a protocol to extend the MSM for simulating the T-jump triggered peptide unfolding dynamics and calculating the 2DIR signals using the direct nonlinear exciton propagation (NEP) method. We use the small trpzip2 hairpin peptide unfolding dynamics as an example. The whole story is summarized briefly in following, and one can find the details in the original manuscripts [55]. To construct MSMs, we first group each of the two sets of conformations, which are fetched from simulation data, into 2,000 microstates at 300 K (T_1) and 350 K (T_2) using a K-centers clustering algorithm. We then used the superlevel-set hierarchical clustering (SHC) algorithm [19] to lump microstates together in further to construct macrostate MSMs and 17-macrostate and 13-macrostate MSMs at 300 and 350 K are obtained, respectively. We simulate the relaxation dynamics after the T-jump by calculating the evolution of populations of metastable states using Eq. 9.1. 2DIR spectra signals for each of the metastable states have been calculated from the structure ensemble obtained from MSM, and we performed a weighted sum of these spectra signals of different states based on their instantaneous populations to obtain an overall signal.

The equilibrium population distribution for the macrostates calculated from the 17-state MSM at 300 K is presented in Fig. 9.4a. The representative structures for all the states with population are shown in Fig. 9.4b. The simulated linear absorption signal shows two main transitions: the low-frequency, stronger transition peaked at $\sim$1,640 cm^{-1} and the high-frequency, weaker transition peaked at $\sim$1,670 cm^{-1}, which is consistent with the experimental result.

And the signal from NMR structure shows a similar feature as in the MSM with a lower intensity for the high-frequency peak. However, the simulated absorptive (KI+KII) 2DIR signals at 300 K using MSM (left) and NMR structure (right) are different (see in Fig. 9.5). In the MSM result, both peaks have an asymmetric feature with the antidiagonal line width broadening at the red end, which agreed with experimental observation. As the comparison, the NMR structure-based result has the symmetric peaks elongated along the parallel direction. The simulated transient Fourier Transform infrared spectroscopy (FTIR) and 2DIR at the different time points after the T-jump are shown in Fig. 9.6. The signal lineshape change performances with time of both transient FTIR and 2DIR are consistent with the fact that the population shifts from the folded state to other unfolded states.

9.4 Modeling Fluorescence Spectroscopy

Fluorescence spectroscopy always remains one of the most important and successful tools for understanding the protein folding. The changes of fluorescence wavelength, intensity, lineshape and lifetime of the intrinsic or extrinsic fluorophores are monitored and assigned to different folding stages [49, 50]. Further unraveling of the fluorescence signals with more molecular level insight usually requires the help from theoretical modeling [3, 21, 37], which is in general complicated due to the entangled and congested nature of the fluorescence transitions as well as the difficulties in simulating the protein folding processes.

The hybrid quantum mechanics and molecular mechanics (QM/MM) method is commonly used to model the fluorescence lineshape in the protein-solvent systems

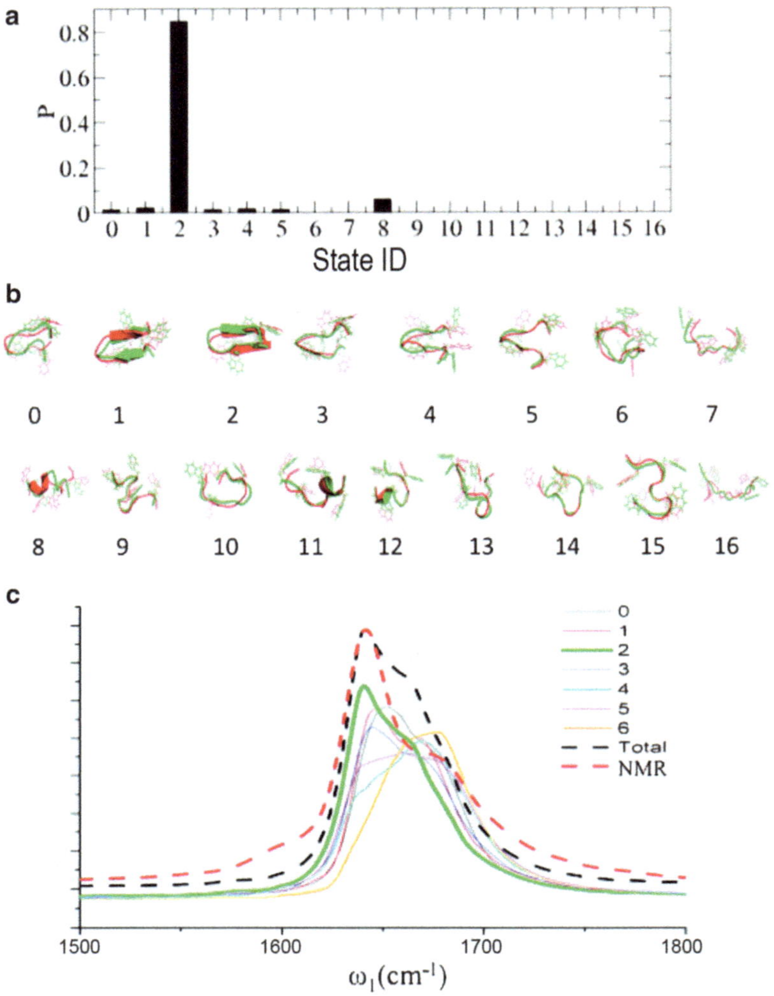

Fig. 9.4 (**a**) Equilibrium populations of 17-state at 300 K computed using MSMs for the trpzip2 peptide. (**b**) Representative structures of 17 metastable states (**c**) The overall 300 K equilibrium linear absorption FTIR signal (*black dash*) and the contributions (*solid lines*) from states with population >1 %, simulated using MSM data. The overall signal is scaled by 1.2 for a better presentation. *Green solid line* gives the contribution from the most significant macro state (state 2). *Red dash line* gives the simulated absorption signal based on the NMR structure, it is scaled to have the same maximum as the MSM overall signal for a better comparison

[17, 27, 33]. Many models can be employed to count in the vibrational feature of the fluorescence spectra for single chromophores, such as the Brownian oscillator model, the displaced harmonic oscillator model and the anharmonic oscillator model [24, 36, 48, 52]. Huang-Rhys factor gained by theoretical Franck-Condon analysis is introduced to evaluate the vibronic coupling. For large molecules, however, it often becomes very difficult to calculate the fluorescence by the quantum chemistry

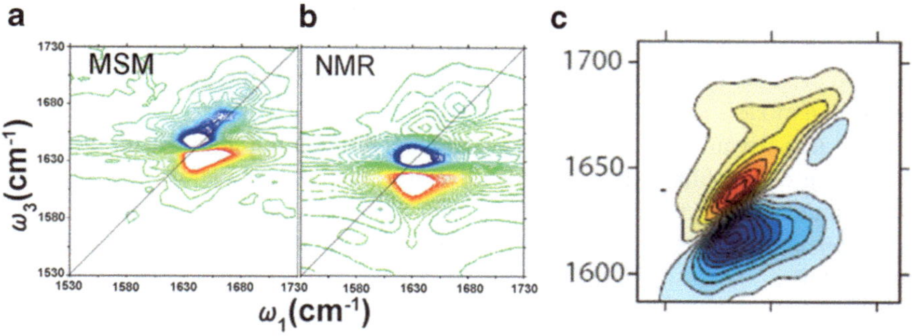

Fig. 9.5 Simulated absorptive 2DIR spectra for the trpzip2 peptide at 300 K, computed from the MSM and nanosecond MD simulations starting from the NMR structure are shown in (**a**) and (**b**) respectively. In (**c**), the experimental results [34] at 298 K are shown

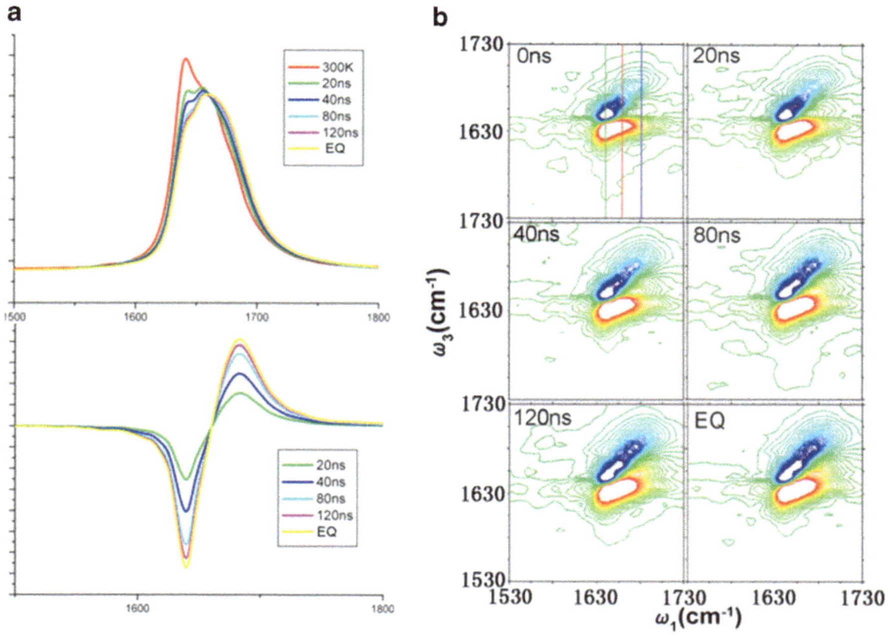

Fig. 9.6 Simulated Transient FTIR and 2DIR at the different time points after the T-jump are shown in (**a**) and (**b**) respectively. For (**a**), the plots on the *lower panel* are the difference spectra with the signal before the T-jump at 300 K

calculation for the whole molecule. For multi-chromophoric systems with weak intermolecular interaction, an alternative and effective approach is to use the Frenkel exciton model and plays an important role in understanding processes concerning the transfer of excitation energy in biological systems. So it is widely used in linear optical response calculation [26, 29]. In these systems, the intramolecular interactions are much stronger than the intermolecular forces. The Hamiltonian of whole system may be written as a sum of individual excitations and their interactions and exciton states are linear combinations of localized excited states. How to evaluate the exciton-exciton coupling accurately is the most important task in exciton model.

The non-trivial work required in sampling the configuration distribution and long timescale folding kinetics adds an extra layer of complexity in simulating the peptide folding fluorescence. Here we try to show the vibrationally-resolved tryptophan band lineshape of trpzip2's fluorescence spectral simulation based on the MSMs.

The trpzip2 system is treated as an tryptophan aggregate, and described by effective Hamiltonians. Considering aggregate consisting of p monomers, and neglecting the weak interaction in the electronic ground state $\left|\phi_1^0, \phi_2^0, \cdots, \phi_p^0\right\rangle = |0\rangle$, one can obtains the ground state Hamiltonian for p-monomer aggregate as

$$H^g(x_1, \cdots, x_p) = |0\rangle \left[\sum_{n=1}^{p} h_n^0(x_n)\right] \langle 0| \tag{9.12}$$

Here h_n^0 is the ground state Hamiltonian of monomer n. The wavefunction for the single excitation state is

$$\left|\Phi_n^1\right\rangle = \left|\phi_1^0 \cdots \phi_n^1 \cdots \phi_p^0\right\rangle \tag{9.13}$$

where, ϕ_n^0 represents the ground state of the nth monomer and ϕ_n^1 is its first excited state. The adjacent excited state configurations interact via coupling elements J, and the excited state nuclear Hamiltonian is a $p \times p$ matrix of the form

$$H^e(x_1, \cdots, x_p) = \begin{pmatrix} h_1^e & J_{1,2} & \cdots & \cdots & J_{1,p} \\ J_{1,2} & h_2^e & \cdots & \cdots & J_{2,p} \\ \cdots & \cdots & \cdots & \cdots & \cdots \\ J_{1,p-1} & J_{2,p-1} & \cdots & h_{p-1}^e & J_{p-1,p} \\ J_{1,p} & J_{2,p} & \cdots & J_{p-1,p} & h_p^e \end{pmatrix} \tag{9.14}$$

where h_n^e represents the excited state Hamiltonian of the nth monomer, $J_{m,n}$ is the coupling between two Frenkel exciton (FE) states localized on the monomers m and n. Under a first-order approximation, $J_{m,n}$ is approximated to be [16],

$$J_{m,n} = \left\langle \Phi_m^1 \left| H \right| \Phi_n^1 \right\rangle \approx \left\langle \phi_m^1 \phi_n^0 \left| V_{mn} \right| \phi_m^0 \phi_n^1 \right\rangle. \tag{9.15}$$

J is split into the following three terms as

$$J_{m,n} = J_{m,n}^{coul} + J_{m,n}^{ex} + J_{m,n}^{overlap} \tag{9.16}$$

where $J^{coul}_{m,n}$, $J^{ex}_{m,n}$, and $J^{overlap}_{m,n}$ are defined as

$$J^{coul}_{m,n} = \int d\vec{r} \int d\vec{r}' \rho^1_m(\vec{r}) \frac{1}{|r - r'|} \rho^1_n(\vec{r}'),$$
$$J^{ex}_{m,n} = \int d\vec{r} \int d\vec{r}' \rho^1_m(\vec{r},\vec{r}') \frac{1}{|r - r'|} \rho^1_n(\vec{r}',\vec{r}) \tag{9.17}$$

For a molecular dimer, $J^{overlap}_{m,n} = -\omega_0 \int d\vec{r} \rho^1_m(\vec{r}) \rho^1_n(\vec{r})$. Here ρ^1_m and ρ^1_n are the transition densities of the first excited state of molecule m and n, respectively.

In the above FE model, a monomer (m) unit is described by a single mode Hamiltonian including the electronic ground state $|\phi^0_m\rangle$ and the first excited state $|\phi^1_m\rangle$,

$$h_m = |\phi^0_m\rangle h^0_m \langle \phi^0_m| + |\phi^1_m\rangle h^e_m \langle \phi^1_m| \tag{9.18}$$

where the ground state and the excited state Hamiltonians can be written, respectively, as

$$h^0_m = -\frac{1}{2}\frac{d^2}{dx^2} + \frac{1}{2}\omega_0^2 x^2, \tag{9.19}$$

$$h^e_m = -\frac{1}{2}\frac{d^2}{dx^2} + \frac{1}{2}\omega_0^2 (x - x_e)^2 + \Delta E_{FE} \tag{9.20}$$

x is the coordinate, and ΔE_{FE} is the 0–0 transition energy. The x_e and ω_0 are the effective displacement and frequency, respectively. In fact, there are $3N - 6$ vibrational modes in one molecule containing N atoms, and each mode has its own ω_k and $x_{e(k)}$. However, due to the large Gaussian broadening arising from complicated environment in the observed spectra, the multiple vibrational behavior is usually not resolved and different vibrations merge into a single vibronic progression feature, which can be described by an effective mode with Huang-Rhys factor [53],

$$S = \frac{\lambda}{\omega_0} = \frac{\sum_k \lambda_k}{\omega_0} \tag{9.21}$$

$$\omega_0 = \sqrt{\frac{\sum_k \omega_k^4 x^2_{e(k)}}{\sum_k \omega_k^2 x^2_{e(k)}}} \tag{9.22}$$

The effective displacement x_e is related to S by $S = \frac{1}{2}\omega_0 x_e^2$. The reorganization energy λ_k of each mode ω_k is $\lambda_k = s_k \omega_k$ and the Huang-Rhys factor s_k for each mode ω_k can also be expressed in terms of the displacement parameter $x_{e(k)}$ as $s_k = \frac{1}{2}\omega_k x^2_{e(k)}$.

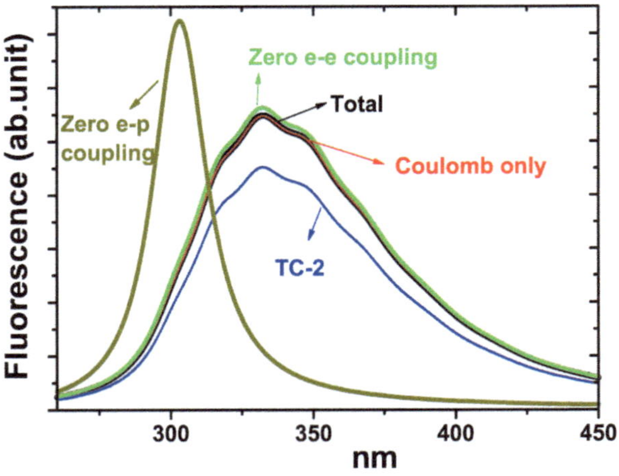

Fig. 9.7 The calculated fluorescence spectra of trpzip2 at 300 K. The *black curve* is the overall fluorescence spectrum; the *blue curve* corresponds to the contribution from Metastable state 2; the *red curve* is calculated with only the coulomb term considered in the exciton-exciton couplings; the *bright green curve* is calculated with zero exciton-exciton coupling; the *light gray* is calculated with no exciton-phonon coupling

With the constructed Hamiltonians H^g and H^e above, the correlation function $C_e(t)$ for emission in linear response spectroscopy is given as

$$C^e(t) = \frac{Tr[e^{-\beta H^e} e^{iH^e t/\hbar} \mu e^{-iH^g t/\hbar} \mu]}{Tr[e^{-\beta H^e}]} \tag{9.23}$$

Thus, the optical emission cross section $\beta(\omega)$ can be obtained

$$\beta(\omega) \propto \omega^3 \int_{-\infty}^{\infty} dt \; \exp(-i\omega t - \gamma|t|) C^e(t) \tag{9.24}$$

Here γ represents the dephasing factor. All the parameters of Hamiltonian are calculated by quantum chemistry based on the representative trpzip2 conformations obtained from previously constructed MSMs.

For trpzip2, 17-state and 13-state MSMs have been built from extensive molecular dynamic simulations at 300 and 350 K, respectively [55]. In the fluorescence simulation, we have selected one representative conformation from each metastable state for the further calculation of the fluorescence signal. The representative conformation has been chosen as the central conformation of the most probable microstate (see Ref. [55] for details of the microstate construction) within each metastable state. For simplicity, here we use TC-n (Typical Conformation n) to label the representative conformation of each mestable state, for example TC-2 is the typical structure for metastable state 2. The vibrationally-resolved fluorescence spectra of trpzip2 can be obtained by summing up the weighted contributions from

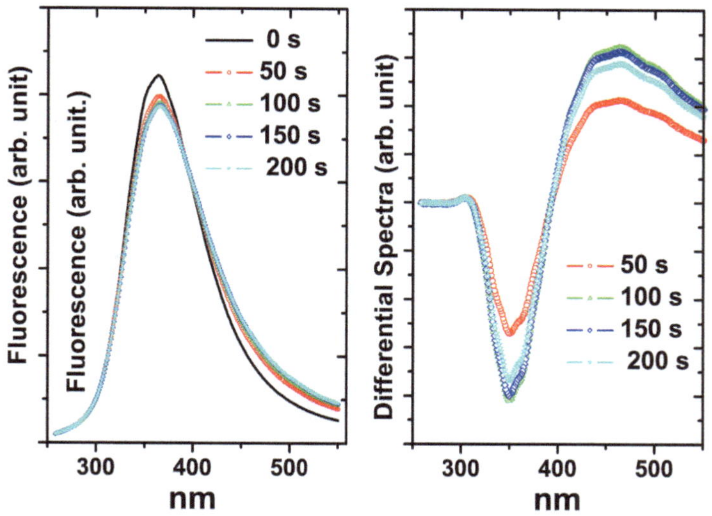

Fig. 9.8 T-jump induced transient fluorescence lineshapes (*left*) and the differential spectra (*right*) between 0 and 200 ns

those typical conformations. Figure 9.7 presents the steady state spectra at 300 K, which nicely resemble the experimental lineshape in Ref. [50]. The maximum of the simulated fluorescence is ~332.6 nm (the black line in Fig. 9.7) while that of the experimental trpzip2 fluorescence is ~351 nm at 297 K [50]. Consistent with the experiment, the simulated spectrum shows a significant asymmetric feature. The lineshape of the spectra at 350 K become more asymmetric, as demonstrated in Fig. 9.8, with a unsymmetrical ratio of ~1.71. The experimental [50] ratio at 347 K's is about 1.88. A 3.4 nm red shift is found between the simulated 300 and 350 K fluorescence signals. This agrees with Gruebele's experiment, in which they reported a 3–5 nm red shift from 297 to 347 K [50]. The simulation thus decently reproduced the experimental tryptophan band fluorescence signal and its temperature dependence. Besides, theoretical analysis show that exciton-exciton couplings between different tryptophan groups almost has no change on the lineshape. the asymmetric feature observed in the fluorescence mainly originates from the vibronic couplings instead of the conformational inhomogeneity.

A temperature jump induced fluorescence signal was also simulated, the populations of metastable states will relax to the equilibrium distribution at T_2 starting from their initial distributions. Based on the population relaxation of each metastable state, the transient fluorescence signals at different moments can be generated. The population change converges at about 200 ns. The lineshape of the spectrum during the T-jump process (see in Fig. 9.8) also has the asymmetric character. At 0 ns, the spectral maximum is 350.6 nm, with unsymmetrical ratio about 1.51. From 0 to 200 ns, the spectral maximum gradually red shifts and the asymmetricity increases. The differential spectra from 0 ns are also plotted in Fig. 9.8. Compared with 0 ns, the intensity of blue tail decreases and that of red tail increases.

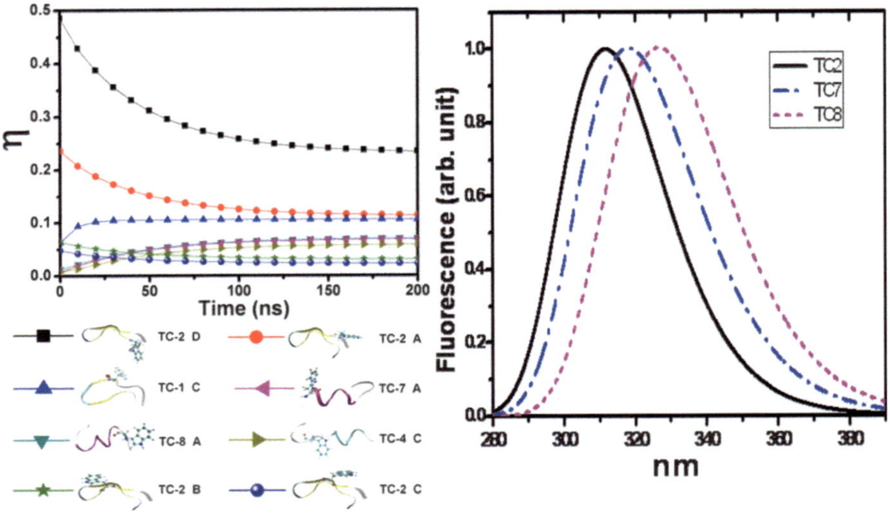

Fig. 9.9 The time dependence of the state population weights η after T-jump (*top*) and the conformations of the tryptophan fluorophores on which the eight most populated ^1L_a states locate (*bottom*). On the *left*, one typical folded state and two typical unfolded states' fluorescence spectra are showed

To reveal the molecular details underlying the trpzip2 unfolding fluorescence, we define $\eta_{ir} = \mathcal{P}_i \mathfrak{B}_{ir}$ for a specific $^1\mathrm{L}_a$ state r (r=A,B,C,D) in a peptide conformation TC-state i. Here, $\mathcal{P}_i$ is the TC-state population, and $\mathfrak{B}_{ir} = \frac{e^{-E_{ir}/kT}}{\sum_{r'} e^{-E_{ir'}/kT}}$ is the Boltzmann weight of the $^1\mathrm{L}_a$ state (emission state of indole-related compounds) with index r (r=A,B,C,D), E_{ir} is the corresponding excitation energy. η_{ir} thus represents the probability of a certain state ir as the initial state of the emission. Changes of η for all the 52 states during the T-jump unfolding process are plotted in the top panel of Fig. 9.9. The locations of the eight most weighted fluorophores are demonstrated at the bottom panel of Fig. 9.9. The two fluorophores with largest η are both on the TC2, a typical folded state. When the temperature changes from 300 to 350 K, η decrease from 48.5 to 23.1 % for TC-2 D, and from 23.6 to 11.3 % for TC-2 A, respectively. And the values for TC-1 C, TC-8 A and TC-7 A increase from 6.3, 1.3 and 0.8 % to 10.6, 7.2 and 7.0 %, respectively. We can also define $\alpha_{ir} = \eta_{ir} D_{ir}^2$, in which D_{ir} is the amplitude of the transition dipoles. α thus represents the contribution of a certain state to the spectrum. Most of the fluorophores on the unfolded configurations, including 1C, 8A, 7A and 4C has significant red shifted transition energies compared with those on the folded configurations, such as 2D and 2A. The analysis above provides a molecular level explanation of the fluorescence change with the rise of temperature: As the temperature increases, the trpzip2 peptide experiences an unfolding process, the populations of blue-end fluorophores on the folded peptides, such as TC-2 D and TC-2 A, decrease and those of red-end

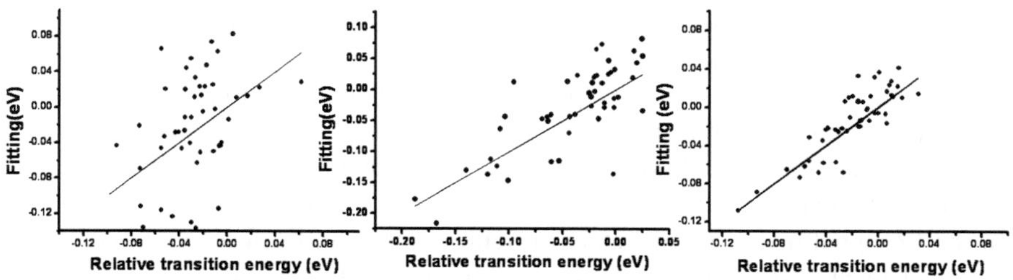

Fig. 9.10 Correlations between the electric potentials and the transition energies only (*left*), between geometry parameters and the transition energies only (*middle*) and between electric potentials plus geometry parameters and the transition energies (*right*)

fluorophores on the unfolded peptides, such as TC-8 A and TC-7 A, increase at the same time, which causes the red shift in the spectrum.

To understand the influence of the environment on the tryptophan fluorescence, we investigated the correlation between the tryptophan transition frequencies and the environmental factors. Calculations are performed on ^1L_a equilibrium states of all 52 3MIs at 350 K. The electric potentials on each of the 10 heavy atoms in 3MI, which generated by the point charge distribution of the environment, are calculated. Based on those 52 calculated data sets, the 3MI transition energy with 10 parameters (10 heavy atoms) are fitted using multiple linear regression analysis. The left panel of Fig. 9.10 shows the poor correlation between the fitted transition energies and the transition energies calculated by quantum chemistry. The ^1L_a state of tryptophan fluorophores is a typical $\pi - \pi*$ conjugated transition on indole ring. Thus the geometry distortion of the 3MI group, which is induced by environment and have effect on electronic delocalization, may have influence on the transition energy as well. We thus carried out another fitting procedure which includes six geometry parameters to describe the mechanical influence of the environment on the transition frequencies. The middle part of Fig. 9.10 gives a better correlation between geometry parameters and transition energies. And then the right part of Fig. 9.10, which include both environmental electric potential and environment induced geometry changes, shows the more largely improved correlation between the fitted and calculated transition energy. This means that the origin of tryptophan fluorescence shows a combined influence from both the environment field and the 3MI geometry distortions.

9.5 Summary and Future Perspective

Based on MSMs, the T-jump triggered long time unfolding related IR, 2DIR and vibrationally-resolved fluorescence of small peptide β-hairpin trpzip2 can be simulated. We can demonstrate that sufficient conformational sampling are crucial for obtaining accurate spectroscopic observables. MSMs provide a good way to simulate the time resolved spectroscopy from the relaxation of the metastable state populations.

Evidently, sufficient accuracy spectroscopic signal simulation for peptides and proteins can push theoretical work become more capable to shed light on interpreting protein folding/unfolding problems. Correspondingly, this will also help to refine the theoretical model. The complexity of the atomistic processes contributing to the IR signal makes it still rather difficult to interpret IR absorption patterns in terms of local structural organisation and atomic motions. Due to computational demand, the limited conformational number certainly trims fluorescence spectra calculation accuracy. Obtaining spectroscopic signals of each state still requires an ensemble average over many protein conformations. There is every reason to believe that developing high efficiency protocol, which combines quantum mechanism with sampling algorithm, still is the future trend for theorists.

Acknowledgements W.Z. gratefully acknowledges the support of the NSFC QingNian Grant 21003117, NSFC Key Grant 21033008 and Instrument in Science and Technological Ministry of China Grant 2011YQ09000505. J.S. acknowledges the support of the NSFC QingNian Grant 21103166.

References

1. Anfinsen CB (1973) Principles that govern the folding of protein chains. Science 181: 223–230
2. Arai S, Hirai M (1999) Reversibility and hierarchy of thermal transition of hen egg-white lysozyme studied by small-angle x-ray scattering. Biophys J 76:2192–2197
3. Bahar I, Lezon TR, Yang LW, Eyal E (2010) Global dynamics of proteins: bridging between structure and function. Annu Rev Biophys 39:23–42
4. Balbach J (2000) Compaction during protein folding studied by real-time nmr diffusion experiments. J Am Chem Soc 122:5887–5888
5. Bowman GR, Huang X, Pande VS (2010) Network models for molecular kinetics and their initial applications to human health. Cell Res 20:622–630
6. Buchete NV, Hummer G (2008) Coarse master equations for peptide folding dynamics. J Phys Chem B 112:6057–6069
7. Chiti F, Taddei N, Baroni F, Capanni C, Stefani M, Ramponi G, Dobson CM (2002) Kinetic partitioning of protein folding and aggregation. Nat Struct Mol Biol 9:137–143
8. Chodera JD, Singhal N, Pande VS, Dill KA, Swope WC (2007) Automatic discovery of metastable states for the construction of markov models of macromolecular conformational dynamics. J Chem Phys 126:155101–155117
9. Dobson CM (2003) Protein folding and misfolding. Nature 426:884–890
10. Dobson CM (2004) Principles of protein folding, misfolding and aggregation. Semin Cell Dev Biol 15:3–16
11. Duan Y, Kollman PA (1998) Pathways to a protein folding intermediate observed in a 1-microsecond simulation in aqueous solution. Science 282:740–744
12. Gaggelli E, Kozlowski H, Valensin D, Valensin G (2006) Copper homeostasis and neurodegenerative disorders (alzheimer's, prion, and parkinson's diseases and amyotrophic lateral sclerosis). Chem Rev 106:1995–2044
13. Greenfield NJ (2007) Using circular dichroism spectra to estimate protein secondary structure. Nat Protoc 1:2876–2890

14. Hartl FU (1996) Molecular chaperones in cellular protein folding. Nature 381:571–580
15. Hartl FU, Hayer-Hartl M (2009) Converging concepts of protein folding in vitro and in vivo. Nat Struct Mol Biol 16:574–581
16. Hsu CP, Fleming GR, Head-Gordon M, Head-Gordon T (2001) Excitation energy transfer in condensed media. J Chem Phys 114:3065–3072
17. Hu H, Yang W (2008) Free energies of chemical reactions in solution and in enzymes with ab initio quantum mechanics/molecular mechanics methods. Annu Rev Phys Chem 59:573–601
18. Huang X, Bowman GR, Bacallado S, Pande VS (2009) Rapid equilibrium sampling initiated from nonequilibrium data. Proc Natl Acad Sci USA 106:19765–19769
19. Huang X, Yao Y, Bowman GR, Sun J, Guibas LJ, Carlsson G, Pande VS (2010) Constructing multi-resolution markov state models (msms) to elucidate rna hairpin folding mechanisms. Pac Symp Biocomput 15:228–239
20. Krimm S, Bandekar J (1986) Vibrational spectroscopy and conformation of peptides, polypeptides, and proteins. Adv Protein Chem 38:181–364
21. Kubelka J, Keiderling TA (2001) Differentiation of β-sheet-forming structures: ab initio-based simulations of IR absorption and vibrational CD for model peptide and protein β-sheets. J Am Chem Soc 123:12048–12058
22. Lyubartsev AP, Martsinovski AA, Shevkunov SV, Vorontsov-Velyaminov PN (1992) New approach to Monte Carlo calculation of the free energy: method of expanded ensembles. J Chem Phys 96:1776–1783
23. Marinari E, Parisi G (1992) Simulated tempering: a new Monte Carlo scheme. Europhys Lett 19:451–458
24. Mukamel S (1995) Principles of nonlinear optical spectroscopy. Oxford University Press, New York
25. Noé F, Fischer S (2008) Transition networks for modeling the kinetics of conformational change in macromolecules. Curr Opin Struct Biol 18:154–162
26. Ohta K, Yang M, Fleming GR (2001) Ultrafast exciton dynamics of j-aggregates in room temperature solution studied by third-order nonlinear optical spectroscopy and numerical simulation based on exciton theory. J Chem Phys 115:7609–7621
27. Onuchic JN, Wolynes PG (2004) Theory of protein folding. Curr Opin Struct Biol 14:70–75
28. Pfuhl M, Driscoll PC (2000) Protein nuclear magnetic resonance spectroscopy in the new millennium. Philos Trans R Soc Lond Ser A 358:513–545
29. Renger T, Marcus RA (2002) On the relation of protein dynamics and exciton relaxation in pigment–protein complexes: an estimation of the spectral density and a theory for the calculation of optical spectra. J Chem Phys 116:9997–10019
30. Segel DJ, Bachmann A, Hofrichter J, Hodgson KO, Doniach S, Kiefhaber T (1999) Characterization of transient intermediates in lysozyme folding with time-resolved small-angle x-ray scattering. J Mol Biol 288:489–499
31. Selkoe DJ (2003) Folding proteins in fatal ways. Nature 426:900–904
32. Semisotnov GV, Rodionova NA, Razgulyaev OI, Uversky VN, Gripas AF, Gilmanshin RI (1991) Study of the "molten globule" intermediate state in protein folding by a hydrophobic fluorescent probe. Biopolymers 31:119–128
33. Senn HM, Thiel W (2009) Qm/mm methods for biomolecular systems. Angew Chem Int Ed 48:1198–1229
34. Smith AW, Lessing J, Ganim Z, Peng CS, Tokmakoff A, Roy S, Jansen TLC, Knoester J (2010) Melting of a β-hairpin peptide using isotope-edited 2D IR spectroscopy and simulations. J Phys Chem B 114:10913–10924
35. Snow CD, Nguyen H, Pande VS, Gruebele M (2002) Absolute comparison of simulated and experimental protein-folding dynamics. Nature 420:102–106
36. Song J, Liang WZ, Zhao Y, Yang JL (2006) Conformational flexibility and its effect on the vibrationally resolved absorption and fluorescence spectra of oligofluorenes. Appl Phys Lett 89:071917
37. Sreerama N, Woody RW (2004) Computation and analysis of protein circular dichroism spectra. Methods Enzymol 383:318–351

38. Sriraman S, Kevrekidis IG, Hummer G (2005) Coarse master equation from bayesian analysis of replica molecular dynamics simulations. J Phys Chem B 109:6479–6484
39. Stefani M, Dobson CM (2003) Protein aggregation and aggregate toxicity: new insights into protein folding, misfolding diseases and biological evolution. J Mol Med 81:678–699
40. Stryer L (1995) Biochemistry, 4th edn. W.H.Freeman, New York
41. Sugita Y, Okamoto Y (1999) Replica-exchange molecular dynamics method for protein folding. Chem Phys Lett 314:141–151
42. Swope WC, Pitera JW, Suits F (2004) Describing protein folding kinetics by molecular dynamics simulations. 1. Theory. J Phys Chem B 108:6571–6581
43. Torii H, Tasumi M (1992) Model calculations on the amide-i infrared bands of globular proteins. J Chem Phys 96:3379
44. Torii H, Tasumi M (1998) Ab initio molecular orbital study of the amide i vibrational interactions between the peptide groups in di- and tripeptides and considerations on the conformation of the extended helix. J Raman Spec 29:81–86
45. Tozzini V (2005) Coarse-grained models for proteins. Curr Opin Struct Biol 15:144–150
46. Venkatramani R, Mukamel S (2002) Correlated line broadening in multidimensional vibrational spectroscopy. J Chem Phys 117:11089–11101
47. Wutrich K (1986) NMR of proteins and nucleic acids Wiley, New York
48. Yan YJ, Mukamel S (1986) Eigenstate-free, green function, calculation of molecular absorption and fluorescence line shapes. J Chem Phys 85:5908–5923
49. Yang WY, Gruebele M (2004) Detection-dependent kinetics as a probe of folding landscape microstructure. J Am Chem Soc 126:7758–7759
50. Yang WY, Pitera JW, Swope WC, Gruebele M (2004) Heterogeneous folding of the trpzip hairpin: full atom simulation and experiment. J Mol Biol 336:241–251
51. Zanni MT, Asplund MC, Hochstrasser RM (2001) Two-dimensional heterodyned and stimulated infrared photon echoes of n-methylacetamide-d. J Chem Phys 114:4579–4590
52. Zhao Y, Knox RS (2000) A brownian oscillator approach to the kennard-stepanov relation. J Phys Chem A 104:7751–7761
53. Zhao Y, Liang WZ, Nakamura H (2006) Semiclassical treatment of thermally activated electron transfer in the intermediate to strong electronic coupling regime under the fast dielectric relaxation. J Phys Chem A 110:8204–8212
54. Zhuang W, Abramavicius D, Mukamel S (2004) Peptide secondary structure determination by three-pulse coherent vibrational spectroscopies: a simulation study. J Phys Chem B 108: 18034–18045
55. Zhuang W, Cui RZ, Silva DA, Huang X (2011) Simulating the T-jump-triggered unfolding dynamics of trpzip2 peptide and its time-resolved IR and two-dimensional IR signals using the markov state model approach. J Phys Chem B 115:5415–5424

Chapter 10
The Dilemma of Conformational Dynamics in Enzyme Catalysis: Perspectives from Theory and Experiment

Urmi Doshi and Donald Hamelberg

Abstract The role of protein dynamics in catalysis is a contemporary issue that has stirred intense debate in the field. This chapter provides a brief overview of the approaches and findings of a wide range of experimental, computational and theoretical studies that have addressed this issue. We summarize the results of our recent atomistic molecular dynamic studies on *cis-trans* isomerase. Our results help to reconcile the disparate perspectives regarding the complex role of enzyme dynamics in the catalytic step and emphasize the major contribution of transition state stabilization in rate enhancement.

Keywords Enzyme dynamics • Accelerated Molecular dynamics • Catalysis • Molecular mechanics • Kramers' rate theory • Conformational dynamics • Cis-trans isomerization/isomerase • Cyclophilin A • Principal component analysis • NMR relaxation dispersion • Protein flexibility • Structure-function • Dynamics-function • Multi-exponential • Kinetics • Free energy barrier

10.1 Introduction

Enzymes are biological catalysts that enhance the rates of a plethora of reactions important for life processes [1]. In the absence of enzymes, the reactions in solution would be 10^{17} to 10^{19} times slower [2]. Along with remarkable selectivity, the speed up of rates by enzymes allows for reactions to occur at timescales of milliseconds to seconds, which is relevant for cellular function [2]. Ever since the discovery of the first enzyme in 1833 [3], the quest for understanding how enzymes are able

U. Doshi (✉) • D. Hamelberg (✉)
Department of Chemistry and the Center for Biotechnology and Drug Design, Georgia State University, Atlanta, GA 30302-3965, USA
e-mail: udoshi@gsu.edu; dhamelberg@gsu.edu

K. Han et al. (eds.), *Protein Conformational Dynamics*, Advances in Experimental Medicine and Biology 805, DOI 10.1007/978-3-319-02970-2_10,

to achieve such proficient catalytic properties has been ongoing. Our knowledge in enzymology in the last four decades has increased to a large extent, leading to several successful industrial applications [3]. These advances, in part, have been based on protein engineering efforts that have exploited the structure-function relation, i.e. focusing on the structural changes upon mutagenesis and its effects on enzyme function [4]. Conceptually, a chemical reaction is visualized as an energy barrier between reactant and product that must be overcome by thermal activation of reactant molecules [2, 5]. The increase in the reaction rates is then, evidently, due to significant reduction of this activation barrier by enzymes. It has been now well-established that enzyme active sites provide favorable environment for electrostatic stabilization of the transition state to a much greater degree relative to that of the ground states [6]. This, therefore, has led to efforts for structural and energetic optimization of active sites. Although other factors such as steric strain, desolvation and entropy have been proposed to play a role in decreasing the activation energy, their effects, even if present, are very small as compared to the dominant contribution arising from transition state stabilization (TSS) [5, 7]. The TSS principle has been exploited in rational approaches for designing and engineering enzymes with better catalytic efficiencies than the naturally occurring ones. However, such attempts have met only moderate success, underscoring our inadequate understanding of the factors responsible for the catalytic power of enzymes and the need for new concepts to explain enzyme function. Since the beginning of this millennium, an increasing number of studies have suggested the role of protein dynamics in enzyme action [8].

The earliest model to understand enzyme specificity emphasized stringent requirement of shape complementarities between enzyme and substrate (the lock and key mechanism [9]) and was deficient in the role for any protein motions. Almost seven decades later, the model proposed by Koshland [10], for the first time, accommodated protein flexibility and suggested that upon encountering the substrate, considerable structural rearrangements took place in the enzyme and the substrate, such that a perfect fit was 'induced'. In this mechanism, substrate binding results in the enzyme visiting those conformations that are not sampled in the free unliganded enzyme [11]. Experimental evidence of structural variability in proteins came later from solution nuclear magnetic resonance (NMR) spectroscopic [12] and X-ray studies that showed that in many instances the structure adopted by an enzyme upon binding its substrate is distinct to that by a free enzyme [13, 14]. Moreover, even in a free enzyme, an ensemble of models yielded from NMR structure determination studies and high B-factors in X-ray crystallography suggested relatively higher flexibility in certain disordered regions of the proteins. One of the earlier implications of protein motions arose from allostery that involved concerted large-scale transitions between at least two conformations or states of a protein and where each conformation (generally interpreted then as a static structure) could be associated with a distinct function [15]. Presence of conformational ensembles was also indicated in the conformational selection view of binding, in which the substrate could choose a subset of enzyme conformers to find the best fit in the complex [16]. Pioneering molecular dynamics simulations had also long before revealed the dynamical nature of biomolecules [17]. It is now commonplace to find several examples of enzymes in literature that undergo, on one hand, substantial conformational changes (sometimes large-scale such as opening and closing of a loop or hinged movements of domains) for substrate binding and product release [18]. On the other hand, thermal fluctuations in a protein may result in an ensemble of conformations that show only minor deviations from the overall native structure, involving a small set of degrees of freedom [19]. Whether such stochastic thermal fluctuations have any functional significance in molecular recognition and catalysis has largely been an unexplored issue, until the last decade. More recently, enzyme flexibility has been suggested to be a crucial determinant of its catalytic efficiency [18, 20–25]. However, the question of how exactly the dynamical motions help in accelerating the catalytic rates has remained elusive, despite intensive research on experimental and theoretical fronts accompanied with unprecedented controversies in the field of enzymology (*vide infra*).

10.2 Nature of Protein Dynamics

Protein dynamics encompass motions that span broad range of timescales and length-scales [26]. The timescales and amplitudes of these motions are governed by the features of the underlying energy landscape [27]. The energy landscape of biomolecules is hyper-dimensional and rugged with multiple energy wells or conformational states (or sub-states) separated by barriers of varying lengths (Fig. 10.1). When conformational states are separated by barriers of several k_BTs (k_B being the Boltzmann constant and T, the temperature) interconversions between them are slower, ranging from microseconds to seconds and involving collective motions of many degrees of freedom, for e.g. allostery, protein folding/unfolding, enzyme catalysis. Within each state, transitions between closely related conformational sub-states constitute the relatively faster fluctuations on the picosecond-nanosecond timescale, for e.g. side-chain rotations or loop motions. Local flexibility at the atomic level within each conformational sub-state includes femtosecond-bond vibrations. Motions may involve a group of atoms located proximally in sequence or space or those that are distal to each other but move in a concerted fashion. The timescale of these motions can fall anywhere in the wide continuum range depending on the barrier heights between the sub-states. The energy landscape represents all the possible states of a protein in a solvent environment. Protein dynamics is coupled to the motions of the surrounding water molecules [28] that rearrange around the protein (hydration shell) involving breakage and formation of hydrogen bonding network [29]. Water dynamics contributes an added level of roughness to the protein energy landscape [29]. When a protein undergoes mutation or binds a substrate/ligand; or there are changes in the external conditions, for e.g. in temperature or solvent composition, the modified system of protein-solvent is then represented by a distinct energy landscape in which the equilibria between the various states may be modified. The conformational ensemble of each

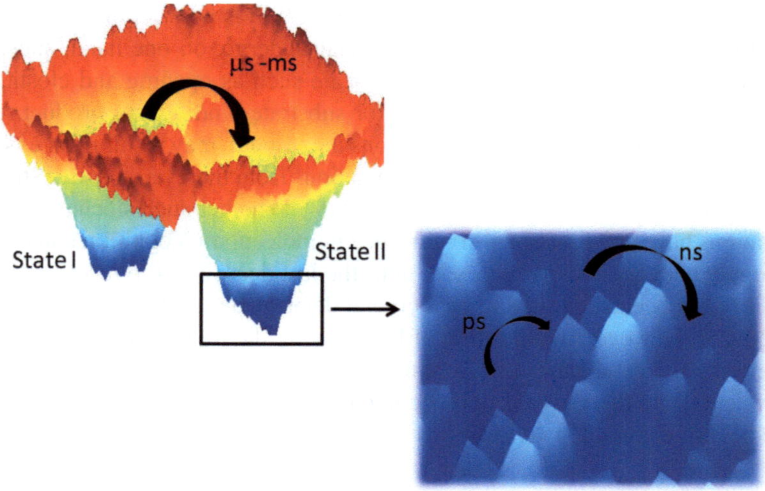

Fig. 10.1 Protein conformational dynamics. Schematic representation of a region of an energy landscape depicting two conformational states separated by a relatively high energy barrier. Transition from one state to another may involve a local change in few degrees of freedom or large-scale with collective motions of many degrees of freedom. Depending upon the barrier height, the transition may take microseconds to milliseconds or even seconds. If we zoom into the bottom of any such state, we may find that the bottom of the well is rugged with multiple valleys separated by barriers of different heights. Interconversions between closely-related sub-states are in the nanosecond timescale whereas faster picoseconds transitions may occur between even more closely-related sub-states (Adapted from Frauenfelder et al. [27, p. 1598] & Henzler-Wildman and Kern [26, p. 964])

state (i.e. State I or II in Fig. 10.1) may generally have structural identity and thermodynamic properties that are different from those of the other. Besides, since fluctuating to another state requires crossing large barriers, the time the system spends in a particular energy well is long enough to allow direct characterization of these states from ensemble- and time-averaged low-resolution bulk experiments. Relaxation to equilibrium conditions is usually monitored spectroscopically following sudden perturbation of the system (for e.g. by temperature, pressure, or pH) to obtain kinetic information about interconversion between states. Probing dynamics in the faster regime is more challenging, partly because of the short lifetimes of the conformational sub-states. From an ensemble-averaged signal, it is not possible to characterize the sub-states and their exchange rates. Advancements in NMR relaxation studies now allow accessing a wide spectrum of timescales from picoseconds to seconds at atomic resolution [30]. One can obtain high-resolution (i.e. site-specific) information about conformational sub-states when their interconversion rates are in the intermediate to rapid regime on the NMR timescale. The Carr-Purcell-Meiboom-Gill relaxation dispersion (CPMG-RD) method, which reduces the line-broadening due to interconversions between sub-states, have been combined with isotope labeling of backbone amides and methyl carbons to probe millisecond dynamics in many free as well as substrate-bound enzymes [31, 32]. The heterogeneous behavior of a single enzyme molecule has been revealed by single molecule fluorescence studies. Enzymatic turnover rates for a single molecule have been shown to fluctuate over a broad range of timescales (from milliseconds to tens of seconds) [33, 34]. Such variation in turnover rates is suggested to be due to interconversions between enzyme conformers. Further, single molecule fluorescence quenching by electron transfer have been used to measure the timescale at which the donor-acceptor distances fluctuate, from which the range of timescales for conformational fluctuations in an entire protein has been inferred [33]. Interestingly, the range of timescales spanned by conformational fluctuations has been found to be similar to that by enzymatic turnover rates. Molecular dynamics (MD) simulations have been greatly instrumental in probing dynamics on the sub-microsecond timescale and providing a complete structural characterization of transient species in atomistic details [17, 35, 36]. The advantage of MD computational studies is that even states with higher energy (with low probability) can be accessed, which is not possible with current experiments. Progress in computational architecture and methodologies has been impressive in the past decade [37–39], permitting routine use of MD alongside experimental investigations. However, accessing long timescale dynamics still presents substantial challenge for atomistic MD studies.

10.3 Computational Modeling of the Effects of Enzyme Dynamics on Catalysis

Computational simulations of enzyme catalysis are challenging from many aspects. An enzyme molecule represented in atomistic details and solvated in explicit solvent involves very many degrees of freedom. Sampling of such large systems is restricted to a few hundreds of microseconds in classical MD simulations. Such studies can help characterize the reactant or the product states from an equilibrium point of view. However, the enzymatic chemical step of transforming the reactant to product, that usually takes milliseconds, cannot be simulated directly yet. In view of the fact that conventional MD trajectories fail to capture the suggested functionally important millisecond dynamic motions, refuting or establishing a role of protein dynamics in catalysis becomes unattainable. To simulate reactions involving bond breaking and formation, quantum mechanical (QM) treatment is required. Also, it is important to model polarization effects caused by environmental dynamics. Due to the high computational cost involved, QM simulations of large enzyme systems are not feasible. Therefore, hybrid methods have been implemented; in which part of the system involved in the chemistry is treated quantum mechanically while the rest of the system is represented with molecular mechanics (MM). Here, we briefly describe some of the currently used approaches for modeling enzyme catalyzed reactions and the kind of information that can be obtained from these methods.

10.3.1 Classical Molecular Dynamics

Although traditional MD is limited in modeling the entire catalytic pathway in continuity, i.e. the progression of the reaction from reactants to products via transition state, useful insights about enzyme dynamics can be gained from analyzing individual atomistic MD trajectories of enzymes in the absence of the substrate and those bound to either the reactant or the product. We have recently demonstrated this point in our model enzyme system, Cyclophilin A (CypA), which belongs to a class of peptidyl prolyl isomerases. CypA catalyzes *cis-trans* isomerization of prolyl peptide bonds [40]. The isomerization reaction does not involve any bond breaking and formation, making it an ideal candidate for classical MD studies. In addition, the location of the transition state is also well-defined (i.e. regions around $\omega \sim 90^{\text{o}}$, where ω is the peptide bond torsional angle between the Xaa-Pro motif in the substrate). Such reactions, in which local conformational change occur at a peptide bond, are associated with several biological switches and have important functional consequences [41]. We carried out one very long normal MD simulations of free CypA and three simulations of CypA bound to the substrate that was restrained in either the *cis*, *trans* or transition state configurations. We then applied principal component analysis of these trajectories, which is a mathematical procedure to reduce the dimensionality of huge data sets and deconvolute the local fast fluctuations from collective large-scale motions [42, 43]. Essentially, a new set of orthogonal coordinates is constructed by diagonalization of the covariance matrix of the atomic coordinates in a trajectory. The resulting set of coordinates or eigenvectors are arranged according to the decreasing order of their corresponding eigenvalues. Eigenvalues represent the variance along the corresponding eigenvectors. The first few eigenvectors have the largest variances such that most of the protein motions can be described by projecting the original data onto these eigenvectors (also called principal components). Figure 10.2 shows the slowest eigenmode projected on the CypA structure. The flexibility observed in the loop regions (motions depicted by the arrows) in the free enzyme is lost in the presence of the substrate when it is in the *cis* or the *trans* state. Interestingly, some of the slowest motions in one of the loops are retained when the substrate is in the transition state configuration. However, the direction of motion of that loop is opposite to the direction observed in the free enzyme. Similarly, comparisons of other eigenmodes in free and substrate-bound enzymes can help in identifying motions that are inherent to the enzyme and those that are important in the transition state. Also, similarity in motions can be determined from calculating dot products between eigenvectors in the reactant-, product-, transition-state-bound enzyme or the *apo* enzyme. In our recent studies, projection of the conformational phase space of the CypA active site residues onto the first three principal components revealed some important features [40, 46]. The conformational space sampled by the active site residues bound to transition-state substrate was seen to be more

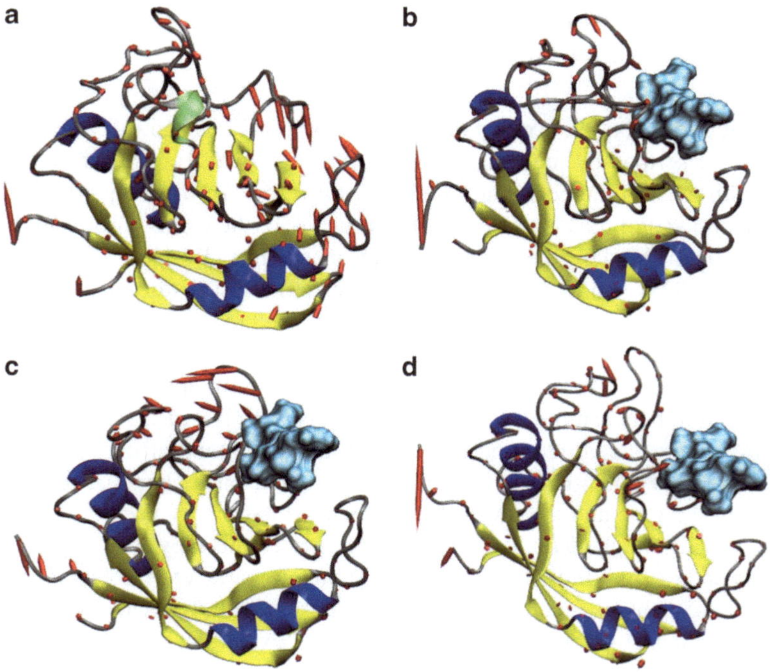

Fig. 10.2 Comparison of the slowest modes in free CypA and CypA-substrate complexes along the reaction coordinate. CypA structure is projected along the slowest eigenmode (*red arrows*) obtained from principal component analysis of trajectories of (**a**) free CypA and CypA in complex with the substrate (*cyan*) in the (**b**) trans, (**c**) transition state and (**d**) *cis* configurations. The lengths of the arrows are proportional to the extent of motion, where the head and tail of the arrows represent the most positive and negative projections, respectively. The slowest motions predominantly involve loops (*gray*) (The figures were prepared using VMD [44] and IED [45])

restricted that those sampled by the enzyme in complex with the ground states. Moreover, each of the three substrate-bound conformational ensembles was a subset of the significantly much broader ensemble sampled by the free CypA. Furthermore, from two-dimensional free energy profiles, we were able to identify the enzyme-substrate interactions coupled to the ω dihedral and, thereby, to the chemical step [46]. Recent computational studies by others on human CypA and its homologs in two other species have shown that reaction-coupled modes (i.e. eigenmodes that are coupled to the maximum displacement of the peptidyl-prolyl ω dihedral of the substrate) are conserved [47]. These comparative studies in non-homologous enzymes that catalyze the same chemistry have further suggested that conformational dynamical motions, which were coupled to the chemical step, facilitated the interactions between the enzyme and the substrate in the active site in a similar manner.

10.3.2 Combined Quantum Mechanical and Molecular Mechanical Methods

Electronic redistribution accompanying formation or breakage of chemical bonds cannot be described by molecular mechanical force fields, which typically model bonds with a harmonic potential. Therefore, for reactions involving bond breaking or formation and/or metals, the electronic structure of the atoms participating in the reaction should be described with quantum mechanics. To make the calculations tractable, the QM treatment, i.e. the detailed electronic structure method is limited to the atoms of the substrate and the enzyme involved in the reactions, whereas the surrounding regions of the enzyme and the solvent are modeled with MM force field [48]. Improved boundary treatments are now available for appropriate modeling of the interaction between the QM and MM regions [49]. Detailed reviews on hybrid QM/MM methods can be found elsewhere [35, 49–53] and articles cited therein. The accuracy of potential energy surfaces computed for a chemical reaction will depend on the type and the level of molecular orbital theory used to describe the QM region. Although *ab initio* methods are most accurate, they are computationally very demanding and limited to only a few atoms. Reparameterized semi-empirical methods and density functional theory are most popularly used due to the low computational cost involved and the possibility of extending to larger systems [35, 49, 52]. To compute potentials of mean force (PMF), potential energy is averaged over ensemble of structures generated from conformational sampling *via* classical MD. Detailed QM calculation is performed on each configuration of normal MD trajectory. MD simulations are implemented in restricted and small overlapping regions along the reaction coordinate using umbrella sampling [54] or free energy perturbation methods [55]. The activation free energies can then be calculated from the PMF's. Another method that is computationally efficient for larger systems and extensively used to calculate the free energy profiles of chemical reactions in solution and enzymes is the empirical valence bond method [56]. A wide range of QM/MM implementations [35, 49, 57, 58] have been carried out to identify enzyme motions deemed important for catalysis [20], estimate rate constants, kinetic isotope effects (KIEs) and their temperature dependence that are consistent with experimental values for several enzymes and their mutants. This has helped to gain insights for the role of protein dynamics in enzyme function at the molecular level. In most cases, equilibrium motions were identified that facilitated the chemistry but not found to be coupled to the chemical step.

10.3.3 Enhanced Sampling Methods

Sampling schemes in which the Hamiltonian is modified by addition of a bias potential are widely employed in either constrained or unconstrained simulations. Umbrella sampling [54] is one of the oldest and commonly used methods to generate PMF along the reaction coordinate and estimate free energy of activation ($\Delta G^{\#}$)

from the PMF profiles. The reaction coordinate is divided into several windows and the phase space is sampled by MD in each window. Such PMF from MM force fields do not include nuclear quantum effects. Therefore, techniques to incorporate the effects of zero point and quantized vibrational energies in the MM-PMF have been proposed using normal mode analysis [49].

Unlike umbrella sampling, accelerated molecular dynamics (aMD) approach [59] does not require restricting the phase space and prior knowledge of the reaction coordinate. aMD has been shown to successfully address the problem of inadequate sampling and accessing long timescale dynamics [60–62]. The main idea behind aMD is to modify the potential energy surface near the minima by adding a bias potential such that the minima are raised and the transition rates out of the basin are increased. A continuous and non negative bias potential is added only when the potential energy falls below a preset threshold boost energy. The prescription for the bias potential maintains the basic shape of the potential and does not allow the forces (i.e. derivative of the modified potential) to be zero at places where the modified potential is equal to the boost energy. The equilibrium properties of the original landscape can be reproduced by a simple reweighting procedure to remove the effects of the bias [59]. The main advantage is that the free energy profiles can be projected onto any choice of variables, not necessarily the progress coordinate. aMD permits the flexibility to boost an entire system or only selective set of degrees of freedom belonging to either one or more molecules in a system. Our recent work has shown that it is also possible to retrieve the kinetics on the original potential by using Kramers' rate theory [60, 63] and establish the relation between roughness of the underlying energy landscape and the effective diffusion coefficient [64]. Accelerated molecular dynamics can be greatly valuable in directly simulating enzyme-catalyzed reactions and their corresponding reference reactions in solution that involve very large activation barriers [62]. Moreover, aMD has the potential to probe the direct effects that millisecond timescale dynamical motions of the enzyme may have on the catalyzed reaction in atomistic details [40]. We have recently employed aMD to simulate the catalyzed and the uncatalyzed *cis-trans* isomerization of a prolyl peptide bond in a realistic enzyme model with lower activation barriers (Sect. 10.5). Improved versions of aMD have been developed by our group in which only rotatable dihedrals i.e. degrees of freedom most relevant for conformational sampling are subjected to acceleration. This has resulted in the reweighted equilibrium properties with significantly less statistical noise than the previous implementations [65].

10.4 Controversy Over the Role of Enzyme Dynamics in Catalysis

It is now broadly accepted that enzyme flexibility is important for function, there exist multiple conformations of an enzyme and enzyme dynamics span multiple timescales. However, whether enzyme dynamics contribute to catalysis or not is an intensely debated contemporary issue. It is not within the scope of this chapter to review the available exhaustive literature on this problem. Therefore, here, we shortly summarize some of the key experimental observations whose interpretations failed to reconcile with computational studies, theoretical predictions for which no compelling experimental evidence exist and disagreements based on semantic issues. Recent reviews and articles [7, 66, 67] and references therein provide detailed discussions on this matter.

NMR relaxation dispersion studies detected amide nitrogens in one of the loop regions in the substrate-bound CypA that underwent chemical exchange during catalysis [31]. Since the global exchange rates of these amides coincided with the catalytic rates, it was suggested that enzyme dynamics directly promoted catalysis [32]. Moreover, conformational exchanges of these same amide nitrogens preexisted in the free enzyme and on the same millisecond-timescale as those observed during substrate turnover [32, 68]. This led to the proposition that catalysis-coupled dynamics is an inherent property of enzymes. Similar conclusions were drawn for dihydrofolate reductase (DHFR), another extensively-studied paradigm to investigate the effects of dynamics on catalysis. The rates of hydride transfer catalyzed by DHFR seemed to be dictated by the dynamics that facilitated the sampling of various conformational states during the course of the catalytic cycle [69]. In subsequent studies on DHFR mutants, millisecond-timescale dynamics observed in the active site loop of the wild-type (wt) was abrogated [70]. The loss of flexibility was concomitant with the decrease in the hydride transfer rates in the mutants. Virtual similarity was also noted in the structural comparisons of active sites in the wt and mutants. Therefore, it was inferred that dynamics is directly coupled with the catalyzed hydride transfer, with the assumption that there was no apparent difference in the electrostatic interactions made in the active sites of wt and mutants [70]. These claims were later contested by computational investigations that showed that the trend of reduced hydride transfer rates in mutants could be reproduced without including any dynamical effects in the model [71]. In agreement to the well-established transition state stabilization principle, it was further suggested that the reduction in the rates was due to an increase in the activation barriers that were predominantly determined by electrostatic preorganization at the active site (as shown earlier by the same group [72]), and not due to the loss of conformational motions in the mutants [71].

Primarily, the experimental observables to monitor tunneling in enzyme-catalyzed hydrogen-transfer reactions are primary KIEs and their temperature dependence. KIEs quantify the isotope dependent differences in the reaction rates (i.e. k_H/k_D). Although the chemical properties of hydrogen isotopes (protium, deuterium and tritium) are the same, differences in their zero-point vibrational energies lead to smaller activation energy and hence faster rate for the transfer of protium than for deuterium or tritium. However, much larger KIEs measured at ambient temperature than expected from a semi-classical description of reaction kinetics (i.e. transition state theory that incorporates quantized vibrational energies of the ground state) are often taken as an indication of quantum tunneling (i.e. passage through the barrier) [73]. Since KIEs are sensitive to the changes in hydrogen donor-acceptor distance and overlap of donor-acceptor wave functions, models invoking a role for protein dynamics in hydrogen transfer reactions have been proposed to explain the unusually large KIEs and their anomalous temperature dependence (or independence) [22]. In case of DHFR, it is argued that if conformational dynamics were directly coupled to the hydride transfer step, it would modify the hydrogen donor-acceptor distance or the wavefunction overlap between the substrate and the product in the mutants. And modifications of these parameters would reflect in the changes in KIEs and their temperature dependence. However, a subsequent experimental investigation found striking similarity in the magnitude of the KIEs and their temperature dependence in the wt and the same DHFR mutants [74] as studied in the previously mentioned experiments [70]. These results re-emphasized that the reaction rates were affected by the changes in the active site electrostatic preorganization (increase in reorganization energy) as a result of mutations and there did not seem any direct coupling between conformational fluctuations and the actual hydride transfer step [74].

The other source of controversy is the fast (femtosecond) vibrational dynamics at the atomic level that are either interpreted as statistical or non-statistical motions. Long-range network of coupled motions on the femtosecond-picosecond timescale have been proposed from detailed computational studies to assist hydride transfer [75, 76] by facilitating tunneling. These motions are considered to be promoting vibrations that bring about fluctuations in equilibrium ensemble [20]. Another set of theoretical and computational studies have suggested the non-statistical vibrational modes of residues distal to the active site of an enzyme to actively drive the reaction over the barrier [24, 77, 78]. Ring polymer molecular dynamics that incorporated zero-point and tunneling effects could capture non-statistical vibrations of only a small set of atoms involved in the hydride transfer reactions [79]. However, such non-statistical dynamics existed only for short length-scales (i.e. up to 4–6 Å of the transferring hydrogen), beyond which no coupling was found with the hydride transfer step [79]. In support of these conclusions, experimental studies on thermophilic DHFR have found no evidence for long-range coupling with the chemical step in distal mutations [80]. Also, computational simulations by other independent groups could reproduce experimental observations without invoking any non-statistical motions [7, 81]. Another contentious aspect of coupling between vibrational dynamics and the chemical step is the disconnect in the timescale of the two processes – typical turnover rates for biochemical reactions are on the millisecond timescale and vibrational motions occur in femtoseconds. In case of hydride transfer, it is argued that the actual cleavage of the C–H bond is fast on the femtosecond-picosecond timescale, thus accommodating catalysis-promoting role for fast vibrations. For that matter, direct contribution of slow millisecond dynamics in CypA-catalyzed *cis-trans* isomerization, which also occurs on the same timescale, is thought to be more convincing. However, as we show further (Sects. 10.5 and 10.6), coincidence of enzyme motions and the chemical step on the same timescale does not guarantee that enzyme motions are driving the chemical step.

To rationalize the observations of large KIEs and their temperature dependence not expected from classical over the barrier kinetics, several mathematical models with simple functional forms have been proposed with a role for promoting vibrations to effect quantum mechanical tunneling. In the full tunneling model, fast femtosecond-promoting vibrations bring about fluctuations in the donor-acceptor distance [22]. This distance sampling enhances the extent of wave function overlap between donor and acceptor, and thereby tunneling. Although convincing interpretations in certain cases of large KIEs and their anomalous temperature dependence (and independence) have been provided by this simple model [73], not all available data could be explained [82]. Another model have suggested 'barrier compressing' vibrations to be instrumental in reducing donor-acceptor distance (i.e. decrease in the width of the barrier), which would lead to efficient tunneling [83, 84]. For this model, however, there has still been lack of support in terms of direct experimental evidence and from computational simulations that have pointed out the breakdown in the dependence of tunneling and KIEs on physically unreasonable short donor-acceptor distances used in the model [81]. If, according to the model, compression in the donor-acceptor distance is accompanied with a decrease in the barrier height and width, and thereby, an increase in tunneling probability, tunneling may become meaningless in cases where the barriers are so small that reaction might as well be able to proceed 'over the barrier'.

Certain amount of confusion and apparent disagreement in the field regarding the role of protein dynamics in catalysis have arose from the definitions of 'catalysis'. Ideally, catalysis is the rate enhancement of a reaction by an enzyme as compared to an uncatalyzed reaction in solution, which follows the same mechanism as the one in the enzyme's active site. Discussions on the choice of another 'reference' reaction can be found in references [7, 67]. Large scale conformational changes are brought about in adenylate kinase by a hierarchy of timescales [18, 68]. The opening of the active site lid is crucial for substrate binding and product release whereas the chemical step takes place in the closed state. The rate determining step is the opening of the lid after the reaction has occurred. It was obvious that protein dynamics impacted catalysis when it was referred to as the overall step [85], whereas others who considered catalysis as the chemical step, did not find any direct coupling between millisecond-timescale dynamics and the chemical step. Such findings were obtained from coarse-grained simulations mapped with the properties of an atomistic model in an explicit solvent [86].

10.5 CypA as a Model to Investigate the Direct Effects of Enzyme Dynamics on Catalysis

As we have seen above, hydride transfer reactions involve nuclear quantum mechanical effects and are more complex to investigate computationally. CypA, on the other hand, catalyzes the reversible *cis-trans* isomerization of peptide bonds preceding proline residues and does not involve any bond formation or bond breaking. There are several advantages for modeling *cis-trans* isomerization reactions: (i) classical MD can be used for simulation; (ii) the choice for the reaction coordinate, i.e. the peptide bond dihedral, ω, is validated; (iii) along with the *cis* and the *trans* (regions around ω ~ 0° and 180°, respectively) the transition state regions are clearly distinguishable (regions around ω ~ 90°); (iv) optimized parameters for peptide bond dihedrals that have reproduced experimentally measured activation barriers and *cis-trans* equilibria are available [87]; (v) CypA is very-well characterized experimentally, making validation of simulations by comparison to experiments possible [88]. However, *cis-trans* isomerization in solution is extremely slow, occurring on the second-timescale. And the catalyzed reaction by CypA takes milliseconds, which is on the timescale still beyond the reach of conventional MD. Since the probabilities of transitions from the *trans* well to the *cis* and vice versa are very small, the system may remain trapped in an energy basin and with no guarantee that even a single transition will be observed from trajectories longer than milliseconds. We, therefore, built a model system in which the potential energy barriers around the peptidyl-prolyl bond were significantly reduced in the CypA-substrate. We summarize below the results of our recent studies on CypA-catalyzed *cis-trans* isomerization [40]. The V_2 force constant of the dihedral potential, $\frac{V_2}{2}[1 + \cos(n\phi - \gamma)]$, in the AMBER force field [89] controls the rotational barrier around the ω bond and modification of V_2 does not perturb *cis-trans* equilibria. A decrease in V_2 allowed adequate sampling of *cis-trans* transitions from conventional MD and to extract reliable kinetics. We carried out atomistic-explicit-solvent classical MD simulations of *cis-trans* isomerization in the free substrate and that bound to CypA. Rotational barriers were systematically increased by setting V_2 to increasing values (i.e. 7, 9, and 11 kcal/mol) in three different sets of simulations. On one hand, the decay of probability of survival in the *trans* well invariably exhibited mono-exponential behavior for the uncatalyzed reactions (Fig. 10.3a–c). The decays for the catalyzed *cis-trans* isomerization reactions, on the other hand, were multi-phasic. The differences in the kinetic behaviors of the uncatalyzed and the catalyzed reactions revealed the nature of the environment that is coupled to the reaction. Aqueous solvent presents a homogeneous environment in which solvent dynamics occurs on a very narrow or perhaps a single timescale and much faster than the timescale of the (uncatalyzed) reaction. In the enzyme, however, the different modes, some on the same timescale and some slightly faster and slower than the catalyzed chemical step, got coupled with the substrate dynamics over the barrier. Fitting the decays of survival probability with multi-exponential functions without deciding *a priori* [90, 91] the number of phases yielded distributions of time constants. Expectedly, with the increase in the rotational barrier, the single time constant for the uncatalyzed isomerization or the distributions of time constants obtained for the catalyzed reaction gradually shifted to slower timescales (Fig. 10.3d–f). However, the average time constant (caption of Fig. 10.4.) for the catalyzed isomerization was always faster than the time constant for the corresponding uncatalyzed reaction. Also, the trend of progressively diminishing relative amplitudes of the faster phases and increasing slower phases was notable. These results suggested the coupling between the chemical step and the dynamics in CypA. As to which enzyme modes would get coupled and affect the chemical step depended on the timescale of the reaction. However, evidence for this coupling did not mean dynamics would bring about enhancement in chemical rates (*vide infra*).

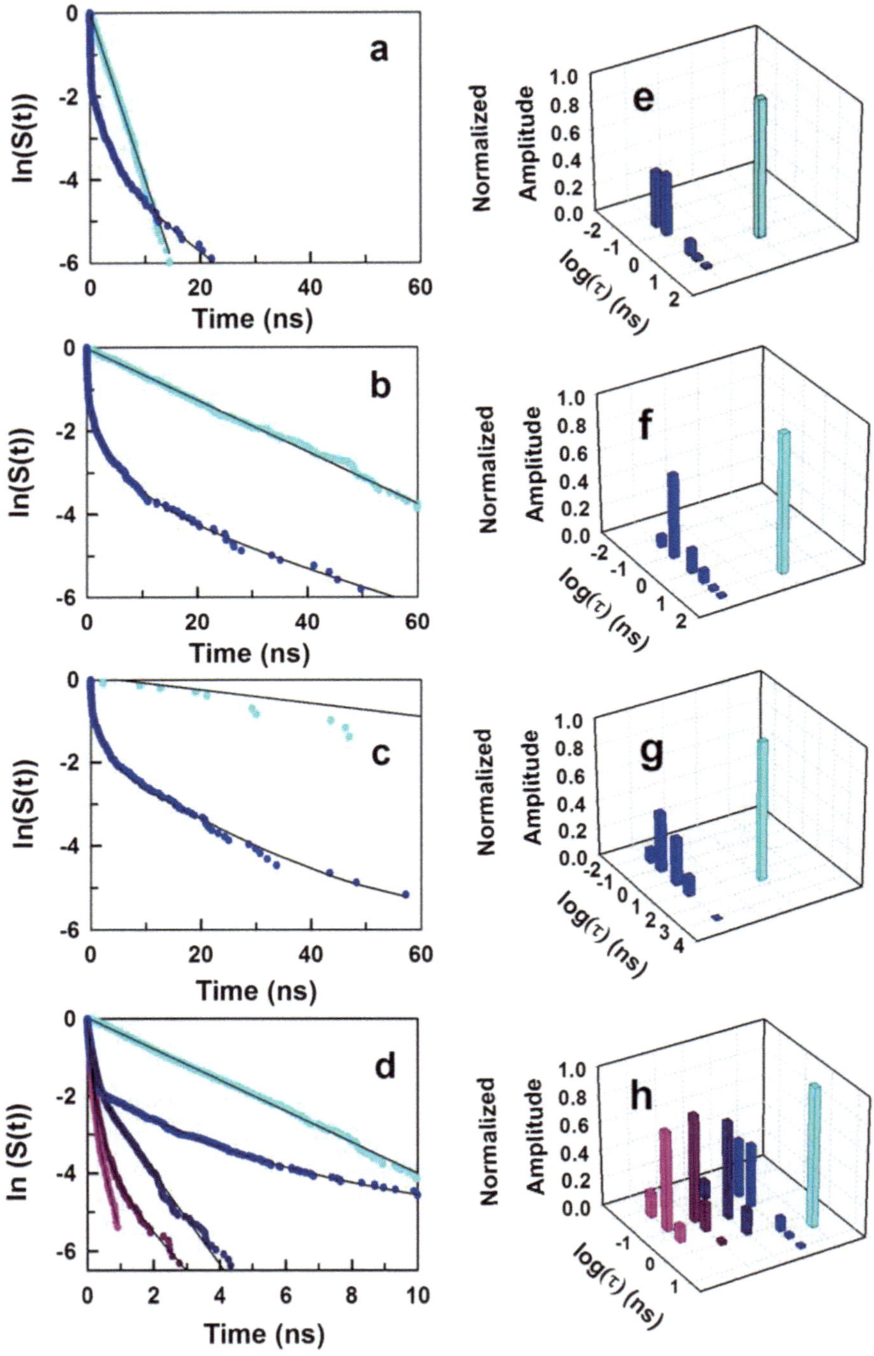

a
b
c
d
e
f
g
h
ln(S(t))
ln (S(t))
Time (ns)
Normalized
Amplitude
log(τ) (ns)

We, further, investigated how changes in the CypA dynamics would affect isomerization rates in the substrate. The idea was to mimic experimental systems wherein mutational changes (i.e. slowing down or abrogation) in enzyme dynamics correlated with diminished rates. However, in our study, we aimed to speed up the dynamics of CypA by subjecting only CypA to accelerated MD in simulations of CypA-substrate complexes. The substrate with a reduced rotational barrier ($V_2 = 7.0$ kcal/mol) was simulated with conventional MD. As the extent of acceleration was increased on CypA, all the enzyme modes sped up, resulting in the coupling of relatively faster modes with the chemical step. The multi-exponential behavior was still present but with decreased number of phases (Fig. 10.3d, h). The speed up in the slower phases caused an increase in the relative contribution of the faster phases, and clearly, the average time constants became faster, increasing the isomerization rates by factors of ~2, 3 and 5 respectively, for the lowest, intermediate and highest levels of acceleration. These results suggested that enzyme dynamics did affect the chemical step, however, to conclude whether the rate enhancement compared to the uncatalyzed reaction was brought about by CypA dynamics or not, we needed to interpret our findings within the framework of a rate theory that would allow teasing out the relative effects from barrier reduction (Sect. 10.6.2).

10.6 Theoretical Frameworks to Model and Interpret the Effects of Dynamics on Catalysis

10.6.1 Transition State Theory

Transition state theory (TST) [92] provides a simple and generalized theoretical framework to predict rate constants for reactions occurring in gas and condensed phases from the following modern mathematical expression [93]: $k = \gamma(T)\,(k_B T/h)\,\frac{Q^{\#}}{Q^{R}}\exp\left(-\Delta G^{\#}/RT\right)$

Fig. 10.3 Kinetics of *cis-trans* isomerization in solution and active site of CypA. (*Left*) Decay of survival probability (on a logarithmic scale) in the *trans* well as a function of time for the uncatalyzed (*cyan*) and the CypA-catalyzed (*blue*) *cis-trans* isomerization when V_2 was set to (**a**) 7.0 kcal/mol (**b**) 9.0 kcal/mol and (**c**) 11.0 kcal/mol in conventional MD. Effects of CypA dynamics on isomerization kinetics are seen in (**d**) when V_2 was set to 7.0 kcal/mol and CypA was subjected to normal MD (*blue*), lower (*dark blue*), intermediate (*violet*) and higher (*magenta*) levels of acceleration in accelerated MD. For comparison, decay of survival probability in the absence of the enzyme (*cyan*) is also plotted. Continuous *black lines* are (mono- or multi-) exponential fits. (*Right*) (**e–h**) Distributions of time constants (one for each phase) corresponding to **a**, **b**, **c** and **d**, respectively, obtained from exponential fits are shown with respect to their amplitudes

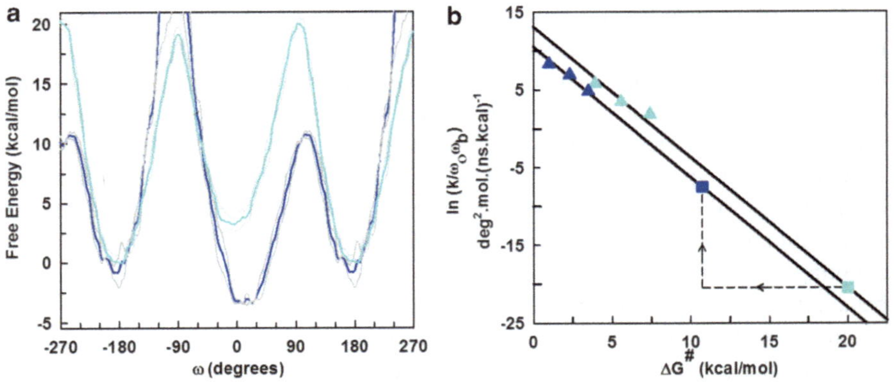

Fig. 10.4 Free energy profiles and prediction of rates of the uncatalyzed and the catalyzed *cis-trans* isomerization from Kramers' plot. (**a**) Reweighted free energy profiles of the uncatalyzed (*cyan*) and the CypA-catalyzed (*blue*) *cis-trans* isomerization generated from accelerated molecular dynamics using the optimized value of 28.0 kcal/mol for V_2 [87]. Upper and lower bounds of error are depicted by *gray lines*. (**b**) Obtained from the linear form of Kramers' rate expression in the high friction regime (Sect. 10.6.2), ln $(k/\omega_o\omega_b)$ is plotted vs. $\Delta G^{\#}$ for the uncatalyzed (*cyan triangles*) and the catalyzed (*blue triangles*) reactions when V_2 was set to 7.0 kcal/mol, 9.0 kcal/mol and 11.0 kcal/mol. Activation free energies, $\Delta G^{\#}$, curvatures of the *trans* basin (ω_b) and barrier region (ω_o) were calculated from potentials of mean force obtained from umbrella sampling. The average rate constant $\boldsymbol{k} = 1/\langle\tau\rangle$, where τ is the average lifetime given by $\langle\tau\rangle = \sum_{i=1}^{n} \boldsymbol{A}_i\tau_i$. A_i and τ_i are the amplitudes and lifetimes of phase i in the exponential function $\boldsymbol{S}\ (\boldsymbol{t}) = \sum_{i=1}^{n} \boldsymbol{A}_i \exp{(\boldsymbol{t}/\tau_i)}$ used for fitting decays in Fig. 10.3. *Black lines* are linear fits with slope = $1/k_BT$. $\Delta G^{\#}$, ω_o, and ω_b for the uncatalyzed (*cyan square*) and the catalyzed (*blue square*) *cis-trans* isomerization with the actual barriers are obtained from plots in (**a**). Assuming linearity at higher values of $\Delta G^{\#}$, the corresponding rate constants are predicted from these Kramers' plots. A decrease in barrier height of ~9 kcal/mol and an effective increase in the catalyzed rate (i.e. after correcting for the reduction in curvatures) are shown as *dashed lines with arrows*

Here, the frequency factor, k_BT/h, is on the order of $\sim 6 \times 10^{12}\ s^{-1}$ at 298 K irrespective of the reaction occurring in the gas or the condensed phase and $\Delta G^{\#}$ is the activation free energy i.e. the free energy difference between the reactant and the transition state. $Q^{\#}$ and Q^R are the partition functions of the transition state and the reactant, respectively. TST allows the incorporation of quantum mechanical effects in $\gamma(T)$, which is the generalized transmission coefficient including the dynamical effects from barrier recrossing, tunneling and non-equilibrium conditions, i.e. deviations from Boltzmann distribution in the phase space [93]. In some implementations, $\Delta G^{\#}$ includes the nuclear quantum effects from quantized vibrations. Transmission coefficients can be evaluated in many different ways as discussed in references [49, 94 and other references therein]. Transmission coefficients calculated for many enzyme-catalyzed reactions have been found to be larger than the reference reactions in solution, thereby contributing to rate enhancement [35].

However, independent theoretical calculations that estimated fluctuations of the solvent coordinates from MD simulations have shown no significant differences in the transmission coefficients of the enzyme-catalyzed and the solution reactions. As compared to the predominant contribution from the exponential dependence of $\Delta G^{\#}$ to the increase in the enzyme-catalyzed rates, the speed up due to larger transmission coefficients, if any, is typically less than an order. In the context of variational TST, if the reaction coordinate is appropriately chosen, the dynamical effects arising from recrossing the transition state hypersurface can become negligible and thus the rate constant can be minimized. The usage of TST for enzymatic reactions have been argued as it does not take into account the direct influence of the surrounding and the accuracy of the rate constant depends on the choice of the dividing hypersurface. Nevertheless, improved variants of TST with ensemble averaging and multidimensional tunneling remain to be the rate theory of choice to rationalize experimentally measured temperature dependence of KIEs and reproduce activation barriers in several wild-type and mutant enzymes, especially those involved in hydride transfer reactions where nuclear quantum effects may be significant [49, 50]. Recently, a model that invoked more than one conformation of the reactant with different transfer rates for product formation under the TST framework has been successful in fitting experimental data on rate constants, KIEs and their temperature dependence for many enzymes [82, 95]. The model assumed certain *a priori* criteria for fitting procedures, one being that the rate of interconversion between the two conformations was much faster than their conversion to the product. Although this simplistic model demonstrated that explaining experimental trend was possible without introducing any direct role for protein motions, it included only two conformations, which is not consistent with the picture implied from single molecule studies, i.e. enzymes have multiple conformations with a wide range of reactivities.

10.6.2 Kramers' Rate Theory

Kramers' rate theory in the overdamped limit has been employed by us in our recent work to interpret the simulation results on CypA [40]. Kramers' framework explicitly allows for the role of the environment in the noise-driven escape over the barrier [96]. On a one-dimensional free energy profile, the rate of escape from the reactant well is given by: $k = (\omega_0\omega_b D_{eff}/2\pi k_B T)\exp(-\Delta G^{\#}/k_B T)$ where ω_o and ω_b are the curvatures of the reactant well and the barrier region, respectively, $\Delta G^{\#}$ is the free energy barrier height and D_{eff} is the effective diffusion coefficient, assumed to be constant along the reaction coordinate. The effects from solvent viscosity or internal friction from the enzyme as well as dynamical effects due to solvent dynamics or enzymatic motions on a wide range of timescales are incorporated into the effective diffusion coefficient. The differences in the energetic roughness in the aqueous environment and the enzyme's active site are also included in D_{eff}, which will, therefore, vary for the uncatalyzed and enzyme-catalyzed reactions. When

Kramers' rate equation is expressed in the logarithmic form and then rearranged, it results in a linear relation: $\ln(k/\omega_0\omega_b) = \ln(D_{eff}/2\pi k_B T) - (1/k_B T)\Delta G^\#$. This means that $\ln(k/\omega_0\omega_b)$ can be plotted vs. $\Delta G^\#$ with a well-defined slope of $1/k_B T$ and D_{eff} can be calculated from the y-intercept. As mentioned in the Sect. 10.5, the rotational barriers around the peptide bond are very high (>20 kcal/mol), prohibiting the use of conventional MD to simulate *cis-trans* isomerization. We, therefore, computed the rates of escape from the *trans* well by setting the rotational barriers to significantly lowered values. Using umbrella sampling, we generated free energy profiles for the uncatalyzed and the CypA-catalyzed *cis-trans* isomerization simulated for each distinct value of the rotational barrier. However, free energy profiles for the actual rotational barrier were obtained by subjecting the substrate to accelerated MD (Fig. 10.4a). Accelerated MD speeds up the transition out of energy well by raising the minima but not modifying the transition state regions (Sect. 10.3.2). Free energies of activation ($\Delta G^\#$) and the curvatures of the *trans* well and the transition state regions were calculated from the free energy profiles. The Kramers' plots for the uncatalyzed and the CypA-catalyzed *cis-trans* isomerization (Fig. 10.4b) in the lower regime of $\Delta G^\#$ were extrapolated to higher values corresponding to the actual reactions. These plots revealed that the effective diffusion coefficient for the uncatalyzed reaction was faster by an order of magnitude. Moreover, the predicted rates for the isomerization with the actual barriers exhibited a speedup of $\sim 10^5$ times for the catalyzed reaction over that in solution, which agreed notably well with experimental estimates. These results provided validation for the use of Kramers' framework in describing enzymatic reactions, despite the notion that Kramers' treatment can be used only for regimes in which the environment relaxes much faster than the chemical reaction. It was clearly seen that catalysis was brought about by a decrease of $\sim$9 kcal/mol in the free energy barrier while the effects of enzyme dynamics opposed the rate enhancement, as reflected in the reduced D_{eff} for the CypA-catalyzed reaction. When CypA dynamics were accelerated, the overall rate constant for *cis-trans* isomerization in the active site had increased. Kramers' approach permitted us to separate the relative contributions from the barrier effects and those from the prefactor. D_{eff} had indeed increased with the speed up of CypA dynamics, however, the overall rate enhancement was hindered by an increase in the free energy barrier heights. The drawback of Kramers' theory is that it cannot take into account quantum mechanical effects, restricting its use for describing reactions involving tunneling. Despite this fact, the effects of disperse enzyme dynamics on the chemical step can be interpreted directly and applied generally across all enzymatic reactions. Unlike the implementation of TST for enzyme reactions in which rates are calculated from activation barriers obtained from equilibrium simulations and estimates of transmission coefficient, in this approach we directly simulated the kinetics of *cis-trans* isomerization and obtained the rates of escape out of an energy basin. The effective diffusion coefficient was then predicted from the information of kinetic rates and free energy barriers. Alternatively, Grote-Hynes theory can be used for analyzing the coupling between slowly relaxing environment as in the enzyme and the reaction dynamics, with the advantage of combining with QM/MM calculations [97].

10.7 Summary

To rationalize outstanding catalytic efficiencies of enzymes, a catalytic-promoting role of enzyme dynamics has been implied from recent experiments and theoretical studies. Establishing enzymatic dynamical effects to rate enhancement has been elusive and contentious. Various computational methodologies and procedures for analyses have been developed to interpret observations from experiments and simulations and further understand at the atomistic level the coupling between enzymatic motions and the reaction occurring at the active site. This has made possible to model chemical reactions that involve bond breaking and formation and simple but very slow isomerization around peptide bonds with large activation barriers. Proposals on how enzyme dynamics may increase catalytic rates via tunneling have been put forth. However, nuclear quantum effects and tunneling, wherever present, are shown to decrease the effective barriers by 2–3 kcal/mol, accounting for a speed up of not more than 1–2 orders of magnitude in enzymatic rates [35]. Also, the presence of tunneling in uncatalyzed reactions in solution, and not only in enzymatic reactions, diminishes its effective contribution in enhancing catalytic rates [94, 98]. Our atomistic MD simulation studies on CypA have provided useful insights and reconciled the opposing perspectives regarding the role of enzyme dynamics in catalysis. Comparing the kinetic behaviors of *cis-trans* isomerization in aqueous solution and the CypA active site suggested that enzyme modes are coupled to and can affect reaction dynamics. Analysis of our results with Kramers' rate theory revealed that enzyme dynamical effects, incorporated in the prefactor, do not increase, but rather opposes the potential enhancement in catalytic rates due to reduction in activation barriers. The predominant factor responsible for the increase in the overall rate is the barrier effect arising from electrostatic stabilization of the transition state.

Acknowledgments We acknowledge support from the National Science Foundation Grant MCB-0953061, the Georgia Research Alliance and Georgia State University.

References

1. Fersht AR (1999) Structure and mechanism in protein science: a guide to enzyme catalysis and protein folding. W.H. Freeman and Company, New York
2. Wolfenden R, Snider MJ (2001) The depth of chemical time and the power of enzymes as catalysts. Acc Chem Res 34:938–945
3. Demirijan DC, Shah PC, Morris-Varas F (1999) Screening for novel enzymes. In: Fessner W-D, Archelas A et al (eds) Topics in current chemistry: biocatalysis – from discovery to application. Springer, Berlin/Heidelberg
4. Fersht AR, Winter GP (2008) Redesigning enzymes by site-directed mutagenesis. In: Porter R, Clark S (eds) Ciba foundation symposium 111 – enzymes in organic synthesis. Wiley, Chichester

5. Pauling L (1946) Molecular architecture and biological reactions. Chem Eng News Arch 24:1375–1377
6. Warshel A (1978) Energetics of enzyme catalysis. Proc Natl Acad Sci USA 75:5250–5254
7. Kamerlin SC, Warshel A (2010) At the dawn of the 21st century: is dynamics the missing link for understanding enzyme catalysis? Proteins 78:1339–1375
8. Benkovic SJ, Hammes-Schiffer S (2003) A perspective on enzyme catalysis. Science 301:1196–1202
9. Lemieux RU, Spohr U (1994) How Emil Fischer was led to the lock and key concept for enzyme specificity. Adv Carbohydr Chem Biochem 50:1–20
10. Koshland DE Jr (1959) Enzyme flexibility and enzyme action. J Cell Comp Physiol 54: 245–258
11. Koshland DE Jr (1960) The active site and enzyme action. Adv Enzymol Relat Subj Biochem 22:45–97
12. Wuthrich K (1995) NMR in structural biology: a collection of papers by Kurt Wüthrich. World Scientific Publishing Co. Pte. Ltd, Singapore
13. Blake CC, Koenig DF, Mair GA et al (1965) Structure of hen egg-white lysozyme. A three-dimensional Fourier synthesis at 2 Angstrom resolution. Nature 206:757–761
14. Johnson LN, Phillips DC (1965) Structure of some crystalline lysozyme-inhibitor complexes determined by X-ray analysis at 6 Angstrom resolution. Nature 206:761–763
15. Monod J, Wyman J, Changeux JP (1965) On the nature of allosteric transitions: a plausible model. J Mol Biol 12:88–118
16. Ma B, Nussinov R (2010) Enzyme dynamics point to stepwise conformational selection in catalysis. Curr Opin Chem Biol 14:652–659
17. McCammon JA, Harvey SC (1987) Dynamics of proteins and nucleic acids. Cambridge University Press, Cambridge
18. Henzler-Wildman KA, Lei M, Thai V et al (2007) A hierarchy of timescales in protein dynamics is linked to enzyme catalysis. Nature 450:913–916
19. McCammon JA, Gelin BR, Karplus M (1977) Dynamics of folded proteins. Nature 267: 585–590
20. Hammes-Schiffer S, Benkovic SJ (2006) Relating protein motion to catalysis. Annu Rev Biochem 75:519–541
21. Masgrau L, Roujeinikova A, Johannissen LO et al (2006) Atomic description of an enzyme reaction dominated by proton tunneling. Science 312:237–241
22. Nagel ZD, Klinman JP (2009) A 21st century revisionist's view at a turning point in enzymology. Nat Chem Biol 5:543–550
23. Nashine VC, Hammes-Schiffer S, Benkovic SJ (2010) Coupled motions in enzyme catalysis. Curr Opin Chem Biol 14:644–651
24. Schwartz SD, Schramm VL (2009) Enzymatic transition states and dynamic motion in barrier crossing. Nat Chem Biol 5:551–558
25. Agarwal PK (2005) Role of protein dynamics in reaction rate enhancement by enzymes. J Am Chem Soc 127:15248–15256
26. Henzler-Wildman K, Kern D (2007) Dynamic personalities of proteins. Nature 450:964–972
27. Frauenfelder H, Sligar SG, Wolynes PG (1991) The energy landscapes and motions of proteins. Science 254:1598–1603
28. Frauenfelder H, Chen G, Berendzen J et al (2009) A unified model of protein dynamics. Proc Natl Acad Sci USA 106:5129–5134
29. Johnson Q, Doshi U, Shen T et al (2010) Water's contribution to the energetic roughness from peptide dynamics. J Chem Theory Comput 6:2591–2597
30. Kleckner IR, Foster MP (2011) An introduction to NMR-based approaches for measuring protein dynamics. Biochim Biophys Acta 1814:942–968
31. Eisenmesser EZ, Bosco DA, Akke M et al (2002) Enzyme dynamics during catalysis. Science 295:1520–1523
32. Eisenmesser EZ, Millet O, Labeikovsky W et al (2005) Intrinsic dynamics of an enzyme underlies catalysis. Nature 438:117–121

33. Min W, English BP, Luo G et al (2005) Fluctuating enzymes: lessons from single-molecule studies. Acc Chem Res 38:923–931
34. English BP, Min W, van Oijen AM et al (2006) Ever-fluctuating single enzyme molecules: Michaelis-Menten equation revisited. Nat Chem Biol 2:87–94
35. McGeagh JD, Ranaghan KE, Mulholland AJ (2011) Protein dynamics and enzyme catalysis: insights from simulations. Biochim Biophys Acta 1814:1077–1092
36. Karplus M, McCammon JA (2002) Molecular dynamics simulations of biomolecules. Nat Struct Biol 9:646–652
37. Friedrichs MS, Eastman P, Vaidyanathan V et al (2009) Accelerating molecular dynamic simulation on graphics processing units. J Comput Chem 30:864–872
38. Shaw DE, Dror RO, Salmon JK et al (2009) Millisecond-scale molecular dynamics simulations on Anton. In: Proceedings of the conference on high performance computing networking, storage and analysis, ACM, Portland, OR, pp 1–11
39. Stone JE, Phillips JC, Freddolino PL et al (2007) Accelerating molecular modeling applications with graphics processors. J Comput Chem 28:2618–2640
40. Doshi U, McGowan LC, Ladani ST et al (2012) Resolving the complex role of enzyme conformational dynamics in catalytic function. Proc Natl Acad Sci USA 109:5699–5704
41. Lee J, Kim SS (2010) An overview of cyclophilins in human cancers. J Int Med Res 38: 1561–1574
42. Amadei A, Linssen AB, Berendsen HJ (1993) Essential dynamics of proteins. Proteins 17: 412–425
43. Levy RM, Srinivasan AR, Olson WK et al (1984) Quasi-harmonic method for studying very low frequency modes in proteins. Biopolymers 23:1099–1112
44. Humphrey W, Dalke A, Schulten K (1996) VMD: visual molecular dynamics. J Mol Graph 14:33–38, 27–38
45. Mongan J (2004) Interactive essential dynamics. J Comput Aided Mol Des 18:433–436
46. McGowan LC, Hamelberg D (2013) Conformational plasticity of an enzyme during catalysis: intricate coupling between cyclophilin A dynamics and substrate turnover. Biophys J 104: 216–226
47. Ramanathan A, Agarwal PK (2011) Evolutionarily conserved linkage between enzyme fold, flexibility, and catalysis. PLoS Biol 9:e1001193
48. Warshel A, Levitt M (1976) Theoretical studies of enzymic reactions: dielectric, electrostatic and steric stabilization of the carbonium ion in the reaction of lysozyme. J Mol Biol 103: 227–249
49. Gao J, Truhlar DG (2002) Quantum mechanical methods for enzyme kinetics. Annu Rev Phys Chem 53:467–505
50. Dybala-Defratyka A, Paneth P, Truhlar DG (2009) Quantum catalysis in enzymes. In: Scrutton NS, Allemann RK (eds) Quantum tunnelling in enzyme-catalysed reactions. The Royal Society of Chemistry, Cambridge
51. Friesner RA, Guallar V (2005) Ab initio quantum chemical and mixed quantum mechanics/molecular mechanics (QM/MM) methods for studying enzymatic catalysis. Annu Rev Phys Chem 56:389–427
52. Senn HM, Thiel W (2007) QM/MM studies of enzymes. Curr Opin Chem Biol 11:182–187
53. Senn HM, Thiel W (2009) QM/MM methods for biomolecular systems. Angew Chem Int Ed Engl 48:1198–1229
54. Torrie GM, Valleau JP (1977) Nonphysical sampling distributions in Monte Carlo free-energy estimation: umbrella sampling. J Comput Phys 23:187–199
55. Zwanzig RW (1954) High-temperature equation of state by a perturbation method. I. Nonpolar gases. J Chem Phys 22:1420–1426
56. Kamerlin SCL, Warshel A (2011) The empirical valence bond model: theory and applications. Wiley Interdiscip Rev Comput Mol Sci 1:30–45
57. Fan Y, Cembran A, Ma S et al (2013) Connecting protein conformational dynamics with catalytic function as illustrated in dihydrofolate reductase. Biochemistry 52:2036–2049

58. Hammes-Schiffer S (2013) Catalytic efficiency of enzymes: a theoretical analysis. Biochemistry 52(12):2012–2020
59. Hamelberg D, Mongan J, McCammon JA (2004) Accelerated molecular dynamics: a promising and efficient simulation method for biomolecules. J Chem Phys 120:11919–11929
60. Doshi U, Hamelberg D (2011) Extracting realistic kinetics of rare activated processes from accelerated molecular dynamics using Kramers' theory. J Chem Theory Comput 7:575–581
61. Markwick PR, Bouvignies G, Blackledge M (2007) Exploring multiple timescale motions in protein GB3 using accelerated molecular dynamics and NMR spectroscopy. J Am Chem Soc 129:4724–4730
62. Hamelberg D, McCammon JA (2009) Mechanistic insight into the role of transition-state stabilization in cyclophilin A. J Am Chem Soc 131:147–152
63. Xin Y, Doshi U, Hamelberg D (2010) Examining the limits of time reweighting and Kramers' rate theory to obtain correct kinetics from accelerated molecular dynamics. J Chem Phys 132:224101
64. Hamelberg D, Shen T, Andrew McCammon J (2005) Relating kinetic rates and local energetic roughness by accelerated molecular-dynamics simulations. J Chem Phys 122:241103
65. Doshi U, Hamelberg D (2012) Improved statistical sampling and accuracy with accelerated molecular dynamics on rotatable torsions. J Chem Theory Comput 8:4004–4012
66. Klinman JP (2013) Importance of protein dynamics during enzymatic C–H bond cleavage catalysis. Biochemistry 52:2068–2077
67. Kohen A (2012) Enzyme dynamics: consensus and controversy. J Biocatal Biotransform 1:1–2
68. Henzler-Wildman KA, Thai V, Lei M et al (2007) Intrinsic motions along an enzymatic reaction trajectory. Nature 450:838–844
69. Boehr DD, McElheny D, Dyson HJ et al (2006) The dynamic energy landscape of dihydrofolate reductase catalysis. Science 313:1638–1642
70. Bhabha G, Lee J, Ekiert DC et al (2011) A dynamic knockout reveals that conformational fluctuations influence the chemical step of enzyme catalysis. Science 332:234–238
71. Adamczyk AJ, Cao J, Kamerlin SC et al (2011) Catalysis by dihydrofolate reductase and other enzymes arises from electrostatic preorganization, not conformational motions. Proc Natl Acad Sci USA 108:14115–14120
72. Liu H, Warshel A (2007) The catalytic effect of dihydrofolate reductase and its mutants is determined by reorganization energies. Biochemistry 46:6011–6025
73. Nagel ZD, Klinman JP (2010) Update 1 of: tunneling and dynamics in enzymatic hydride transfer. Chem Rev 110:PR41–PR67
74. Loveridge EJ, Behiry EM, Guo J et al (2012) Evidence that a 'dynamic knockout' in *Escherichia coli* dihydrofolate reductase does not affect the chemical step of catalysis. Nat Chem 4:292–297
75. Agarwal PK, Billeter SR, Rajagopalan PT et al (2002) Network of coupled promoting motions in enzyme catalysis. Proc Natl Acad Sci USA 99:2794–2799
76. Wong KF, Selzer T, Benkovic SJ et al (2005) Impact of distal mutations on the network of coupled motions correlated to hydride transfer in dihydrofolate reductase. Proc Natl Acad Sci USA 102:6807–6812
77. Antoniou D, Basner J, Nunez S et al (2006) Computational and theoretical methods to explore the relation between enzyme dynamics and catalysis. Chem Rev 106:3170–3187
78. Saen-Oon S, Quaytman-Machleder S, Schramm VL et al (2008) Atomic detail of chemical transformation at the transition state of an enzymatic reaction. Proc Natl Acad Sci USA 105:16543–16548
79. Boekelheide N, Salomon-Ferrer R, Miller TF 3rd (2011) Dynamics and dissipation in enzyme catalysis. Proc Natl Acad Sci USA 108:16159–16163
80. Loveridge EJ, Tey LH, Behiry EM et al (2011) The role of large-scale motions in catalysis by dihydrofolate reductase. J Am Chem Soc 133:20561–20570
81. Kamerlin SC, Mavri J, Warshel A (2010) Examining the case for the effect of barrier compression on tunneling, vibrationally enhanced catalysis, catalytic entropy and related issues. FEBS Lett 584:2759–2766

82. Glowacki DR, Harvey JN, Mulholland AJ (2012) Protein dynamics and enzyme catalysis: the ghost in the machine? Biochem Soc Trans 40:515–521
83. Hay S, Johannissen LO, Sutcliffe MJ et al (2010) Barrier compression and its contribution to both classical and quantum mechanical aspects of enzyme catalysis. Biophys J 98:121–128
84. Hay S, Scrutton NS (2012) Good vibrations in enzyme-catalysed reactions. Nat Chem 4: 161–168
85. Karplus M (2010) The role of conformation transitions in adenylate kinase. Proc Natl Acad Sci USA 107:E71
86. Pisliakov AV, Cao J, Kamerlin SC et al (2009) Enzyme millisecond conformational dynamics do not catalyze the chemical step. Proc Natl Acad Sci USA 106:17359–17364
87. Doshi U, Hamelberg D (2009) Re-optimization of the AMBER force field parameters for peptide bond (Omega) torsions using accelerated molecular dynamics. J Phys Chem B 113:16590–16595
88. Kern D, Kern G, Scherer G et al (1995) Kinetic analysis of cyclophilin-catalyzed prolyl cis/trans isomerization by dynamic NMR spectroscopy. Biochemistry 34:13594–13602
89. Cornell WD, Cieplak P, Bayly CI et al (1995) A 2nd generation force-field for the simulation of proteins, nucleic-acids, and organic-molecules. J Am Chem Soc 117:5179–5197
90. Provencher SW (1976) An eigenfunction expansion method for the analysis of exponential decay curves. J Chem Phys 64:2772–2777
91. Provencher SW (1976) A Fourier method for the analysis of exponential decay curves. Biophys J 16:27–41
92. Eyring H (1935) The activated complex in chemical reactions. J Chem Phys 3:107–115
93. Garcia-Viloca M, Gao J, Karplus M et al (2004) How enzymes work: analysis by modern rate theory and computer simulations. Science 303:186–195
94. Olsson MH, Parson WW, Warshel A (2006) Dynamical contributions to enzyme catalysis: critical tests of a popular hypothesis. Chem Rev 106:1737–1756
95. Glowacki DR, Harvey JN, Mulholland AJ (2012) Taking Ockham's razor to enzyme dynamics and catalysis. Nat Chem 4:169–176
96. Kramers HA (1940) Brownian motion in a field of force and diffusion model of chemical reactions. Physica (Utrecht) 7:284–304
97. Castillo R, Roca M, Soriano A et al (2008) Using Grote-Hynes theory to quantify dynamical effects on the reaction rate of enzymatic processes. The case of methyltransferases. J Phys Chem B 112:529–534
98. Truhlar DG (2010) Tunneling in enzymatic and nonenzymatic hydrogen transfer reactions. J Phys Org Chem 23:660–676

Chapter 11
Exploiting Protein Intrinsic Flexibility in Drug Design

Suryani Lukman, Chandra S. Verma, and Gloria Fuentes

Abstract Molecular recognition in biological systems relies on the existence of specific attractive interactions between two partner molecules. Structure-based drug design seeks to identify and optimize such interactions between ligands and their protein targets. The approach followed in medicinal chemistry follows a combination of careful analysis of structural data together with experimental and/or theoretical studies on the system. This chapter focuses on the fact that a protein is not fully characterized by a single structure, but by an ensemble of states, some of them represent "hidden conformations" with cryptic binding sites. We highlight case studies where both experimental and computational methods have been used to mutually drive each other in an attempt to improve the success of the drug design approaches.

Advances in both experimental techniques and computational methods have greatly improved our physico-chemical understanding of the functional mechanisms in biomolecules and opened a debate about the interplay between molecular structure and biomolecular function. The beautiful static pictures of protein structures may have led to neglecting the intrinsic protein flexibility, however we are entering a new era where more sophisticated methods are used to exploit this ability of macromolecules, and this will definitely lead to the inclusion of the notion in the pharmaceutical field of drug design.

S. Lukman • C.S. Verma • G. Fuentes (✉)
Bioinformatics Institute (BII), Agency for Science, Technology and Research (A*STAR), 30 Biopolis Street #07-01, Matrix Building, 138671 Singapore, Singapore
e-mail: chandra@bii.a-star.edu.sg; gfuentes@bii.a-star.edu.sg; gloria.fuentes@riken.jp

K. Han et al. (eds.), *Protein Conformational Dynamics*, Advances in Experimental Medicine and Biology 805, DOI 10.1007/978-3-319-02970-2_11,

Abbreviations

ATP	adenosine triphosphate
FDA	Food and Drug Administration (US)
CBF	core binding factor
CPMG	Carr–Purcell Meiboom–Gill relaxation dispersion
CSP	chemical shift perturbations
DHFR	dihydrofolate reductase
FBDD	fragment-based drug design
FEP	free energy perturbation
FTIR spectroscopy	Fourier transform infrared spectroscopy
GTP	guanosine triphosphate
GPCR	protein-coupled receptor
HIV	human immunodeficiency virus
HSQC	heteronuclear single quantum correlation
HTS	high-throughput screening
LBD	ligand binding domains
LES	locally enhanced sampling
LIE	linear interaction energy
MAPK	mitogen activated protein kinase
MBP	maltose binding protein
MD	molecular dynamics
MM-PB(GB)SA	molecular mechanics Poisson-Boltzmann (Generalized Born) Surface Area
MRC	multiple receptor conformations
NADPH	nicotinamide adenine dinucleotide phosphate
NMA	normal mode analysis
NMR	nuclear magnetic resonance
NOE	nuclear Overhauser effect
PBP	periplasmic binding protein
PI(3)K	phosphoinositide-3-OH kinase
PRE	paramagnetic relaxation enhancement
RCS	relaxed complex scheme
RDC	residual dipolar coupling
REMD	replica exchange molecular dynamics
SAR	structure activity relationship
SAXS	small-angle X-ray scattering
SBDD	structure-based drug design
TAMD	temperature-accelerated MD
TMD	targeted MD

11.1 Introduction: Including the Importance of Flexibility of Both Receptor and Ligand

For some time, it was thought that all functions that proteins perform stem predominantly from the three-dimensional structures that they adopt. Because these structures are stabilized by a large network of weak interactions, they are easily rearranged by thermal motion at biologically relevant temperatures. This provides proteins an intrinsically dynamism that is best understood by considering proteins as an ensemble of inter-converting conformers [1, 2]; in contrast to the static interpretation that has been generated from representing a single conformation as determined by crystallography. Furthermore, a large body of experimental and computational studies has conclusively highlighted that many biochemical and cellular processes are intimately coupled to the structural fluctuations of proteins. Thus, the explicit description of the dynamical behavior of these molecules is required to relate the structure of biomolecular systems with their function in the biological context.

The range of motions covered by protein dynamics is extraordinarily large both in space and time domains, ranging from 10^{-11} to 10^{-8} m, and from femtoseconds to hours, respectively [3]. It has been generally assumed that fast and local motions can be fundamental in ligand binding and enzymatic catalysis, while slower and global motions modulate allostery and conformational transitions; even longer timescales are characteristic of protein folding and association. Conformational flexibility extends to different levels in the spectrum of motions starting with small local adjustment such as re-orientation of side chains or even backbones, continuing with medium-range movement loops and local alterations in secondary structures, to large-scale conformational changes of structural motifs or domains, including the extreme case of disorder-to-order transitions.

This intrinsic dynamical property of proteins is of extremely importance in the field of ligand-receptor recognition, and thus it has direct consequences in the drug design. Numerous evidences have revealed the limitations of the lock-and-key theory in taking into account the conformational changes upon ligand binding, and in consequence novel mechanisms to account for the inherent plasticity of biomolecules have been elaborated, such as the induced-fit model or the existence of an ensemble of pre-equilibrated conformations, as proposed in the population-shift mechanism or also known as conformational selection [4]. Neglecting any flexibility at the binding site, although has shown some success, in general it has led to failures in predicting protein recognition events and in successfully docking ligands with protein receptors [5].

Currently, it has become a common practice to combine both experimental and computational methods to unravel the role of conformational dynamics associated with the binding of a ligand to a protein. Experimental data can reveal the type of motions inherent to the ligand recognition and the type of mechanism that the system is more likely to use in such an event. This highly system-dependent information can then be used to determine the most suitable computational methods to describe the intrinsic conformational diversity that characterize the system. Reproducing the inherent flexibility of biomolecules has thus become one of the most challenging issues in molecular modeling and simulation studies, as it has direct implications in our understanding of the structure-function relationships, even in areas such as virtual screening and structure-based drug discovery. A small number of current computational and experimental methods that can address this diversity in flexibility are detailed in the next section.

11.2 Experimental Methods Reporting Protein Conformational Changes

There is a large number of experimental techniques that can unravel the implications of conformational dynamics in the protein structure and function, as well as few others characterizing the molecular forces that govern the recognition event (see Fig. 11.1). Here we focus on the first type although in certain cases, these same techniques can complementarily provide information for both types of phenomena.

Infrared and Raman spectroscopies measure the vibrational frequencies of a group of bonded atoms in a molecule. There is a high degree of synergy amongst these experimental probes in the information that they provide. The frequencies are determined by the masses of atoms, the force constants and the geometry of the molecule [6]. This makes possible the structural determination of parameters, such as bond orders, bond lengths and angles, ionization state of ionizable moieties for a molecular group from its vibrational frequencies. Changes are observed when the vibrational frequencies are affected, like it occurs when new interactions are formed between molecular groups upon binding of a small molecule to a protein, such as hydrogen bonding and other bond polarizing electrostatic interactions, and/or geometry distortion from steric clashes. The accuracy of determining these parameters from vibrational frequencies is very high however it is not yet feasible to determine the full structure of a protein from its vibrational spectrum, basically due to spectral crowding. Recent developments in Fourier transform infrared (FTIR) spectroscopy and Raman difference techniques place them as powerful tools for the investigation of protein-ligand interactions [7], and they have made possible to measure the vibrational spectra of any small molecule when it is bound to a protein. In such an experiment, a protein spectrum is measured in the apo and ligand-bound states, and the subtraction of the two yields the spectrum of the bound ligand.

Inelastic neutron scattering captures the diffusive component of protein internal motions by determining the vibrational density of states (frequency distribution). However, this description is complexed due to the large variety of co-existing motions within the protein, where groups of atoms undergo a plethora of continuous

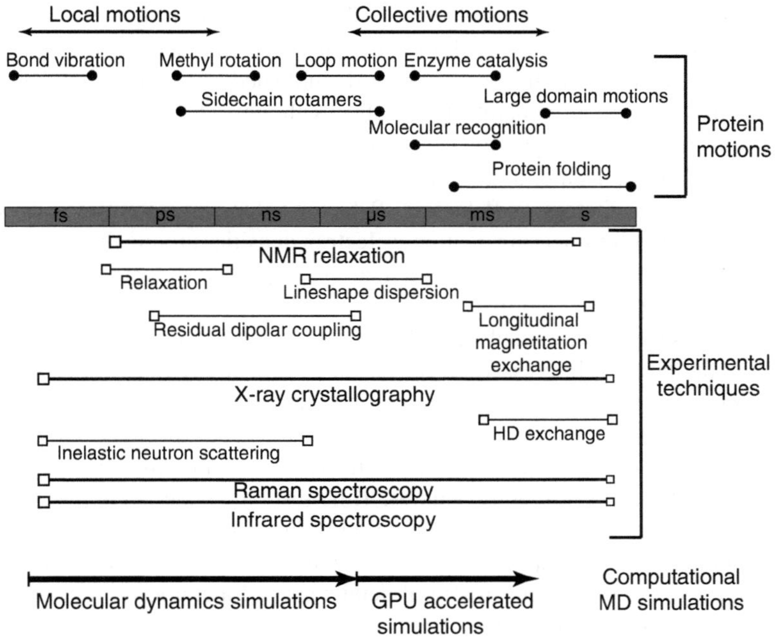

Fig. 11.1 Experimental and computational techniques to study protein motions. A timescale line is drawn as reference ranging from femtoseconds to seconds. Above such a line, the protein motions have been annotated according to their timescale; while below it, the computational methods reporting on the flexibility within the different regions in time are indicated

or jump-like diffusion. Neutron spectroscopy permits the investigation of motions in the time range from 10^{-14} to 10^{-9} s (depending on the timescale different methods are available such as time-of-flight, backscattering, spin echo techniques) [8, 9]. Neutron scattering also provides an unique opportunity for comparing the dynamics of protein and hydration water due to the large incoherent cross section between hydrogen and deuterium, and the fact that hydrogen atoms are distributed 'quasi-homogeneously' in proteins. In most cases, the hydrogenated protein is measured in H_2O and in D_2 technique [10]. They found that the vibrations of the complex are significantly softer relative to the unbound protein, meaning that the protein-ligand association can be significantly more flexible than the apo protein, that resulting in free-energy change that contributes significantly around 3 orders of magnitude more to the binding equilibrium. This was surprising since it has been frequently accepted that the complexation of the ligand will somehow rigidify the complex.

X-Ray crystallography is the most established and accurate method for determination of the three-dimensional structure of a protein. The output data of such a method corresponds to a single set of coordinates of all atoms in the molecule. X-ray crystallography is highly sensitive to the experimental conditions under which it was performed. Thus, in spite of the difficulty and time-consuming process, often for a large number of interesting molecules, there are several structures available characterized under different conditions, collectively constituting a conformational ensemble that contains relevant biological motions and defines the system. A representative example in this line is the work published by Grant and collaborators on kinesins [11]. But in general, it was always thought that the data and models coming out from crystallographic experiments were quite static, mainly providing snapshots of the molecule, and with a glimpse of the system flexibility reflected in the temperature factors or B-factors. B-factor values, as determined crystallographically, represent smearing of atomic electron densities around their equilibrium positions and have been associated as a measure of protein dynamics, flexibility of amino acids and protein stability.

However, several time-resolved X-ray methods have been recently developed following the burst of impressive advances on intense X-ray radiation at the third-generation synchrotron sources. In general all these methods aim to follow in real time the functional conformational changes a protein goes through. Among them, the most popular are time-resolved Laue diffraction and intermediate trapping (or so-called kinetic crystallography) studies. Time-resolved wide-angle X-ray scattering has also emerged for the characterization of large-scale global conformational changes in proteins in a liquid environment, while time-resolved X-ray absorption studies are used to determine structural details around metal ions (for a review of these methods see the work published by Westenhoff and collaborators [12]).

The second most common method of determining the structure of a protein is **nuclear magnetic resonance (NMR) spectroscopy**. Although, this method is in general less accurate than X-ray crystallography and limited to small to medium sized proteins, it is unique in that it can simultaneously describe directly the motions that proteins undergo as well as their structure, and consequently it yields great insight into how proteins 'work' [13]. Another advantage of using NMR structures is that the final solution is not a single structure but a family of structures derived from the set of experimental conditions that best represent the system. Although this family is usually composed of 10–50 structures, this number can be made as large as necessary by deriving more structures that satisfy the NMR experimental constraints. The progress in NMR-based approaches and their impact in deciphering the nature of protein dynamics have been extensively reviewed [14–16]. This type of spectroscopy is uniquely suited to study many of these dynamical processes, because site-specific information can be extracted for a large variety of motions that span many timescales (see Fig. 11.1). When analyzing fast dynamics (ps to ns), the measurements of NMR relaxation rates provide the parameters required to characterize these motions, due to the sensitivity of internal motions towards fluctuations of the local magnetic fields. The most common approach used to get insight into the motional processes is the model-free from Lipari-Szabo that expresses the amplitude of these motions as order parameters, S2 [17]. For slower motions (μs to ms), which normally involve larger conformational transitions with transient intermediates, relaxation dispersion is an amenable tool. In this technique, the additional line-broadening of the NMR signals caused by the conformational exchange between two states can provide information about the relative populations for the different intermediates, their rate of interconversion and additionally the chemical shifts between these exchanging species. The basic NMR experiment is the so-called Carr–Purcell Meiboom–Gill (CPMG) relaxation dispersion [18], that relies on the application of a variable number of refocusing pulses during a fixed interval of time.

Another technique frequently used for determining dynamics spanning a broader timescale is based on the measurement of Residual Dipolar Coupling (RDCs) [19]. The loss of structural information due to the random tumbling of a molecule can be recovered if the protein is dissolved in an anisotropic solution, such as liquid crystalline media. Under these conditions, the macromolecule becomes weakly aligned so the magnetic dipolar interactions do not average to zero and residual dipolar coupling constants can be measured between proximal pair of nuclei. They are a rich source of structural information since they do report on the orientation of bond vectors with respect to the magnetic field. Dipolar coupling data *per se* do not report on the timescale of the motions, however this information is extracted by comparison of the per-residue generalized order parameters estimated using the 3D GAF model of dynamics with those parameters extracted from spin relaxation experiments. The synergism between NMR spectroscopy and molecular dynamics simulations is of great importance due to their high complementarity [20]. On one hand, NMR provides quantitative data on dynamical processes, but these data does not unambiguously describe the motions; on the other hand, MD simulations describe atomic motions at high detail but they are highly dependent on the force fields and model used. Combination of both can reveal a deeper understanding of the dynamics of the system, with a physical and quantitative description of the motions involved.

NMR spectroscopy, apart of providing structural and dynamical information on the protein receptor, offers a great potential in drug discovery since it can reveal information about protein-ligand interactions at atomic resolution [21]. During the past decade, novel NMR screening methods have been developed either to increase the sensitivity and/or to reduce the protein consumption in an attempt to push the technique towards a high-throughput use. A brief description of some new NMR screening techniques applied to drug discovery has been given in very good reviews [22, 23]. NMR screening methods can be divided into two main categories: target-based screening relying on observing the target resonances, and ligand-based screening, which comprises methods that rely on the detection of an altered hydrodynamic property of the ligand or the transfer of an NMR signal between target and ligand. In screening methods that observe the macromolecular target, the parameters that are typically monitored are the chemical shifts. Heteronuclear Single Quantum Correlation (HSQC) experiment exploits the differences in chemical shifts on the two-dimensional correlation spectra of the target upon titration of a ligand. The NMR assay is simple, less time consuming compared with other acquisition methods, applicable to any class of compound, with no upper limit in affinity and highly informative in the identification of the ligand binding site. The SAR (Structure Activity Relationship) by NMR technique makes use of ligand-induced chemical shift perturbations (CSPs) in the protein target to localize the binding sites [24] and it is extensively used in fragment-based drug design (FBDD). Alternatively, experiments that rely on observation of ligand resonances can be used for protein targets of virtually any size, but are restricted in the ligand's binding affinity range; although this scenario is changing due to the introduction of cryo-cooled NMR probe technology. For these methods, the choice of NMR parameters is more diverse, few of them rely on the detection of an altered hydrodynamic property of the ligand, such as transferred NOESY, relaxation-edited or diffusion-edited NMR experiments; while few other will follow the evolution on the transfer of an NMR signal between target and ligand, including saturation transfer difference and WaterLOGSY experiments [25].

Although NMR based screening is only one of the more recent additions to the bag of tools used in drug discovery [26], it promises to revolutionize the field, mainly for the amount of detailed structural information it can provide, ranging from revealing the binding site on the target (HSQC screening), the conformation of the bound ligand (transfer NOE), and additionally it can also supply information facilitating the precise docking of the ligand to the protein's binding pocket (isotope-filtered NOESY).

Small-angle X-ray scattering (SAXS) has become a widely streamline tool in structural molecular biology for low-resolution structural characterization of macromolecules in solution [27–30]. The major advantage of SAXS compared to other structural characterization techniques is that it can be performed under a wide variety of solution environments, including near physiological conditions, and for a wide range of molecular sizes. Additionally, it provides unique information about overall structure and conformational changes of native individual proteins, functional complexes, flexible macromolecules and assembly processes. In a recent study, Fenton and collaborators have used a combination of protein fluorescence and SAXS to monitor and characterize different conformational changes associated with pyruvate kinase triggered by the binding of different allosteric inhibitors and the effect that they elicit on the binding of its substrate [31]. The authors, apart of revealing the intrinsic motions of the system, have been able to propose a novel view in the regulation of the enzyme. Allostery does not derive from a conformational transition upon the binding of a single ligand, but both substrate and inhibitors mutually modulate how the other molecule binds to the enzyme, in line with a linked-equilibrium mechanism and against previously proposed two-state model [32].

In a normal scenario, the identification and characterization of the dynamics in a system will require the combination of different experimental techniques. *E. coli* dihydrofolate reductase (DHFR) is a popular target for drug design against microbial infections. DHFR is one of the most extensively studied enzymes at the structural and dynamical level. The protein progresses through its catalytic cycle transiting into two different states for its Met20 loop, "closed" and "occluded" states. The bacterial enzyme bound to a quinazoline derivative exhibits conformational dynamics, both in the protein and the small molecule, as it was shown by NMR line-broadening properties. On the other hand, crystallographic studies revealed a closed conformation for the complex DHFR-NADPH-ligand, with electron density poorly defined in the loop region, and portions of cofactor and inhibitor, suggesting multiple binding poses and high mobility. NMR chemical shift and RDCs experiments corroborated the closed conformation for the Met20 loop and suggested a different solution preferred orientation for the inhibitor and cofactor. From NMR spectroscopy and X-ray crystallography, the compound was found to bind in an unorthodox orientation but switch internally to drive a dynamic conformational loop change in the protein. The two methods used jointly are highly complementary, and both are necessary to develop a full, accurate picture of this small molecule complex [33].

11.3 Computational Methods to Disclose Flexibility in Proteins

In the modern drug discovery, all steps are closely linked with some computational methods [34]. There exists a high interest on the *in silico* protocols with the ability to predict the binding mode and affinity of small molecules to biomolecular targets of pharmaceutical interest. This is the case in structure-based drug design (SBDD) approaches, where the knowledge of the 3D structure of the target is exploited to design small molecules that tightly bind into the active site and modulate the biological function. The proper incorporation of protein flexibility in the prediction of binding poses and affinities for small compounds has attracted increasing attention recently in current structure-based drug design projects such as virtual screening and protein-ligand docking [35, 36]. Incorporating protein flexibility would allow the expansion of the conformational and chemical space of the hit molecules. Diversity in ligand-binding mechanisms and associated conformational changes make difficult to treat dynamic features of the receptor during docking protocols, and this has been the main reason for neglecting this factor. We discuss here some of the most commonly used computational tools developed to characterize the conformational flexibility in ligand-receptor complexes, focusing on those that are used in the examples presented in the last part of the chapter. There is a large number of published works in the field [35, 37–41].

In spite of the approach taken, drug design procedures need to predict and simulate the intrinsic flexibility of proteins, trying to find the best compromise between accuracy, reliability and computational resources available. Several docking algorithms and programs have incorporated different levels, both implicitly or explicitly, for receptor flexibility during or prior the virtual screening experiments. The less time and computational consuming algorithms simulate the possible movements of the side chains in the active site by a variety of approaches, such as the implementation of soft potentials [42], the use of rotameric conformation libraries, or even just local energy minimization because of the dominance of small side chain rotations during ligand binding [43]. These receptor-ligand docking methods are less computationally demanding than simulation-based free energy methods, for that reason they are the preferred ones when large libraries of compounds need to be tested in order to identify novel chemotypes for a particular target. There are different types of scoring methods available with distinct level of sophistication, although they can predict correctly binding poses for few cases, the identification of new hits from large databases of putative candidates has been found [44]. However, they present few limitations and problems, especially in the accurate prediction of binding energies and the low correlation between experimental and predicted binding affinities, as well as the treatment of protein flexibility, only partially incorporated in the docking scheme.

One approach that it is becoming popular is the use of receptor ensemble-based methods. The molecular docking on multiple receptor conformations (MRC) approach can be used with both experimental structures from different crystal structures, an NMR ensemble or computationally generated conformations. Molecular dynamics (MD) simulations have been extensively used to obtain these conformations that describe the conformational space of proteins [45, 46]. The right choice of the conformations that best represent the full spectrum of the receptor flexibility, which depends highly on the quality of the conformational sampling, is a key for the success of virtual screening experiments. Several strategies have been already proposed together with some recommendations in order to determine how to select the right subset of receptor conformations to use in the docking protocol [47, 48]. This approach has been implemented in the drug industry to screen large databases and predict roughly the binding affinity of these compounds for biologically relevant targets, complementing high-throughput screening (HTS) methodology.

Until very recently and given the intrinsic limitations of conventional computational tools, only events occurring in short timescales could be reproduced at a high accuracy level through all-atom techniques such as molecular dynamics simulations (see Fig. 11.1). Larger structural rearrangements demand the use of enhanced sampling methods relying on modified descriptions of the biomolecular system or the potential surface [49]. These techniques offer an alternative to bridge the detailed intermolecular interactions with larger spatial and longer scale motions. The simplest and crudest approach is to manipulate the energy function used in the MD simulation by applying a restrained potential, such as in targeted MD (TMD) [50] or umbrella sampling [51]. Besides, the energy can as well be altered by using additional energy terms into the potential energy. Such it is the case in the accelerated MD, where a new term is used to boost the potential and help the system to escape from a local minimum [52]. Another option will be the modification of the MD simulation protocol, such as it is done in approaches like replica exchange molecular dynamics (REMD) [53] and locally enhanced sampling (LES) [54].

One of the most challenging tasks is to devise computational methods and protocols that are able to yield better agreement between *in silico* and experimental results using reasonable computer times. Free energy methods have been proposed in order to increase this accuracy in predicting the binding affinities and thus they have been used extensively in the past for the estimation of the binding free energy of small-molecule drugs to proteins [55–57]. There are mainly two different options when referring to the computational resources required. On one hand, the end-point methods like free energy perturbation (FEP), linear interaction energy (LIE), and molecular mechanics Poisson-Boltzmann (Generalized Born) Surface Area (MM-PB(GB)SA) have been shown to give highly accurate estimates of the relative binding energy, and they could be used to improve the *in silico* scoring protocols used in the pharmaceutical industry. The free energy perturbation method can predict the absolute as well as relative binding free energy of ligands using thermodynamical cycles [58, 59]. In spite of the high computational cost associated with the method, FEP provides good predictions when a small series of compounds with similar structures are compared. This makes this technique most useful in lead optimization process. Linear interaction energy method [60], although based on different assumptions than FEP, represents a valid alternative but it is still too slow to be used in high-throughput virtual screening for a large number of candidates. MD simulations are not only used as a pre-technique to generate models for docking but for post-processing to estimate more accurate free binding energy of small molecules. This protocol in which docking poses are conformationally refined by molecular dynamics (MD) followed by prediction of the binding free energy by MM-PB(GB)SA is becoming extremely successful [61, 62].

On the other hand, methods based on a physical path, especially umbrella sampling and metadynamics, although highly demanding in resources, have the advantages of the prediction of the full free energy profile along the binding pathway, which can lead to the estimation of kinetics of binding and unraveling of intermediate states that can report new venues in drug design [63].

11.4 How to Incorporate Experimental and Computation Information on Protein Flexibility in the Designing of New Drugs

There are different approaches and protocols that incorporate computational methods such as molecular dynamics simulations in drug discovery to account for the flexibility of the receptor; however these flexibility-function studies are very rare in the current pharmaceutical research. As a consequence, there is still a communication breakdown between research on protein dynamics and drug design [64], probably due to the demanding pressure for high-throughput methodology.

As mentioned earlier, molecular recognition and, in particular, drug binding are highly dynamical processes, where a large spectrum of responses are involved. On one extreme of the scale, protein motions are limited and the ligand fits into a fairly static binding pocket. For such systems with reduced flexibility, the required additional degrees of freedom can be include by allowing side chains of residues in the binding site to be flexible throughout the search process, by exhaustive search or by using a rotameric library. This induce-fit case is easy to be incorporated in the docking schemes [65].

However, in some other cases, the ligand binds and stabilizes only a subset of many conformations sampled by its dynamical receptor; in agreement with the conformational selection hypothesis, where the flexibility required for the binding is encoded intrinsically in the apo form of the protein. The easiest approach will be to use a combination of experimental structures from different sources and/or conditions, if available, constituting a set of conformers that describes the dynamical behavior of the system. If only one structure is known, computational methods can be used to introduce this variability by adding conformational fluctuations in terms of collective motions. These degrees of freedom have been obtained through performing normal mode analysis using a fully atomistic representation of the protein [66] or an elastic-network model [67]. Another possibility is to perform canonical MD simulations to generate a proper ensemble of conformations for the docking protocol [36]. This idea has been incorporated in virtual screening under the relaxed complex scheme (RCS), where each possible ligand is docked to multiple protein conformations [68], and after that ideally, a holo MD simulation can be carried out. Some systems display novel conformational changes upon ligand binding that are not typically sampled when the ligand is absent.

In order to dissect the different flexibility schemes a protein adopt when coupled to ligand binding, and thus characterize the system and decide upon the computational protocol to carry out, experimental techniques such as X-ray crystallography, NMR experiments, and kinetic measurements are utilized. They can help in the determination of whether the ligand-bound protein conformations pre-exist in the ligand-free state of the protein (population-shift mechanism or conformational selection), or in the identification of intermediate states, potentially pointing toward an induced-fit mechanism. However, in most of the cases, experiments normally suggest the co-existence of both the population-shift and induced-fit mechanisms for ligand binding to a protein, and the decision of the approach to take becomes challenging. Furthermore, these flexibility-function studies can point to new modes of drug action that would be invisible to traditional drug design strategies that consider a static structure of the receptor. Simulations can also displayed motions that reveal the formation of cryptic and allosteric binding sites by exploring further cavities or even revealing transient pockets [69]. Following this idea, motions at one site can be inhibited by the binding at a remote site of selective inhibitors [70, 71]. Similarly modeling protein conformations associated with specific cellular functions might enable the discovery of conformationally selective ligands [72, 73]. In order to include these notions in the drug design protocol, it is necessary to run very long MD simulations or use enhanced sampling methods in order to achieve the required long-range effects involved in the identification of these sites (see previous section in this chapter).

In a different front and apart of this intra-network communication analysis within a protein, the flexibility-function studies can be expanded to protein complexes [74], when a dimeric association is used in the simulation. This opens up a new avenue for developing drugs, such as small molecules, peptidomimetics, and stapled peptides [75, 76], that can pharmacologically disrupt aberrant protein-protein interactions in the complexes.

From a more knowledge-based point of view, when studying protein dynamics upon the binding of a ligand, apart of the analysis of the effects caused and the understanding of the underlying mechanism, a large amount of atomistic information is earned, such as putative hydrogen bonds, level and nature of the residue contributions in the binding, providing a cautionary note for structure-based drug design [61, 77]. This information is not easy to extract in a high-throughput manner, since the analysis to perform will highly depend on the nature of the system; a deep knowledge on the protein can help to unravel the entwined molecular interactions that assemble this functional unit, and reduce its dimensionality. Although the field is on track, further work still needs to be done for the complete automation of the use of MD and other computational techniques in areas such as virtual screening and structure-based drug discovery.

11.5 Examples of Cases Where Flexibility Has Been Taken into Account

In the last section of this chapter, we illustrate the crucial role that intrinsic flexibility of protein structures plays in mediating ligand recognition through representative examples. Ideally, both experimental and computational approaches are mutually used complementing each other and thus symbiotically driving the design of drugs to the protein targets of interest.

11.5.1 Cryptic Pockets and Resistance Mechanism in HIV Integrase

Human immunodeficiency virus (HIV) integrase is one of three virally encoded enzymes required for HIV replication; the other two enzymes are a protease and a reverse transcriptase. HIV integrase pathogenically functions as an enzyme that inserts the viral genetic material into human chromosome, hence the inhibition of HIV integrase (using e.g. small molecule drugs) is essential to treat this viral infection. The integration is achieved by first creating reactive cytosine-adenosine 3′-hydroxyl ends (by cleaving off two nucleotides from the viral cDNA in cytoplasm), and after transport to the nucleus, transferring the hydroxyl ends on the human chromosomal DNA strand.

The first successful use of MD simulations to generate an ensemble of diverse receptor conformations for HIV integrase led to an FDA-approved drug, raltegravir [78]. Before this landmark study, the HIV integrase was deemed to be difficult for structure-based drug design. The diverse conformations generated in the study enabled the discovery of a previously unknown binding trench adjacent to the protein active site. This trench is formed only transiently and went undetected using conventional experimental methods. However, the finding of this channel was influential to the development of the drug [79] that works by inhibiting the strand transfer reaction in the nucleus [80]. Raltegravir was 1st FDA approved to target HIV integrase in 2007, and until Oct 2012, it is the only approved HIV integrase inhibitor, with a good activity against multi-drug resistant HIV-1 strains [81].

McCammon group further developed a method to produce a more accurate model of the active site of HIV integrase based on a restrained MD that ensures correct mono-dentate interactions between the carboxylate groups of the residues at the DDE motif and the two active-site Mg2+ ions [82]. This method has also been used to get insights into the effect of G140S/Q148H raltegravir-resistant double mutant of HIV integrase. Raltegravir adopts both the primary mode and a flipped mode of binding for the wild-type HIV integrase; while in the resistant-double mutant, the primary mode of binding is less accessible to raltegravir and the flipped binding mode is not observed.

11.5.2 Generation of Multiple Conformations for HIV Protease

HIV protease is known to undergo large conformational changes upon binding of natural or drug ligands. Based on the hypothesis of conformational selection (introduced earlier on in this chapter), the ligand-bound conformational states are less populated in the absence of ligands. To account for intrinsic flexibility of HIV protease, MD simulations of apo HIV protease have been performed to generate multiple receptor structures for docking [83]. They showed that the incorporation of protein flexibility enabled them to identify the correct HIV protease inhibitors and the binding modes.

11.5.3 Allosteric Sites in GTPase

GTPase proteins are members of a large superfamily, also known as Ras, consisting of the Ras, Rho, Rab, Ran, and Arf family [84, 85]. GTPase proteins bind and hydrolyze guanosine triphosphate (GTP) through a tight regulation mediated by the binding of guanine nucleotide exchange factors (GEFs) and GTPase-activating proteins (GAPs). Normally, Ras proteins regulate cytoplasmic signaling that mediates gene expression, cell proliferation, and differentiation. Rho proteins are involved in the regulation of actin organization, cell motion, and cell shape. Mutations in Ras are associated with over 25 % of diverse human tumors and leukemia [86, 87]. However, the structures of wild-type and mutant Ras are highly similar as observed by X-ray crystallography, hence rendering traditional structure-based drug design challenging.

Recently, Ras and Rho family members have been shown to possess intrinsic flexibility in the absence and presence of nucleotide ligand [88–91]. This intrinsic flexibility of the GTPase proteins has been explored using MD and accelerated MD simulations, and has led to the identification of their pattern of correlated motions, allosteric coupling between (1) the nucleotide-binding site and the (2) distant loops and membrane interacting C-terminal region [88, 90]. This has opened up new avenues to selectively target allosteric sites, instead of targeting the highly conserved and polar nucleotide-binding site [92]. A later study also using MD simulations reveals that acetylation of a lysine residue on the recently identified allosteric site results in conformational changes and an increased flexibility of Ras; the altered flexibility inhibits Ras from interacting with GEFs and hence the transforming/oncogenic activity of Ras, as verified using biochemical analysis [49]. This computational work highlights the importance of exploiting post-translational modification conformations in therapeutics.

11.5.4 Fragment Based-Screening on Transcription Factor p53

p53 is a transcription factor that regulates a wide variety of genes involved in repair, apoptosis, senescence and metabolism [93–95] in response to stress, e.g. DNA damage, telomere erosion, and hypoxia [96]. p53 is the most widely mutated protein in human cancer, with more than 25,000 mutations reported so far, which makes it a weak point in the complexity of the cells.

MD simulations have successfully been used to characterize the flexibility of a cavity in an oncogenic mutant Y220C of p53 DNA binding domain (DBD) [97]. The authors also employed fragment based screening to examine the druggability of this flexible cavity, followed by crystallizing the complex of fragment hits and mutant to elucidate the binding mode. Their work shows that the fragment hits reduce the dynamics of this flexible cavity. These fragments can then be used to design anti-cancer therapeutic small molecules that bind to the cavity and re-stabilize (hence rescue) the mutant in an allosteric manner. An interesting note on their procedure of MD simulations is the use of isopropanol solvent, based on the rationale that the isopropanol mimics a ligand possessing both polar and non-polar properties that enable it to bind to the Y220C cavity through hydrogen bonding and hydrophobic interactions.

11.5.5 Inhibition of Protein-Protein Interaction Between p53 and MDM2

MDM2 is an E3 ubiquitin ligase that regulates the levels of p53 through binding to the N-terminal transactivation domain and the DBD of p53, and stimulates the ubiquitination of p53 for subsequent proteasomal degradation [98, 99]. To drive the apoptosis of cancer cells, p53 can be reactivated through the inhibition of p53-MDM2 interaction. Moreover, in cancer cells, MDM2 is over-expressed and results in the loss of p53 activity. MD simulations have been performed to generate multiple structures of p53-MDM2, hence accounting for intrinsic flexibility of MDM2 in its complexed form; the set of diverse MDM2 structures has been used for virtual screening that succeeds in identifying two compounds with higher affinity than the natural p53 peptide [100]. These small molecules have distinct scaffolds from nutlin, a well-know inhibitor of p53-MDM2 interaction, bringing the advantage of a more diverse scaffolds of therapeutic small molecules that can be useful to overcome resistance against existing drugs. Moreover, the N-terminal transactivation domain of p53 that inserts into a hydrophobic cleft of MDM2, has inspired the design and synthesis of stapled peptides that bind well to MDM2 and readily enter cells [75]. Efforts to further improve the binding affinity of stapled peptides to MDM2 using MD simulations highlight the requirement for higher helicity and reduced solvent exposure from the stapled peptides [76].

11.5.6 NMR Ensemble of Transcription Factor CBF

The transcription core binding factor (CBF) is important in blood cell development and often implicated in acute myeloid leukemia. The CBF comprises a protein-protein complex formed by core binding factor beta (CBFbeta) and Runx1. CBFbeta increases the affinity of Runx1 towards DNA and protects Runx1 against ubiquitin tagging and the subsequent proteasomal degradation [101]. In leukemia, CBFbeta aberrantly fuses to coiled-coil region of smooth muscle myosin protein (SMMHC) [102]. Unlike the wild-type CBFbeta, the fused CBFbeta binds to Runx1 over tightly and results in the deregulation of the CBF function and the cell cycle progression associated with leukemia. Therefore, the binding between fused CBFbeta and Runx1 is a potential target for inhibitory small molecules with anti-cancer therapeutic properties.

Before screening for small molecules that can inhibit the interaction between CBFbeta and Runx1, the protein intrinsic flexibility of CBFbeta, determined by multiple NMR structures of CBFbeta [103] has been used for virtual screening to identify top 35 hits, which were posteriorly validated using NMR spectroscopy. Interestingly, the identified lead compounds do not bind directly to the CBFbeta-Runx1 interface, suggesting that they act through allosteric or non-competitive mechanism. The allosteric mechanism possibly compensates for the lack of significant curvature at the CBFbeta-Runx1 interface. Then, the virtual screening was followed by compound synthesis and cell assays, which successfully revealed the first small molecule inhibitors of Runx1-CBFbeta interaction with the ability to inhibit the proliferation of the human leukemia cell line ME1 [104].

11.5.7 Understanding Selective Drugs for Abl Kinase Inhibition Using Experimental and Computational Information

Almost 2 % of human genes encode protein kinases, which use ATP to phosphorylate their substrates and are involved in different biological processes such as cell growth, movement, and apoptosis. Next to the ATP-binding site there is a DFG motif, which is highly conserved and often mutated in human cancers

[105]. When the aspartate residue in the DFG motif points into the ATP-binding site (known as adopting the DFG-in conformation), the aspartate can coordinate an ATP-bound magnesium ion. In many kinases, the active DFG-in conformation is more favored than the inactive DFG-out conformation as the former has higher stability. Previously, the anti-cancer drug imatinib (Gleevec, Novartis) was known to inhibit the Abl kinase by selectively targeting a specific DFG conformation, but its mechanism was unclear especially at the level of atomistic detail. Crystallographic studies mostly capture the two distinct DFG conformations [106, 107] but not their transitional mechanism which hampers the understanding of the DFG flip. To address this challenge, long MD simulations have shown that imatinib selectively stabilizes the DFG-out conformation in a pH-dependent manner through electrostatic changes intrinsic to the kinase catalytic cycle. This testable mechanism was verified experimentally using the Abl-imatinib binding assays [108]. The investigations of the DFG conformations of Abl kinase and its inhibition by imatinib are an excellent example where both experimental and computational methods have been used to generate ideas and results that mutually drive the subsequent studies. This study of imatinib-Abl kinase can also inspire the development of selective drugs targeting a particular kinase through understanding of mechanism at atomic level details.

11.5.8 Large Conformational Changes in Periplasmic Binding Protein (PBP) Studied by Accelerated MD

PBPs are expressed by Gram-negative bacteria to function in chemotaxis and to transport various nutrient molecules including amino acids, ions, vitamins, and carbohydrates. Structurally, all PBPs share two-domain architecture with a central inter-domain ligand binding cleft. This architecture allows PBPs to utilize Venus-flytrap mechanism to undergo large scale hinge-bending motions for trapping ligand (upon its binding) at the inter-domain cleft [109].

Well-studied PBPs include bacterial maltose binding protein (MBP), due to its use as an affinity tag for protein expression and purification. Although crystallographic structures suggest the possibility of only two conformations of apo/open and holo/closed, NMR paramagnetic relaxation enhancement (PRE) measurements on apo MBP suggest major open and minor semi closed conformational states [110]. To verify the existence of semi closed conformation of apo MBP, accelerated MD simulations [111, 112] have been used to sample the conformational landscape of this protein, whose conformational transitions may take up to microseconds and milliseconds, and they indeed confirmed the existence of such transient but stable semi closed conformation.

Although MBP is not a drug target on its own, the understanding of conformational dynamics of this protein is a critical starting point for drug design because PBPs and the extracellular ligand binding domains (LBD) of some mammalian receptors have a high structural homology. These receptors include NMDA receptor and G protein-coupled receptor (GPCR) [113], which are of major pharmaceutical interest. The identification of hidden yet stable conformations of the LBD of these receptors can be a promising target for conformationally selective drugs.

In another study, temperature-accelerated MD (TAMD) simulations have been combined with the Cα atom-based elastic network normal mode analysis (NMA) to study the large scale conformational changes in MBP [114]. A combination of few low-frequency modes can describe the entire conformational changes and a single low-frequency mode can rationalize the individual functional conformational change along the transition pathway generated using TAMD, whose trajectories are not trapped in the local phase space. The characterization of the full entire conformational changes provides the diversity in structures that optimistically will lead towards an effective drug design. This study also serves as an inspiration for future studies aimed to explore protein intrinsic flexibility, through integrating insights obtained from a combination of MD simulations, NMA, and experimental methods.

11.5.9 Conformationally Selective Inhibition of Mismatch Repair Protein

Deregulation at the decision point where, upon DNA damage, the cell should either activate the DNA repair machinery or induce cell death, results in cancer and failure of anti-cancer therapy. The MSH2/MSH6 protein complex acts as a sensor for both DNA damage and mismatches, and then a recruiter for additional proteins that function in either DNA repair or cell death depending on the conformation of the MSH2/MSH6. Molecular modeling has been used to obtain the death conformation of the MSH2/MSH6, which is distinguishable from their repair conformation [63]. The results from this computational work have facilitated the identification and synthesis of conformationally selective small molecules, such as reserpine and its analog rescinnamine [115].

11.5.10 Physical Path-Based Free Energy Methods

The mitogen activated protein kinase (MAPK) p38 is an interesting target mainly due to its implication in several signaling mechanisms and the regulation of a wide range of cellular processes, from proliferation to cell survival and apoptosis, among many others [116]. The interest in oncology for the development of p38 inhibitors arises from its role as a tumor suppressor, with the downregulation of the cell cycle progression and the induction of apoptosis [117]. Using metadynamics in combination with few other computational algorithm, Gervasio and collaborators have designed a blind protocol to estimate accurately the binding profile of a series of p38 inhibitors [63]. In addition to the correct prediction of the affinities with their approach, the full characterization of the full binding pathway, exploring all important intermediates and transition states, provides information about the flexibility of the protein in the binding of different inhibitors. For several of the inhibitors there is an increased in protein flexibility that allows the molecules to establish alternative interactions, expanding the range of binding poses that they might adopt. Some of them vary in the localization within the pocket, such as how buried they are, while for an inhibitor a "flipped" conformation has been observed along the binding pathway explored in their work. This structural and dynamical information could potentially be used in the lead optimization process, expanding the current use of computational methods in the design of drugs.

11.6 Conclusion

The understanding of the biological timescales and scope of protein dynamics, the implications of these conformational changes into the biological functions, and the effective design of drugs that inhibit these important roles must come from the combination of the atomistic information provided by computational simulations and experimental studies.

In the recent years, advances in computational techniques and especially in the hardware potency have made possible the use of *in silico* methods as a key tool in simulating structural and functional properties of biological macromolecules. The advance in modern graphical processing units (GPUs) have opened new venues for MD simulations, extending the timescale they can reach and dropping the price and expanding the accessibility of these machines in fields as drug discovery process. This dramatic reduction in the cost of MD simulations and the length of them make this technique an interesting tool to include in the arsenal of biological research and biomedicine. Indeed, these computational methods are widespread in the initial stages of drug discovery and their importance in guiding the rational design has increased in last few decades. The field is evolving towards the tight integration of more expensive and time-consuming experiments with faster and cheaper computational simulations.

However, although routine MD on biological systems is becoming popular; the complementarity with the experimental data needs to be scrutinized carefully. Experimental methods can help us to identify the motion regimes the protein follows in the binding event, which it will determine the kind of computational approach to take in order to capture these motions and predict new small molecules that inhibit more efficiently the protein conformational dynamics of the target. A lot of progress has been done in this direction, but still there is no a general and automatic recipe on how to treat the systems, and expert contribution is required to complete a full description of the flexibility of the protein. Making use of this information promises to initiate new and intensive studies in the field of predicting the most successful inhibitors for our current and future biological targets.

References

1. Kar G, Keskin O, Gursoy A, Nussinov R (2010) Allostery and population shift in drug discovery. Curr Opin Pharmacol 10:715–722
2. Salsbury FR (2010) Molecular dynamics simulations of protein dynamics and their relevance to drug discovery. Curr Opin Pharmacol 10:738–744
3. McCammon JA (1999) Protein dynamics. Rep Prog Phys 47:1
4. Boehr DD, Nussinov R, Wright PE (2009) The role of dynamic conformational ensembles in biomolecular recognition. Nat Chem Biol 5:789–796
5. Gunasekaran K, Nussinov R (2007) How different are structurally flexible and rigid binding sites? Sequence and structural features discriminating proteins that do and do not undergo conformational change upon ligand binding. J Mol Biol 365:257–273
6. Callender R, Deng H (1994) Nonresonance Raman difference spectroscopy: a general probe of protein structure, ligand binding, enzymatic catalysis, and the structures of other biomacromolecules. Annu Rev Biophys Biomol Struct 23:215–245
7. Nienhaus K, Nienhaus GU (2011) Ligand dynamics in heme proteins observed by Fourier transform infrared-temperature derivative spectroscopy. Biochim Biophys Acta 1814: 1030–1041
8. Bu Z, Neumann DA, Lee SH, Brown CM, Engelman DM, Han CC (2000) A view of dynamics changes in the molten globule-native folding step by quasielastic neutron scattering. J Mol Biol 301:525–536
9. Russo D, Pérez J, Zanotti J-M, Desmadril M, Durand D (2002) Dynamic transition associated with the thermal denaturation of a small beta protein. Biophys J 83:2792–2800
10. Balog E, Becker T, Oettl M, Lechner R, Daniel R et al (2004) Direct determination of vibrational density of states change on ligand binding to a protein. Phys Rev Lett 93:028103
11. Grant BJ, McCammon JA, Caves LS, Cross RA (2007) Multivariate analysis of conserved sequence–structure relationships in kinesins: coupling of the active site and a tubulin-binding sub-domain. J Mol Biol 368:1231–1248
12. Westenhoff S, Nazarenko E, Malmerberg E, Davidsson J, Katona G, Neutze R (2010) Time-resolved structural studies of protein reaction dynamics: a smorgasbord of X-ray approaches. Acta Crystallogr A 66:207–219
13. Baldwin AJ, Kay LE (2009) NMR spectroscopy brings invisible protein states into focus. Nat Chem Biol 5:808–814
14. Mittermaier A, Kay LE (2006) New tools provide new insights in NMR studies of protein dynamics. Science 312:224–228
15. Mittermaier AK, Kay LE (2009) Observing biological dynamics at atomic resolution using NMR. Trends Biochem Sci 34:601–611
16. Kleckner IR, Foster MP (2011) An introduction to NMR-based approaches for measuring protein dynamics. Biochim Biophys Acta 1814:942–968
17. Lipari G, Szabo A (1982) Model-free approach to the interpretation of nuclear magnetic resonance relaxation in macromolecules. 1. Theory and range of validity. J Am Chem Soc 104:4546–4559
18. Palmer AG, Grey MJ, Wang C (2005) Solution NMR spin relaxation methods for characterizing chemical exchange in high-molecular-weight systems. Methods Enzymol 394:430–465
19. Lange OF, Lakomek NA, Farès C, Schröder GF, Walter KF et al (2008) Recognition dynamics up to microseconds revealed from an RDC-derived ubiquitin ensemble in solution. Science 320:1471–1475
20. Markwick PR, Bouvignies G, Blackledge M (2007) Exploring multiple timescale motions in protein GB3 using accelerated molecular dynamics and NMR spectroscopy. J Am Chem Soc 129:4724–4730
21. Pellecchia M, Sem DS, Wüthrich K (2002) NMR in drug discovery. Nat Rev Drug Discov 1:211–219

22. Fernández C, Jahnke W (2004) New approaches for NMR screening in drug discovery. Drug Discov Today Technol 1:277–283
23. Pellecchia M, Bertini I, Cowburn D, Dalvit C, Giralt E et al (2008) Perspectives on NMR in drug discovery: a technique comes of age. Nat Rev Drug Discov 7:738–745
24. Shuker SB, Hajduk PJ, Meadows RP, Fesik SW (1996) Discovering high-affinity ligands for proteins: SAR by NMR. Science 274:1531–1534
25. Fielding L (2007) NMR methods for the determination of protein–ligand dissociation constants. Prog Nucl Magn Reson Spectrosc 51:219–242
26. Lepre CA, Moore JM, Peng JW (2004) Theory and applications of NMR-based screening in pharmaceutical research. Chem Rev 104:3641–3676
27. Neylon C (2008) Small angle neutron and X-ray scattering in structural biology: recent examples from the literature. Eur Biophys J 37:531–541
28. Svergun DI (2010) Small-angle X-ray and neutron scattering as a tool for structural systems biology. Biol Chem 391:737–743
29. Jacques DA, Trewhella J (2010) Small-angle scattering for structural biology–expanding the frontier while avoiding the pitfalls. Protein Sci 19:642–657
30. Mertens HD, Svergun DI (2010) Structural characterization of proteins and complexes using small-angle X-ray solution scattering. J Struct Biol 172:128–141
31. Fenton AW, Williams R, Trewhella J (2010) Changes in small-angle X-ray scattering parameters observed upon binding of ligand to rabbit muscle pyruvate kinase are not correlated with allosteric transitions. Biochemistry 49:7202–7209
32. Lee JC (2008) Modulation of allostery of pyruvate kinase by shifting of an ensemble of microstates. Acta Biochim Biophys Sin (Shanghai) 40:663–669
33. Carroll MJ, Gromova AV, Miller KR, Tang H, Wang XS et al (2011) Direct detection of structurally resolved dynamics in a multiconformation receptor-ligand complex. J Am Chem Soc 133:6422–6428
34. Jorgensen WL (2004) The many roles of computation in drug discovery. Science 303: 1813–1818
35. B-Rao C, Subramanian J, Sharma SD (2009) Managing protein flexibility in docking and its applications. Drug Discov Today 14:394–400
36. Lin JH (2011) Accommodating protein flexibility for structure-based drug design. Curr Top Med Chem 11:171–178
37. May A, Zacharias M (2005) Accounting for global protein deformability during protein-protein and protein-ligand docking. Biochim Biophys Acta 1754:225–231
38. Zacharias M (2010) Accounting for conformational changes during protein-protein docking. Curr Opin Struct Biol 20:180–186
39. Spyrakis F, BidonChanal A, Barril X, Luque FJ (2011) Protein flexibility and ligand recognition: challenges for molecular modeling. Curr Top Med Chem 11:192–210
40. Fuentes G, Dastidar SG, Madhumalar A, Verma CS (2011) Role of protein flexibility in the discovery of new drugs. Drug Dev Res 72:26–35
41. Taboureau O, Baell JB, Fernández-Recio J, Villoutreix BO (2012) Established and emerging trends in computational drug discovery in the structural genomics era. Chem Biol 19:29–41
42. Jiang F, Kim SH (1991) "Soft docking": matching of molecular surface cubes. J Mol Biol 219:79–102
43. Zavodszky MI, Kuhn LA (2005) Side-chain flexibility in protein-ligand binding: the minimal rotation hypothesis. Protein Sci 14:1104–1114
44. Villoutreix BO, Eudes R, Miteva MA (2009) Structure-based virtual ligand screening: recent success stories. Comb Chem High Throughput Screen 12:1000–1016
45. Dodson GG, Lane DP, Verma CS (2008) Molecular simulations of protein dynamics: new windows on mechanisms in biology. EMBO Rep 9:144–150
46. Durrant JD, McCammon JA (2011) Molecular dynamics simulations and drug discovery. BMC Biol 9:71
47. Bottegoni G, Rocchia W, Rueda M, Abagyan R, Cavalli A (2011) Systematic exploitation of multiple receptor conformations for virtual ligand screening. PLoS One 6:e18845

48. Rueda M, Bottegoni G, Abagyan R (2010) Recipes for the selection of experimental protein conformations for virtual screening. J Chem Inf Model 50:186–193
49. Yang MH, Nickerson S, Kim ET, Liot C, Laurent G et al (2012) Regulation of RAS oncogenicity by acetylation. Proc Natl Acad Sci USA 109:10843–10848
50. Schlitter J, Engels M, Krüger P (1994) Targeted molecular dynamics: a new approach for searching pathways of conformational transitions. J Mol Graph 12:84–89
51. Bartels C, Karplus M (1998) Probability distributions for complex systems: adaptive umbrella sampling of the potential energy. J Phys Chem B 102:865–880
52. Hamelberg D, Mongan J, McCammon JA (2004) Accelerated molecular dynamics: a promising and efficient simulation method for biomolecules. J Chem Phys 120:11919
53. Sugita Y, Okamoto Y (1999) Replica-exchange molecular dynamics method for protein folding. Chem Phys Lett 314:141–151
54. Stultz CM, Karplus M (1998) On the potential surface of the locally enhanced sampling approximation. J Chem Phys 109:8809
55. Woo HJ, Roux B (2005) Calculation of absolute protein-ligand binding free energy from computer simulations. Proc Natl Acad Sci USA 102:6825–6830
56. Fidelak J, Juraszek J, Branduardi D, Bianciotto M, Gervasio FL (2010) Free-energy-based methods for binding profile determination in a congeneric series of CDK2 inhibitors. J Phys Chem B 114:9516–9524
57. Kondo HX, Okimoto N, Morimoto G, Taiji M (2011) Free-energy landscapes of protein domain movements upon ligand binding. J Phys Chem B 115:7629–7636
58. Bash PA, Field MJ, Karplus M (1987) Free energy perturbation method for chemical reactions in the condensed phase: a dynamic approach based on a combined quantum and molecular mechanics potential. J Am Chem Soc 109:8092–8094
59. Rao SN, Singh UC, Bash PA, Kollman PA (1987) Free energy perturbation calculations on binding and catalysis after mutating Asn 155 in subtilisin. Nature 328:551–554
60. Aqvist J, Marelius J (2001) The linear interaction energy method for predicting ligand binding free energies. Comb Chem High Throughput Screen 4:613–626
61. Fuentes G, Scaltriti M, Baselga J, Verma CS (2011) Synergy between trastuzumab and pertuzumab for human epidermal growth factor 2 (Her2) from colocalization: an in silico based mechanism. Breast Cancer Res 13:R54
62. Joseph TL, Lane DP, Verma CS (2012) Stapled BH3 peptides against MCL-1: mechanism and design using atomistic simulations. PLoS One 7:e43985
63. Saladino G, Gauthier L, Bianciotto M, Gervasio FL (2012) Assessing the performance of metadynamics and path variables in predicting the binding free energies of p38 inhibitors. J Chem Theory Comput 8:1165–1170
64. Peng JW (2009) Communication breakdown: protein dynamics and drug design. Structure (London, England: 1993) 17:319
65. Nabuurs SB, Wagener M, de Vlieg J (2007) A flexible approach to induced fit docking. J Med Chem 50:6507–6518
66. Zacharias M, Sklenar H (1999) Harmonic modes as variables to approximately account for receptor flexibility in ligand–receptor docking simulations: application to DNA minor groove ligand complex. J Comput Chem 20:287–300
67. May A, Zacharias M (2008) Protein-ligand docking accounting for receptor side chain and global flexibility in normal modes: evaluation on kinase inhibitor cross docking. J Med Chem 51:3499–3506
68. Amaro RE, Baron R, McCammon JA (2008) An improved relaxed complex scheme for receptor flexibility in computer-aided drug design. J Comput Aided Mol Des 22: 693–705
69. Eyrisch S, Helms V (2007) Transient pockets on protein surfaces involved in protein-protein interaction. J Med Chem 50:3457–3464
70. Kamerlin SC, Rucker R, Boresch S (2007) A molecular dynamics study of WPD-loop flexibility in PTP1B. Biochem Biophys Res Commun 356:1011–1016

71. Perryman AL, Lin JH, McCammon JA (2009) HIV-1 protease molecular dynamics of a wild-type and of the V82F/I84V mutant: possible contributions to drug resistance and a potential new target site for drugs. Protein Sci 13:1108–1123
72. Drotschmann K, Topping RP, Clodfelter JE, Salsbury FR (2004) Mutations in the nucleotide-binding domain of MutS homologs uncouple cell death from cell survival. DNA Repair (Amst) 3:729–742
73. Salsbury FR, Clodfelter JE, Gentry MB, Hollis T, Scarpinato KD (2006) The molecular mechanism of DNA damage recognition by MutS homologs and its consequences for cell death response. Nucleic Acids Res 34:2173–2185
74. Salsbury FR (2010) Effects of Cisplatin binding to DNA on the dynamics of the E. coli MutS dimer. Protein Pept Lett 17:744–750
75. Bernal F, Tyler AF, Korsmeyer SJ, Walensky LD, Verdine GL (2007) Reactivation of the p53 tumor suppressor pathway by a stapled p53 peptide. J Am Chem Soc 129:2456–2457
76. Joseph TL, Lane D, Verma CS (2010) Stapled peptides in the p53 pathway: computer simulations reveal novel interactions of the staples with the target protein. Cell Cycle 9: 4560–4568
77. Foulkes-Murzycki JE, Scott WRP, Schiffer CA (2007) Hydrophobic sliding: a possible mechanism for drug resistance in human immunodeficiency virus type 1 protease. Structure 15:225–233
78. Schames JR, Henchman RH, Siegel JS, Sotriffer CA, Ni H, McCammon JA (2004) Discovery of a novel binding trench in HIV integrase. J Med Chem 47:1879–1881
79. Hazuda DJ, Anthony NJ, Gomez RP, Jolly SM, Wai JS et al (2004) A naphthyridine carboxamide provides evidence for discordant resistance between mechanistically identical inhibitors of HIV-1 integrase. Proc Natl Acad Sci USA 101:11233–11238
80. Summa V, Petrocchi A, Bonelli F, Crescenzi B, Donghi M et al (2008) Discovery of raltegravir, a potent, selective orally bioavailable HIV-integrase inhibitor for the treatment of HIV-AIDS infection. J Med Chem 51:5843–5855
81. Croxtall JD, Keam SJ (2009) Raltegravir: a review of its use in the management of HIV infection in treatment-experienced patients. Drugs 69:1059–1075
82. Perryman AL, Forli S, Morris GM, Burt C, Cheng Y et al (2010) A dynamic model of HIV integrase inhibition and drug resistance. J Mol Biol 397:600–615
83. Meagher KL, Carlson HA (2004) Incorporating protein flexibility in structure-based drug discovery: using HIV-1 protease as a test case. J Am Chem Soc 126:13276–13281
84. Wennerberg K, Rossman KL, Der CJ (2005) The Ras superfamily at a glance. J Cell Sci 118:843–846
85. Rojas AM, Fuentes G, Rausell A, Valencia A (2012) The Ras protein superfamily: evolutionary tree and role of conserved amino acids. J Cell Biol 196:189–201
86. Reddy EP, Reynolds RK, Santos E, Barbacid M (1982) A point mutation is responsible for the acquisition of transforming properties by the T24 human bladder carcinoma oncogene. Nature 300:149–152
87. Prior IA, Lewis PD, Mattos C (2012) A comprehensive survey of Ras mutations in cancer. Cancer Res 72:2457–2467
88. Grant BJ, Gorfe AA, McCammon JA (2009) Ras conformational switching: simulating nucleotide-dependent conformational transitions with accelerated molecular dynamics. PLoS Comput Biol 5:e1000325
89. Gorfe AA, Grant BJ, McCammon JA (2008) Mapping the nucleotide and isoform-dependent structural and dynamical features of Ras proteins. Structure 16:885–896
90. Grant BJ, McCammon JA, Gorfe AA (2010) Conformational selection in G-proteins: lessons from Ras and Rho. Biophys J 99:L87–L89
91. Lukman S, Grant BJ, Gorfe AA, Grant GH, McCammon JA (2010) The distinct conformational dynamics of K-Ras and H-Ras A59G. PLoS Comput Biol 6:e1000922
92. Grant BJ, Lukman S, Hocker HJ, Sayyah J, Brown JH et al (2011) Novel allosteric sites on Ras for lead generation. PLoS One 6:e25711

93. Vogelstein B, Lane D, Levine AJ (2000) Surfing the p53 network. Nature 408:307–310
94. Li T, Kon N, Jiang L, Tan M, Ludwig T et al (2012) Tumor suppression in the absence of p53-mediated cell-cycle arrest, apoptosis, and senescence. Cell 149:1269–1283
95. Vousden KH, Prives C (2009) Blinded by the light: the growing complexity of p53. Cell 137:413–431
96. Dai C, Gu W (2010) p53 post-translational modification: deregulated in tumorigenesis. Trends Mol Med 16:528–536
97. Basse N, Kaar JL, Settanni G, Joerger AC, Rutherford TJ, Fersht AR (2010) Toward the rational design of p53-stabilizing drugs: probing the surface of the oncogenic Y220C mutant. Chem Biol 17:46–56
98. Haupt Y, Maya R, Kazaz A, Oren M (1997) Mdm2 promotes the rapid degradation of p53. Nature 387:296–299
99. Shimizu H, Burch LR, Smith AJ, Dornan D, Wallace M et al (2002) The conformationally flexible S9-S10 linker region in the core domain of p53 contains a novel MDM2 binding site whose mutation increases ubiquitination of p53 in vivo. J Biol Chem 277:28446–28458
100. Bowman AL, Nikolovska-Coleska Z, Zhong H, Wang S, Carlson HA (2007) Small molecule inhibitors of the MDM2-p53 interaction discovered by ensemble-based receptor models. J Am Chem Soc 129:12809–12814
101. Perry C, Eldor A, Soreq H (2002) Runx1/AML1 in leukemia: disrupted association with diverse protein partners. Leuk Res 26:221–228
102. Castilla LH, Wijmenga C, Wang Q, Stacy T, Speck NA et al (1996) Failure of embryonic hematopoiesis and lethal hemorrhages in mouse embryos heterozygous for a knocked-in leukemia gene CBFB-MYH11. Cell 87:687–696
103. Huang X, Peng JW, Speck NA, Bushweller JH (1999) Solution structure of core binding factor beta and map of the CBF alpha binding site. Nat Struct Biol 6:624–627
104. Gorczynski MJ, Grembecka J, Zhou Y, Kong Y, Roudaia L et al (2007) Allosteric inhibition of the protein-protein interaction between the leukemia-associated proteins Runx1 and CBFbeta. Chem Biol 14:1186–1197
105. Greenman C, Stephens P, Smith R, Dalgliesh GL, Hunter C et al (2007) Patterns of somatic mutation in human cancer genomes. Nature 446:153–158
106. Nagar B, Hantschel O, Young MA, Scheffzek K, Veach D et al (2003) Structural basis for the autoinhibition of c-Abl tyrosine kinase. Cell 112:859–871
107. Young MA, Shah NP, Chao LH, Seeliger M, Milanov ZV et al (2006) Structure of the kinase domain of an imatinib-resistant Abl mutant in complex with the Aurora kinase inhibitor VX-680. Cancer Res 66:1007–1014
108. Shan Y, Seeliger MA, Eastwood MP, Frank F, Xu H et al (2009) A conserved protonation-dependent switch controls drug binding in the Abl kinase. Proc Natl Acad Sci USA 106:139–144
109. Mao B, Pear MR, McCammon JA, Quiocho FA (1982) Hinge-bending in L-arabinose-binding protein. The "Venus's-flytrap" model. J Biol Chem 257:1131–1133
110. Tang C, Schwieters CD, Clore GM (2007) Open-to-closed transition in apo maltose-binding protein observed by paramagnetic NMR. Nature 449:1078–1082
111. Bucher D, Grant BJ, McCammon JA (2011) Induced fit or conformational selection? The role of the semi-closed state in the maltose binding protein. Biochemistry 50:10530–10539
112. Bucher D, Grant BJ, Markwick PR, McCammon JA (2011) Accessing a hidden conformation of the maltose binding protein using accelerated molecular dynamics. PLoS Comput Biol 7:e1002034
113. Felder CB, Graul RC, Lee AY, Merkle HP, Sadee W (1999) The Venus flytrap of periplasmic binding proteins: an ancient protein module present in multiple drug receptors. AAPS PharmSci 1:7–26

114. Vashisth H, Brooks CL (2012) Conformational sampling of maltose-transporter components in Cartesian collective variables is governed by the low-frequency normal modes. J Phys Chem Lett 3:3379–3384
115. Vasilyeva A, Clodfelter JE, Gorczynski MJ, Gerardi AR, King SB et al (2010) Parameters of reserpine analogs that induce MSH2/MSH6-dependent cytotoxic response. J Nucleic Acids 162018–162030
116. Pearson G, Robinson F, Gibson TB, Xu B-E, Karandikar M et al (2001) Mitogen-activated protein (MAP) kinase pathways: regulation and physiological functions. Endocr Rev 22:153–183
117. Bulavin DV, Fornace AJ (2004) p38 MAP kinase's emerging role as a tumor suppressor. Adv Cancer Res 92:95–118

Chapter 12
NMR and Computational Methods in the Structural and Dynamic Characterization of Ligand-Receptor Interactions

Michela Ghitti, Giovanna Musco, and Andrea Spitaleri

Abstract The recurrent failures in drug discovery campaigns, the asymmetry between the enormous financial investments and the relatively scarce results have fostered the development of strategies based on complementary methods. In this context in recent years the rigid lock-and-key binding concept had to be revisited in favour of a dynamic model of molecular recognition accounting for conformational changes of both the ligand and the receptor. The high level of complexity required by a dynamic description of the processes underlying molecular recognition requires a multidisciplinary investigation approach. In this perspective, the combination of nuclear magnetic resonance spectroscopy with molecular docking, conformational searches along with molecular dynamics simulations has given new insights into the dynamic mechanisms governing ligand receptor interactions, thus giving an enormous contribution to the identification and design of new and effective drugs. Herein a succinct overview on the applications of both NMR and computational methods to the structural and dynamic characterization of ligand-receptor interactions will be presented.

Keywords Allostery • Conformational sampling • Enhanced sampling • Elastic Network Model/ENM • Essential Dynamics/ED • Docking • Drug discovery • HTS/high through put screening • Integrin antagonists • Ligand observed • Molecular dynamics/MD • Monte Carlo • NMR/nuclear magnetic resonance • NOE nulear overhauser effect • Normal Mode Analysis/NMA • Target observed • SAR structure activity relationship • trNOE/transfer nuclear overhauser effect • Receptor flexibility

M. Ghitti • G. Musco (✉) • A. Spitaleri
Dulbecco Telethon Institute c/o Biomolecular NMR Laboratory S. Raffaele Scientific Institute, Via Olgettina 58, 20132 Milan, Italy
e-mail: musco.giovanna@hsr.it

K. Han et al. (eds.), *Protein Conformational Dynamics*, Advances in Experimental Medicine and Biology 805, DOI 10.1007/978-3-319-02970-2_12,

12.1 Introduction

Drug discovery is a complex and expensive process, requiring approximately 12 years and costing >800 million dollars to develop one new medicine from the earliest stages of discovery until it is available for treating patients. At the end of this time-consuming and challenging endeavour only few drugs, after having been successfully tested in the clinics, are able to reach the market. High-throughput screening (HTS), though being one of the primary pharmaceutical methods for the identification of lead compounds, display a high false positive rate. Several inhibitors are not routinely translated into new drugs because they often have undesirable side effects leading to the delay or failure of drug discovery projects [1]. The reasons of this shortcoming are diverse, including unclear target biology, inappropriate leads, poor potency or selectivity of the discovered drug, lack of efficacy, unexpected animal toxicity and unwanted drug-like properties. It is therefore of paramount importance to identify lead compounds that interact with the target in a biologically relevant mechanism without inducing adverse or paradoxical modes of action. The recurrent failures and the asymmetry between the enormous investments (in terms of time and financial resources) and the relatively scarce results have therefore fostered the development of strategies based on complementary methods (e.g. bioinformatics, computational chemistry, cell biology, medicinal chemistry, enzymology, molecular biology, protein chemistry, genomics, proteomics, metabolomics, structural biology) to drive the drug discovery and development processes in a more efficient and productive way [2]. In this context, in recent years the original lock-and-key binding concept originally introduced by Emil Fischer in 1894, in which a frozen ligand accommodates into a static receptor, had to be revisited in favour of more dynamic models of molecular recognition, able to account for conformational changes of both the ligand and the receptor. The high level of complexity required by a dynamic description of molecular recognition necessitates a multidisciplinary investigation approach. In this perspective, the combination of nuclear magnetic resonance (NMR) spectroscopy with molecular docking, conformational searches along with molecular dynamics simulations has given new insights into the dynamic mechanisms governing ligand-receptor interactions, thus giving an enormous contribution to the identification and design of new and effective therapeutic drugs [3, 4]. This chapter is meant to give a succinct overview on the application of both NMR and computational methods to the structural and dynamic characterization of ligand-receptor interactions. It will be organized in four sections:

I. We will briefly outline the strength and the weakness of NMR in ligand screening (Sect. 12.2);
II. We will discuss the fundamental role of an exhaustive conformational search in the selection of suitable ligands for docking calculations (Sect. 12.3);
III. We will highlight the importance of collective protein dynamics in the design of drugs targeting allosteric receptors (Sect. 12.4);

IV. Finally, we will focus on a case study, in which the combination of experimental and computational techniques resulted in a successful strategy in the identification of real integrin αvβ3 antagonists (Sect. 12.5).

12.2 NMR-Based Screening

One of the most powerful aspects of NMR spectroscopy relies on its ability to characterize at atomic detail protein-ligand interactions under physiological conditions. Especially in those situations in which several biophysical techniques might fail because the interactions are inherently weak and transient, or because the protein-ligand complex does not crystallize, NMR can play a unique role in the molecular characterization. In this regard, the vast range of applicability of the method, which is able to investigate affinity constants spanning from nM to mM values, can potentially contribute to improve the efficiency of drug discovery programs. NMR spectroscopy has therefore become a well-established tool both for screening techniques in leads identification and for the examination of structure activity relationship (SAR) [3].

Herein we will briefly summarize the most popular NMR-based screening techniques, that are traditionally classified in target-observed and ligand-observed techniques, depending whether the binding event is detected monitoring the changes in the NMR parameters of the ligand or of the target (protein, nucleic acid), respectively. The major parameters being sensitive to ligand binding include: chemical shifts, relaxation and translational diffusion properties, intermolecular cross-relaxations. Depending on the kind of experiments site-specific information or simple binding assessment can be obtained.

12.2.1 Target-Observed Techniques

Target-observed methods are usually based on the acquisition and comparison of ^{1}H-^{13}C or ^{1}H-^{15}N heteronuclear single quantum coherence (HSQC) spectra of the ^{13}C and ^{15}N isotopically enriched target in the presence or absence of unlabelled ligand. If the spectral assignment of the protein is known, binding can be proved monitoring the chemical shift displacements (CSD) of the target upon ligand binding. CSD is a highly sensitive tool for proving interactions, for mapping binding sites and detecting residues which are directly interacting with the ligand or that are indirectly affected by the association [5]. Information obtained from CSD can be then used to filter docking solutions or even to drive the docking, thus limiting the conformational search problem (see Sect. 12.3) [5]. One of the most popular applications of chemical shift mapping is the "SAR-by-NMR" methodology in which small organic molecules that bind to proximal subsites of a protein are identified, optimized and linked together to produce high-affinity ligands [6]. Thanks to NMR intrinsic high sensitivity to detect weak interactions, the method is successfully applied in fragment based drug discovery, whose main concept consists in building up high-affinity ligands in a modular way, starting from the identification of low affinity binders through the screening of a compound library. The chemical substructures (MW < 300 Da), sufficient to elicit a minimal yet specific and localized interaction with the target are then merged to generate higher affinity ligands guided by structural information. In contrast to ligand-based experiments (see Sect. 12.2.2), observation of the target molecule is not restricted to an affinity limit, however the throughput of these assays is intrinsically low, due to the relatively large amounts of recombinant labelled target (in the mg range) and to the relatively lengthy experimental time needed. Target-observed experiments also suffer from one important limitation, represented by the molecular weight of the target, which is restricted to an upper limit of approximately 40–50 kDa (larger molecular sizes are accessible in case of oligomer proteins). One major problem for larger macromolecules resides in the fast relaxation rates, reflecting on turn in poor spectra quality. These detrimental effects can be partly overcome by special techniques, such as TROSY (transverse relaxation optimized spectroscopy), CRINEPT (cross-correlated relaxation-enhanced polarization transfer) at high magnetic fields, deuteration and selective labelling [7]. One seminal example in which the power of target based methods has been exploited up to its size limits is represented by Malate Synthase (723 residues) in which the chemical shift changes of $^{1}H_{N}$, ^{15}N and $^{13}C'$ nuclei upon binding of pyruvate have been mapped onto the three-dimensional structure of the molecule [8]. Another successful case of TROSY-based NMR experiments is represented by Methyl-TROSY-based NMR spectroscopy performed on 20S archaeal proteasome from Thermoplasma acidophilium, which provided evidence for a novel class of 20S proteasome inhibitors [9]. Nevertheless, it should be pointed out that this kind of NMR studies on supramolecular structures are still relatively rare in the literature, because they require large investments in terms of sample preparation and acquisition time, thus limiting their vast applicability.

12.2.2 Ligand-Observed Techniques

Ligand-observed NMR screenings monitor the NMR spectrum of a ligand under free and bound conditions and are based on the concept that upon binding to a macromolecule, the apparent molecular weight and the hydrodynamic radius of the small molecule change substantially by several orders of magnitude. The dynamic nature of ligand receptor interactions influences the appearance of the NMR spectra, several different NMR experiments can be therefore performed to detect and quantify these changes. Importantly, ligand-observed experiments are only applicable on complexes that are in the fast exchange regime on the NMR chemical shift difference timescale, i.e. when the difference in chemical shift between the free and bound form of the ligand is considerably smaller then the exchange constant k_{ex}. This condition requires high dissociation rate constants (k_{off}) and usually applies, as a rough rule of thumb, when the ligand binds with a $K_d > 10^{-5}$ M, assuming a diffusion limited k_{on} rate constant ($k_{on} \sim 10^7$–$10^8 s^1 m^1$). Through this fast exchange mechanism it is possible to obtain information on the low population bound state by analysing the resonances from the free form in exchange with the bound form. In this situation the observed NMR parameters for the ligand will reflect the population weighted average of the free and bound form. Therefore, when a small molecule binds to its target it will adopt the dynamical properties of the larger molecular-weight protein. The spectroscopic and dynamic properties of a small molecule tumbling in solution, which typically include slow relaxation rates, fast diffusion and fast Brownian motions, vanishing or weakly positive Nuclear Overhauser Effects (NOE) cross-peaks, will drastically change when binding occurs, resulting in high relaxation rates, reduced diffusion and Brownian motions, and development of negative NOEs. These distinct differences imply that changes in the ligand NMR spectral parameters can be monitored to assess target binding.

As ligand-observed NMR techniques rely on the rapid and efficient transfer of spectral characteristics between the free and the bound state of a ligand, they have the enormous advantage to require only small amounts of purified unlabelled receptor (pM-μM concentrations), which is order of magnitudes lower compared to the quantities required by protein detected methods. Thanks to the reduced burden on protein expression and purification these methods are routinely used to interrogate receptor-ligand interactions. Moreover, as a large difference between the molecular weight of the small compounds and the target molecule is required, the size of the receptor does not constitute a limiting factor. As the approach becomes even more sensitive and effective with increasing molecular weights, the target may be even immobilized, bound to lipid vesicles, or bound on the surface of the cells (see Sect. 12.5.2). An important drawback affecting ligand based approaches, as compared to target based methods, consists in their inability to a priori localize the binders on the receptor. Additionally, a major caveat affecting all these approaches relying on the rapid exchange of the NMR properties between the free and the bound form, originates from the fact that they are all biased towards weakly binding ligands and large ligand molar excesses. As a consequence, high ligand concentrations with respect to the target may start to occupy low affinity non specific binding sites, giving rise to false positives. Nevertheless, the use of simple one-dimensional ^{1}H spectra and the ability to screen mixtures without deconvolution, fosters the application of ligand based experiments in high throughput screenings.

In the following we will summarize the most popular ligand-based NMR experiments. For a detailed discussion of the theoretical and practical aspects of the single experiments we encourage the reader to refer to several seminal reviews, that illustrate in detail the methods and their applications [10–13].

12.2.2.1 Transverse Relaxation Rates

One of the most well-established class of NMR binding assays relies on the comparison of the ligand's relaxation rate in the presence and absence of the target. Relaxation reflects the hydrodynamic radius and rotational tumbling rate (τ_c) of species in solution [11]. When a small molecule binds to a large molecule, the small molecule transiently possesses similar NMR properties as the large molecule, thus assuming fast transverse relaxation rates, which on turn lead to spectral line broadening of the small molecule signal. The simplest experiment requires the acquisition of a ligand spectrum (or mixture of compounds) in the absence and presence of a protein target. Enhanced transverse relaxation of ligand(s) upon the addition of the protein target, reflecting in line broadening of the ligand(s) resonances, indicates the transient formation of a bound complex. One of the earliest example of applications of transverse relaxation for the efficient screening of large libraries of compounds has been presented by Fesik and co-workers to detect ligands that bind to the FK506 binding protein [14].

12.2.2.2 Diffusion Experiments

Comparison of the ligand's diffusion coefficient, describing the translational mobility of the molecule, with or without the protein target, can be also used to prove intermolecular interactions between protein targets and ligands. Most diffusion filters are based on pulsed field gradient (PFG) stimulated echo (STE) experiments [15, 16]. They follow the same basic principles as relaxation experiments but rely on differences in translational instead of rotational motion. Briefly, in the case of a ligand in fast exchange between the free and receptor-bound states, the observed translational diffusion coefficient (D_{obs}) is given by: $D_{obs} = D_{free}\ \chi_{free} + D_{bound}\ (1-\chi_{free})$, where χ_{free} is the mole fraction of the ligand in the free state and D_{free} and D_{bound} are the translational diffusion coefficients for the free and receptor-bound ligand, respectively. The use of diffusion-edited NMR spectroscopy for screening compound libraries was first illustrated by Lin et al. [17, 18]. In this study, a change in the diffusion rate of DL-isocitric lactone upon binding to hydroquinine 9-phenanthryl ether was used to identify the ligand binding in the presence of a mixture of non binding compounds.

12.2.2.3 Saturation Transfer Difference (STD)

Saturation Transfer Difference (STD) was introduced in 1999 by Meier and Meyer in a seminal paper describing the study of the interaction of wheat germ agglutinin with saccharides [19]. Several other STD experiments for analysing mixtures of putative ligands have since been reported, using a broad range of targets, including transmembrane receptors on whole cells [20, 21]. In STD experiments, a 1D steady-state NOE experiment is measured for a ligand (or a library of ligands) in the presence of a small amount of target, usually at a ligand:protein ratio of about 100:1. The method relies on the selective saturation of receptor protons by irradiating regions of the ^{1}H NMR spectrum (for example the aliphatic region of the spectrum, between −1 and 2 ppm) that are usually not occupied by resonances from small organic molecules. Due to the effective spin diffusion of the receptor, saturation quickly propagates across the entire macromolecule. If the ligand(s) binds to the receptor, saturation will be transferred to the ligand(s), whereby ligand protons which are near in space to the receptor will be mostly affected by saturation. As a result of the saturation transfer the intensity of the ligand signals will be attenuated. Subtraction of the saturated spectrum from the reference spectrum without saturation, generates the STD spectrum which contains only signals of the binding ligands. STD can be also used to determine the binding epitope of the ligand by exploiting the fact that STD signal intensities (I_{STD}) are not equal for the different protons in the ligand [22].

12.2.2.4 Water Ligand Observation by Gradient Spectroscopy (WaterLOGSY)

WaterLOGSY is an experiment strictly related to STD, whereby the selective saturation of the protein is achieved by irradiation of water protons [23]. The transfer of magnetization occurs from bulk water to the ligand via the receptor through multiple pathways including: (i) long-lived water molecules that are within the binding pocket, (ii) solvent exchangeable protons of the binding pocket, (iii) remote exchangeable protons that propagate their magnetization state across the receptor via spin diffusion. Differently from STD, WaterLOGSY is not suitable for epitope mapping, but is similarly well suited for competition experiments to confirm the binding site of screening hits and to estimate binding affinity. Alternatively, a known ligand with relatively weak affinity can be used as a reporter or spy molecule: screening is performed with the known ligand present in all samples, and a decrease in binding of this reporter indicates competition by a fragment in the mixture.

12.2.2.5 Transferred NOE (trNOE)

In trNOE experiments reversible protein-ligand complexes can be examined under chemical equilibrium, monitoring the changes in nuclear spin relaxation of the ligand in the presence of a sub-stoichiometric amount of target. The technique is based upon transfer of nuclear spin relaxation of a small ligand from the bound to the free state, provided that the ligand-protein dissociation constants are sufficiently high ($k_{off} > 300\ s^{-1}$ and $K_d > 10^{-7}$ M). The trNOE experiment [24] relies on the different tumbling time (τ_c) of either free or bound ligand. Small ligands (MW < 1,000 Da), which have small τ_c, usually develop weak positive NOE at a slow rate, whereas large receptors, which show large τ_c, are characterized by a rapid development of strong negative NOE [25]. When a ligand binds in fast

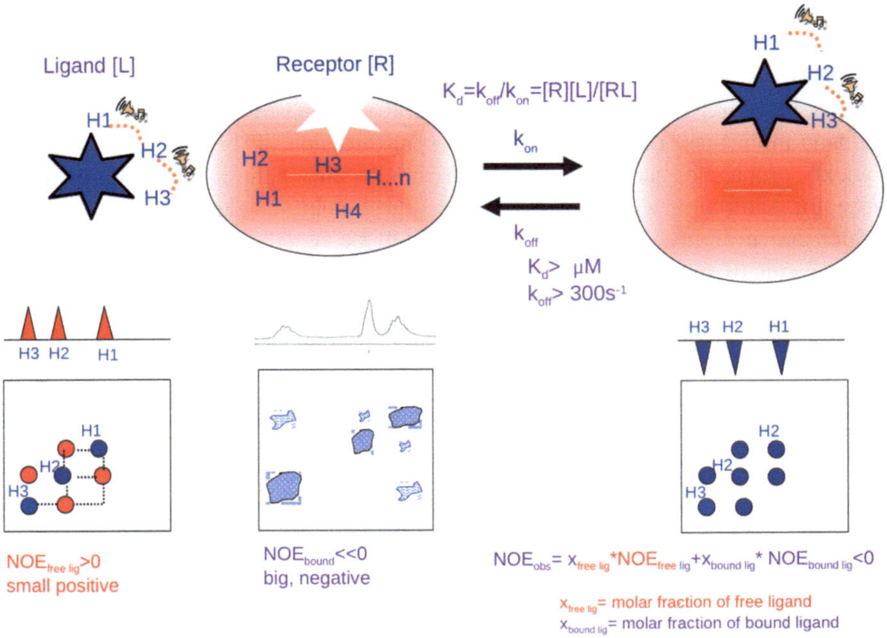

Fig. 12.1 Schematic representation of trNOE experiment The ligand and the receptor are represented as a *blue star* and *red oval*, respectively. The NOE effect between protons is represented with the symbol of the music (representing the radiofrequency). The sharp small positive NOEs of the free ligand are represented with *red cross-peaks* outside the diagonal, the NOEs of the receptor are represented as *broad blue peaks* outside the diagonal. The negative trNOEs of the ligand when bound to the receptor are represented as *blue cross-peaks*

exchange with the receptor, it transiently adopts the tumbling time of a large molecule during its bound life-time; thus, the motional characteristics of the bound state are carried into solution and detected using the signals of the free ligand. In particular, as discussed elsewhere [25, 26] the averaged cross-relaxation rate $<\sigma_{ij}>$, which is responsible of the NOE intensity build-up in trNOE experiments in fast exchange regime, is given by $<\sigma_{ij}> = X^F\sigma^F{}_{ij} + X^B\sigma^B{}_{ij}$, whereby X^F and X^B are the molar fractions of free and bound ligand; and $\sigma^F{}_{ij}$ and $\sigma^B{}_{ij}$ are the cross-relaxation rates of free and bound ligand, respectively. As long as the inequality $|X^B\sigma^B{}_{ij}| > |X^F\sigma^F{}_{ij}|$ applies, $<\sigma_{ij}>$ will be dominated by the bound state. Binding compounds will be therefore characterized by NOE cross-peaks that have changed sign in the presence of the receptor, whereas non-binders will show no change in the presence of the receptor and display either zero or negative cross-peaks with respect to the diagonal of the NOESY spectrum (Fig. 12.1). Several methods have been developed that allow a quantitative interpretation of trNOEs and, thus, yield reliable information about the bioactive conformation of bound ligand. trNOE has been successfully applied to detect binding within the framework of a lead generation method (SHAPES), based on a limited but diverse library of small compound scaffolds whose shapes are commonly found in known therapeutic agents [27].

12.2.2.6 Interligand NOEs for Pharmacophore Mapping (INPHARMA) and Interligand Nuclear Overhauser Effect (ILOE)

Once a ligand has been confirmed to bind to a protein by methods such as trNOE, two recently developed techniques may be used to facilitate identification of the bound ligand's orientation inside the receptor binding site. The protein mediated INPHARMA method is based on the observation of interligand, spin diffusion mediated, trNOE data, between two ligands, binding competitively and weakly to a macromolecular receptor. Interligand NOEs are then used to obtain information on the binding mode of one ligand with respect to the other. The interligand NOEs between a known binder and a new compound are then used to determine the orientation of the new complex without the need for complete structure identification. The method has been successfully applied to a mixture of epothilone A and baccatin III [28] in the presence of tubulin, representing the first observation of protein mediated NOEs between two ligands that bind competitively and consecutively to the same target molecule.

A second method, which should not be confused with the previous one, is the ILOE approach, which uses trNOEs to identify small molecule ligands bound to the protein target simultaneously in close proximity to each other. Interligand NOEs are then used to provide information about the orientation of ligands, which can be then linked to yield a high affinity lead compound in the correct conformation [29]. Following an NMR-based approach SAR by interligand NOE method, Becattini et al. were able to identify two chemical fragments that bind on the surface of Bid, a proapoptotic member of the Bcl-2 family [30].

12.2.2.7 Ligand Fluorine Chemical Shift Perturbation

Although most NMR screening experiments focus mainly on ^{1}H, the ^{19}F nucleus presents unique properties that render it a highly effective probe for NMR screening. First, ^{19}F is usually incorporated in drugs to enhance their pharmacokinetic. Second, the absence of endogenous ^{19}F in biological molecules allows direct observation of ligand spectra, without the need of relaxation filters and/or difference spectroscopy to eliminate receptor or large solvent signals. Third, ^{19}F is present at 100 % natural abundance with a sensitivity comparable to that of ^{1}H. Finally, the chemical shift range of ^{19}F is much larger than that of ^{1}H (900 ppm) implying high sensitivity of the ^{19}F chemical shift to local changes. Chemical shift changes of the fluorinated molecule in the bound state can be very large, thus allowing the detection of very weak binding. For example, in the fluorine chemical shift anisotropy and exchange for screening (FAXS) strategy, developed by Dalvit and co-workers, useful screening information can be gained by looking at the ^{19}F relaxation of a small library of compounds. In this approach, the relaxation properties of a small set of ^{19}F "spy" compounds report on the binding of a larger set of higher affinity binders via competitive displacement. The large ^{19}F chemical shift dispersion and low fragment concentration enable screening of large mixtures thus offering high throughput [31, 32].

12.3 The Ligand Flexibility: the Importance of an Exhaustive Conformational Sampling in Docking Calculations

Molecular docking is a computational method to investigate intermolecular complexes formed between two or more constituent molecules. It comprises the process of generating a model of a complex based on the known three-dimensional structures of its components, i.e. the receptor (protein, or nucleic acids) and the ligand (a peptide, an protein, a small molecule), free or bound to other species [33]. The docking procedure consists in the search for the precise ligand conformations and orientations (usually referred as docking poses) within a given target protein, when the structure of the protein is known or modelled. Fast approximate mathematical methods (so called scoring functions) are used to predict the binding affinity between two molecules and to rank the docking poses. Pioneered during the early 1980s [34], molecular docking is still a field of intense research, as it represents a fundamental component in many drug discovery programs [35] and a primary tool for the virtual screening of large chemical libraries [36].

The typical system described in docking calculations usually includes the ligand, the receptor and the solvent molecules. Because of the enormous number of degrees of freedom associated to the solvent molecules, they are normally excluded from the calculations, or implicitly modelled in the scoring functions. However, the number of degrees of freedom associated to both the ligand and receptor still remains computationally untreatable. The dimensionality of the problem can be further reduced through different approximations, allowing for a more time effective sampling of the conformational space. The most basic one is the rigid-body approximation, that treats both the ligand and the receptor as rigid entities. However, this constitutes a strong approximation, as both the ligand and the protein generally undergo structural rearrangements upon complex formation, thus requiring the introduction of flexibility in docking algorithms [37]. This aspect is particularly relevant in those cases in which a flexible ligand is docked into a known three-dimensional binding site. As a matter of fact, unrealistic high-energy descriptions of the ligand conformer can lead to wrong conclusions on the ligand-receptor interactions, or small changes in the ligand input conformation can cause drastic differences in the geometries and the scores of the docking poses [38]. At present several programs exist, including HADDOCK, Gold, Autodock, MOE, Glide, FlexX and Surflex, that try to account for a certain level of flexibility for the ligand and in some cases also for the receptor [39–43]. However, as pointed out by Tirado-Rives and Jorgensen when describing the binding of flexible ligands to HIV-1 reverse transcriptase [44], insufficient conformational sampling can compromise the efficiency of current docking methodology in the ranking of diverse compounds in high-throughput virtual screening. In this context, it should be pointed out that the identification of relevant ligand conformations that might affect binding affinity is often challenging for standard spectroscopic and diffraction techniques, as it is virtually impossible to experimentally characterize the rapid transition from one minimum to the other. In this scenario detailed conformational searches of ligands prior docking calculations might be a successful strategy to improve the accuracy and the prediction power of docking calculations. In the next section we will briefly discuss the methods for the conformational searches outside the docking framework, but still in the context of protein-ligand interactions analysis.

12.3.1 Replica-Exchange Molecular Dynamics (REMD)

Molecular Dynamics (MD) simulation is the most popular method used to sample the molecular conformational space [45]. Based on the integration of Newton's law, it allows to compute time–dependent properties and to follow individual particles' motion along time [46]. Assuming that the ergodic hypothesis holds, an infinitely long MD trajectory should be able to sample the entire conformational space. Nevertheless, at room temperature the probability of crossing high energy barriers is often too small to be observed during a finite MD simulation. Most likely, even with several hundred nanoseconds of simulations, the system might be confined to limited regions of the conformational space. A solution usually applied to overcome the limited sampling efficiency of MD simulations at room temperature consists in raising the simulation temperature. The additional kinetic energy available at higher temperature allows the crossing of high energy barriers, thus ensuring a wider sampling of the conformational space. This methodology is the basis of two computational approaches, simulated annealing (SA) and parallel tempering (also named replica-exchange molecular dynamics). SA consists in heating up the system in order to jump out from the initial local minimum to explore other minima [47]. The heating step is followed by a gradual cooling, which allows the system to slowly settle down to a lower energy minimum. This method is widely exploited for the local structural optimization of polypeptides, that have broad and energy rough surfaces requiring extremely long simulations to find the global minimum. SA combined to Rosetta protein modelling suite proved to perform very well in the design of the bound conformation of the C-terminal portion of the RGS14 GoLoco motif peptide when bound to the $G\alpha_{i1}$ receptor [48].

REMD [49] is based on the run of multicopy MD simulations randomly initialized, at different temperatures. The conformations are then exchanged at different temperatures following the Metropolis criterion (see Sect. 12.3.2). The strength and robustness of this method allows to sample both low and high energy configurations [50]. In drug discovery focusing on cardiovascular and metabolic disease extensive REMD has been for example exploited in the design of a new antagonist of the apelin APJ receptor, a class A G-protein-coupled receptor (GPCR) working as co-receptor for HIV cellular entry [51]. This work showed that the peptides promoting a β-turn at the RPRL motif displayed a good affinity for the APJ receptor.

Okumura et al. [52] compared the computational efficiency of the traditional constant temperature MD with REMD in a series of inhibitors of HIV-1 reverse transcriptase. Herein the authors showed that the conformational populations are accurately estimated by both methods, however, REMD converged at a faster rate, especially for one ligand (rilpivirine), which is characterized by multiple stable states separated by high-free energy barriers. Moreover, they showed that for small drug-like molecules with energetic barriers separating the stable states, the use of REMD with Weighted Histogram Analysis Method (WHAM) is an efficient computational approach for estimating the contribution of ligand conformational reorganization to binding affinities.

12.3.2 Monte Carlo (MC) Search Methods

Monte Carlo search methods are stochastic techniques based on the use of random numbers and probability statistics to sample the conformational space [53]. A Monte Carlo search consists of two steps: (1) generation of a new trial conformation, and (2) decision whether the new conformation will be accepted or rejected. The trial conformation is usually accepted or rejected according to a temperature-dependent probability of the Metropolis type $p = \min[1, e^{-\beta \Delta U}]$ where ΔU is the difference in potential energy between the trial and starting conformations, $\beta = 1/kT$, k is the Boltzmann constant and T is the temperature. MC methods have a significant advantage over MD methods, as they use a simpler energy function that does not require any sort of derivative information. In addition, MC methods are more efficient in stepping energy barriers, thus allowing more complete conformational searches.

Several docking programs are directly interfaced with MC-based algorithms; combinatorial small molecule growth (CombiSMoG) for example introduces the philosophy of combinatorial synthesis into computational drug design combining a knowledge-based potential with Monte Carlo ligand growth algorithm [54]. The method was successfully applied to design picomolar inhibitors of human carbonic anhydrase II [54]. Macromodel is another popular commercial software capable to perform Monte Carlo search of small molecule [55]. It was used in [56], were the conformational minima of a small set of HIV reverse transcriptase inhibitors have been located using a Metropolis Monte Carlo simulation. Herein the conformational space was sampled exploring a set of rotatable bonds contributing to the conformational flexibility of the molecule. Another example of MC search combined to docking is represented by the design of a new highly flexible inhibitor against acyl-CoA [57]. ConfGen is a conformational search program, that, similarly to Macromodel, efficiently generates bioactive conformations exploiting MC simulation [58]. This approach has been used in the development of novel IKKβ inhibitors with IC50 values lower than 10 μM [59]. MC search is also implemented in Molecular Operating Environment (MOE) software, a fully integrated commercial drug discovery software package [60]. MOE offers three methods for conformers generation: systematic search, stochastic search, and low mode molecular dynamics. OMEGA is another commercial conformers generation tool that uses a systematic, knowledge-based approach to generate ligand conformers [61]. The program generates initial 3D structures from a library of fragments, then it exhaustively enumerates all rotatable torsions using pre-defined libraries, and finally it samples this large conformational space using geometric and energy criteria. As an example, potential inhibitors of HIV-1 protease have been designed by the combination of conformational search with Omega and docking with Autodock [62].

Finally, it is worth pointing out that it is quite common after a MC search (or MD sampling) to define structural similarities between conformations [63] using clustering approaches. Although this method can provide valuable insights into the structural diversity of conformations, it may end up with a collection of clusters poorly related to the actual energetics of the system. An alternative to MC to perform a more reliable exhaustive conformational analysis is based on free energy surface calculations and will be described in the following section.

12.3.3 Enhanced Conformational Sampling Methods Based on Bias Potentials

The methods presented in the previous sections often fail to generate reliable equilibrium conformations because of the rugged and complex nature of the Free-Energy Surface (FES) that is accessible to the system. As a consequence, computational sampling is often relegated to some local, unrealistic minima, which could compromise subsequent docking studies. Conformational sampling and transitions are too slow to occur spontaneously in fully atomic MD simulations. These long time-scales originate from relatively high free energy barriers between metastable states, hampering efficient sampling of conformational space in conventional MD calculations. The well-recognized limitations of sampling in atomistic dynamics have led to many innovative alternatives to enhance the coverage of the thermally accessible conformational space and to capture *rare events* (events that might happen on a long timescale) [64]. One trick that is commonly used to address this problem is to add a potential bias in order to force the *rare event* to occur. In this context, several techniques, including the local-elevation method [65], taboo search [66], the Wang–Landau method [67], adaptive force bias [68] conformational flooding [69], umbrella sampling [70], weighted histogram techniques [71], transition state theory and path sampling [72] and free energy guided sampling [73] have been developed to address the sampling problem. In this context, Metadynamics (MetaD) [74] has emerged as a powerful coarse-grained non-Markovian molecular-dynamics approach for the acceleration of rare events and the efficient and rapid computation of multidimensional free energy surfaces as a function of a restricted number of degrees of freedom, named collective variables (CVs). Differently from other sampling methods, in which the calculation of FES requires an additional step (such as WHAM [67]), MetaD directly provides a good estimate of the free energy of the system projected into the CVs.

The principal advantages of MetaD are the following: (a) it is able to escape local minima by overcoming large free energy barriers and (b) it allows to reconstruct the FES in the space of the chosen CVs. This is possible because the CVs evolve under a continuous addition of a history dependent potential energy, built as a sum of repulsive Gaussian terms, that forces the dynamics to visit previously unexplored regions of the conformational space and discourages the system from returning to these regions. Hereby, the system can escape minima along low free energy paths and can explore other minima in the free energy landscape, thus allowing an enhanced sampling of the conformational space. The efficiency and reliability of MetaD strongly depends on which CVs are chosen as reaction coordinates for a particular mechanism. There is no a priori rule to define the correct set of CVs, because of their strong dependence on the physical and chemical properties of the system. However, their choice should satisfy the following properties:

1. CVs should be able to distinguish between the initial, intermediate and final state.
2. CVs should describe all slow events relevant to the process under investigation that cannot be sampled within the typical scale of the simulation.
3. Their number should be kept to a minimum.

MetaD simulations have been exploited to predict equilibria in a variety of different molecules. For instance, iduronic acid, unlike most other monosaccharides, can adopt different ring conformations, depending on the context of the molecular structure. Accurate modelling of this building block is essential for understanding the role of glycosaminoglycans and other glycoconjugates [75]. Exploration of a conformational space of eight drug-like molecules, including all major classes of diseases such as antivirals, anticancer, and lifestyle drugs, has been evaluated by MetaD enhanced molecular dynamics with the weighted holistic invariant molecular (WHIM) descriptors, which does not require a prior knowledge of the accessible conformations [76]. Combination of biasing potentials and traditional alchemical free energy techniques, in particular Thermodynamic Integration (TI) and local elevation/umbrella sampling (LE/US) methods, allowed to explore the conformational equilibrium of highly flexible ligands such as guanosine-5′-triphosphate (GTP) and 8-substituted GTP analogues [77]. Moreover LE/US has been also proposed to study in water the relative free energies and interconversion barriers of β-d-glucopyranose ring, a fundamental building block of a series of drugs [78].

12.4 The Receptor Flexibility

Presently, the vast majority of molecular docking applications considers the ligand conformational flexibility either during docking calculations [40, 42] or using libraries of conformers [41, 79]. Conversely, the flexibility of the receptor is usually neglected as the number of degrees of freedom which should be considered in the calculations is extremely computationally demanding. However, this approximation constitutes a major drawback in docking calculations, as it does not account for conformational adjustments in response to environmental biological stimuli. The frontiers of computer aided drug discovery should be therefore expanded to account for both the inherent motions of the system and for potential induced fit phenomena. Several research groups have started to consider protein flexibility into structure based drug design, as summarized in detail in several recent reviews [37, 80, 81]. This represents a major progress in the field, increasing the discrimination power between binders and non-binders (the so called enrichment factor) and improving the ability to predict the correct binding poses and ligand induced conformational changes. However, at present there is no reliable, easy-to-use docking/screening protocol that accounts for all protein conformational changes. Thus, in several pharmaceutical applications protein flexibility is still neglected, conceivably limiting the success rate of drug discovery campaign. This important issue should be therefore carefully considered in those cases in which the receptor displays a substantial conformational transition upon ligand binding [82]. Diverse experimental methods exist to elucidate collective motions, including NMR, X-ray crystallography, cryo-Electron Microscopy, as well as single-molecule fluorescence or electron-transfer measurements. In the last two decades considerable efforts have been also dedicated to integrate experimental data with computational techniques. The comprehension of the allosteric mechanisms, through which a local structural perturbation has distant dynamic long-range effects, might be crucial to give new insights into the protein function and its regulation mechanisms [83, 84]. In the following we will briefly summarize the most popular *in silico* methods to characterize proteins dynamic in the context of ligand-induced conformational changes.

12.4.1 Normal Mode Analysis (NMA)

NMA method [85] is based on the assumption that: (1) any given equilibrium system fluctuates around a single well defined conformation and (2) the nature of these thermally induced fluctuations can be calculated assuming a simple harmonic potential. NMA determines the independent harmonic modes of the molecule, whereby each single mode comprises the concerted motions of many atoms. The standard application of NMA to a large biological molecule is computationally expensive, however several sophisticated numerical techniques exist to extract the lowest frequency modes of large molecules. Despite the intrinsic approximations required by the method, including the use of a simple harmonic potential, solvent exclusion and the inability to model multiple minima, NMA succeeded in determining functionally relevant motions in several biological systems [86], such as lysozome, crambin and ribonuclease [87], myosin, NtrC, hemoglobin [88, 89], DNA-dependent polymerase [90], and several others [91, 92]. In all these cases, the lowest frequency modes compared well with the experimentally conformational changes observed upon ligand binding. Of note, Gramicidin A (GA) was the first membrane protein examined by NMA, thus extending the range of applicability of the method and opening new perspectives in the field. Extensive computational studies on GA have been performed since then [93, 94]. For example NMA analysis was able to give new insights into GA gating mechanism and slow conformational transitions, showing that its major motions consist in a counter-rotation of the two helices around the pore axis, accompanied by a slight expansion of the channel mouths at the EC and CP ends. Notably, the global modes of motion predicted by the NMA models were consistent with experiments [95, 96]. Recently, NMA has been also applied to flexible docking problems. Cavasotto at al., inspired by the current representation of the ligand-receptor binding process, presented a normal-mode based methodology to incorporate receptor flexibility in ligand docking and virtual screening of cAMP-dependent protein kinase [97]. In 2012, Kruger et al., developed the NMSim web server (http://www.nmsim.de) implementing a three-step approach for multiscale modelling of protein conformational changes [98]. Efrat Mashiach et al. [99] presented FiberDock, a new method for docking refinement, which models backbone flexibility by an unlimited number of normal modes. Finally, coarse-grained normal modes have been shown to be useful for the rapid prediction of functional sites [100].

12.4.2 Elastic Network Model (ENM)

Although NMA has been used over three decades to study intrinsic flexibility of proteins, interest in this approach increased after the development of NMA based on Elastic Network Model, which proved to be accurate and robust despite its simplified physical model and force field description [101]. The network representation adopted in ENMs takes advantage of principles deriving from both NMA and spectral graph theory to obtain analytical solutions for equilibrium dynamics, that can be readily implemented in efficient computational algorithms. The model was recently exploited for the development of ElNémo: a normal mode web server for protein movement analysis (http://igs-server.cnrs-mrs.fr/elnemo/index.html) [94]. The first simplified model, the Gaussian Network Model (GNM) [102] represents a protein structure as a network of nodes (alpha-carbons) and elastic springs. In GNM all the fluctuations and inter residues distances are *gaussianly* distributed around their equilibrium coordinates. When this model is applied to coarse grained proteins' description, it shows significant agreement with experimental crystallographic B-factors for several proteins. Further extension of the model by Atilgan et al. [103] included information of the direction of motions exploiting Anisotropic Network Model (ANM).

Recently, the ENM was successfully applied to study large-scale conformational transitions in the maltose-binding protein and in the nucleotide binding domains of a maltose-transporter [104] or to analyze immunological relevant proteins such as HIV gp120 plasticity in complexes with CD4 binding fragments, CD4 mimetic proteins, and various antibody fragments [105]. Also the type and the extent of conformational changes undergone upon activation of rhodopsin have been extensively examined by various experiments and computations, including GNM and ANM studies [106, 107]. In this case, the lowest ANM modes derived from the model correctly predicted 93 % of the effects of 119 rhodopsin mutants [107].

12.4.3 Molecular Dynamics (MD)

In the absence of sufficient experimental data, the method that mostly contributes to the understanding of biomolecular flexibility is MD simulation. Simulations in explicit solvent can model flexibility of both ligand and receptor in a realistic way, taking into account the fundamental role of water-mediated interactions. However, the high computational cost combined with the long time-scale of conformational changes strongly limits the applicability range of the method. Nevertheless, MD simulations have been successfully and widely utilized in drug design and development. One paradigmatic example highlighting the invaluable role of MD in rational drug discovery is represented by HIV integrase. In 1999 the solved structure of HIV integrase in complex with an inhibitor provided a platform for the drug development of a different class of inhibitors. In this context MD simulations were able to successfully predict more than one possible orientation of HIV ligands binding [108, 109], as later confirmed by crystallography studies [110, 111]. Another seminal example of the useful insights offered by MD simulations in structure based drug design was reported in 2011, when Buch et al. performed on graphics processing unit (GPU)-based infrastructures 495 MD simulations of 100 ns each to simulate the complete binding process of the inhibitor benzamidine to β-trypsin [112]. The approach allowed the identification of the lowest energy binding mode of a ligand to a receptor. Monitoring of the binding process at an atomic resolution can be potentially assist the development of drugs able to control and modulate the ligand-receptor recognition process. In this context, Shaw and collaborators formulated a mechanism for the flipping of a conserved motif of Abl tyrosine kinases combining microsecond MD simulations with crystallographic and kinetic experiments. Importantly, the conformation of this motif was crucial to discriminate between active/inactive kinase conformations. Their results led to the identification of a class of potent inhibitors of both Src and Abl that recognize the inactive kinase conformations [113]. Recently, Skjærven et al. reported another successful application of unbiased MD simulation on the chaperonine GroEL. Multiple 100 ns MD simulations revealed a pre-existing equilibrium between the unliganded closed T-state and the fully open R-state, even in the absence of bound nucleotide. This study provided a model for the structure-dynamic relationship of GroEL folding machine, supplying atomic insights into the interactions potentially important for the large scale conformational transitions driven by ATP binding and hydrolysis [114].

12.4.4 Principal Component Analysis (PCA)

MD simulations generate an overwhelming amount of information contained in the trajectory of atomic coordinates. The graphic visualization of such a trajectory reveals the tremendous complexity of protein motion, but it is insufficient in giving useful insights into the overall dynamics. To simplify the intricate network of motions it is possible to identify concerted fluctuations with large amplitude performing Principal Component Analysis on a large number of configurations extracted from MD trajectory [115] or on a set of experimental structures (see Fig. 12.4 in Sect. 12.5). PCA is a widely used statistical technique to retrieve dominant patterns and representative distributions from noisy data. The idea is to simplify the description of a complex system from a multidimensional to a reduced dimensionality space, spanned by only few principal components (PCs), thus elucidating the principal/dominant features underlying the observed data. This method is based on the notion that by far the largest fraction of positional fluctuations in proteins occurs along only a small subset of collective degrees of freedom. The subset of largest-amplitude variables forms a set of generalized internal coordinates that can be used to describe the dynamics of a protein. Often a small subset of 5–10 % of the total number of degrees of freedom yields a remarkably accurate approximation. In other words, PCA is a multi-dimensional linear least square fit procedure in the configurational space. Mathematically, it is based on the calculation and diagonalization (after a fitting procedure to remove the translational and rotational motions in the trajectories) of the positional covariance matrix of atomic fluctuations [115] to yield collective variables that are sorted according to their contributions to the total mean-square fluctuation. For studies aiming to relate large scale motions to function, it is possible to reduce the computational effort by selecting only backbone or alpha-carbon atoms for PCA; this analysis is often named Essential Dynamics analysis (ED). Hence, PCA identifies displacements of groups of residues and emphasizes amplitude and direction of dominant protein motions, providing a reliable method to extrapolate collective functional motions from relatively short MD trajectories [116]. As a matter of fact it has been successfully exploited to describe motions occurring over longer timescale [117], such as opening and closing events or large conformational transitions occurring in enzymes and regulatory proteins. For example, ED analysis was used to capture the early stages of the gating process in a potassium channel [118] or to reveal a gating-like conformational change in the catalytic loop of a HIV-1 integrase [119, 120], or to suggest that conformational selection, rather than induced-fit, is the dominant mechanism in the molecular recognition dynamics of ubiquitin [121]. Lou et al. applied PCA to analyze the resulting trajectory from MD simulations of adenylate kinase [122]. Herein the computational results were discussed in light of experimental data revealing substantial fluctuations characterizing the motion of adenylate kinases in solution.

To conclude, one of the major challenging task for computational drug design consists in the prediction of large domain motions. This objective requires an important change in our mind-set, as increasing evidences clearly show that the comforting idea of a ligand perfectly adapting inside a static protein structure is now outdated. The approach, that searches through an ensemble of conformations the one which best accommodates the ligand, is still a strong biological approximation. However, to date it constitutes an acceptable compromise between affordable computational efforts and reliability of the results. Nevertheless, in the near future, it is expected that calculations that will fully account for protein receptor flexibility and dynamics, though computationally demanding, may likewise become routine, leading to significantly improved computer-aided identification of effective drugs. Moreover, presently underexploited target classes, such as ion channels, nuclear hormone receptors or transporters, whose functions are strictly related to their structural flexibility will be object of future drug design studies. Similarly, the molecular description of allosteric modulation will provide new opportunities to subtly regulate biological processes and will be likely object of future drug discovery campaigns.

12.5 Integrin αvβ3-Drugs Interaction: A Case Study for Ligands and Receptor Dynamics

In this section we will focus on a case study where the binding and the conformational properties of both the ligands and the receptor have been analyzed by the combined use of spectroscopic, biochemical and computational techniques. The synergy between the different methods resulted in a successful strategy for the identification of a real integrin αvβ3 antagonist. This example is particularly suited to illustrate several issues related to:

i. The conformational characterization of flexible ligands and the allosteric effects induced on the receptor upon ligand binding;
ii. The characterization of receptor-ligand molecular interactions in their natural membrane environment.

12.5.1 The Combination of Metadynamics and Docking Calculations: A Successful Strategy for Structure Based Drug Design

Recent advances in cancer therapy include molecules interfering with angiogenesis and moieties that recognize specific receptors expressed onto the tumour endothelium and/or cells, thus allowing the ligand-directed targeted delivery of various drugs and particles to tumours. In this context, integrins play a pivotal role regulating cellular functions crucial for the initiation, progression and metastasis

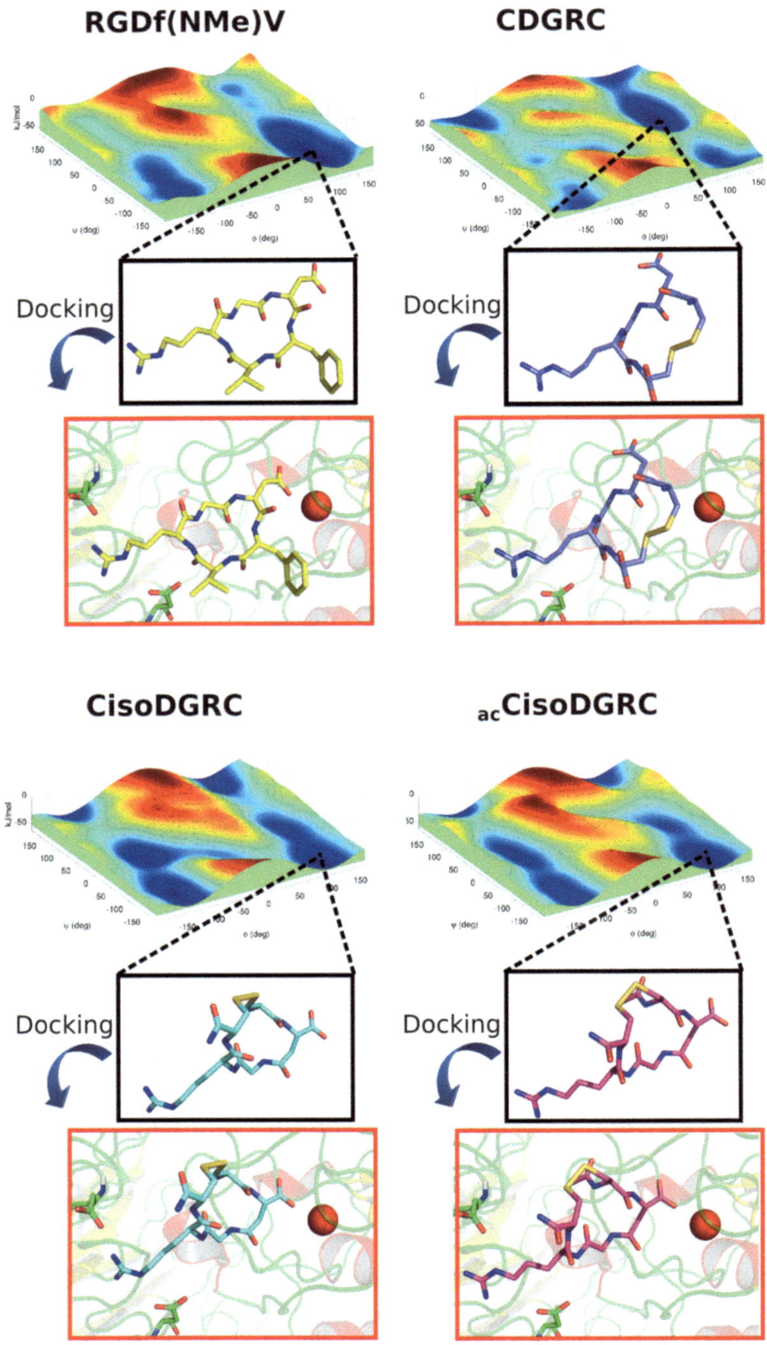
RGDf(NMe)V
CDGRC
Docking
Docking
CisoDGRC
acCisoDGRC
Docking
Docking

of solid tumours [123]. Herein, integrin αvβ3 exerts a key function in endothelial cell survival and migration, as it is an important transmembrane adhesion receptor highly expressed during angiogenesis. It has therefore gained attention as attractive therapeutic target in anti-angiogenic cancer therapy. The sequence Arg-Gly-Asp (RGD), which is contained in natural αvβ3 interactors, such as vitronectin, fibronectin, fibrinogen, osteopontin, and tenascin, is by far the most prominent ligand to promote specific cell adhesion through stimulation. This sequence is therefore attractive as a lead for the development of integrin antagonists. Recent biochemical studies showed that deamidation of the NGR sequence gives rise to isoDGR, a new αvβ3-binding motif [124]. As a novel class of peptidic integrin ligands it paves the way to drug-design studies focusing on the synthesis and characterization of a new generation of isoDGR-based cyclopeptides. [125–127]. IsoAspartic acid is a β-aminoacid, which induces high flexibility in isoDGR-containing macrocycles, thus augmenting the range of accessible inter-converting conformations which are difficult to be characterized by the common spectroscopic techniques. Nevertheless, an accurate and reliable determination of the accessible conformations of macrocycles containing the isoDGR signature conformation is a prerequisite for a reliable docking screening. Exploring the conformational space of peptides with sufficient detail is computationally very demanding and often beyond the reach, even for state-of-the-art atomistic molecular simulations techniques. As previously discussed, MetaD has emerged as a powerful coarse-grained non-Markovian molecular-dynamics approach for the acceleration of rare events and the efficient and rapid computation of multidimensional free energy surfaces as a function of a restricted number of collective variables. In this case MetaD, combined to docking calculations was successfully exploited to evaluate *in silico* the different binding properties of the cyclopeptides CisoDGRC, CDGRC (a non-binder) and RGDf(NMe)V to αvβ3. In particular, MetaD was applied to exhaustively and rapidly characterize the conformational equilibrium of flexible ligands prior docking calculations. The combination of MetaD and docking allowed to discriminate *in silico* binders from non-binders [126]. In this work it was demonstrated that MetaD performed on Gly φ and ψ angles reliably describes the free energy surface of a relevant set of RGD, DGR and isoDGR-containing cyclopeptides, thus allowing the scrutiny of their intrinsic conformational equilibria and the quantitative estimation of the population of the conformers (Fig. 12.2). In addition, these MetaD-generated conformations well agreed with NMR-derived experimental data performed on the

Fig. 12.2 Free energy surfaces (FES, kJ/mol) of RGDf(NMe)V, CDGRC, CisoDGRC and $_{ac}$CisoDGRC reconstructed by well-tempered MetaD using central Gly φ and ψ angles as collective variables (CVs). For each cyclopeptide, a representative structure extracted from the corresponding FES global minimum is shown in licorice in the *black box*. In the *red boxes*, the ligand-αvβ3 binding site, as calculated by the docking program HADDOCK are shown. The ligand is represented in licorice, αvβ3 is represented in cartoon, the side chains of αvβ3 directly involved in the binding (ASP218 and ASP150) and $Mn2^{+}$ MIDAS cation are shown with *green* licorice and *red* sphere, respectively (Modified from [125]) free molecules. It is worth noting that MetaD was successfully applied to predict the effect of chemical modifications on the cyclopeptides conformational equilibrium. This prediction power is particularly relevant for macrocycles in which conformational heterogeneity can be exploited to fine tune the ligand selectivity and affinity towards a specific receptor. As a matter of fact, it was demonstrated that MetaD could be successfully applied to predict the conformational effects of N-terminal acetylation of the macrocycle CisoDGRC ($_{ac}$CisoDGRC) and to generate reliable structural models that were docked inside αvβ3. In particular, MetaD predicted that N-terminal acetylation should remove "unproductive" conformations which should result in an increased affinity of $_{ac}$CisoDGRC over CisoDGRC (Fig. 12.2). Importantly, the computational predictions were validated in vitro through binding experiments by conventional flow cytometry analysis by testing both living cells and recombinant αvβ3. The results have been further confirmed performing binding and competition experiments acquiring 2D trNOE spectra on living cells [126, 128] (see Sect. 12.5.2). Overall the combination of docking and conformational sampling through MetaD allowed to discriminate among binding and non-binding cyclopeptides, contributing to define descriptors for good ligands and to rapidly discard *in silico* "unproductive" ligands. Overall, these findings provide support for applying MetaD/docking to improve the rationale design of isoDGR-based new diagnostic and therapeutic agents along with the rapid and accurate screening of peptide libraries. Finally, it is conceivable that coupling MetaD to docking can be successfully exploited in other ligand-receptor systems following the identification of appropriate CVs for ligand conformational ensemble characterization. It is worth noting that such an approach may be well exploited in the field of peptide-targeted agents, which represent an emerging frontier in angiogenesis, where the availability of *in silico* methods for rapid and reliable screening of targeted compounds is critically needed before entering chemical synthesis and binding experiments.

12.5.2 Characterization of Receptor-Ligand Molecular Interactions in the Natural Membrane Environment: trNOE Experiments Interrogate and Rank αvβ3-Ligand Interactions in Living Human Cancer Cells

A crucial contribution to the efficacy of targeting approaches relies on the characterization of receptor-ligand molecular interactions in their natural membrane environment. However, this is an inherently difficult goal to achieve. Usually reductionist approaches are adopted, where binding experiments are mostly performed on recombinant purified proteins. Nevertheless, these studies are often hampered by the limited availability of the target receptor. In addition, because of the intricate network of macromolecules simultaneously exerting different biological activities at membrane level, binding assays using purified proteins often fail to reflect the true nature of the cellular environment. Hence, drug-discovery studies may benefit from binding assays performed in physiological conditions, which might require accessory proteins contributing to ligand-receptor interactions. Herein, solution nuclear magnetic resonance spectroscopy, because of its non invasive nature, is increasingly utilized [4]. In this framework trNOE methods were successfully exploited directly on patient-derived intact cancer cells to prove the selective binding of various cyclic ligands including CRGDC, RGDf(NMe)V, CisoDGRC (the three peptides are ligands for αvβ3) and CDGRC (negative control). Binding was also proven for the peptide CNGRC, a ligand of aminopeptidase N (CD13), another membrane-spanning surface protein playing a pivotal role in tumour growth and metastatic spread. The selected cell lines were melanoma ($\alpha v\beta 3^{+}CD13^{-}$) and a non-small lung carcinoma cell lines ($\alpha v\beta 3^{+}CD13^{+}$) which display different phenotypes for αvβ3 and CD13. Briefly, in this experiment the spectrum of the free ligand and in presence of five millions cells are acquired. The free ligand shows positive NOEs (opposite sign with respect to the diagonal), in the presence of the cells, if it binds to the receptor, it transiently adopts the tumbling time of the receptor attached to the cell and can hereby transfer the negative NOE (same sign with respect to the diagonal) of the protein complex to the population of the free molecule. If the ligand does not bind, the cross-peaks outside the diagonal will maintain the positive sign (Fig. 12.3). The method allows using different cell lines, with different receptors, which can be also silenced with siRNA techniques to prove recognition specificity. Only very small amount of receptors are needed to prove binding (in the picomolar range). Non specific binding can be straightforwardly established by competitive binding with stronger ligands performing competition experiments thus defining an affinity ranking of different ligands in a physiological context [128]. This method was also applied to validate the MetaD hypothesis, which had predicted an increased affinity of $_{ac}$CisoDGRC over CisoDGRC (see Sect. 12.5.1). In conclusion, trNOE experiments performed directly on living cells allows to follow ligand-receptor interactions with receptors involved in tumour angiogenesis directly in a natural cellular environment that may confer relevant biological structural conformations that cannot be duplicated in vitro with single components. Furthermore, this method might have large general applications in drug discovery studies, because it can be easily exploited in other cellular systems to rapidly screen libraries of ligands and to contribute to drug design.

12.5.3 The Role of Essential Dynamics in the Study of αvβ3 Allostery Upon Ligand Binding

As member of the integrin family αvβ3 relays signals bi-directionally across the plasma membrane between the extracellular ligand binding site and the cytoplasmic domains. Signal transfer is allosterically coupled to three major equilibrium conformational states, including the (1) inactive bent state (low affinity); (2) intermediate extended state with a closed headpiece, and (3) active extended form with an

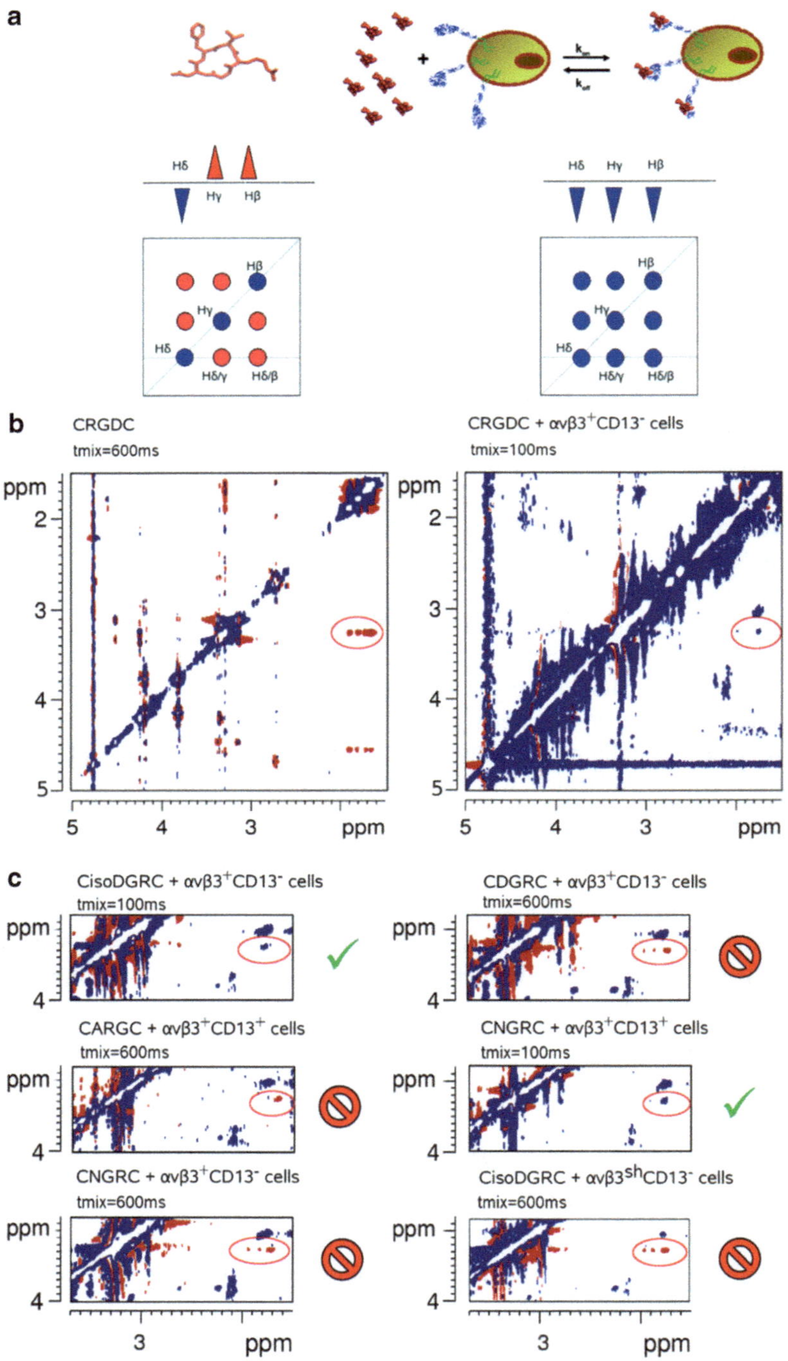
a
k_on
k_off
Hδ
Hγ
Hβ
Hδ/γ
Hδ/β
b
CRGDC
tmix=600ms
CRGDC + αvβ3+CD13- cells
tmix=100ms
ppm
c
CisoDGRC + αvβ3+CD13- cells
tmix=100ms
CDGRC + αvβ3+CD13- cells
tmix=600ms
CARGC + αvβ3+CD13+ cells
tmix=600ms
CNGRC + αvβ3+CD13+ cells
tmix=100ms
CNGRC + αvβ3+CD13- cells
tmix=600ms
CisoDGRC + αvβ3shCD13- cells
tmix=600ms

open headpiece (high affinity) [129]. Although controversy remains concerning the level of complexity in integrin allostery, it is generally recognized that the out-in signalling following ligand binding and the consequent switch from inactive to active state is accompanied by an outwards movement of the β-hybrid domain, characterized by a swing-out angle varying between 10° and 80° [129, 130] (Fig. 12.4). Out-in activation of αvβ3 by natural ligands occurs through the recognition of the tripeptidic motif RGD. This sequence, as mentioned before, is becoming a lead for developing integrin antagonists. However, in this context it should be pointed out that a major drawback with integrin antagonists is their potential to activate the receptor. In fact, all clinically approved β3 integrin antagonists act as partial agonists, as they can activate signalling through allosteric changes in the β chain, generating aberrant integrin signalling. Pharmaceutical integrin antagonists designed to block adhesive protein binding may actually promote conformational changes and receptor clustering, thus converting integrin in a multifaceted signalling machine. This might have been the case for example for oral RGD based αIIbβ3-inhibitors which showed a significant increase in mortality in patients treated with these reagents [131, 132]. One potential explanation for this extraordinary and costly failure in drug development might rely on the fact that, binding of natural ligands and of some small-molecule "antagonists" to integrins might induce exposure of neoepitopes referred to as ligand-induced binding site (LIBS) epitopes [133]. The exposure of LIBS is a consequence of conformational changes which indicate that the "antagonists" paradoxically induce the active conformation, causing outside-in signalling and subsequent receptor activation [134, 135]. Hence RGD mimetics fit into the traditional binding pocket of the receptor, competing with its natural ligands but, at the same time, act as partial agonist, inducing a response of the receptor. Therefore, although therapeutic targeting of integrins is highly attractive, the mimicking of ligands may result in an intrinsic paradoxical integrin receptor activation and may therefore not be the ideal strategy for integrin inhibition [134, 136]. In the development process of new αvβ3 inhibitors, studies aiming at the characterization of the receptor allosteric events induced by ligand binding can therefore offer crucial information on the potential benefits and drawbacks elicited by the lead molecule.

Fig. 12.3 trNOE on living human cancer cells. (**a**) Schematic representation of trNOE experiment performed on cell surface bound receptors. (**b**) Representative spectra of a free ligand (*left*), of free cells (*center*), of ligand bound to the cell surface receptor (*right*). The spectrum of the free ligand shows positive NOEs (opposite sign with respect to the diagonal). During ligand–receptor binding, the ligand transiently adopts the tumbling time of the receptor and can transfer the negative NOE (same sign with respect to the diagonal) of the protein complex to the population of the free molecule (**c**) selected region of trNOE experiments performed on different ligands and on difference cell lines (indicated on the *top* of the panel), *red circles* indicate the cross-peaks deriving from correlations of the arginine δ/γ and δ/β protons of the ligand; ligands interacting with the cell lines are indicated with a green "v" symbol, whereas not interacting ligands are indicated with the "stop" symbol (Modified from [128])

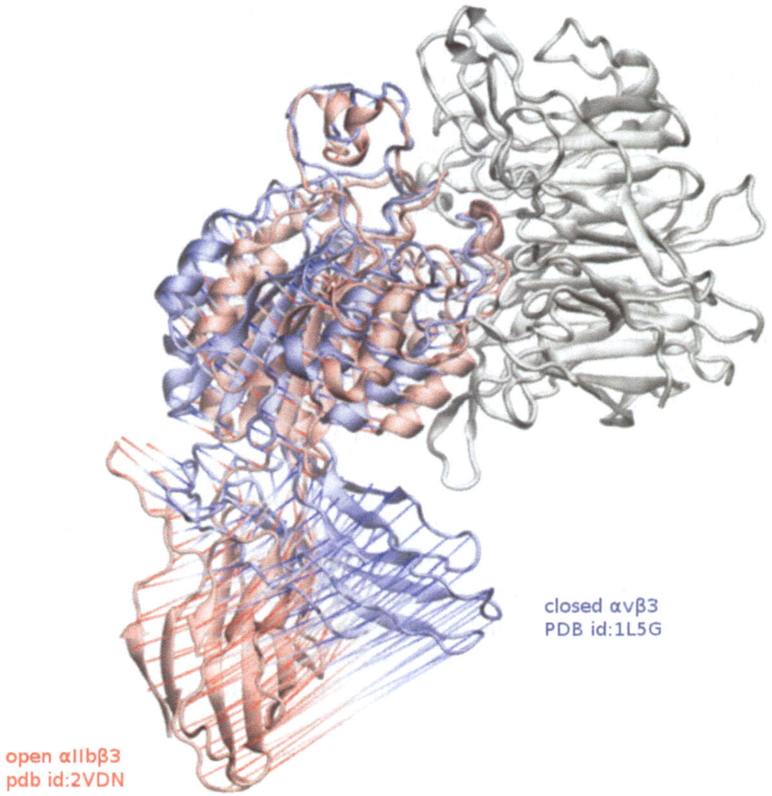

Fig. 12.4 Principal Component Analysis shows integrin headpiece allosteric transition from the closed to the open state. Porcupine visualization of the first dominant eigenvector (corresponding to the swing out movement of integrin hybrid domain) obtained from PCA on a set of crystallographic structures. Two integrin structures representing the closed (PDB code: 1L5G) and the open (PDB code: 2VDN) state are shown in cartoon representation with the α chain coloured in Silver, the β3 of the closed and open conformation are coloured in *Pastel blue* and *Metallic Pastel red*, respectively

In the framework of a drug-discovery project focusing on the development of αvβ3 ligands based on the isoDGR sequence we performed all atoms MD simulations on αvβ3 in complex with RGDf(NMe)V and with isoDGR containing cyclopeptides to characterize the receptor conformational changes induced by different ligands. Calculations showed that both cyclopeptides anchor to the αv and β3 extracellular domains through an electrostatic clamp that exploits similar though not identical interaction patterns. On one hand, the Arg guanidinium groups of the ligands are engaged in stable salt-bridges with the carboxylate of D218 and/or of D150 within the αv β-propeller domain. On the other hand, their Asp/isoAsp carboxylates bind to the I-like domain of the β3 subunit, coordinating the MIDAS ion via a carboxylate oxygen. Herein, we observed relevant differences in the

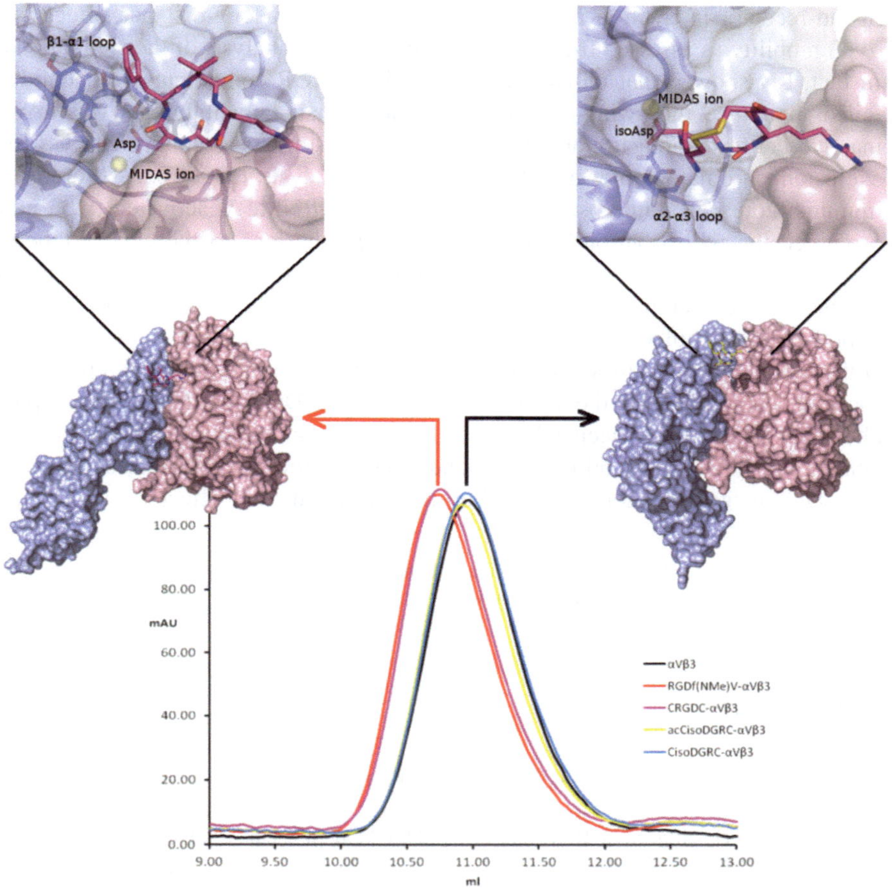

Fig. 12.5 Differences in the ligand-receptor interaction patterns induce different long range conformational changes on αvβ3. On the *top* is shown a representative binding mode of RGD (*left*) and isoDGR (*right*) containing cyclopeptides, extracted from MD simulations. The carboxylate group of Asp/isoAsp residue of the ligands (represented with sticks) coordinates the MIDAS cation and two different regions of integrin β3 chain. In agreement with Essential Dynamics analysis, gel filtration profiles confirm different mobility of free αvβ3 or with saturating amounts of RGDf(NMe)V, CRGDC, CisoDGRC and $_{ac}$CisoDGRC (on the bottom). RGD ligands induce opening of the receptor (*left structure*), as assessed by the smaller elution volume as compared to isoDGR ligands, which maintain αvβ3 an inactive conformation (*right structure*) (Modified from [137])

coordination pattern of the second carboxylate oxygen: in isoDGR containing cyclopeptides this oxygen interacts with the α2-α3 loop in the β3 subunit, whereas, the very same oxygen in RGDf(NMe)V stably binds to the β1-α1 loop of the β3 subunit (Fig. 12.5) [137]. Remarkably, this latter interaction, playing a relevant role as trigger of the β3 swing-out mechanism [138], is barely present in the simulations with isoDGR containing cyclopeptides. Importantly, differences in the ligand-receptor interaction patterns resulted in different long-distance effects on αvβ3 mobility as assessed from the essential dynamics analysis of the simulations. In fact, ED showed that unlike RGDf(NMe)V, isoDGR-containing cyclopeptides failed to induce the swing-out of the hybrid domain, maintaining αvβ3 in its inactive conformation. Importantly, the *in silico* results were confirmed by size-exclusion chromatography and flow cytometry analysis (Fig. 12.5). Moreover, immunofluorescence microscopy showed that RGDf(NMe)V activated αvβ3 and promoted the redistribution of αvβ3 from focal adhesions to the cell periphery, an event critical for cell migration. In contrast, isoDGR containing cyclopeptides do not induce accumulation of αvβ3 at the cell border, further supporting the hypothesis that isoDGR containing cyclopeptides compete with ligand binding without inducing integrin activation. Altogether, these findings hold major promises for drug design, based on the intrinsic ability of the isoDGR motif to block receptor allosteric activation. Conceivably, isoDGR based drugs might replace the current generation of integrin-binding compounds, representing a promising solution in designing integrin antagonists, devoid of intrinsic paradoxical effects.

Acknowledgements The authors wish to thank Fondazione Telethon and the Italian Association against Cancer (AIRC) for continuous support.

References

1. Macarron R (2006) Critical review of the role of HTS in drug discovery. Drug Discov Today 11:277–279
2. Giersiefen H, Hilgenfeld R, Hillisch A (2003) Modern methods of drug discovery: an introduction. EXS 93:1–18
3. Powers R (2009) Advances in nuclear magnetic resonance for drug discovery. Expert Opin Drug Discov 4:1077–1098
4. Pellecchia M, Bertini I, Cowburn D et al (2008) Perspectives on NMR in drug discovery: a technique comes of age. Nat Rev Drug Discov 7:738–745
5. Williamson MP (2013) Using chemical shift perturbation to characterise ligand binding. Prog Nucl Magn Reson Spectrosc 73:1–16
6. Shuker SB, Hajduk PJ, Meadows RP et al (1996) Discovering high-affinity ligands for proteins: SAR by NMR. Science 274:1531–1534
7. Takeuchi K, Wagner G (2006) NMR studies of protein interactions. Curr Opin Struct Biol 16:109–117
8. Tugarinov V, Kay LE (2003) Quantitative NMR studies of high molecular weight proteins: application to domain orientation and ligand binding in the 723 residue enzyme malate synthase G. J Mol Biol 327:1121–1133
9. Kay LE (2011) Solution NMR spectroscopy of supra-molecular systems, why bother? A methyl-TROSY view. J Magn Reson 210:159–170
10. Lepre CA (2011) Practical aspects of NMR-based fragment screening. Methods Enzymol 493:219–239
11. Lepre CA, Moore JM, Peng JW (2004) Theory and applications of NMR-based screening in pharmaceutical research. Chem Rev 104:3641–3676
12. Dalvit C (2009) NMR methods in fragment screening: theory and a comparison with other biophysical techniques. Drug Discov Today 14:1051–1057

13. Goldflam M, Tarrago T, Gairi M et al (2012) NMR studies of protein-ligand interactions. Methods Mol Biol 831:233–259
14. Hajduk PJ, Olejniczak ET, Fesik SW (1997) One dimensional relaxation- and diffusion-edited NMR methods for screening compounds that bind to macromolecules. J Am Chem Soc 119:12257
15. Price SW (1997) Pulsed-field gradient nuclear magnetic resonance as a tool for studying translational diffusion: Part 1. Concepts Magn Reson 9:299–366
16. Price SW (1998) Pulsed-field gradient nuclear magnetic resonance as a tool for studying translational diffusion: Part II. Experimental aspects. Concepts Magn Reson 10:197–197
17. Lin M, Shapiro MJ, Wareing JR (1997) Diffusion-edited NMR ± affinity NMR for direct observation of molecular interactions. Biophys J 119:5249–5250
18. Lin M, Shapiro MJ, Wareing JR (1997) Screening mixtures by affinity NMR. J Org Chem 62:8931
19. Mayer M, Meyer B (1999) Characterization of ligand binding by saturation transfer difference NMR spectroscopy. Angew Chem Int Ed 38:1784–1788
20. Claasen B, Axmann M, Meinecke R et al (2005) Direct observation of ligand binding to membrane proteins in living cells by a saturation transfer double difference (STDD) NMR spectroscopy method shows a significantly higher affinity of integrin alpha(IIb)beta3 in native platelets than in liposomes. J Am Chem Soc 127:916–919
21. Mari S, Serrano-Gómez D, Cañada FJ, Corbí AL, Jiménez-Barbero J (2005) 1D-STD NMR experiments on living cells. The DC-SIGN/oligomannose interaction. Angew Chem Int Ed 44:298
22. Angulo J, Nieto PM (2011) STD-NMR: application to transient interactions between biomolecules-a quantitative approach. Eur Biophys J 40:1357–1369
23. Dalvit C, Fogliatto G, Stewart A et al (2001) WaterLOGSY as a method for primary NMR screening: practical aspects and range of applicability. J Biomol NMR 21:349–359
24. Clore GM, Gronenborn AM (1982) Theory and applications of the transferred nuclear Overhauser effect to the study of the conformations of small ligands bounds to proteins. J Magn Reson 48:402–417
25. Neuhaus D, Williamson MP (2000) The NOE in structural and conformational analysis. Methods in stereochemical analysis, vol 24. Wiley-VCH, New York
26. Post CB (2003) Exchange-transferred NOE spectroscopy and bound ligand structure determination. Curr Opin Struct Biol 13:581–588
27. Fejzo J, Lepre CA, Peng JW et al (1999) The SHAPES strategy: an NMR-based approach for lead generation in drug discovery. Chem Biol 6:755–769
28. Sanchez-Pedregal VM, Reese M, Meiler J et al (2005) The INPHARMA method: protein-mediated interligand NOEs for pharmacophore mapping. Angew Chem Int Ed Engl 44:4172–4175
29. Becattini B, Pellecchia M (2006) SAR by ILOEs: an NMR-based approach to reverse chemical genetics. Chemistry 12:2658–2662
30. Becattini B, Culmsee C, Leone M et al (2006) Structure-activity relationships by interligand NOE-based design and synthesis of antiapoptotic compounds targeting Bid. Proc Natl Acad Sci USA 103:12602–12606
31. Vulpetti A, Hommel U, Landrum G et al (2009) Design and NMR-based screening of LEF, a library of chemical fragments with different local environment of fluorine. J Am Chem Soc 131:12949–12959
32. Dalvit C, Fagerness PE, Hadden DT et al (2003) Fluorine-NMR experiments for high-throughput screening: theoretical aspects, practical considerations, and range of applicability. J Am Chem Soc 125:7696–7703
33. Rognan D (2013) Proteome-scale docking: myth and reality. Drug Discov Today Technol 10:e403–e409
34. Kuntz ID, Blaney JM, Oatley SJ et al (1982) A geometric approach to macromolecule-ligand interactions. J Mol Biol 161:269–288

35. Meng XY, Zhang HX, Mezei M et al (2011) Molecular docking: a powerful approach for structure-based drug discovery. Curr Comput-Aided Drug Des 7:146–157
36. Kitchen DB, Decornez H, Furr JR et al (2004) Docking and scoring in virtual screening for drug discovery: methods and applications. Nat Rev Drug Discov 3:935–949
37. Lexa KW, Carlson HA (2012) Protein flexibility in docking and surface mapping. Q Rev Biophys 45:301–343
38. Feher M, Williams CI (2009) Effect of input differences on the results of docking calculations. J Chem Inf Model 49:1704–1714
39. Morris GM, Huey R, Lindstrom W et al (2009) AutoDock4 and AutoDockTools4: automated docking with selective receptor flexibility. J Comput Chem 30:2785–2791
40. Dominguez C, Boelens R, Bonvin AM (2003) HADDOCK: a protein-protein docking approach based on biochemical or biophysical information. J Am Chem Soc 125:1731–1737
41. de Vries SJ, van Dijk AD, Krzeminski M et al (2007) HADDOCK versus HADDOCK: new features and performance of HADDOCK2.0 on the CAPRI targets. Proteins 69:726–733
42. Jones G, Willett P, Glen RC (1995) Molecular recognition of receptor sites using a genetic algorithm with a description of desolvation. J Mol Biol 245:43–53
43. Chemical CGI (2012) Molecular operating environment (MOE) 2013.08. Chemical Computing Group Inc., Montreal
44. Tirado-Rives J, Jorgensen WL (2006) Contribution of conformer focusing to the uncertainty in predicting free energies for protein-ligand binding. J Med Chem 49:5880–5884
45. Schlick T, Collepardo-Guevara R, Halvorsen LA et al (2011) Biomolecular modeling and simulation: a field coming of age. Q Rev Biophys 44:191–228
46. Karplus M, McCammon JA (2002) Molecular dynamics simulations of biomolecules. Nat Struct Biol 9:646–652
47. Kirkpatrick S, Gelatt CD Jr, Vecchi MP (1983) Optimization by simulated annealing. Science 220:671–680
48. Sammond DW, Bosch DE, Butterfoss GL et al (2011) Computational design of the sequence and structure of a protein-binding peptide. J Am Chem Soc 133:4190–4192
49. Sugita Y, Okamoto Y (1999) Replica-exchange molecular dynamics method for protein folding. Chem Phys Lett 314:141–151
50. Ostermeir K, Zacharias M (2013) Advanced replica-exchange sampling to study the flexibility and plasticity of peptides and proteins. Biochim Biophys Acta 1834:847–853
51. Macaluso NJ, Pitkin SL, Maguire JJ et al (2011) Discovery of a competitive apelin receptor (APJ) antagonist. ChemMedChem 6:1017–1023
52. Okumura H, Gallicchio E, Levy RM (2010) Conformational populations of ligand-sized molecules by replica exchange molecular dynamics and temperature reweighting. J Comput Chem 31:1357–1367
53. Frenkel D, Smit B (1996) Understanding molecular simulation: from algorithms to applications. Academic, San Diego
54. Grzybowski BA, Ishchenko AV, Kim CY et al (2002) Combinatorial computational method gives new picomolar ligands for a known enzyme. Proc Natl Acad Sci USA 99:1270–1273
55. Mohamadi F, Richards NG, Guida WC et al (1990) Macromodel—an integrated software system for modeling organic and bioorganic molecules using molecular mechanics. J Comput Chem 11:440–467
56. Forti F, Cavasotto CN, Orozco M et al (2012) A multilevel strategy for the exploration of the conformational flexibility of small molecules. J Chem Theory Comput 8:1808–1819
57. Doi T, Muraoka T, Ohshiro T et al (2012) Conformationally restricted analog and biotin-labeled probe based on beauveriolide III. Bioorg Med Chem Lett 22:696–699
58. Watts KS, Dalal P, Murphy RB et al (2010) ConfGen: a conformational search method for efficient generation of bioactive conformers. J Chem Inf Model 50:534–546
59. Huang JJ, Wu XW, Jia JM et al (2013) Novel IKKbeta inhibitors discovery based on the co-crystal structure by using binding-conformation-based and ligand-based method. Eur J Med Chem 63C:269–278

60. Anonymous (2008) MOE; Chemical Computing Group: 1010 Sherbrooke St. W, Suite 910, Montreal, Quebec, Canada H3A 2R7
61. Hawkins PC, Nicholls A (2012) Conformer generation with OMEGA: learning from the data set and the analysis of failures. J Chem Inf Model 52:2919–2936
62. Soliman MH (2013) A hybrid structure/pharmacophore-based virtual screening approach to design potential leads: a computer-aided design of South African HIV-1 subtype C protease inhibitors. Drug Dev Res 74:283–295
63. Levy Y, Becker OM (2001) Energy landscapes of conformationally constrained peptides. J Chem Phys 114:993–1009
64. Mitsutake A, Mori Y (2013) Enhanced sampling algorithms. Methods Mol Biol 924:153–195
65. Huber T, Torda AE, van Gunsteren WF (1994) Local elevation: a method for improving the searching properties of molecular dynamics simulation. J Comput Aided Mol Des 8:695–708
66. Cvijovicacute D, Klinowski J (1995) Taboo search: an approach to the multiple minima problem. Science 267:664–666
67. Wang F, Landau DP (2001) Efficient, multiple-range random walk algorithm to calculate the density of states. Phys Rev Lett 86:2050–2053
68. Darve E, Rodriguez-Gomez D, Pohorille A (2008) Adaptive biasing force method for scalar and vector free energy calculations. J Chem Phys 128:144120
69. Grubmuller H (1995) Predicting slow structural transitions in macromolecular systems: conformational flooding. Phys Rev E Stat Phys Plasmas Fluids Relat Interdiscip Topics 52:2893–2906
70. Patey GN, Valleau JP (1975) Monte-Carlo method for obtaining interionic potential of mean force in ionic solution. J Chem Phys 63:2334–2339
71. Ferrenberg AM, Swendsen RH (1988) New Monte Carlo technique for studying phase transitions. Phys Rev Lett 61:2635–2638
72. Bolhuis PG, Chandler D, Dellago C et al (2002) Transition path sampling: throwing ropes over rough mountain passes, in the dark. Annu Rev Phys Chem 53:291–318
73. Zhou T, Caflisch A (2012) Free energy guided sampling. J Chem Theory Comput 8: 2134–2140
74. Laio A, Parrinello M (2002) Escaping free-energy minima. Proc Natl Acad Sci USA 99:12562–12566
75. Oborsky P, Tvaroska I, Kralova B et al (2013) Toward an accurate conformational modeling of iduronic acid. J Phys Chem B 117:1003–1009
76. Spiwok V, Hlat-Glembova K, Tvaroska I et al (2012) Conformational free energy modeling of druglike molecules by metadynamics in the WHIM space. J Chem Inf Model 52:804–813
77. Garate JA, Oostenbrink C (2013) Free-energy differences between states with different conformational ensembles. J Comput Chem 34:1398–1408
78. Hansen HS, Hunenberger PH (2011) A reoptimized GROMOS force field for hexopyranose-based carbohydrates accounting for the relative free energies of ring conformers, anomers, epimers, hydroxymethyl rotamers, and glycosidic linkage conformers. J Comput Chem 32:998–1032
79. McGaughey GB, Sheridan RP, Bayly CI et al (2007) Comparison of topological, shape, and docking methods in virtual screening. J Chem Inf Model 47:1504–1519
80. Wong CF, McCammon JA (2003) Protein flexibility and computer-aided drug design. Annu Rev Pharmacol Toxicol 43:31–45
81. Carlson HA, McCammon JA (2000) Accommodating protein flexibility in computational drug design. Mol Pharmacol 57:213–218
82. Seeliger D, de Groot BL (2010) Conformational transitions upon ligand binding: holo-structure prediction from apo conformations. PLoS Comput Biol 6:e1000634
83. Berendsen HJ, Hayward S (2000) Collective protein dynamics in relation to function. Curr Opin Struct Biol 10:165–169
84. Bahar I, Lezon TR, Bakan A et al (2010) Normal mode analysis of biomolecular structures: functional mechanisms of membrane proteins. Chem Rev 110:1463–1497

85. Brooks B, Karplus M (1985) Normal modes for specific motions of macromolecules: application to the hinge-bending mode of lysozyme. Proc Natl Acad Sci USA 82: 4995–4999
86. Brooks B, Karplus M (1983) Harmonic dynamics of proteins: normal modes and fluctuations in bovine pancreatic trypsin inhibitor. Proc Natl Acad Sci USA 80:6571–6575
87. Levitt M, Sander C, Stern PS (1985) Protein normal-mode dynamics: trypsin inhibitor, crambin, ribonuclease and lysozyme. J Mol Biol 181:423–447
88. Petrone P, Pande VS (2006) Can conformational change be described by only a few normal modes? Biophys J 90:1583–1593
89. Xu C, Tobi D, Bahar I (2003) Allosteric changes in protein structure computed by a simple mechanical model: hemoglobin TR2 transition. J Mol Biol 333:153–168
90. Delarue M, Sanejouand YH (2002) Simplified normal mode analysis of conformational transitions in DNA-dependent polymerases: the elastic network model. J Mol Biol 320: 1011–1024
91. Tama F, Sanejouand YH (2001) Conformational change of proteins arising from normal mode calculations. Protein Eng 14:1–6
92. Dietzen M, Zotenko E, Hildebrandt A et al (2012) On the applicability of elastic network normal modes in small-molecule docking. J Chem Inf Model 52:844–856
93. Roux B (2002) Computational studies of the gramicidin channel. Acc Chem Res 35:366–375
94. Suhre K, Sanejouand YH (2004) ElNemo: a normal mode web server for protein movement analysis and the generation of templates for molecular replacement. Nucleic Acids Res 32:W610–W614
95. Lu HP (2005) Probing single-molecule protein conformational dynamics. Acc Chem Res 38:557–565
96. Harms GS, Orr G, Montal M et al (2003) Probing conformational changes of gramicidin ion channels by single-molecule patch-clamp fluorescence microscopy. Biophys J 85:1826–1838
97. Cavasotto CN, Kovacs JA, Abagyan RA (2005) Representing receptor flexibility in ligand docking through relevant normal modes. J Am Chem Soc 127:9632–9640
98. Kruger DM, Ahmed A, Gohlke H (2012) NMSim web server: integrated approach for normal mode-based geometric simulations of biologically relevant conformational transitions in proteins. Nucleic Acids Res 40:W310–W316
99. Mashiach E, Schneidman-Duhovny D, Peri A et al (2010) An integrated suite of fast docking algorithms. Proteins 78:3197–3204
100. Ming D, Cohn JD, Wall ME (2008) Fast dynamics perturbation analysis for prediction of protein functional sites. BMC Struct Biol 8:5–15
101. Bahar I, Erman B, Haliloglu T et al (1997) Efficient characterization of collective motions and interresidue correlations in proteins by low-resolution simulations. Biochemistry 36: 13512–13523
102. Tirion MM (1996) Large amplitude elastic motions in proteins from a single-parameter, atomic analysis. Phys Rev Lett 77:1905–1908
103. Atilgan AR, Durell SR, Jernigan RL et al (2001) Anisotropy of fluctuation dynamics of proteins with an elastic network model. Biophys J 80:505–515
104. Vashisth H, Brooks CL 3rd (2012) Conformational sampling of maltose-transporter components in Cartesian collective variables is governed by the low-frequency normal modes. J Phys Chem Lett 3:3379–3384
105. Korkut A, Hendrickson WA (2012) Structural plasticity and conformational transitions of HIV envelope glycoprotein gp120. PLoS One 7:e52170
106. Rader AJ, Anderson G, Isin B et al (2004) Identification of core amino acids stabilizing rhodopsin. Proc Natl Acad Sci USA 101:7246–7251
107. Isin B, Rader AJ, Dhiman HK et al (2006) Predisposition of the dark state of rhodopsin to functional changes in structure. Proteins 65:970–983
108. Perryman AL, Forli S, Morris GM et al (2010) A dynamic model of HIV integrase inhibition and drug resistance. J Mol Biol 397:600–615

109. Schames JR, Henchman RH, Siegel JS et al (2004) Discovery of a novel binding trench in HIV integrase. J Med Chem 47:1879–1881
110. Maertens GN, Hare S, Cherepanov P (2010) The mechanism of retroviral integration from X-ray structures of its key intermediates. Nature 468:326–329
111. Hare S, Gupta SS, Valkov E et al (2010) Retroviral intasome assembly and inhibition of DNA strand transfer. Nature 464:232–236
112. Buch I, Giorgino T, De Fabritiis G (2011) Complete reconstruction of an enzyme-inhibitor binding process by molecular dynamics simulations. Proc Natl Acad Sci USA 108:10184–10189
113. Seeliger MA, Ranjitkar P, Kasap C et al (2009) Equally potent inhibition of c-Src and Abl by compounds that recognize inactive kinase conformations. Cancer Res 69:2384–2392
114. Skjaerven L, Grant B, Muga A et al (2011) Conformational sampling and nucleotide-dependent transitions of the GroEL subunit probed by unbiased molecular dynamics simulations. PLoS Comput Biol 7:e1002004
115. Amadei A, Linssen AB, Berendsen HJ (1993) Essential dynamics of proteins. Proteins 17:412–425
116. Lange OF, Grubmuller H (2006) Can principal components yield a dimension reduced description of protein dynamics on long time scales? J Phys Chem B 110:22842–22852
117. Cheng X, Ivanov I, Wang H et al (2007) Nanosecond time scale conformational dynamics of the human α7 nicotinic acetylcholine receptor. Biophys J 93:2622–2634
118. Grottesi A, Domene C, Hall B et al (2005) Conformational dynamics of M2 helices in KirBac channels: helix flexibility in relation to gating via molecular dynamics simulations. Biochemistry 44:14586–14594
119. Brigo A, Lee KW, Iurcu Mustata G et al (2005) Comparison of multiple molecular dynamics trajectories calculated for the drug-resistant HIV-1 integrase T66I/M154I catalytic domain. Biophys J 88:3072–3082
120. Lee MC, Deng J, Briggs JM et al (2005) Large-scale conformational dynamics of the HIV-1 integrase core domain and its catalytic loop mutants. Biophys J 88:3133–3146
121. Lange OF, Lakomek NA, Fares C et al (2008) Recognition dynamics up to microseconds revealed from an RDC-derived ubiquitin ensemble in solution. Science 320:1471–1475
122. Lou H, Cukier RI (2006) Molecular dynamics of apo-adenylate kinase: a principal component analysis. J Phys Chem B 110:12796–12808
123. Desgrosellier JS, Cheresh DA (2010) Integrins in cancer: biological implications and therapeutic opportunities. Nat Rev Cancer 10:9–22
124. Curnis F, Longhi R, Crippa L et al (2006) Spontaneous formation of L-isoaspartate and gain of function in fibronectin. J Biol Chem 281:36466–36476
125. Spitaleri A, Mari S, Curnis F et al (2008) Structural basis for the interaction of isoDGR with the RGD-binding site of alphavbeta3 integrin. J Biol Chem 283:19757–19768
126. Spitaleri A, Ghitti M, Mari S et al (2011) Use of metadynamics in the design of isoDGR-based alphavbeta3 antagonists to fine-tune the conformational ensemble. Angew Chem Int Ed Engl 50:1832–1836
127. Frank AO, Otto E, Mas-Moruno C et al (2010) Conformational control of integrin-subtype selectivity in isoDGR peptide motifs: a biological switch. Angew Chem Int Ed Engl 49: 9278–9281
128. Mari S, Invernizzi C, Spitaleri A et al (2010) 2D TR-NOESY experiments interrogate and rank ligand-receptor interactions in living human cancer cells. Angew Chem Int Ed Engl 49:1071–1074
129. Xiao T, Takagi J, Coller BS et al (2004) Structural basis for allostery in integrins and binding to fibrinogen-mimetic therapeutics. Nature 432:59–67
130. Puklin-Faucher E, Vogel V (2009) Integrin activation dynamics between the RGD-binding site and the headpiece hinge. J Biol Chem 284:36557–36568
131. Meadows TA, Bhatt DL (2007) Clinical aspects of platelet inhibitors and thrombus formation. Circ Res 100:1261–1275

132. Quinn MJ, Byzova TV, Qin J et al (2003) Integrin alphaIIbbeta3 and its antagonism. Arterioscler Thromb Vasc Biol 23:945–952
133. Du X, Gu M, Weisel JW et al (1993) Long range propagation of conformational changes in integrin alpha IIb beta 3. J Biol Chem 268:23087–23092
134. Bassler N, Loeffler C, Mangin P et al (2007) A mechanistic model for paradoxical platelet activation by ligand-mimetic alphaIIb beta3 (GPIIb/IIIa) antagonists. Arterioscler Thromb Vasc Biol 27:e9–e15
135. Du XP, Plow EF, Frelinger AL et al (1991) Ligands "activate" integrin alpha IIb beta 3 (platelet GPIIb-IIIa). Cell 65:409–416
136. Ahrens I, Peter K (2008) Therapeutic integrin inhibition: allosteric and activation-specific inhibition strategies may surpass the initial ligand-mimetic strategies. Thromb Haemost 99:803–804
137. Ghitti M, Spitaleri A, Valentinis B et al (2012) Molecular dynamics reveal that isoDGR-containing cyclopeptides are true alphavbeta3 antagonists unable to promote integrin allostery and activation. Angew Chem Int Ed Engl 51:7702–7705
138. Zhu J, Zhu J, Negri A et al (2010) Closed headpiece of integrin alphaIIbbeta3 and its complex with an alphaIIbbeta3-specific antagonist that does not induce opening. Blood 116:5050–5059

Chapter 13
Molecular Dynamics Simulation of Membrane Proteins

Jingwei Weng and Wenning Wang

Abstract Membrane proteins play crucial roles in a range of biological processes. High resolution structures provide insights into the functional mechanisms of membrane proteins, but detailed biophysical characterization of membrane proteins is difficult. Complementary to experimental techniques, molecular dynamics simulations is a powerful tool in providing more complete description of the dynamics and energetics of membrane proteins with high spatial-temporal resolution. In this chapter, we provide a survey of the current methods and technique issues for setting up and running simulations of membrane proteins. The recent progress in applying simulations to understanding various biophysical properties of membrane proteins is outlined.

Keywords Membrane protein • MD simulation • Force fields • Lipid bilayer • Coarse-grained method • Enhanced sampling method • Free energy calculation

Abbreviations

MD	molecular dynamics
LJ	Lennard-Jones
APL	area per lipid
DPPC	dipalmitoylphosphatidylcholine

J. Weng
Department of Chemistry, Fudan University, Shanghai, China
e-mail: jwweng@fudan.edu.cn

W. Wang (✉)
Department of Chemistry, Fudan University, Shanghai, China

Institute of Biomedical Sciences, Fudan University, Shanghai, China
e-mail: wnwang@fudan.edu.cn

K. Han et al. (eds.), *Protein Conformational Dynamics*, Advances in Experimental Medicine and Biology 805, DOI 10.1007/978-3-319-02970-2_13,

POPC	palmitoyloleoylphosphatidylcholine
CG	coarse-grained
PC	phosphatidylcholine
EM	energy minimization
PME	Particle Mesh Ewald
RF	reaction field
SMD	steered MD
AFM	atomic force microscopy
PMF	potential of mean force
TMD	targeted MD
ABF	adaptive biasing force
CV	collective variable
ENM	elastic network model
NMA	normal mode analysis
ANM	anisotropic network model
VSD	voltage sensor domain
GPCR	G protein coupled receptor
AQP	aquaporin
EGFR	epidermal growth factor receptor
DOR	delta opioid receptor

13.1 Introduction

Membrane proteins constitute 20–30 % of all proteins encoded by genes in most genomes [1]. They are involved in crucial physiological processes of life, including transporting various ions and molecules across membranes, transducing energy, sensing and sending chemical signals, regulating intracellular vesicular transport etc. Therefore, membrane proteins often serve as important drug targets. It has been estimated that about 60 % drug targets for human diseases are membrane proteins [2].

Understanding the biological function and the underlying molecular mechanism of membrane proteins rely heavily on their high-resolution 3D structures. Compared to soluble proteins, the structure determination of membrane protein is much more difficult. The first crystal structure of membrane protein was determined in 1985 [3]. In the following two decades, only ~100 unique membrane protein structures were reported. But this number is growing exponentially, and up to date ~400 unique membrane protein structures have been solved (http://blanco.biomol.uci.edu/mpstruc/listAll/list). The crystal structures often provide clues to unraveling the working mechanism of membrane proteins, but proteins often cycle between multiple conformational/functional states during the biological processes and the structural information of different states of the same membrane protein is limited. On the other hand, when structures of multiple states of the membrane protein are determined, it is still difficult for experimental techniques to reveal the transition mechanism between these states, which require the dynamic description of these processes with high spatial and temporal resolutions [4].

Computer simulation, specifically molecular dynamics (MD) simulation, provides such a tool of exploring the membrane protein dynamics with atomistic and femtosecond resolutions. MD simulation of biomolecular system is based on classic Newton's equation of motion and empirical potential energy functions. There is a long history of application of MD simulations to biomolecular systems starting with the simulation of BPTI in 1977 [5]. Since then MD simulation has been widely applied and become a common technique to study the structure-function relationship of biomolecules. The application of MD simulation to membrane proteins, however, has a relatively short history due to the limited available protein structures and the immature lipid force fields. In recent years, developments of the force fields for lipids largely improve the description of the lipid bilayer and the protein-lipid interactions. On the other hand, a more challenging aspect of membrane protein simulation is the "time gap" between the simulations and the functional relevant process of membrane protein. For a complex system of membrane protein, the typical atomic MD simulation timescale is generally limited to 10^{-8} to 10^{-7} s by the computer resources accessible for most research groups. On the other hand, unfortunately, most of the biologically interesting processes occur at microsecond to millisecond or even longer timescales. Recently, the development of high-performance computer facilities and advances in computational algorithms begin to progressively bridging the gap. The methodology advances mainly focus on coarse-grained force fields and various enhanced sampling methods. With these advances of computer hardware and software, many important biologically relevant phenomena of membrane proteins have been investigated, such as the conformational changes, ligand binding, ion conductance and substrate transport, receptor oligomerization and assembly in membrane etc. In this chapter, we will first introduce the most commonly used techniques in setting up and simulate the membrane protein systems, and then give a brief review of some applications of MD simulation for membrane proteins.

13.2 Force Fields

For all molecular dynamics simulations, two important issues must always be kept in mind: (1) whether the interactions between atoms are accurately described and (2) whether the simulation is long enough for the system to sufficiently sample its conformational phase space. Reliable dynamics and thermodynamics properties can be obtained only when both issues are well satisfied. In this section, we will first introduce the empirical force fields which are currently used to describe atom-atom interactions in membrane protein simulations.

13.2.1 General Issues of Force Fields

In classical MD simulations, atoms are often reduced to point-like particles. The interactions between the particles are typically modeled by sums of pairwise or multibody potentials including bond stretching, angle bending, torsional twisting, out-of-plane bending, Lennard-Jones (LJ) interactions and Coulomb interactions. A general form of the potential energy function can be written as,

$$
\begin{aligned}
V(r) = & \sum_{bonds} k_b(b-b_0)^2 + \sum_{angles} k_\theta(\theta-\theta_0)^2 \\
& + \sum_{dihedrals} k_\phi\,(1+\cos(n\phi-\phi_0)) + \sum_{impropers} k_\psi(\psi-\psi_0)^2 \\
& + \sum_{\substack{non\text{-}bonded\\ pairs(i,j)}} 4\varepsilon_{ij}\left[\left(\frac{\sigma_{ij}}{r_{ij}}\right)^{12} - \left(\frac{\sigma_{ij}}{r_{ij}}\right)^{6}\right] + \sum_{\substack{non\text{-}bonded\\ pairs(i,j)}} \frac{q_i q_j}{\varepsilon_D r_{ij}}
\end{aligned}
$$

although there may be some extra terms in certain force fields. The potential energy function V is dependent on position vector r of all particles, from which the inter-particle distance b and r_{ij}, the angle θ, the dihedral angle φ and the improper dihedral angle ψ in the expression are derived. The parameters in bonded terms, including the force constants k_b, k_θ, k_φ, k_ψ and the equilibration distance b_0, angle θ_0, improper angle ψ_0, dihedral phase angle φ_0 and multiplicity n, and those in non-bonded terms, including the LJ well depth ε_{ij}, the collision diameter σ_{ij}, and the partial particle charges q_i and q_j, are all dependent on the particle type involved in each term. ε_D is the dielectric constant. The functional form of potential energy and the set of parameters constitute a force field. The parameters in force fields are derived from a combination of experimental data and quantum mechanical calculations [6]. Parameterized force fields are computationally efficient and allow for simulation of biomolecules with hundreds of thousands of atoms for hundreds of nanoseconds.

A good force field should provide satisfactory agreement with all available experimental data and a well determined parameter set is crucial to its accuracy. In parameter development, a basic assumption is often adopted that the particles bearing similar chemical environment can share the same parameters (partial charges are sometimes treated more specifically). For example, backbone carbonyl groups and amino groups in proteins are often regarded to be the same to the groups in N-methylacetamide, and methyl groups in amino acid side chains are treated equally with those in alkanes. This assumption greatly reduces the number of parameters as all particles involved are now reduced to a few particle types and the same parameter set can be transferred between particles of the same type, thereby simplifies the parameter optimization procedure. In practice, a large biomolecule is usually divided into appropriate model molecules of about ten heavy atoms. Then parameter optimization can be conducted individually for each small molecule by fitting to its quantum mechanical calculation results and experimental data. The resulted parameters are directly transferred to the original biomolecules and further verified by running simulations for complex systems and comparing the results with experimental data. The parameters may be fine-tuned afterwards to better fit all used data.

Though the transferability is an approximation to real nature, it has produced most of the commonly used force fields today, such as CHARMM [7], AMBER [8, 9], GROMOS [10] and OPLS [11], and each of them reaches a satisfactory agreement with experimental data. In the absence of more detailed rules, every force field follows its own philosophy in practice. The differences in functional formula, particle type and model molecule definition, quantum mechanical calculation methods and experimental data used, and partial charges fitting strategy etc. lead to very different parameters, even for similar particle types. Because of the differences in philosophy, arbitrarily mixing parameters from different force fields would lead to loss of accuracy, and is generally not recommended. Another important difference is the particular water model utilized in each force field. CHARMM and AMBER are paired with the TIP3P model [12], while GROMOS uses the SPC model [13]. A mismatch between protein force field and water model would lead to several kJ/mol discrepancies between computational and experimental values [14].

Different levels of particle details are adopted in various force fields, categorized as 'all-atom', 'united-atom' and 'coarse-grained' (Fig. 13.1). All-atom force fields are most popular, including the CHARMM, OPLS and AMBER force fields. They treat every atom explicitly as one particle, including hydrogen atoms, and provide the most detailed description of biomolecules. United-atom force fields adopt a similar particle definition except aliphatic carbons. One carbon and its associated hydrogen atoms are combined as one particle, and the computational cost of simulating the hydrogens could be saved. Coarse-grained force fields use larger particle units. The whole amino acid side chain or even several intact residues can be represented by a single particle and the collective physiochemical properties of the individual atoms are endowed to it. Due to the reduced number of particles and faster dynamics, coarse-grained models could simulate systems with orders of magnitude increase in time scale and system size.

13.2.2 Lipid Force Fields

Lipids are amphiphilic molecules consisted of charged or polar head groups and hydrophobic acyl chains. In organisms, they often assemble noncovalently to form a highly heterogeneous system called lipid bilayer, which varies largely in density and polarity on a typically ∼5 nm length scale. Lipid bilayers are often fluid-like, showing evident dynamic properties in a wide range of time and spatial scale, ranging from tail twisting, lateral diffusion, and flip-flop etc. It is quite challenging to make a lipid force field since an elaborate balance between hydrophobicity and hydrophilicity must be achieved to describe its complicated dynamic behavior.

The lipid bilayer dynamics also causes the sampling problem in lipid parameterization. As a pure lipid bilayer usually fluctuates on a multinanosecond time scale [15], hundreds of nanoseconds are usually required to equilibrate the structure,

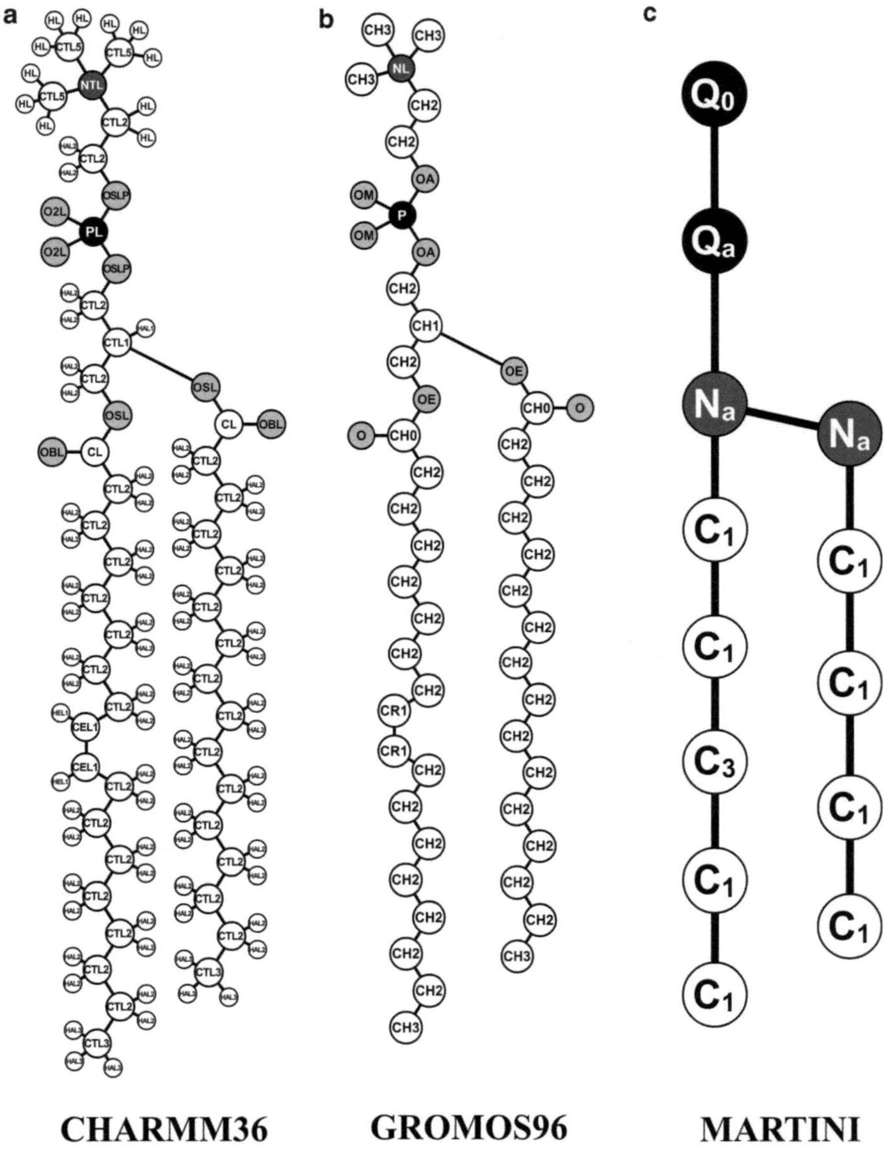

Fig. 13.1 Particle definition of palmitoyloleoylphosphatidylcholine (POPC) in (**a**) all-atom, (**b**) united-atom and (**c**) coarse-grained force fields. The particle types are indicated

and to sample the major states for accurate statistical averages [16]. Motions of ions are also slowed down near the lipids. Binding of sodium or calcium to lipids requires tens or one hundred nanosecond equilibration time [17]. To obtain a reliable parameter set, large amount of computational resources are required.

Another obstacle in parameterization comes from experimental techniques. Because of the mobile nature of lipids, the most powerful techniques utilized in lipid studies including X-ray and neutron diffraction, solid state nuclear magnetic resonance and infrared spectroscopy, usually provide indirect information which is interpreted based on models and assumptions. Structural information of lipids is more limited and less reliable compared with proteins. Parameterization can be supplemented by quantum mechanics calculations, but experimental data are still indispensable for validation. A consensus for parameter validation of lipids in the early days is using the average area per lipid (APL). APL could be obtained from X-ray or neutron diffraction and volumetric experiments, and compared with the calculated results obtained from a constant pressure-zero tension simulation by dividing the total surface area of membrane using half of the total number of lipids. Recently, a larger list of properties is proposed for validation, including the structure factors (APL and bilayer thickness), carbon-deuterium order parameter, NMR spin lattice relaxation times, elastic moduli, electron density profile and lipid translational diffusion constants etc. [18].

Last decade has witnessed significant progress in lipid force field development and the commonly used protein force fields have published their compatible lipid parameters. The all-atom force field CHARMM has updated several times these years. The CHARMM27r (C27r) parameter set [19] improved the behavior on dipalmitoylphosphatidylcholine (DPPC) bilayers over C27 set [20] by revising the torsion potential energies for short alkanes. The latest version C36 [21] made further progress in providing correct surface area for both saturated and unsaturated chains by modifying selected torsional, LJ, and partial charge parameters. The agreement with experiments in other properties was also improved. Another popular atomistic force field AMBER was less frequently used for lipid simulation as it did not have a specific lipid force field for a long time. Its general parameter set GAFF was an alternative though it tended to underestimate average APL [22, 23]. A new parameter set Slipids was recently developed for several saturated and unsaturated phospholipids using AMBER philosophy [24, 25]. It could accurately describe the structural properties of lipid bilayers under a range of temperatures, and the NMR order parameter profiles and the scattering form factors are also well reproduced.

United-atom lipid force fields can be traced back to Berger parameter set [26], though the GROMOS force field is currently the more rigorous one. GROMOS suffered from unneglectable disagreement with experiments in electron density profile and APL in its early versions 45A3 [27] and its performance has been greatly improved these years. GROMOS 53A6 set [28] increased the van der Waals radius between the choline methyl groups and the non-ester phosphate oxygens and well reproduced the structural and hydration properties of common phosphatidylcholine lipids which have varying length and unsaturation degree of the acyl tails [18]. The latest version 54A7 [29] made some small modifications by adding a new atom type for headgroup etc., and provided the best agreements with the experimental results in GROMOS series, especially in the ordering of the choline and glycerol moieties and the orientation of the headgroup dipole.

Coarse-grained (CG) force fields are very powerful in studying lipid dynamics for its high-efficiency in sampling. A simple lipid CG model called MARTINI [30, 31] has become very popular these years. It adopted a four-to-one mapping procedure by combining approximately four heavy atoms into a coarse-grained particle (Fig. 13.1). This model enabled rapid calculations on lipid dynamics such as self-assemble, phase separation and membrane fusion, which usually span microsecond timescales.

Currently, parameterization of phosphatidylcholine (PC) lipids is quite satisfying, but these lipids only constitute a part of all biologically relevant lipids. Works on parameters of other lipids such as cholesterols and sphingolipids are still on the way. A compatible comprehensive lipid parameter set could be expected in the future.

13.2.3 Compatibility of Protein and Lipid Force Fields

Membrane proteins exist in a very heterogeneous environment. Within about 5 nm's space, they would experience very different physicochemical environment ranging from a strongly polar water phase, a concentrated electrolyte solution, an ordered hydrophobic matrix and a disordered hydrophobic matrix [32]. A good protein parameter set should behave well in all these environments to ensure a reliable description of protein-lipid interactions. Direct experimental measurements on protein-bilayer systems are always preferred as target data for parameterization, however, the information obtained is often too complicated to interpret or too global for fine tuning of parameters. Several simplified systems are used instead for direct comparison between simulations and experiments.

The Radzicka-Wolfenden system is most commonly used. Proteins are modeled as small molecule analogs of amino acid side chains and bilayers are represented by biphasic systems of water and cyclohexane [33, 34]. When systems reach equilibration, the concentration is measured experimentally in each phase and the partition coefficient determines the free energy of transfer. As the same property can also be obtained from computation, a direct comparison between simulations and experiments can be used to test [35, 36] and optimize [28] force field parameters.

Though the Radzicka-Wolfenden system provides a bridge between simulations and experiments, it could be doubted in two aspects: first, side chains by themselves do not represent all aspects of protein structure, especially the backbone; second, the isotropic solvents do not resemble the complex environment of lipid bilayer. More realistic systems using pure lipid bilayer, peptide, or even proteins have been proposed in experiments [37], though simulations are still left behind.

Currently, a reasonable choice of force field for protein-bilayer simulation is to employ the same set of force fields for either system to keep the consistency in parameterization philosophy. However, it should be kept in mind that these parameters are not as convincing in protein-bilayer systems as they are used individually. Case-by-case verifications on these parameters are still needed. As experimental techniques on membrane protein systems are improving, force fields may be further trained and improved to attain a better agreement between simulations and experiments.

13.3 Preparing the Simulation System

In this section we will present the technical details in setting up membrane protein systems for MD simulations. MD simulations need to start from initial coordinates of the system. Biologically meaningful production simulations are obtained only when the initial coordinates are "correct". A common procedure for system setting-up usually starts with a high-resolution structure of membrane protein. The starting structure will first be repaired and correctly protonated, then it is embedded into a bilayer membrane and solvated. When the restraints on the system are gradually removed in the equilibration process, the initial coordinate for production simulation is finally obtained.

13.3.1 Preparing the Protein Structure Coordinates

To setup a membrane protein simulation system, one needs the starting structure coordinates of the protein. Membrane structure coordinates solved by X-ray crystallography or NMR are all deposited to the Protein Data Bank and can be obtained freely from the website: http://www.rcsb.org. The downloaded PDB files of the membrane protein structures usually contain atoms belonging to amino acid residues as well as heteroatoms constituting ligands, water molecules, or even some detergents in the precipitation buffer. The protein coordinates are commonly reserved. If the ligands and the crystal water molecules are biologically relevant, e.g. important for protein structure stability, they should also be retained.

The experimental structures are usually not "complete". Hydrogen atoms are often invisible by crystallography. Some side chains in proteins may be unresolved due to the low resolution diffraction data or their intrinsic high flexibility. Some residues, usually at the loop region, may be even totally missing. The coordinates of the missing atoms can be guessed with structure building tools, such as psfgen (implemented in NAMD package) or pdb2gmx (implemented in GROMACS package). Or, if there are too many missing atoms, especially the backbone atoms, homology modeling programs such as Modeller [38] or online servers such as SWISS-MODEL Workspace [39] (http://swissmodel.expasy.org/) may be utilized to repair them. The protonation states of the titratable residues are often kept in their default states. But when proteins show evident pH-dependent behavior, the protonation states could be of physiological importance. The states could be predicted using an empirically based method PROPKA [40] (http://propka.ki.ku.dk/) or a more physically based method H++ [41] (http://biophysics.cs.vt.edu/). Possible disulfide bonds should also be searched.

When all the missing atoms are generated and all the titration states are determined, certain force field could be assigned to the protein using the structure building tools. An energy minimization (EM) process is recommended afterwards to remove any bad contact hidden in crystal structures or introduced in the repairing process.

13.3.2 Preparing the Lipid Bilayer

Pure phospholipids bilayers are often used in simulations to model cell membranes. In early times, the initial coordinates of a lipid bilayer were often generated *ab initio*. Tools such as CHARMM membrane builder server (http://www.charmm-gui.org/?doc=input/membrane_only&step=1) would stack lipid by lipid to build up an integral bilayer. An alternative way to obtain the initial coordinates is to download pre-equilibrated bilayers from internet. Lipidbook database [42] provides an easier way from which various membrane structures attached in the published works can be found (http://lipidbook.bioch.ox.ac.uk/). It is important to make sure that the downloaded bilayer models have the proper size for the specific membrane proteins one desires to simulate. As periodic boundary conditions are usually employed, the size of the bilayer must be large enough to avoid interactions between the protein and its periodic images, though a larger membrane would be more time-consuming. After the modification, further equilibration is recommended. This process usually requires dozens of nanoseconds MD simulations and can be monitored simply by area per lipid (APL), as mentioned in Sect. 13.2.2.

13.3.3 Embedding Protein in a Lipid Bilayer

With a refined membrane protein and an equilibrated membrane model at hand, the next step is to embed the protein in the bilayer. The embedding process aims to provide a protein-lipid complex with an unperturbed protein structure, a well-structured bilayer and a reasonable protein-lipid interface. Though several hundred nanoseconds MD simulation would be enough for lipids to spontaneously assemble into bilayer around membrane proteins [43], the computational cost for most membrane protein systems is still very expensive, even with coarse-grained force fields. Many time-saving strategies are developed, which often build up a protein-lipid complex in vacuo, and then solvate and equilibrate the system.

Though there are different strategies for protein embedding, an initial location of the protein relative to the bilayer is always the prerequisite for all the methods. The positioning process is largely guided by hydrophobicity and hydrophilicity match between the protein and the bilayer. All the membrane integrated proteins have a transmembrane spanning region characterized by the absence of charged residues at the molecular surface, usually capped with Tyr and Trp residues at both edges of the region corresponding to the lipid-solvent interfaces. The protein must be reoriented and translated so that its transmembrane spanning region could be well matched up with the hydrophobic tails of the bilayer and the Tyr and Trp residues of the protein inserted into the headgroup/tail interfacial region. The procedure could be conducted manually with molecular graphics tools, such as VMD. OPM database provide a more precise way for the positioning [44]. It is based on a computational approach developed to predict the transmembrane region of the protein and to evaluate the transfer energy of protein from water to membrane. The optimized spatial arrangement of proteins could be found using the approach. Searching within the database could be easily executed using proteins' name or PDB ID and the optimized structure coordinates can be downloaded readily (http://opm.phar.umich.edu/).

After the determination of the initial location, the protein is superimposed with the bilayer. Embedding protocols would be utilized to remove the overlapping lipids and create a proper protein-bilayer interface. There are various kinds of strategies at hand which can be roughly categorized into two types: the bilayer-biased methods and the interface-biased methods.

13.3.3.1 The Bilayer-Biased Methods

The bilayer-biased methods try to retain the pre-equilibrated bilayer structure as much as possible. Though the lipids close to the protein would be removed or modified when the protein-lipid interface is created, the farther ones are almost undisturbed. A typical method of this class is to simply delete the overlapping lipids within a certain distance cut-off. The applied cut-off length could be 5–6 Å for protein-phosphorus distances or 0.8–1.6 Å for the minimal distance between any two atoms on protein and lipids. Though this method is extremely fast and well preserves the bilayer structure, it often produces a rather rugged protein-lipid interface due to the highly disordered nature of bilayer lipids. The lipids are often either too close to the protein or too far from it. A long and fussy equilibration process will be required to remove the protein-lipid clashes and fill up the gap between them. The process may take dozens of nanoseconds as lipid diffusion takes place on a comparable time scale.

To improve the protein-lipid interface, methods specialized in creating compatible cavity in bilayer for proteins are proposed. These methods first use the three-dimensional description of the protein shape to define the cavity. Regular geometric solids such as a cylinder [45] or a finer Connolly solvent-accessible surface [46] are acceptable. The cavity is then emptied by removing all the overlapping lipids using a protein-phosphorus distance cutoff and further refined by a weak repulsive potential which would drive out all the lipids atoms, e.g. the hydrophobic tails. The resulted cavity would better fit the protein with hardly any clash between them. This method could be implemented with mdrun_hole in GROMACS package. Other programs such as GRASP [47] may also be required to generate the solvent-accessible surface for GROMACS to use. A drawback of the method is that many parameters need to be specified. As no universal setting is available, trial-and-error is often required [48]. Another problem is that the Connolly surface only concerns van der Waals interactions, and the protein-lipid interface needs further optimization to include electrostatic effects.

The protein-growing strategy proposed recently well settles the above problems [49]. As in its initial location, protein is first laterally scaled down and a cavity in bilayer is created by deleting all the overlapping lipids to accommodate the compressed protein. Then the protein is gradually scaled up to its original size in a short MD simulation. As the protein grows, the nearby lipids are gradually pushed away and optimally form a protein-bilayer interface without disturbing of the overall structure of the bilayer. As the bilayer structure is still close to its natural state, the equilibration run could be shorter. For proteins with a large sectional area, some lipids would have to travel a long distance during the growing stage and the bilayer structure may be evidently disturbed. A moderate scaling-down ratio (such as 0.5) would circumvent the problem. The protein-growing method is easy to use even for proteins with very irregular shape and its computational tool g_membed is already available in the GROMACS suite.

13.3.3.2 The Interface-Biased Methods

In contrast to the bilayer-biased methods, the interface-biased methods focus on creating an elaborate protein-lipid interface, though large-scale perturbation is cast on the bilayer structure. This class of methods could be traced back to the "ab initio" approach [50] implemented in the CHARMM package. The method gains its name as the membrane around the protein is constructed lipid by lipid. Each added amphiphatic molecule is randomly chosen from a pre-equilibrated lipid libraries and its location is decided according to the distribution of equilibrated bilayers. A rigid-body conformational search is conducted afterwards. The process includes randomly rotating around membrane normal and translating along membrane plane to minimize protein-lipid clashes or lipid-lipid clashes. Though an optimized protein-lipid interface is guaranteed, this method evidently messes up the bilayer structure. An equilibration process is needed after the construction to re-equilibrate the bilayer.

Shrinking approach [48] is another interface-biased method but with more moderate perturbation on the bilayer structure. The pre-equilibrated bilayer is first expanded by laterally translating lipid molecules in the membrane plane. A scaling-up factor of 4 is usually appropriate in practice. After superimposing the protein, the overlapping lipids are deleted and then the lipid bilayer is scaled back to the normal density progressively. The possible clashes are eliminated by a subsequent energy minimization (EM) procedure. The EM procedure would ensure the formation of appropriate protein-lipid and lipid-lipid packing, while the protein are restrained to avoid artificial conformational changes. A scaling factor of 0.95 for each shrinking step works well in most cases and a total of 26 steps ends up with a protein-bilayer complex structure very close to equilibration state.

13.3.4 Solvation

As most embedding procedure is carried out in vacuo, the system is solvated with water molecules and ions afterwards. The geometry-based solvation strategy used for globular proteins could be directly applied here. A pre-equilibrated water box is first superimposed onto the protein-bilayer system, and truncated to fit the system. Then the overlapping solvent molecules are removed based on a certain cutoff distance. As the equilibration of solvent molecules is generally fast both in the bulk and at the solvent-lipid interface [51, 52], several hundreds of picoseconds are usually enough to relax the system and form optimal protein-solvent and lipid-solvent interfaces. In some cases, water molecules might be misplaced in the hydrophobic core of bilayer during the solvation process, which is thermodynamically unfavorable. It would be better to manually delete these water molecules before the equilibration process using molecular graphics tools, otherwise an equilibration process of several dozens of nanoseconds might be required for the water molecules to spontaneously diffuse away. There are also cases when vacuum bubbles appear in the system after a short equilibration run. This is often due to the improper cutoff distance employed in the solvation procedure, in which too many solvent molecules are deleted. The problem can be fixed by using a smaller cut-off distance in solvation or by performing a short isothermal-isobaric MD simulation to resize the simulation box by restraining the protein atoms at the same time. Addition of ions could be carried out by randomly replacing water molecules to reach a physiological salt concentration (typically ~0.15 mol/L). The added ions are usually kept a certain distance (~0.5 nm) away from the solute so that they can freely diffuse to the surface of the solute in the equilibration process. The system may look like this after solvation (Fig. 13.2).

13.3.5 Equilibration

After solvation there might still be defects in the system, especially at the protein-lipid, water-lipid and water-protein interfaces. Equilibration of several nanoseconds is often necessary to relax the system and fix defects. It is worth noting that if there is a big problem in the protein-bilayer complex, e.g. the protein is placed with a wrong orientation or at a wrong depth relative to the bilayer, the equilibration process may not fix it.

A typical equilibration process can often be divided into two stages. Solvent can be equilibrated first as water molecules and ions usually move fast. Position restraints are cast on both protein and lipid in this stage to avoid unreasonable distortions in the structures due to the defects introduced in the system setup procedure. Several hundreds of picoseconds are often enough to obtain an optimized solvent configuration. After the equilibration of solvent, both lipid and solvent are allowed to freely move while restraints are cast on protein atoms only. An

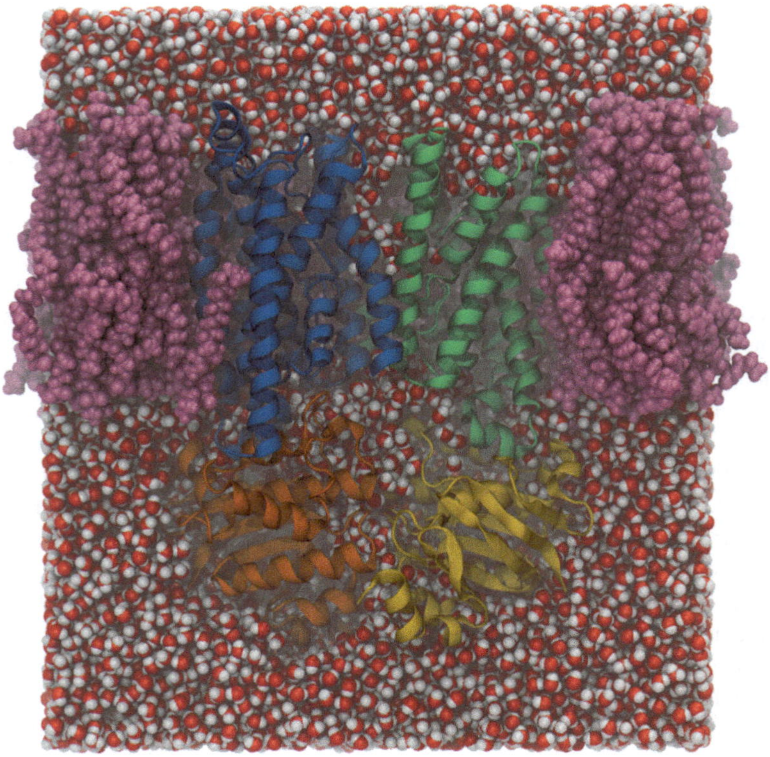

Fig. 13.2 Membrane protein system after solvation. The vitamin B_{12} importer BtuCD is represented by *ribbons* with its transmembrane domains buried in the bilayer, the lipid molecules are represented by *spheres in grey*, and the water molecules are represented by *spheres in black and grey*. The lipid and the water molecules between protein and viewer are removed for clarity

isothermal-isobaric simulation using semi-isotropic pressure coupling algorithm (see Sect. 13.4.1 for further details) is recommended so that the volume and the surface tension could be relaxed simultaneously. An energy minimization process or a short isothermal-isobaric simulation using isotropic algorithm may be used before the semi-isotropic process to obtain a 'good' initial structure. The length of the process depends on the system size and the embedding method. As mentioned above, the protein-lipid complex constructed with the simple "delete lipids within a cut-off" method may require long time simulation to reach equilibrium, while the shrinking method or the protein-growing method may minimize the equilibration time. As previous simulation study demonstrated that 10–20 ns are required to equilibrate pure DPPC bilayers, at least a few nanoseconds simulation is recommended to equilibrate the protein-bilayer system. APL could be used to monitor the process. Equilibration is approximately reached when the APL curve turns flat. Then the restraints on protein could be gradually diminished followed by the production run.

13.4 Simulation Methods

13.4.1 All-Atom Molecular Dynamics (MD) Simulation

All-atom MD simulation method treats all atoms explicitly as point particles in the system including protein, lipid bilayer and water, providing a natural environment of the membrane protein and mimicking the experimental conditions. The inter-atomic interactions are described by parameterized potential energy functions, also known as force fields (see Sect. 13.2). In all-atom MD simulations, the Newton equation of motion is integrated by numerical finite difference methods over small time steps. At each time step, the total force on each particle is calculated according to the coordinates of all the particles in the system and the force field parameters, and then the forces are used to integrate the equation to predict the positions of the particles for the next step. To avoid numerical instability, the time step of integration is limited typically to 1–2 fs, which is the time scale of the highest-frequency vibration in the system (covalent bond stretching). As a result, the motions of the system are described by a series of snapshots (conformations) of the system (known as trajectory) at femtosecond resolution. Various properties of interest can be evaluated using the trajectories based on the principles of statistical thermodynamics.

The simulation parameters for membrane protein systems are generally the same as those for globular protein systems. As the isothermal-isobaric algorithm is becoming dominant for membrane proteins, one of the major differences between two kinds of systems is the type of pressure coupling [15]. Globular proteins are generally isotropic due to their free rotation in solvent, so isotropic pressure coupling is always used which couples the contributions in x, y and z directions by using a single proportional scaling factor. But for membrane protein systems, the membrane surface (on x-y plane) is never equivalent to the membrane normal (along z direction). If three directions are forced to couple together, the fluctuations in the surface would be largely restrained and the surface tension would be inappropriately specified. To allow surface fluctuations, semi-isotropic pressure coupling should be used. In this case, the pressure contribution in z direction is decoupled from those in x and y direction, so the surface could get rid of the restraint and its motion would be driven by its own nature.

Another important issue is the treatment of non-bonded interactions. Biomolecules and membranes are often highly charged and involve massive electrostatic interactions inside of them. Because of the slow decay of Coulomb potential, these interactions are usually significant at large distances. A simple truncation strategy to remove long-range interactions beyond certain distance (typically between 1.5 and 2 nm) could considerably reduce the computational cost, but leads to major distortions in simulation systems, especially in bilayers [53]. Two methods are now commonly used to treat the long-range problem: Particle Mesh Ewald (PME) techniques [54, 55] and reaction field (RF) approaches [56]. PME is the most popular and preferred method at the time. By infinitely replicating the simulation box in all three directions, PME could efficiently sums up all of the interactions in this periodic system, and include long-range electrostatic interactions. RF approaches provide an alternative for the calculation. They improve the truncation strategy by introducing a correction term to electrostatic force, which is obtained by analytically solving the Poisson-Boltzmann equation of the response of a uniform dielectric media. Both methods have their drawbacks. PME artificially enhances the periodicity of the system, while RF estimates heterogeneous system properties using a homogenous model [57]. In practice, a safe choice between PME and RF is to employ the same method as the force field developers use: PME for CHARMM, OPLS and AMBER force fields and RF approaches for the GROMOS force fields. Thus, the accuracy of the force fields can be preserved.

Calculation of long-range LJ force may also get renewed attention. Though a truncation at ~1 nm is routinely used for LJ force in current simulations, a recent work showed that neglecting the long-range terms led to 50 % overestimation of the isothermal compressibility for bulk heptane, and approximately halved the calculated surface tension of alkane/vapor interfaces [58]. As long-range LJ correction in heterogeneous systems is very different from that in homogeneous systems, further efforts will be needed to examine the importance of long-range corrections, to test the correction algorithms, and to incorporate the algorithms into the highly parallelized MD simulation codes.

Currently all-atom MD simulations are generally limited to nanosecond to microsecond timescale due to the small integration time step of femtosecond. Recent advances in computer hardware and development of software keep breaching this limit and simulations of hundreds of microseconds have been reported [59]. However, these kinds of simulations are not normal in the community yet and even this timescale is still far short compared to those of many biologically relevant events for membrane proteins. In addition, simulation of the complex system of membrane protein is demanding in respect of the statistical sampling quality. A useful strategy for production run is to start multiple trajectories with either multiple starting structures or using the same starting structure with different initial velocities. A long sampling study on rhodopsin showed that multiple trajectories could provide better agreement with experiment instead of using a single one [60].

13.4.2 Coarse-Grained (CG) MD Simulation

One way to bridge the gap between the timescales of all-atom MD simulations and the slow biological processes is to use CG force field as mentioned in Sect. 13.2.2. One of the popular CG force field MATINI enables the simulation to take much longer time steps (approximately ten times longer than the all-atom MD simulation), providing 2–3 orders of magnitude speedup to extend the simulation timescales to microsecond. However, because of the coarseness of the model, hydrogen bonds were poorly described when MARTINI was extended to proteins. Extra restraints had to be cast on backbone atoms to stabilize the structure of proteins and large-scale conformational changes were excluded in simulations [61]. One strategy to overcome this difficulty is to switch between different coarse-graining levels in a single simulation. When the system is in the coarse-grained level, it samples efficiently in the phase space, and as the system switches to the fine-grained level, structural refinement is allowed and atomic details are provided. This strategy is known as multiscale simulation method, and has been widely applied to membrane protein systems [62, 63].

13.4.3 Enhanced Sampling Methods

Instead of coarse-graining the force fields, another solution to bridging the simulation-experimental time gap is using enhanced sampling of the configuration space of the system. It is not possible to give a comprehensive review of enhanced sampling methods here. We briefly introduce several enhanced sampling methods that are widely used in membrane protein simulations.

The steered MD (SMD) simulation method [64] applies external forces to a selected group of protein atoms, which are anchored to a spring of elastic constant k and pulled from the initial position at a velocity v in order to accelerate a certain type of conformational change. Initially SMD simulation was performed to interpret the data from atomic force microscopy (AFM) experiments. Although SMD simulation is based on non-equilibrium sampling, the trajectories can also be used to calculate potential of mean force (PMF) along specific coordinates based on Jarzynski's identity [65]. It should be noted that when using SMD to calculate PMF very long simulation time at slow pulling velocity is needed to ensure the proper sampling of the system. Similar with SMD, targeted MD (TMD) simulation [66] exerts external force on protein to drive the protein from an initial conformational state to a targeted one. When two conformational states are known for a membrane protein, TMD simulation can be applied to explore the transition process between the two structures.

Some enhanced sampling methods are designed to improve sampling for the calculation of PMF along predefined reaction coordinates. Umbrella sampling is the oldest approach of these methods, in which the computation along the reaction coordinate is divided into subintervals (windows) and in each window a static biasing potential is used to further improve the sampling. These windows are then analyzed together and the PMF profile can be reconstructed using reweighting algorithms such as WHAM method [67, 68]. The more recently developed methods of adaptive biasing force (ABF) [69–71] and metadynamics [72, 73] use history dependent biasing force or potential to enhance sampling along certain reaction coordinates. These methods enable the construction of a multidimensional free energy surface as a function of a set of reaction coordinates or collective variables (CVs). The key issue of applying these methods in computing PMF of a complex system as membrane protein is the proper choice of the CVs, which should represent the degrees of freedom of the slowest conformational motions in the system ensuring that the sampling of all other dimensions orthogonal to the reaction coordinates converges quickly during the simulation time.

13.4.4 Elastic Network Model

Elastic network model (ENM) [74] is a coarse-grained version of normal mode analysis (NMA), which provides information of the vibrational modes intrinsically accessible to a protein structure. In ENM, the protein structure is represented as a network of nodes (C_α atoms) connected by elastic springs within a specific cutoff distance. The optimal cutoff distance for the most widely used ENM anisotropic network model (ANM) is suggested to be 18 Å, and a uniform force constant is used in ENM. In recent years, ENM is widely used to explore the collective conformational motions of proteins based on the observation that the global modes elucidated by ENM are functionally significant. This method is also extensively applied to study membrane protein due to its low computational cost [75]. In the application of ENM to membrane protein, the lipid bilayer and solvent are usually not taken into account. It has been shown that the global modes of membrane protein are essentially dictated by the overall shape of the protein while the environment has little effect.

13.5 Typical Biological Questions of Membrane Protein Explored by Simulation Methods

The first MD simulation study of membrane protein of bacteriorhodopsin was reported in 1995 [76]. Since then, MD simulation and related techniques have been successfully applied to unravel structure-function relationship of various membrane proteins, such as ion channels, aquaporins, transporters, receptors, ATP-synthases and so on. Some excellent reviews of this field have been published recently [77–85]. Here, we are not aiming to a comprehensive survey, but present some examples to showcase the applicability of MD simulation to membrane protein study, focusing on several typical questions of membrane protein biophysics.

13.5.1 Conformational Change

Majority of membrane proteins undergo large-scale conformational changes during the biological processes. For example, the voltage sensor domain (VSD) of the voltage-gated potassium channel changes its conformation in response to the changes of membrane potential (voltage) to control the activation-deactivation of the channel [86]. The transporter proteins switch between outward-facing and inward-facing conformations to translocate substrates across cell membrane [87], and the membrane receptors alter their conformations upon extracellular ligand binding. Most of these conformational changes involve slow inter-domain motion/rearrangement that span long time scales from tens of microseconds to millisecond and beyond. Therefore, it is generally unfeasible for the direct equilibrium all-atom MD simulation to sample the conformational space with enough statistical accuracy at the level of the nowadays computer resources. However, numerous previous studies [78, 88] have shown that equilibrium all-atom simulations at sub-microsecond time scales can provide useful information of the conformational dynamics of the physiologically relevant states of membrane protein, such as the local conformational changes, the tendency of the conformational transition as well as inter-residue interactions crucial in the conformational movement.

In the MD simulation studies of the voltage-gated potassium channel Kv1.2, standard MD simulations of $\sim$1 μs under a hyperpolarizing membrane potential revealed the secondary structure transformation of the S4 helix from α- to 3_{10}-helix with a 120° rotation, although the later step of S4 downward translation toward deactivated state was not observed [89]. To explore the entire transition process from activated to deactivated states of Kv1.2, combined use of unbiased and biased MD simulations revealed three intermediate states of the voltage sensor domain [90]. Recently, impressive extended all-atom MD simulations of Kv1.2 channel over 200 μs were reported [91]. The entire conformational change processes during deactivation and activation of the channel were observed and a complete mechanistic model of voltage gating has been proposed. Interestingly, the main observations in the previous equilibrium and biased nonequilibrium simulations are in good agreement with this extended long time MD simulation study.

For G protein coupled receptors (GPCRs), another family of membrane proteins of biological importance, long-time scale all-atom MD simulations have also been invested to explore the conformational changes [92], but the entire conformational transition from inactive to active state is still out of reach. Provasi and Filizola combined biased MD and metadynamics simulation to calculate the PMF profile connecting two experimental end structures [93]. The conformational change pathway was initially obtained from the biased MD simulation trajectories, and subsequently used in the path collective variable-based metadynamics simulations. Two pathways and four metastable states were identified along the conformational transition [93].

Elastic network model was also employed to study the conformational change of membrane proteins, including ion channels, receptors and transporters. The large-scale conformational changes of membrane proteins can be captured in a few low-frequency normal modes, although these analyses lack the atomistic details. We direct the readers for a recent detailed survey by Bahar [75].

13.5.2 Ion and Ligand Permeability

Transporting substances across membrane is the main function of membrane channels and transporters. The permeation of ions and ligands is therefore a key topic in membrane protein simulations. The most extensively studied system is the potassium channel. To understand the process of K^+ permeation through the selective filter of KcsA channel, early study calculated the PMF of three ions using umbrella sampling [94], revealing concerted pathways for K^+ permeation and the most stable binding site of K^+. SMD simulation was also employed to compute the PMF of K^+ permeation through KcsA [95], generally in agreement with the findings of the umbrella sampling study. An interesting study [96] that compared the performance of three methods, i.e. SMD, umbrella sampling and metadynamics in calculating the PMF of K^+ permeation indicated that the PMFs obtained are qualitatively similar but metadynamics was recommended for its computational efficiency and accuracy. The K^+ conductance of Kv1.2 channel was studied using direct all-atom MD simulations at microsecond timescales, revealing a detailed conduction mechanism supporting the Hodgkin-Keynes "knock-on" model [97]. This straightforward MD simulation also allowed direct observation of the time limiting step of K^+ dehydration.

Water channel aquaporins (AQPs) belong to another family of membrane proteins extensively characterized both by experiments [98] and simulations [82]. The process of water permeation through aquaporins takes places at nanosecond timescale. Therefore, direct all-atom simulations are adequate to characterize the dynamics of translocation and calculate the free energy profile of water permeation. Simulation studies have shown that the free energy barrier for water permeation through AQPs is about 3 kcal/mol [99, 100]. AQPs can transport other molecule than water, but the permeation rates are much slower. Permeation of glycerol through aquaglyceroporin GlpF was studied using enhanced sampling methods SMD [101] and ABF [102], giving permeation free energy barriers of 7.3 and 8.7 kcal/mol respectively, in agreement with the experimental value of 9.6 ± 1.5 kcal/mol [103]. It is interesting to note that calculation with umbrella sampling gave a lower energy barrier of 3.2 kcal/mol [99]. Larger ligand molecules have intrinsic conformational flexibilities, thereby introduce additional degrees of freedom in the simulation which may render difficulties in convergence of the sampling.

13.5.3 Dimerization of Receptors

Membrane bound receptors is responsible for signaling cross the cell membrane. These receptors have ectodomains that bind extracellular ligands, single transmembrane helix, and intracellular signaling domains. Membrane bound receptors (e.g. integrin, epidermal growth factor receptor (EGFR)) are often activated through dimerization, which involves the dimerization of transmembrane helices of the two monomers. Therefore TM helices dimerization has been extensively studied with various computational approaches. CG MD simulations were used to investigate the glycophorin A dimer assembly from separate monomers in lipid bilayer, with the resultant dimer structure in good agreement with experimentally determined one [104]. The PMF of glycophorin A dimerization were also calculated, showing that the dimerization is assisted by lipid-induced interactions [105, 106]. Recently, extended all-atom MD simulations of EGFR showed that the self-assembly of the TM helix dimer in lipid bilayer took place on timescale of 50–75 μs [107].

GPCRs, which bear seven transmembrane helices, have also been shown to form dimers and/or oligomers in lipid bilayer. Umbrella sampling and CG simulation were used to calculate the free energy profile of delta opioid receptor (DOR) dimer formation and by using a diffusion limited model of dimerization the lifetime of this dimer was estimated to be 4.4 s [93]. A similar simulation with CG metadynamics obtained a shorter lifetime of 0.2 s [108]. Obviously this time scale is inaccessible to direct MD simulations.

References

1. Krogh A, Larsson B, von Heijne G et al (2001) Predicting transmembrane protein topology with a hidden Markov model: application to complete genomes. J Mol Biol 305:567–580
2. Overington JP, Al-Lazikani B, Hopkins AL (2006) How many drug targets are there? Nat Rev Drug Discov 5:993–996
3. Deisenhofer J, Epp O, Miki K et al (1985) Structure of the protein subunits in the photosynthetic reaction centre of *Rhodopseudomonas viridis* at 3A resolution. Nature 318:618–624
4. Henzler-Wildman K, Kern D (2007) Dynamic personalities of proteins. Nature 450:964–972
5. McCammon JA, Gelin BR, Karplus M (1977) Dynamics of folded proteins. Nature 267:585–590
6. Guvench O, MacKerell AD Jr (2008) Comparison of protein force fields for molecular dynamics simulations. Methods Mol Biol 443:63–88
7. MacKerell AD, Bashford D, Bellott M et al (1998) All-atom empirical potential for molecular modeling and dynamics studies of proteins. J Phys Chem B 102:3586–3616
8. Case DA, Cheatham TE 3rd, Darden T et al (2005) The Amber biomolecular simulation programs. J Comput Chem 26:1668–1688
9. Ponder JW, Case DA (2003) Force fields for protein simulations. Adv Protein Chem 66:27–85
10. Scott WRP, Hunenberger PH, Tironi IG et al (1999) The GROMOS biomolecular simulation program package. J Phys Chem A 103:3596–3607
11. Jorgensen WL, Maxwell DS, TiradoRives J (1996) Development and testing of the OPLS all-atom force field on conformational energetics and properties of organic liquids. J Am Chem Soc 118:11225–11236
12. Jorgensen WL, Chandrasekhar J, Madura JD et al (1983) Comparison of simple potential functions for simulating liquid water. J Chem Phys 79:926–935
13. Berendsen HJC, Postma JPM, van Gunsteren WF et al (1981) Interaction models for water in relation to protein hydration. In: Pullman B (ed) Intermolecular forces. Reidel, Dordrecht, pp 331–342
14. Hess B, van der Vegt NF (2006) Hydration thermodynamic properties of amino acid analogues: a systematic comparison of biomolecular force fields and water models. J Phys Chem B 110:17616–17626

15. Anezo C, de Vries AH, Holtje HD et al (2003) Methodological issues in lipid bilayer simulations. J Phys Chem B 107:9424–9433
16. Wohlert J, Edholm O (2006) Dynamics in atomistic simulations of phospholipid membranes: nuclear magnetic resonance relaxation rates and lateral diffusion. J Chem Phys 125:204703
17. Bockmann RA, Grubmuller H (2004) Multistep binding of divalent cations to phospholipid bilayers: a molecular dynamics study. Angew Chem Int Ed 43:1021–1024
18. Poger D, Mark AE (2010) On the validation of molecular dynamics simulations of saturated and cis-monounsaturated phosphatidylcholine lipid bilayers: a comparison with experiment. J Chem Theory Comput 6:325–336
19. Klauda JB, Brooks BR, MacKerell AD et al (2005) An ab initio study on the torsional surface of alkanes and its effect on molecular simulations of alkanes and a DPPC bilayer. J Phys Chem B 109:5300–5311
20. Feller SE, MacKerell AD (2000) An improved empirical potential energy function for molecular simulations of phospholipids. J Phys Chem B 104:7510–7515
21. Klauda JB, Venable RM, Freites JA et al (2010) Update of the CHARMM all-atom additive force field for lipids: validation on six lipid types. J Phys Chem B 114:7830–7843
22. Jojart B, Martinek TA (2007) Performance of the general amber force field in modeling aqueous POPC membrane bilayers. J Comput Chem 28:2051–2058
23. Rosso L, Gould IR (2008) Structure and dynamics of phospholipid bilayers using recently developed general all-atom force fields. J Comput Chem 29:24–37
24. Jambeck JPM, Lyubartsev AP (2012) An extension and further validation of an all-atomistic force field for biological membranes. J Chem Theory Comput 8:2938–2948
25. Jambeck JPM, Lyubartsev AP (2012) Derivation and systematic validation of a refined all-atom force field for phosphatidylcholine lipids. J Phys Chem B 116:3164–3179
26. Berger O, Edholm O, Jahnig F (1997) Molecular dynamics simulations of a fluid bilayer of dipalmitoylphosphatidylcholine at full hydration, constant pressure, and constant temperature. Biophys J 72:2002–2013
27. Chandrasekhar I, Kastenholz M, Lins RD et al (2003) A consistent potential energy parameter set for lipids: dipalmitoylphosphatidylcholine as a benchmark of the GROMOS96 45A3 force field. Eur Biophys J Biophys Lett 32:67–77
28. Oostenbrink C, Villa A, Mark AE et al (2004) A biomolecular force field based on the free enthalpy of hydration and solvation: the GROMOS force-field parameter sets 53A5 and 53A6. J Comput Chem 25:1656–1676
29. Schmid N, Eichenberger AP, Choutko A et al (2011) Definition and testing of the GROMOS force-field versions 54A7 and 54B7. Eur Biophys J Biophys Lett 40:843–856
30. Marrink SJ, de Vries AH, Mark AE (2004) Coarse grained model for semiquantitative lipid simulations. J Phys Chem B 108:750–760
31. Marrink SJ, Risselada HJ, Yefimov S et al (2007) The MARTINI force field: coarse grained model for biomolecular simulations. J Phys Chem B 111:7812–7824
32. Marrink SJ, Berendsen HJC (1994) Simulation of water transport through a lipid-membrane. J Phys Chem 98:4155–4168
33. Radzicka A, Wolfenden R (1988) Comparing the polarities of the amino-acids – side-chain distribution coefficients between the vapor-phase, cyclohexane, 1-octanol, and neutral aqueous-solution. Biochemistry 27:1664–1670
34. Wolfenden R (2007) Experimental measures of amino acid hydrophobicity and the topology of transmembrane and globular proteins. J Gen Physiol 129:357–362
35. Maccallum JL, Tieleman DP (2003) Calculation of the water-cyclohexane transfer free energies of neutral amino acid side-chain analogs using the OPLS all-atom force field. J Comput Chem 24:1930–1935
36. Villa A, Mark AE (2002) Calculation of the free energy of solvation for neutral analogs of amino acid side chains. J Comput Chem 23:548–553
37. MacCallum JL, Tieleman DP (2011) Hydrophobicity scales: a thermodynamic looking glass into lipid-protein interactions. Trends Biochem Sci 36:653–662

38. Sali A, Blundell TL (1993) Comparative protein modeling by satisfaction of spatial restraints. J Mol Biol 234:779–815
39. Kiefer F, Arnold K, Kunzli M et al (2009) The SWISS-MODEL Repository and associated resources. Nucleic Acids Res 37:D387–D392
40. Sondergaard CR, Olsson MHM, Rostkowski M et al (2011) Improved treatment of ligands and coupling effects in empirical calculation and rationalization of pK(a) values. J Chem Theory Comput 7:2284–2295
41. Anandakrishnan R, Aguilar B, Onufriev AV (2012) H++3.0: automating pK prediction and the preparation of biomolecular structures for atomistic molecular modeling and simulations. Nucleic Acids Res 40:W537–W541
42. Domanski J, Stansfeld PJ, Sansom MSP et al (2010) Lipidbook: a public repository for force-field parameters used in membrane simulations. J Membr Biol 236:255–258
43. Scott KA, Bond PJ, Ivetac A et al (2008) Coarse-grained MD simulations of membrane protein-bilayer self-assembly. Structure 16:621–630
44. Lomize MA, Lomize AL, Pogozheva ID et al (2006) OPM: orientations of proteins in membranes database. Bioinformatics 22:623–625
45. Shen L, Bassolino D, Stouch T (1997) Transmembrane helix structure, dynamics, and interactions: multi-nanosecond molecular dynamics simulations. Biophys J 73:3–20
46. Faraldo-Gomez JD, Smith GR, Sansom MS (2002) Setting up and optimization of membrane protein simulations. Eur Biophys J 31:217–227
47. Nicholls A, Sharp KA, Honig B (1991) Protein folding and association: insights from the interfacial and thermodynamic properties of hydrocarbons. Proteins Struct Funct Bioinform 11:281–296
48. Kandt C, Ash WL, Tieleman DP (2007) Setting up and running molecular dynamics simulations of membrane proteins. Methods 41:475–488
49. Wolf MG, Hoefling M, Aponte-Santamaria C et al (2010) g_membed: efficient insertion of a membrane protein into an equilibrated lipid bilayer with minimal perturbation. J Comput Chem 31:2169–2174
50. Woolf TB, Roux B (1994) Molecular dynamics simulation of the gramicidin channel in a phospholipid bilayer. Proc Natl Acad Sci USA 91:11631–11635
51. Luzar A, Chandler D (1996) Hydrogen-bond kinetics in liquid water. Nature 379:55–57
52. Luzar A, Chandler D (1996) Effect of environment on hydrogen bond dynamics in liquid water. Phys Rev Lett 76:928–931
53. Patra M, Karttunen M, Hyvonen MT et al (2003) Molecular dynamics simulations of lipid bilayers: major artifacts due to truncating electrostatic interactions. Biophys J 84:3636–3645
54. Darden T, York D, Pedersen L (1993) Particle mesh Ewald – an N.log(N) method for Ewald sums in large systems. J Chem Phys 98:10089–10092
55. Essmann U, Perera L, Berkowitz ML et al (1995) A smooth particle mesh Ewald method. J Chem Phys 103:8577–8593
56. Tironi I, Sperb R, Smith P et al (1995) A generalized reaction field method for molecular-dynamics simulations. J Chem Phys 102:5451–5459
57. Bockmann RA, Caflisch A (2005) Spontaneous formation of detergent micelles around the outer membrane protein OmpX. Biophys J 88:3191–3204
58. Klauda JB, Venable RM, MacKerell AD et al (2008) Considerations for lipid force field development. In: Feller SE (ed) Computational modeling of membrane bilayers. Academic, London, pp 1–48
59. Dror RO, Dirks RM, Grossman JP et al (2012) Biomolecular simulation: a computational microscope for molecular biology. Annu Rev Biophys 41:429–452
60. Grossfield A, Feller SE, Pitman MC (2007) Convergence of molecular dynamics simulations of membrane proteins. Proteins Struct Funct Bioinform 67:31–40
61. Monticelli L, Kandasamy SK, Periole X et al (2008) The MARTINI coarse-grained force field: extension to proteins. J Chem Theory Comput 4:819–834
62. Stansfeld PJ, Sansom MSP (2011) From coarse grained to atomistic: a serial multiscale approach to membrane protein simulations. J Chem Theory Comput 7:1157–1166

63. Ayton GS, Voth GA (2007) Multiscale simulation of transmembrane proteins. J Struct Biol 157:570–578
64. Isralewitz B, Gao M, Schulten K (2001) Steered molecular dynamics and mechanical functions of proteins. Curr Opin Struct Biol 11:224–230
65. Jarzynski C (1997) Nonequilibrium equality for free energy differences. Phys Rev Lett 78:2690–2693
66. Schlitter J, Engels M, Kruger P et al (1993) Targeted molecular-dynamics simulation of conformational change – application to the T <−−−> R transition in insulin. Mol Simul 10:291–308
67. Roux B (1995) The calculation of the potential of mean force using computer-simulations. Comput Phys Commun 91:275–282
68. Souaille M, Roux B (2001) Extension to the weighted histogram analysis method: combining umbrella sampling with free energy calculations. Comput Phys Commun 135:40–57
69. Darve E, Pohorille A (2001) Calculating free energies using average force. J Chem Phys 115:9169–9183
70. Henin J, Chipot C (2004) Overcoming free energy barriers using unconstrained molecular dynamics simulations. J Chem Phys 121:2904–2914
71. Darve E, Rodriguez-Gomez D, Pohorille A (2008) Adaptive biasing force method for scalar and vector free energy calculations. J Chem Phys 128:144120
72. Laio A, Parrinello M (2002) Escaping free-energy minima. Proc Natl Acad Sci USA 99:12562–12566
73. Barducci A, Bussi G, Parrinello M (2008) Well-tempered metadynamics: a smoothly converging and tunable free-energy method. Phys Rev Lett 100:020603
74. Atilgan AR, Durell SR, Jernigan RL et al (2001) Anisotropy of fluctuation dynamics of proteins with an elastic network model. Biophys J 80:505–515
75. Bahar I, Lezon TR, Bakan A et al (2010) Normal mode analysis of biomolecular structures: functional mechanisms of membrane proteins. Chem Rev 110:1463–1497
76. Edholm O, Berger O, Jahnig F (1995) Structure and fluctuations of bacteriorhodopsin in the purple membrane: a molecular dynamics study. J Mol Biol 250:94–111
77. Stansfeld PJ, Sansom MS (2011) Molecular simulation approaches to membrane proteins. Structure 19:1562–1572
78. Lindahl E, Sansom MS (2008) Membrane proteins: molecular dynamics simulations. Curr Opin Struct Biol 18:425–431
79. Johnston JM, Filizola M (2011) Showcasing modern molecular dynamics simulations of membrane proteins through G protein-coupled receptors. Curr Opin Struct Biol 21:552–558
80. Khalili-Araghi F, Gumbart J, Wen PC et al (2009) Molecular dynamics simulations of membrane channels and transporters. Curr Opin Struct Biol 19:128–137
81. Arinaminpathy Y, Khurana E, Engelman DM et al (2009) Computational analysis of membrane proteins: the largest class of drug targets. Drug Discov Today 14:1130–1135
82. Wang Y, Shaikh SA, Tajkhorshid E (2010) Exploring transmembrane diffusion pathways with molecular dynamics. Physiology 25:142–154
83. Shaikh SA, Li J, Enkavi G et al (2013) Visualizing functional motions of membrane transporters with molecular dynamics simulations. Biochemistry 52:569–587
84. Grossfield A (2011) Recent progress in the study of G protein-coupled receptors with molecular dynamics computer simulations. Biochim Biophys Acta (BBA) Biomembr 1808:1868–1878
85. Maffeo C, Bhattacharya S, Yoo J et al (2012) Modeling and simulation of ion channels. Chem Rev 112:6250–6284
86. Sigworth FJ (2003) Structural biology: life's transistors. Nature 423:21–22
87. Jardetzky O (1966) Simple allosteric model for membrane pumps. Nature 211:969–970
88. Gumbart J, Wang Y, Aksimentiev A et al (2005) Molecular dynamics simulations of proteins in lipid bilayers. Curr Opin Struct Biol 15:423–431

89. Bjelkmar P, Niemela PS, Vattulainen I et al (2009) Conformational changes and slow dynamics through microsecond polarized atomistic molecular simulation of an integral Kv1.2 ion channel. PLoS Comput Biol 5:e1000289
90. Delemotte L, Tarek M, Klein ML et al (2011) Intermediate states of the Kv1.2 voltage sensor from atomistic molecular dynamics simulations. Proc Natl Acad Sci USA 108:6109–6114
91. Jensen MO, Jogini V, Borhani DW et al (2012) Mechanism of voltage gating in potassium channels. Science 336:229–233
92. Rosenbaum DM, Zhang C, Lyons JA et al (2011) Structure and function of an irreversible agonist-β2 adrenoceptor complex. Nature 469:236–240
93. Provasi D, Filizola M (2010) Putative active states of a prototypic G-protein-coupled receptor from biased molecular dynamics. Biophys J 98:2347–2355
94. Berneche S, Roux B (2001) Energetics of ion conduction through the K+ channel. Nature 414:73–77
95. Gwan JF, Baumgaertner A (2007) Cooperative transport in a potassium ion channel. J Chem Phys 127:045103
96. Piccinini E, Affinito F, Brunetti R et al (2007) Exploring free-energy profiles through ion channels: comparison on a test case. J Comput Electron 6:373–376
97. Jensen MO, Borhani DW, Lindorff-Larsen K et al (2010) Principles of conduction and hydrophobic gating in K+ channels. Proc Natl Acad Sci USA 107:5833–5838
98. Carbrey JM, Agre P (2009) Discovery of the aquaporins and development of the field. Handb Exp Pharmacol 190:3–28
99. Hub JS, de Groot BL (2008) Mechanism of selectivity in aquaporins and aquaglyceroporins. Proc Natl Acad Sci USA 105:1198–1203
100. Wang Y, Tajkhorshid E (2010) Nitric oxide conduction by the brain aquaporin AQP4. Proteins Struct Funct Bioinform 78:661–670
101. Jensen MO, Park S, Tajkhorshid E et al (2002) Energetics of glycerol conduction through aquaglyceroporin GlpF. Proc Natl Acad Sci USA 99:6731–6736
102. Henin J, Tajkhorshid E, Schulten K et al (2008) Diffusion of glycerol through *Escherichia coli* aquaglyceroporin GlpF. Biophys J 94:832–839
103. Borgnia MJ, Agre P (2001) Reconstitution and functional comparison of purified GlpF and AqpZ, the glycerol and water channels from *Escherichia coli*. Proc Natl Acad Sci USA 98:2888–2893
104. Psachoulia E, Fowler PW, Bond PJ et al (2008) Helix-helix interactions in membrane proteins: coarse-grained simulations of glycophorin a helix dimerization. Biochemistry 47:10503–10512
105. Janosi L, Prakash A, Doxastakis M (2010) Lipid-modulated sequence-specific association of glycophorin A in membranes. Biophys J 99:284–292
106. Sengupta D, Marrink SJ (2010) Lipid-mediated interactions tune the association of glycophorin A helix and its disruptive mutants in membranes. Phys Chem Chem Phys 12:12987
107. Shan Y, Arkhipov A, Kim ET et al (2013) Transitions to catalytically inactive conformations in EGFR kinase. Proc Natl Acad Sci USA 110:7270–7275
108. Johnston JM, Aburi M, Provasi D et al (2011) Making structural sense of dimerization interfaces of delta opioid receptor homodimers. Biochemistry 50:1682–1690

Chapter 14
Free-Energy Landscape of Intrinsically Disordered Proteins Investigated by All-Atom Multicanonical Molecular Dynamics

Junichi Higo and Koji Umezawa

Abstract We introduce computational studies on intrinsically disordered proteins (IDPs). Especially, we present our multicanonical molecular dynamics (McMD) simulations of two IDP-partner systems: NRSF–mSin3 and pKID–KIX. McMD is one of enhanced conformational sampling methods useful for conformational sampling of biomolecular systems. IDP adopts a specific tertiary structure upon binding to its partner molecule, although it is unstructured in the unbound state (i.e. the free state). This IDP-specific property is called "coupled folding and binding". The McMD simulation treats the biomolecules with an all-atom model immersed in an explicit solvent. In the initial configuration of simulation, IDP and its partner molecules are set to be distant from each other, and the IDP conformation is disordered. The computationally obtained free-energy landscape for coupled folding and binding has shown that native- and non-native-complex clusters distribute complicatedly in the conformational space. The all-atom simulation suggests that both of induced-folding and population-selection are coupled complicatedly in the coupled folding and binding. Further analyses have exemplified that the conformational fluctuations (dynamical flexibility) in the bound and unbound states are essentially important to characterize IDP functioning.

Keywords Molecular dynamics • Multicanonical • All-atom model • Free-energy landscape • Coupled folding and binding

J. Higo (✉)
Institute for Protein Research, Osaka University, 3-2 Yamadaoka, Suita, Osaka 565-0871, Japan
e-mail: higo@protein.osaka-u.ac.jp

K. Umezawa
Graduate School of Advanced Science and Engineering, Waseda University, Okubo 3-4-1, Shinjuku-Ku, Tokyo 169-8555, Japan

K. Han et al. (eds.), *Protein Conformational Dynamics*, Advances in Experimental Medicine and Biology 805, DOI 10.1007/978-3-319-02970-2_14,

14.1 Introduction

Intrinsically disordered proteins (IDPs) or intrinsically disordered regions (IDRs) are structurally disordered in the unbound state (free state) and adopt a well-defined tertiary structure upon binding to their partner molecules [9, 55, 62]. Thus, folding and binding are coupled indivisibly. This property proposes a new scheme for the structure-activity relationship as opposed to the well-known "lock and key" scheme, and is referred to as "coupled folding and binding" [9]. IDPs and IDRs exist in various eukaryotic genomes [46, 60], and play important roles in physiological processes, such as cellular signal transduction, protein phosphorylation, molecular assemblies, transcription, and translation regulation [9, 13, 24]. Therefore, IDPs and IDRs are thought to be potential drag targets [38]. A single IDP or IDR, as a hub protein, can interact with various targets exhibiting various biological functions [8, 47, 48]. Based on the biological significance, IDP database have been established: DisProt [53], GTOP [11], DICHOT [10], MobiDB [7], IDEAL [12] and D^2P^2 [44].

It is difficult to experimentally detect various molecular conformations that temporally appear in coupled folding and binding. A computer simulation is a suitable technique to persuade such transient conformations. A coarse-grained model provides large-scale molecular motions in a short computing time because the IDP and the partner are highly simplified and the surrounding solvent is ignored. Taking this advantage, the coarse-grained model proposed an interesting binding mechanism for IDPs: a "fly-casting mechanism" [52]. The unfolded conformation of IDP in the unbound state has a greater interaction radius than a well-packed structure does, and this increases the binding rate constant to capture the partner molecule. However, a coarse-grained simulation proposed an objection to the fly-casting mechanism [22]: The unfolded conformation has slower translational diffusion than the compact one (i.e. the unfolded state does not accelerate the binding). Instead, they argued that the structural flexibility of IDP in an encounter–complex state reduces the free-energy barrier between the encounter complex and the final native complex. In other words, an induced-folding (or induced-fit) mechanism [39, 54] dominates the coupled folding and binding. A population-selection (or population-shift) is known as another binding mechanism [2, 29, 63], which may be opposite to the induced-folding mechanism. Okazaki and Takada studied which the induced-folding or the population-selection is dominant in coupled folding and binding by using a coarse-grained model [45]. They argued that the binding mechanism shifts from population-selection to induced-folding with increasing the strength of the IDP-partner interactions and/or with an increment of the interaction radius. The coarse-grained model was also effectively used to study binding kinetics, where parameters used in the coarse-grained model were modulated to reproduce experimental rate constants [51].

The above-mentioned coarse-grained models are categorized in a simplified protein model, the Gō-like model [14], where the interactions are usually designed so that a target structure (i.e. the native structure or the native complex structure) has the global energy minimum: In other words, the protein conformation fluctuates under a target-structure-oriented bias. Therefore, the coarse-grained model may skip over some important intermediates in the coupled folding and binding process in exchange for the simplification of the model. We emphasize that an all-atom simulation is useful to link the experiments and the coarse-grained models.

Although time consuming, the all-atom simulation of biomolecules in an explicit solvent provides a conformational ensemble of the system at an atomic resolution. A conventional MD (i.e. canonical MD) simulation was used to investigate the structural dynamics for the complex of an IDP, the C-terminal segment of FCP1, and the partner protein, the winged-helix domain of RAP74 [61]. The simulation has shown that FCP1 retains disorder in complex with RAP74, which was also examined by an NMR experiment [36]. The large flexibility in the bound state suggests that the chain-entropy loss upon binging is small. Such structural disorder in the bound state has been generally termed "fuzzy" [57]. Chen has used an all-atom model with a continuum solvent to investigate the conformations of the C-terminal segment of p53 from the bound state with the partner protein, S100B(ββ), to the unbound sate, and found that the unbound state of p53 contains a native-like bound conformation [3]. Chen suggested that those residual conformations in the unbound state control thermodynamically the entropic cost for binding, and that it is not evident for population-selection mechanism in this case because p53 is unfolded at an intermediate state between the bound and free states, which supports the fly-casting mechanism. A multi-scale sampling technique, where an all-atom system is combined with a coarse grained system to enhance the conformational sampling, has shown that non-native interactions between an IDP and its partner facilitate binding by reducing the entropic cost to reach the final form in the free-energy landscape [59]. Therefore, this work supports the induced-folding mechanism. A conventional MD of an IDR, the N-terminal tail of the p53 protein, supports the population-selection scheme [23]. Anchor residues in the tail, which are buried in the IDP-partner interface in the complex, frequently adopt the bound form even in the unbound state. A protein sortase has an intrinsically disordered loop, and the N-terminal of the loop is structured upon binding to a signal peptide and the C-terminal is done upon binding to a calcium ion. A multi-scale enhanced sampling simulation of this system has proposed that the N-terminal tends to exhibit population-selection, whereas the C-terminal does the induced-folding [41]. Interestingly, the signal-peptide binding and the calcium-ion binding enhance cooperatively the structure formation of the disordered loop. These works suggest that the mechanism for coupled folding and binding has multifaceted aspects.

As above, not only the coarse-grained model but also the all-atom models have uncovered various pictures for coupled folding and binding. Natural explanation for coupled folding and binding is: each IDP/IDR has its own mechanism. Another explanation is that the induced-folding and population-selection are mixed in an IDP-partner system, and then either mechanism appears on the front responding to a situation. Below we describe our simulation results of two IDP-partner systems: NRSF–mSin3 and pKID–KIX systems. In the initial configuration of simulation, the IDPs (NRSF and pKID) and the partner molecules (mSin3 and KIX) were put distant to each other in an explicit solvent, and the IDP conformations were disordered. Then, a multicanonical MD (McMD) simulation, which is explained briefly blow, was performed to enhance conformational sampling of the IDP-partner systems.

14.2 Multicanonical MD

14.2.1 Methods

Protein folding and protein-ligand binding have been studied individually or separately. Because each of folding and binding accompanies large configurational motions, a powerful sampling method is required for the computational study. In coupled folding and binding, on the other hand, folding and binding are coupled indivisibly. Thus computational clarification of this phenomenon is a challenging task. We have attacked this task using an enhanced conformational sampling method, McMD simulation developed by Nakajima et al. [42].

Historically, the multicanonical algorithm was developed first to investigate statistical properties of a two-dimensional Potts model on lattices based on Monte Carlo (MC) sampling [1]. The MC-based method was applied to biological systems [16, 34] and extended to molecular dynamics [17, 42]. The version by Nakajima et al., denoted as McMD simulation in this chapter, executes the sampling in a Cartesian coordinate space. Adoption of the Cartesian coordinates made the multicanonical method extendable readily to a flexible multi-molecular system in explicit solvent. Below we briefly explain the McMD method. Methodological details are given in a review [18].

At a temperature T, the potential energy E of a system fluctuates, and a long MD simulation yields an energy distribution $P_c(E,T)$. This distribution, characterized by E and T, is called a "canonical energy distribution". Because a single energy value is assigned to a conformation, E is a function of atomic coordinates of the system:

$$E = E\left(\boldsymbol{r}_1, \boldsymbol{r}_2, \ldots, \boldsymbol{r}_N\right), \tag{14.1}$$

where $\boldsymbol{r}_i$ is the position of atom i expressed as $\boldsymbol{r}_i = [x_i, y_i, z_i]$, where x_i, y_i, and z_i are its x-, y-, and z-coordinates, respectively. The parameter N is the number of atoms in the system (i.e. atoms in biomolecules and solvent molecules). In conventional MD (canonical MD), the force acting on atom i is given as

$$\boldsymbol{f}_i = -\frac{\partial E}{\partial x_i}\boldsymbol{e}_x - \frac{\partial E}{\partial y_i}\boldsymbol{e}_y - \frac{\partial E}{\partial z_i}\boldsymbol{e}_z, \tag{14.2}$$

where $\boldsymbol{e}_x$, $\boldsymbol{e}_y$, and $\boldsymbol{e}_z$ are unit vectors parallel to the x, y, and z axes, respectively. A simulation, which is performed by solving Newtonian equations of motion with using Eq. 14.2 under constant temperature and volume, is called the canonical MD simulation, and $P_c(E,T)$ is computed from time-average of E along the simulation trajectory.

Here we suppose that $P_c(E,T)$ is known accurately in advance. Statistical mechanics ensures that the density of states $n(E)$ of the system is given by the following equation: $n(E) \propto P_c(E,T)\exp[E/RT]$, where R is the gas constant (we assume that energy is measured in an kcal/mol unit). Here let us introduce another potential energy E_{mc} (called "multicanonical energy") as

$$E_{mc} = E + RT \ln [P_c (E, T)] \propto RT \ln [n(E)] \tag{14.3}$$

Then, we define the force acting on atom i as

$$\begin{aligned} \boldsymbol{f}_i^{mc} &= -\frac{\partial E_{mc}}{\partial x_i}\boldsymbol{e}_x - \frac{\partial E_{mc}}{\partial y_i}\boldsymbol{e}_y - \frac{\partial E_{mc}}{\partial z_i}\boldsymbol{e}_z \\ &= -\left[\frac{\partial E}{\partial x_i}\boldsymbol{e}_x - \frac{\partial E}{\partial y_i}\boldsymbol{e}_y - \frac{\partial E}{\partial z_i}\boldsymbol{e}_z\right]\frac{\partial E_{mc}}{\partial E} \\ &= \frac{\partial E_{mc}}{\partial E}\boldsymbol{f}_i. \end{aligned} \tag{14.4}$$

Then, the MD simulation at T with Eq. 14.4 for the force evaluation is called "multicanonical MD (McMD)". The simulation trajectory provides the following energy distribution:

$$P_{mc} (E, T) \propto n(E) \exp\left[-\frac{E_{mc}}{RT}\right] = \frac{n(E)}{n(E)} = const. \tag{14.5}$$

Equation 14.5 ensures that the energy distribution from a long McMD run converges to a flat distribution. Although the system consists of a variety of atoms (i.e. atoms in biomolecules, water molecules, and ions), the single term $-\partial E_{mc}/\partial E$ is multiplied to forces of any atom for performing McMD. In other words, replacement of $\boldsymbol{f}_i$ by $-\partial E_{mc}/\partial E \times \boldsymbol{f}_i$ in an MD computer program converts canonical MD to McMD.

However, we do not know an accurate function form for $P_c(E,T)$ a priori. Then, McMD is done iteratively, through which the accuracy of $P_c(E,T)$ increases and the multicanonical energy distribution gradually converges to a flat function in a wide energy range: $P_{mc}(E,T) \rightarrow const.$ See the review [18] for detailed procedure. After reaching the flat distribution, we perform the last McMD simulation (production run), during which a number of conformations are stored. We denote the ensemble of stored conformations in the production run as Q_{mc}.

Because McMD produces the flat energy distribution in the wide energy range, the conformations in Q_{mc} have various energies, some of which correspond to high-temperature conformations and some of which to low-temperature ones. Therefore, when we select conformations from Q_{mc} whose energies are probable at a room temperature T_{room}, the ensemble of the selected conformations is a canonical ensemble at T_{room}, denoted as $Q_c(T_{\text{room}})$. Similarly, when conformations probable at another temperature T' are taken from Q, the generated ensemble is a canonical ensemble at $T' : Q_c(T')$. By this way, we can generate the canonical ensemble at arbitrary temperature.

With increasing the system size, the complexity of the system and the volume of conformational space increase drastically. Accordingly, the simulation length required for sampling increases. To expand the applicability of McMD to such a large and complex system, we developed a trajectory-parallelization method of McMD (trivial trajectory-parallelization of McMD; TTP-McMD) [20], where a number of McMD runs are executed from various initial conformations of simulation. Importantly, the ensemble of the multiple trajectories is equivalent to a long single trajectory of McMD [27]. TTP-McMD was used to compute the free-energy landscape of IDP systems [21, 58]. This method is mentioned again later.

14.2.2 Free-Energy Landscape

Now we have the conformational ensemble $Q_c(T_{\text{room}})$. Biophysically interesting characteristics in protein folding or binding may be related to some structural parameters of conformations stored in $Q_c(T_{\text{room}})$. These parameters may be such as an end-to-end distance, a radius of gyration, a solvent accessible surface area, an inter-atomic distance, a number of inter-residue contacts (quantities listed up to here are those related to folding), and an inter-molecular distance, a number of inter-molecular residue-residue contacts, and an interface area (these are related to molecular binding). Or one can define an arbitrary parameter to analyze the conformational ensemble. Assuming a one-, two-, or three-dimensional space by those parameters, we can generate a distribution function of $Q_c(T_{\text{room}})$ in the space. For instance, a three-dimensional distribution function $P(s_1,s_2,s_3;T_{\text{room}})$ is defined by counting the number of conformations detected in a small volume $\Delta s_1 \times \Delta s_2 \times \Delta s_3$ at a position $[s_1,s_2,s_3]$, where s_i is one of the parameters. When one or two parameters are selected, the distribution function is denoted as $P(s_1;T_{\text{room}})$ or $P(s_1,s_2;T_{\text{room}})$. Benefit for generating a low-dimensional space is visibility.

A potential of mean force, *PMF*, is a quantity categorized into free energy, which is defined as

$$PMF\,(s_1, s_2, s_3; T_{room}) = -RT_{room} \ln\left[P\,(s_1, s_2, s_3; T_{room})\right]. \tag{14.6}$$

Because the probability is large at a site where snapshots are frequently detected, *PMF* is low at the site. This means that the structure/configuration at this site is thermodynamically stable. Patterns of *PMF* in the conformational space provide an image of a "free-energy landscape". Generation of the landscape at a different temperature T' is achieved by replacing T_{room} by T' in Eq. 14.6.

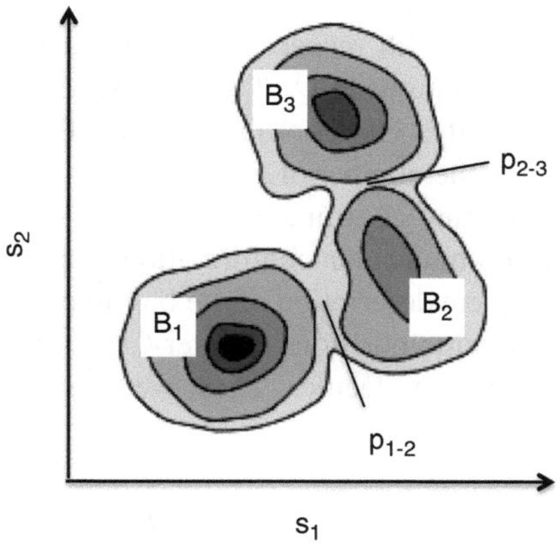

Fig. 14.1 Scheme for free-energy landscape illustrated two-dimensionally by parameters s_1 and s_2. The darker the tone, the lower the free energy (i.e. potential of mean force; *PMF*). Three free-energy basins, B_1, B_2, and B_3, are shown. *Saddle point* p_{1-2} separates B_1 and B_2, and p_{2-3} does B_2 and B_3

One may obtain a free-energy landscape as illustrated in Fig. 14.1, which is a two-dimensional representation for the landscape. A region with low *PMF* at T_{room} is called a "free-energy basin" (or simply "basin"). In this figure, the basin B_1 is the global free-energy minimum. The global minimum may correspond to the native structure (or native complex structure) at T_{room} if the force filed parameters are accurate enough and the simulation length is long enough. An important quantity to characterize the free-energy landscape is a free-energy barrier, of which the height is the *PMF* value at the saddle point p_{1-2} or p_{2-3}. We can presume that an inter-basin transition between B_1 and B_2 occurs directly via passing the saddle point p_{1-2} and one between B_2 and B_3 via p_{2-3}. Contrarily, a transition between B_1 and B_3 occurs indirectly. To argue the conformational change for this indirect transition, one may pick up conformations from B_1, B_2, B_3, p_{1-2}, and p_{2-3}.

If the free-energy landscape is presented as Fig. 14.1, we can specify pathways readily. However, a structural parameter, mentioned above, may mislead in identifying the free-energy barriers. This is because largely different protein conformations may have the same parameter value. We have experienced that a free-energy barrier identified in a conformational space vanishes completely in another space [21, 31]. A universal method to unerringly identify the free-energy basins and barriers is desired. However, no such a universal method exists.

Here we explain a method to naturally construct the conformational space without prejudice. We start with the high-dimensional space, although it is invisible. This space specifies full details of the system in each snapshot of $Q_c(T_{\mathrm{room}})$. The configuration of the system in a snapshot is expressed by a vector as:

$$\boldsymbol{X} = [x_1, y_1, z_1, x_2, y_2, z_2, \ldots, x_N, y_N, z_N]\,. \tag{14.7}$$

The $\boldsymbol{X}$ may consist of coordinates of a protein and a ligand when ligand–protein binding is studied. The conformational space is a $3N$-dimensional space, and the conformations stored in $Q_c(T_{\mathrm{room}})$ distribute in the space. Our study usually focuses on the biomolecules (i.e. protein and ligand). Then, to reduce the degrees of freedom with respect to mutual translation and rotation of the biomolecules, the conformations in $Q_c(T_{\text{room}})$ are superimposed onto a reference conformation given arbitrarily. The coordinates used for analyses are those after the superposition, where the number of independent coordinates is $3N-6$ substantially.

Internal coordinates, such as a set of inter-atomic distances, can also express the biomolecular conformational space. Note that our analyses may not require complete description of the biomolecular configuration. For instance, we can select Cα-atomic positions or inter-Cα-atomic distances to specify the overall structures of the biomolecules. Then, we present the system configuration by those selected quantities as

$$\boldsymbol{q} = [q_1, q_2, q_3, \ldots, q_M], \tag{14.8}$$

where M is the number of the selected quantities. Although M is smaller than $3N-6$, the M-dimensional space is still high dimensional ($M \gg 3$) generally. Then, we encounter again a difficulty to analyze the conformational distribution because the M-dimensional distribution $P(\boldsymbol{q};T_{\text{room}})$ is invisible. A commonly used strategy for analyzing the high-dimensional distribution is to reduce the space dimensionality without loosing important features of the high-dimensional distribution.

We have frequently used principal component analysis (PCA) to view the free-energy landscape because PCA defines coordinate axes, along which the conformational distribution $P(\boldsymbol{q};T_{\text{room}})$ has large variances. For simplicity, we suppose that the full space is two dimensional (q_1 and q_2), and the full distribution is as Fig. 14.2a. Suppose that this two-dimensional distribution is invisible for us. Then, we define new axes $\boldsymbol{v}_1$ and $\boldsymbol{v}_2$ as in Fig. 14.2. Suppose that the one-dimensional distribution projected on $\boldsymbol{v}_1$ or $\boldsymbol{v}_2$ is visible because of the space-dimensionality reduction. The $\boldsymbol{v}_1$ is an axis where the projected distribution $P(v_1;T_{\text{room}})$ widely spread and the basins are distinguishable (Fig. 14.2b). Along the axis $\boldsymbol{v}_2$, in contrast, the distribution $P(v_2;T_{\text{room}})$ is narrow and basins may be merged into one (figure not shown). From visual comparison of Figs. 14.2a, b, one may state that the reduction of the space-dimensionality, $P(\boldsymbol{q};T_{\text{room}}) \rightarrow P(v_1;T_{\text{room}})$, is not useful because $P(\boldsymbol{q};T_{\text{room}})$ discriminates more clearly the basins and barriers than $P(v_1;T_{\text{room}})$ does. However, remember that the total dimension of the full space is M (or $3N$). Then we need to select an adequate set of axes to construct the free-energy landscape with a smaller dimension than four. Otherwise the basins and barriers are invisible as mentioned above.

Now we explain how to compute such axes using PCA. We first calculate a variance-covariance matrix C from the conformational ensemble $Q(T_{\text{room}})$ as

$$C_{mn} = < q_m q_n >_{Troom} - < q_m > < q_n >_{Troom} \tag{14.9}$$

where C_{mn} is the matrix element (m,n), and $< \ldots >_{\text{Troom}}$ is the ensemble average over the conformations in $Q(T_{\text{room}})$. Diagonalization of C provides M eigenvectors

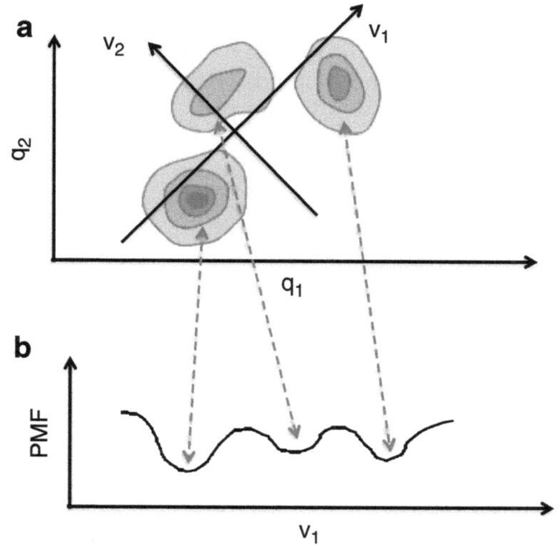

Fig. 14.2 (**a**) Scheme of the full-dimensional distribution, although the conformational space is provided two-dimensionally ($\boldsymbol{q} = [q_1, q_2]$) for simplicity. The distribution $P(\boldsymbol{q};T_{\text{room}})$ is converted to potential of mean force as $PMF(\boldsymbol{q};T_{\text{room}}) = -RT_{\text{room}}\ln[P(\boldsymbol{q};T_{\text{room}})]$ (Eq. 14.6). Two axes $\boldsymbol{v}_1$ and $\boldsymbol{v}_2$ may be derived from principal component analysis (PCA) (see text for details). (**b**) One-dimensional distribution projected on $\boldsymbol{v}_1$ converted to potential of mean force as $PMF(v_1;T_{\text{room}}) = -RT_{\text{room}}\ln[P(v_1;T_{\text{room}})]$. *Broken-line arrows* present correspondence of three basins between panels (**a**) and (**b**)

and eigenvalues. An eigenvector $\boldsymbol{v}$ and an eigenvalue λ are paired satisfying an equation $C\boldsymbol{v} = \lambda\boldsymbol{v}$. We assign an index k to each of $\boldsymbol{v} - \lambda$ pairs so that the k-th eigenvector and eigenvalue are denoted as $\boldsymbol{v}_k$ and λ_k, respectively, and arrange the eigenvectors in descending order of pairing eigenvalues. The eigenvectors satisfy equations $C\boldsymbol{v}_k = \lambda_k\boldsymbol{v}_k$ and $\boldsymbol{v}_i \cdot \boldsymbol{v}_j = \delta_{ij}$. The latter equation assures that the eigenvectors can construct M-dimensional orthogonal axes, and the former does that the variance of conformational distribution along a larger-eigenvalue eigenvector is larger than that along a smaller-eigenvalue eigenvector. Then, the position of a snapshot in the M-dimensional space is expressed as

$$\boldsymbol{s} = [s_1, s_2, \ldots, s_M], \qquad (14.10)$$

where the element s_i is expressed as

$$s_i = \boldsymbol{v}_i \cdot (\boldsymbol{q} - <\boldsymbol{q}>_{T\,room}). \qquad (14.11)$$

We refer to the M-dimensional space constructed by the eigenvectors as "PC space". Projecting the conformations in $Q_c(T_{\text{room}})$ on the PC space using Eq. 14.10, we generate a conformational distribution. By projecting the conformations on a subspace constructed by a few (one, two or three) eigenvectors, we can generate a visible low-dimensional distribution. These eigenvectors should be those with large conformational variance (i.e. large eigenvalues) to capture important features (basins and barriers) in the free-energy landscape. Picking $\boldsymbol{v}_1$, $\boldsymbol{v}_2$ and $\boldsymbol{v}_3$, a snapshot is simply expressed in the three-dimensional PC subspace as

$$\boldsymbol{s} = [s_1, s_2, s_3] . \tag{14.12}$$

We have experienced that free-energy barriers detected in the three-dimensional PC space disappear in a low-dimensional space constructed by well-used structural parameters such as the radius of gyration or the accessible surface area [21, 31].

14.2.3 Our McMD Research

We have applied McMD to polypeptide folding and ligand–protein flexible docking. So far, we have computed the free-energy landscape of a 40-residue $\alpha+\beta$ protein [25] and a 57-residue protein consisting of two long helices [26], where the initial simulation conformations were fully disordered in an explicit solvent. A flexible docking of lysozyme and sugar in an explicit solvent was performed, where lysozyme and sugar were set distant to each other in the initial configuration [33]. McMD produced a free-energy landscape, in which the lowest free-energy basin was assigned to the native complex structure and the lowest basin was separated from the other minor basins by free-energy barriers.

To expand the applicability of McMD to a more complicated system, recently we have developed a trajectory–parallelization method [20], where multiple McMD runs are initiated from various conformations. Then, a long trajectory is generated with simply connecting the multiple trajectories. Importantly, the integrated long trajectory is theoretically equivalent to a single McMD trajectory because the detailed balance is satisfied at the connection points of the multiple trajectories [27]. Because the initial conformations spread widely in the conformational space, the integrated trajectory covers a larger region than a real single trajectory, even though the integrated-trajectory length is equal to or shorter than the real single trajectory. This method, called "trivial trajectory parallelization (TTP)", was used for the McMD sampling of IDP systems.

To computationally investigate large configurational motions, force field is essentially important. We have developed an AMBER-hybrid force field [32] expressed as $E(w) = (1 - w)E_{94} + wE_{96}$, where E_{94} and E_{96} are the AMBER parm94 [4] and parm96 [35] force fields, respectively, and the parameter w is the mixing rate. We computed the free-energy landscape of short peptides by McMD simulations and quantum chemical calculations with varying w from 0 to 1, and found that $E(0.75)$ produces the best agreement of the free-energy landscape between the two computation methods. Furthermore, in McMD simulations with $E(0.75)$, a peptide with a helical propensity folds into a helix, while a peptide with a β-hairpin propensity forms a β-hairpin. The force field for a water molecule is TIP3P [30]. A cell-multipole expansion method [6] was used to compute long-range electrostatic interactions, and the net charge of the system was neutralized by introducing Cl^- and four Na^+ ions. A spherical boundary condition was used for generating the solvent environment. A computer program, Presto ver. 3 [40], was used for performing McMD.

14.3 Free-Energy Landscape of Coupled Folding and Binding

So far, as mentioned in the Introduction section, we have computed the free-energy landscape of two IDP systems: NRSF–mSin3 and pKID–KIX. Remember that NRSF and pKID are IDPs, and mSin3 and KIX the partners. For simplicity, the paper on the NRSF–mSin3 system [21] is referred to as “the NRSF-mSin3 paper”, and the paper on the pKID–KIX system [58] as “the pKID-KIX paper”. The structure of the partner molecule was weakly restrained around the pdb structure to prevent the partner from unfolding during McMD, although IDP was completely flexible and allowed translational and rotational motions. We emphasize that the force fields for proteins, solvent and ions are the same and that the simulation protocol is also the same between the two systems. Details for the simulation protocol is described in the papers. In the initial conformations of simulation, the IDPs were set to be distant from the partners in an explicit solvent and the IDP conformations were disordered: See Fig. 1c of the NRSF-mSin3 paper, and Fig. 1b of the pKID-KIX paper. During McMD, IDPs fluctuated around the partner molecules adopting a variety of complex forms. The computed free-energy landscapes (see Fig. 5 of the NRSF-mSin3 paper and Fig. 4 of the pKID-KIX paper) consisted of various structural clusters, where one of them corresponded to the native-like complex and the other to non-native clusters. We discussed the conformational pathways for coupled folding and binding. The non-native clusters provided a variety of IDP conformations in both the bound and unbound states.

The size of cluster i in the free-energy landscape at temperature T is proportional to the free energy $F_i(T)$ of the cluster as

$$F_i(T) = -RT \ln\left[Z_i(T)\right], \tag{14.13}$$

where Z_i is a partition function of cluster i. By setting the cluster border in the conformational space in an appropriate way, we can compute Z_i:

$$Z_i = \int_{\text{cluster } i} P\left(s_1, s_2, s_3; T\right) ds_1 ds_2 ds_3, \tag{14.14}$$

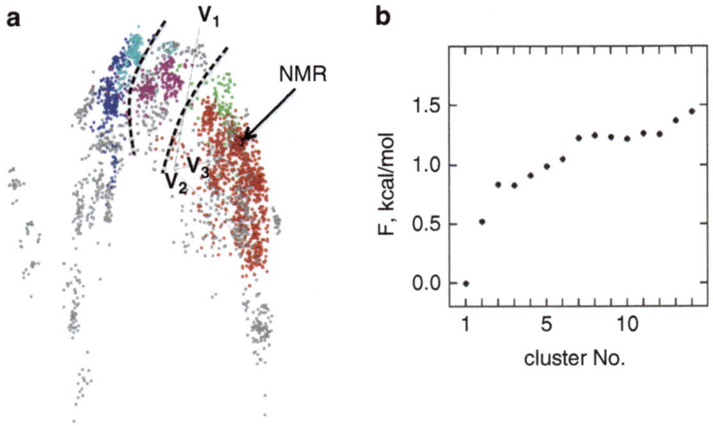

Fig. 14.3 (**a**) Conformational distribution of the NRSF–mSin3 system at 300 K, displayed in a three-dimensional PC subspace, where coordinate axes $\boldsymbol{v}_1$, $\boldsymbol{v}_2$, and $\boldsymbol{v}_3$ are eigenvectors with the largest, second largest, and third largest eigenvalues. A *small sphere* represents a sampled conformation probable at 300 K. The first cluster (i.e. the lowest free-energy cluster) consists of *red-colored spheres*. The second, third, fourth and fifth clusters (i.e. the second, third, fourth and fifth lowest free-energy clusters) are colored in *blue*, *magenta*, *green*, and *cyan*, respectively. *Gray spheres* belong to other minor clusters. The *arrow* indicates the position of the NMR complex structure (native complex). *Broken lines* show free-energy barriers separating three super-clusters. See paper [21] for detailed characterization of the free-energy barriers. (**b**) Free energy, *F*, assigned to each cluster at 300 K (Eq. 14.13). The cluster ordinal numbers on the x-axis are arranged so that the smaller the number the lower the free energy. The free energy for the largest cluster is set to zero. The NMR complex structure is involved in the largest cluster

where the integral is taken over the volume of the cluster. The larger the Z_i, the lower the free energy of the cluster i. A free-energy difference between clusters i and j is given as

$$\Delta F(T) = F_j - F_i = -RT \ln\left[Z_j(T)/Z_i(T)\right]. \quad (14.15)$$

The free-energy landscape (conformational distribution) of the NRSF–mSin3 system is shown in Fig. 14.3a. Importantly, that the lowest free-energy cluster, shown in red in the figure, involves the native complex (see also Figs. 7 and 8 of the NRSF-mSin3 paper). In other words, the computed lowest free-energy cluster corresponds to the experimental structure [43]. We have identified two free-energy barriers (broken lines in Fig. 14.3a), which divide the distribution into three super-clusters. Subsequent analyses showed that passing a barrier corresponds to a specific motion of NRSF from hairpin-like conformations to helical conformations and the other to a change of the molecular orientation of NRSF in the mSin3 cleft. Figure 14.3b plots the free-energy values of the clusters.

McMD simulation of the unbound NRSF has produced a rugged free-energy landscape consisting of various structures involving α and β secondary-structure elements. Thus the unbound NRSF is disordered in solution as a whole. This multiple-structure property has been found in other peptide systems [19, 28].

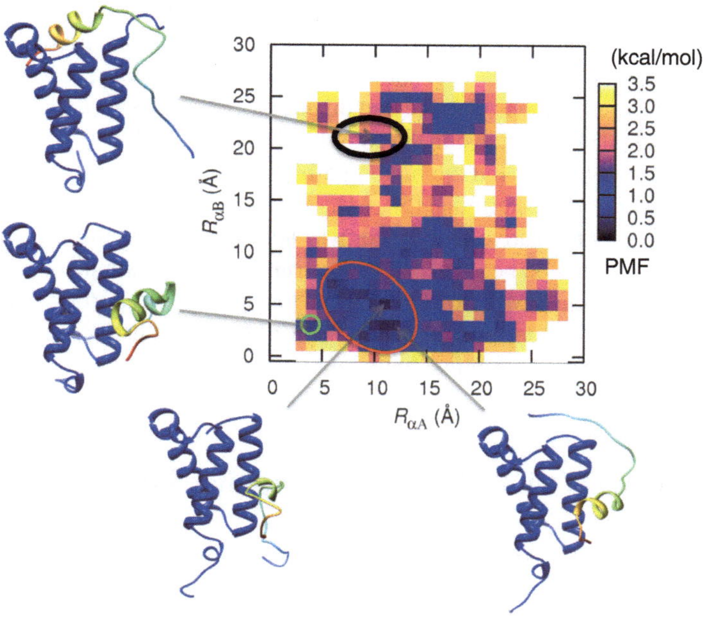

Fig. 14.4 Potential of mean force (PMF) of the pKID-KIX system at 315 K mapped on the plane of $R_{\alpha A}$ and $R_{\alpha B}$. The quantities $R_{\alpha A}$ and $R_{\alpha B}$ are defined as follows: $R_{\alpha A} = \left| \vec{R}_{\alpha A} - \vec{R}^{NMR}_{\alpha A} \right|$ and $R_{\alpha B} = \left| \vec{R}_{\alpha B} - \vec{R}^{NMR}_{\alpha B} \right|$, where $\vec{R}_{\alpha A}$ and $\vec{R}_{\alpha B}$ respectively represent the center positions of the N-terminal and C-terminal halves of pKID in a snapshot, and $\vec{R}^{NMR}_{\alpha A}$ and $R^{NMR}_{\alpha B}$ those in the reference structure (i.e. the NMR model), respectively. The lowest PMF value was set to zero. This figure also displays some complex structures picked from the PMF map, where KIX is represented by the *blue ribbon* and pKID by *rainbow* (*blue* N-terminal and *red* C-terminal). *Green* and *red circles* represent the native-like and largest clusters, respectively. The *black circle* represents a cluster where pKID bind to the MLL binding sites (see text for details)

Interestingly, most of these conformations in the unbound state could be found in the bound state (see Fig. 10 of the NRSF-mSin3 paper). This point is discussed again later.

In the NMR structure of the pKID–KIX complex (PDB ID: 1kdx), each of the N- and C-terminal halves of pKID adopts helix and the two helices are spaced by non-helical two amino-acid residues, a proline and a phosphorylated serine. Interestingly, our McMD simulation has shown that each half has a helical propensity even in the unbound state, whereas the unbound pKID is, as a whole, disordered. Detailed analyses have shown that the helix propensity for the N-terminal half is larger than that for the C-terminal half (Fig. 2 of the pKID–KIX paper). This result agrees with an experimental result for the unbound pKID [49]. In the bound state, the free-energy landscape has provided a native-like cluster (a green circle in Fig. 14.4). However, the largest cluster was not this native-like cluster but one indicated by the red circle. In the largest cluster, the C-terminal half was helical whereas the N-terminal half was disordered. Furthermore, in the native-like cluster the N-terminal half of pKID was somewhat fragile, whereas the C-terminal half adopts a well-ordered helix. We note that this fragility of the N-terminal half agrees with experimental observations: In the complex the N-terminal helix is more flexible than the C-terminal helix, and the binding affinity of the N-terminal half to KIX is weaker than that of the C-terminal [50, 64].

The reason why the native-like cluster was not the largest in the computed free-energy landscape may be because pKID was truncated in the simulation: the pKID sequence used for the NMR experiment was longer than the portion deposited to PDB, and we used the deposited portion for McMD. The unstructured regions of pKID may stabilize the C-terminal helix more.

14.4 Binding of pKID to the MLL Binding Site of KIX

An NMR study [5] has shown that KIX binds to two protein segments: a segment taken from the activation domain of the mixed lineage leukemia (MLL) transcription factor and a segment taken from the cMyb transcription factor. The NMR structure (PDB ID: 2agh) shown that both segments adopt helix upon binding to KIX. Here we refer to this complex structure as MLL–KIX–cMyb. The helical C-terminal half of pKID in the largest cluster and the native-like cluster (red- and green-circle clusters in Fig. 14.4, respectively) was similar to the c-Myb segment structure. The MLL-binding site of KIX is far from the pKID-binding site. We have checked whether the McMD simulation produce a similar complex structure with the MLL segment in the MLL–KIX–cMyb complex, and found that the pKID structure taken from the black-circle cluster in Fig. 14.4 was similar with the MLL segment structure (Fig. 14.5).

It is worthwhile to note that in the presence of MLL, pKID binds to KIX with the twofold higher affinity than pKID in the absence of MLL [15]. Our simulation suggests that MLL facilitates the pKID binding to the genuine binding site via blocking the MLL binding site.

14.5 Coupled Folding and Binding

Scheme 14.1 is a schematic representation of the free-energy landscape for coupled folding and biding of the NRSF–mSin3 system, obtained from McMD. The bound state involves a variety of complex forms. Then, a non-native complex moves in the free-energy landscape and is stabilized when reaching the native-like form. Some non-native complexes may overcome the free-energy barriers to reach the native-like complex.

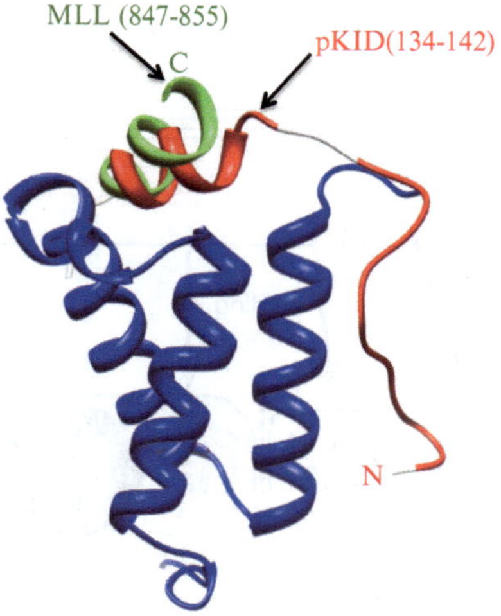

Fig. 14.5 Structures of pKID (*red*) and MLL segment (*green*) bound to KIX (*blue*). The N- and C-terminal halves of pKID are shown in *red*, and two residues spacing the halves are shown in while. The MLL segment structure was taken from an NMR structure (PDB ID: 2agh) of the MLL–KIX–cMyb complex, where the c-Myb segment was removed. Character "*N*" indicates the N-terminal of pKID, and "*C*" the C-terminal of the MLL segment. The numbers "134-142" and "847-855" are the residue ordinal numbers for the segments used in the original PDB files

As mentioned in Introduction, there are two representative mechanisms to explain molecular binding: induced-folding and population-selection. Which of the two mechanisms does work in coupled folding and binding for the current systems? Remember that the unbound NRSF fluctuates among various conformations, and most of these NRSF conformations are found in the bound state. Therefore, if a temporally formed helical conformation binds to mSin3, this process supports the population-selection mechanism. On the other hand, other non-helical NRSF conformations are also bindable to mSin3, and yield various non-native complex structures (encounter complexes). Once the non-native complex is formed, the conformation moves in the free-energy landscape, and finally reaches the most thermodynamically stable cluster (the native-like cluster). This process supports the induced-folding mechanism. The free-energy landscape has shown that multiple pathways are possible between the non-native and native complexes. Thus, the induced-folding mechanism has an entropic advantage against the population-selection mechanism, in which a small number of pathways are possible. Finally, we presume that the induced-folding mechanism is dominant for the NRSF–mSin3 system.

The pKID-KIX paper reports that in the unbound state, the C-terminal half of pKID contains a nascent helix with a smaller content than the N-terminal does (Fig. 2 of the pKID–KIX paper). This result agrees qualitatively with an experiment [49]. Contrarily, in the bound state, the largest cluster consists of conformations

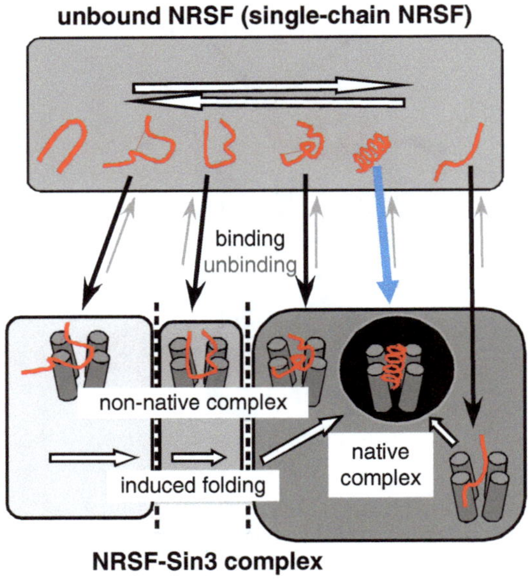

Scheme 14.1 Schematic representation of the free-energy landscape for coupled folding and biding for the NRSF–mSin3 system. The *upper half* represents the unbound NRSF (*red* string) fluctuating among various conformations. These conformations are bindable to the partner molecule, mSin3. The *lower half* represents the complex state. Only when the helical NRSF in the unbound state binds to mSin3 (four-cylinder model), the native complex is formed immediately (*blue colored arrow*), where the population-selection mechanism works. Other complex forms are non-native. Then, the non-native complexes move in the free-energy landscape, some of which overcome free-energy barrier(s) shown by *broken lines* to reach the native complex form. This movement is categorized to the induced-fit mechanism. We do not exclude a possibility of dissociation, which is shown by *shaded arrows*

where the C-terminal half is helical whereas the N-terminal half is flexible and somewhat disordered. This result is also consistent with experiments [50, 64]. The folding of the C-terminal half is explained with two alternative scenarios. One is: the wide distribution in Fig. 14.4 demonstrates that various encounter complexes are formable in the bound state. Then, the helix is induced during the conformational fluctuations with keeping the contacts to KIX. This scenario accords with the induced-folding mechanism. The other scenario (population-selection) is: the nascent helical structure of pKID formed in the unbound state is selected by the interactions with KIX. Thus, the largest cluster is formed in an early phase of complex formation. Because the multicanonical algorithm provides an equilibrium distribution [18], we cannot select deterministically which mechanism is dominant.

The N-terminal half of pKID may have a different binding mechanism than the C-terminal half. The N-terminal half has a structural variety in the largest cluster. Thus the interactions between the N-terminal half and KIX are weak even after the C-terminal half has folded into helix. As notified above, the fragility of the N-terminal half has been observed in the experiments [50, 64]. Therefore, it is likely that the

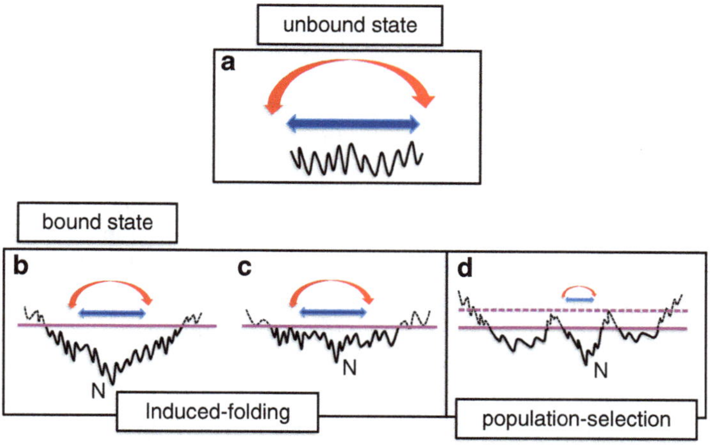

Scheme 14.2 Free-energy landscapes for the unbound state of IDP (**a**), IDP-partner bound state with induced-folding (**b**) and (**c**), and IDP-partner bound state with population-selection (**d**). Character "*N*" represents the position of the native complex. *Red-colored and blue-colored curved arrows* indicate amounts of rotational and translational motions of IDP: the larger the arrows, the larger the entropy. *Magenta-colored solid lines* indicate a free-energy level of the unbound state. The landscapes above the *magenta lines* are shown by *black broken lines. Magenta broken line* is the free-energy level in a more condensed IDP-partner solution

N-terminal-half folding stabilizes supplementarily the native complex. The coarse-grain model by Okazaki and Takada [45] has proposed that a weak IDP-partner interaction leads the binding mechanism to population-shift. Then, we presume that the N-terminal half is controlled by the population-selection mechanism.

Scheme 14.2 illustrates free-energy landscapes for an IDP-partner system. The unbound IDP has a wide conformational variety (chain entropy) in the thermal fluctuations (Scheme 14.2a). Furthermore, the unbound IDP has large translation and rotation entropies because the IDP can move freely in solution. The free-energy landscape for the bound state is shown in Scheme 14.2b–d. In the induced-folding mechanism (Scheme 14.2b, c), once a non-native encounter complex is formed, the complex can reach the native complex without dissociation. The landscape of Scheme 14.2b may have a faster rate to reach the native form than that of Scheme 14.2c does. To maintain binding, the free energy for the non-native complex should be lower than that of the unbound state. Otherwise the non-native complexes may be dissociated before reaching the native form, which is essentially the same as the population-selection mechanism.

In the population-selection mechanism (Scheme 14.2d), a non-native complex cannot overcome a free-energy barrier(s) even though the free energy of the non-native complex is lower than that of the unbound state. Then the non-native complex is dissociated most likely. A chance for reaching the native form is assigned to encounter complexes whose conformations are similar to the native form. Here we note that the free-energy level of the unbound state moves according to the solute

(IDP and partner) concentration: With increasing the concentration, the level rises. Then, population-selection switches with induced-folding as indicated by the broken line in Scheme 14.2d.

The entropy change upon binding is an important issue. The unbound IDP can vary the molecular orientation and translation freely (red and blue arrows in Scheme 14.2a). In the induced-folding mechanism, a variety of orientation and translation as well as various encounter-complex forms are possible. Contrarily, in the population-selection mechanism, the orientation and translation as well as the encounter complex form are restricted considerably: Red and blue arrows in Scheme 14.2b, c are larger than those in Scheme 14.2d. Therefore, the entropy loss for population-selection is larger than that for induced-folding.

We should note that the conformational variety of the unbound IDP also affects the binding mechanism. When a native-like form (i.e. the form in the native complex) has a large probability in the unbound state, this situation increases the advantage of population-selection. If the unbound state is restricted in small volumes in the conformational space (or the unbound state consists of some restricted conformational clusters), then the entropy of the unbound state is small. Then the entropy cost $-T\Delta S = -T(S_{\text{bound}} - S_{\text{unbound}})$ upon binding is also small. In other words, the restricted unbound state facilitates IDP-partner binding in either the induced-folding or population-selection mechanism.

The "conformational flexibility" is an important keyword to characterize IDP for both of induced-folding and population-selection. The flexibility in the bound state is more important in induced-folding than in population-selection. I.e. in the induced-folding mechanism the bound-IDP conformation fluctuates largely without dissociation, and then the entropic loss upon binding is small. This small entropic loss may ease the complex formation because a small enthalpy gain can compensate the small entropic loss. In population-selection, contrarily, the large entropy loss upon binding should be compensated by a large enthalpy gain.

Our all-atom McMD simulations of the NRSF–mSin3 system has suggested that the main mechanism for coupled folding and binding of this system is induced-folding. However, the existence of free-energy barriers in the bound state and the existence of helical conformations in the unbound state suggest that population-selection works as a secondary mechanism.

Last in this chapter, we note that the coarse-grained model is also useful to rationally explain coupled folding and binding, as done by Okazaki and Takada [45], and Terakawa and Takada [56]. Recently, Matsushita and Kikuchi [37] have proposed a novel picture for intrinsic disorder, where structural frustrations are designed for inducing the intrinsic disorder. We believe that both the all-atom and coarse-grained models are useful to understand the mechanism of IDP-partner binding.

Acknowledgements Both authors were supported by a Grant-in-Aid for Scientific Research on Innovative Areas (21113006) from the Ministry of Education, Culture, Sports, Science and Technology (MEXT) Japan. J.H. was supported by grants from the New Energy and Industrial Technology Development Organization (NEDO) Japan.

References

1. Berg BA, Neuhaus T (1992) Multicanonical ensemble: a new approach to simulate first-order phase transitions. Phys Rev Lett 68:9–12
2. Bosshard HR (2001) Molecular recognition by induced fit: how fit is the concept? News Physiol Sci 16:171–173
3. Chen J (2009) Intrinsically disordered p53 extreme C-terminus binds to S100B(ββ) through "fly-casting". J Am Chem Soc 131:2088–2089
4. Cornell WD, Cieplak P, Bayly CI, Gould IR, Merz KM, Ferguson DM, Spellmeyer DC, Fox T, Caldwell JW, Kollman PA (1995) A second generation force field for the simulation of proteins, nucleic acids, and organic molecules. J Am Chem Soc 117:5179–5197
5. De Guzman RN, Goto NK, Dyson HJ, Wright PE (2006) Structural basis for cooperative transcription factor binding to the CBP coactivator. J Mol Biol 355:1005–1013
6. Ding H-Q, Karasawa N, Goddard WA (1992) Atomic level simulations on a million particles: the cell multipole method for Coulomb and London nonbond interactions. J Chem Phys 97:4309–4315
7. Domenico TD, Walsh I, Martin AJM, Tosatto SCE (2012) MobiDB: a comprehensive database of intrinsic protein disorder annotations. Bioinformatics 28:2080–2081
8. Dunker AK, Cortese MS, Romero P, Iakoucheva LM, Uversky VN (2005) Flexible nets. The roles of intrinsic disorder in protein interaction networks. FEBS J 272:5129–5148
9. Dyson HJ, Wright PE (2005) Intrinsically unstructured proteins and their functions. Nat Rev Mol Cell Biol 6:197–208
10. Fukuchi S, Homma K, Minezaki Y, Gojobori T, Nishikawa K (2009) Development of an accurate classification system of proteins into structured and unstructured regions that uncovers novel structural domains: its application to human transcription factors. BMC Struct Biol 9:26
11. Fukuchi S, Homma K, Sakamoto S, Sugawara H, Tateno Y, Gojobori T, Nishikawa K (2009) The GTOP database in 2009: updated content and novel features to expand and deepen insights into protein structures and functions. Nucleic Acids Res 37:D333–D337
12. Fukuchi S, Sakamoto S, Nobe Y, Murakami DS, Amemiya T, Hosoda K, Koike R, Hiroaki H, Ota M (2012) IDEAL: intrinsically disordered proteins with extensive annotations and literature. Nucleic Acids Res 40:D507–D511
13. Fuxreiter M, Tompa P, Simon I, Uversky VN, Hansen JC, Asturias FJ (2008) Malleable machines take shape in eukaryotic transcriptional regulation. Nat Chem Biol 4:728–737
14. Go N (1983) Theoretical studies of protein folding. Annu Rev Biophys Bioeng 12:183–210
15. Goto NK, Zor T, Martinez-Yamout M, Dyson HJ, Wright PE (2002) Cooperativity in transcription factor binding to the coactivator CREB-binding protein (CBP). J Biol Chem 277:43168–43174
16. Hansmann UHE, Okamoto Y (1993) Prediction of peptide conformation by multicanonical algorithm: new approach to the multiple-minima problem. J Comput Chem 14:1333–1338
17. Hansmann UHE, Okamoto Y, Eisenmenger F (1996) Molecular dynamics, Langevin, and hybrid Monte Carlo simulations in multicanonical ensemble. Chem Phys Lett 259:321–330
18. Higo J, Ikebe J, Kamiya N, Nakamura H (2012) Enhanced and effective conformational sampling of protein molecular systems for their free energy landscapes. Biophys Rev 4: 27–44
19. Higo J, Ito N, Kuroda M, Ono S, Nakajima N, Nakamura H (2001) Energy landscape of a peptide consisting of α-helix, 3_{10}-helix, β-turn, β-hairpin, and other disordered conformations. Protein Sci 10:1160–1171
20. Higo J, Kamiya N, Sugihara T, Yonezawa Y, Nakamura H (2009) Verifying trivial parallelization of multicanonical molecular dynamics for conformational sampling of a polypeptide in explicit water. Chem Phys Lett 473:326–329
21. Higo J, Nishimura Y, Nakamura H (2011) A free-energy landscape for coupled folding and binding of an intrinsically disordered protein in explicit solvent from detailed all-atom computations. J Am Chem Soc 133:10448–10458

22. Huang Y, Liu Z (2009) Kinetic advantage of intrinsically disordered proteins in coupled folding–binding process: a critical assessment of the "fly-casting" mechanism. J Mol Biol 393:1143–1159
23. Huang Y, Liu Z (2011) Anchoring intrinsically disordered proteins to multiple targets: lessons from N-terminus of the p53 protein. Int J Mol Sci 12:1410–1430
24. Iakoucheva LM, Brown CJ, Lawson JD, Obradovic Z, Dunker AK (2002) Intrinsic disorder in cell signaling and cancer-associated proteins. J Mol Biol 323:573–584
25. Ikebe J, Kamiya N, Shindo H, Nakamura H, Higo J (2007) Conformational sampling of a 40-residue protein consisting of α and β secondary-structure elements in explicit solvent. Chem Phys Lett 443:364–368
26. Ikebe J, Standley DM, Nakamura H, Higo J (2011) Ab initio simulation of a 57-residue protein in explicit solvent reproduces the native conformation in the lowest free-energy cluster. Protein Sci 20:187–196
27. Ikebe J, Umezawa K, Kamiya N, Sugihara T, Yonezawa Y, Takano Y, Nakamura H, Higo J (2011) Theory for trivial trajectory parallelization of multicanonical molecular dynamics and application to a polypeptide in water. J Comput Chem 32:1286–1297
28. Ikeda K, Galzitskaya OV, Nakamura H, Higo J (2003) β-hairpins, α-helices, and the intermediates among the secondary structures in the energy landscape of a peptide from a distal β-hairpin of SH3 domain. J Comput Chem 24:310–318
29. James LC, Tawfik DS (2003) Conformational diversity and protein evolution–a 60-year-old hypothesis revisited. Trends Biochem Sci 28:361–368
30. Jorgensen WL, Chandrasekhar J, Madura JD, Impey RW, Klein ML (1983) Comparison of simple potential functions for simulating liquid water. J Chem Phys 79:926–935
31. Kamiya N, Higo J, Nakamura H (2002) Conformational transition states of beta-hairpin peptide between the ordered and disordered conformations in explicit water. Protein Sci 11:2297–2307
32. Kamiya N, Watanabe YS, Ono O, Higo J (2005) AMBER-based hybrid force field for conformational sampling of polypeptides. Chem Phys Lett 401:312–317
33. Kamiya N, Yonezawa Y, Nakamura H, Higo J (2008) Protein-inhibitor flexible docking by a multicanonical sampling: native complex structure with the lowest free energy and a free-energy barrier distinguishing the native complex from the others. Proteins 70:41–53
34. Kidera A (1995) Enhanced conformational sampling in Monte Carlo simulations of proteins: applications to a constrained peptide. Proc Natl Acad Sci USA 92:9886–9889
35. Kollman PA, Dixon RW, Cornell WD, Chipot C, Pohorille A (1997) The development/application of a 'minimalist' organic/biochemical molecular mechanic force field using a combination of Ab initio calculations and experimental data. In computer simulations of biological systems. Comput Simul Biomol Syst Theor Exp Appl 3:83–96
36. Lawrence CW, Showalter SA (2012) Carbon-detected ^{15}N NMR spin relaxation of an intrinsically disordered protein: FCP1 dynamics unbound and in complex with RAP74. J Phys Chem Lett 3:1409–1413
37. Matsushita K, Kikuchi M (2013) Frustration-induced protein intrinsic disorder. J Chem Phys 138:105101
38. Metallo SJ (2010) Intrinsically disordered proteins are potential drug targets. Curr Opin Chem Biol 14:481–488
39. Monod J, Wyman J, Changeux JP (1965) On the nature of allosteric transitions: a plausible model. J Mol Biol 12:88–118
40. Morikami K, Nakai T, Kidera A, Saito M, Nakamura H (1992) PRESTO (PRotein Engineering SimulaTOr): a vectorized molecular mechanics program for biopolymers. Comput Chem 16:243–248
41. Moritsugu K, Terada T, Kidera A (2012) Disorder-to-order transition of an intrinsically disordered region of sortase revealed by multiscale enhanced sampling. J Am Chem Soc 134:7094–7101
42. Nakajima N, Nakamura H, Kidera A (1997) Multicanonical ensemble generated by molecular dynamics simulation for enhanced conformational sampling of peptides. J Phys Chem 101:817–824

43. Nomura M, Uda-Tochio H, Murai K, Mori N, Nishimura Y (2005) The neural repressor NRSF/REST binds the PAH1 domain of the Sin3 corepressor by using its distinct short hydrophobic helix. J Mol Biol 354:903–915
44. Oates ME, Romero P, Ishida T, Ghalwash M, Mizianty MJ, Xue B, Dosztányi Z, Uversky VN, Obradovic Z, Kurgan L, Dunker AK, Gough J (2013) D^2P^2: database of disordered protein predictions. Nucleic Acids Res 41(Database issue):D508–D516
45. Okazaki K, Takada S (2008) Dynamic energy landscape view of coupled binding and protein conformational change: induced-fit versus population-shift mechanisms. Proc Natl Acad Sci USA 105:11182–11187
46. Oldfield CJ, Cheng Y, Cortese MS, Brown CJ, Uversky VN, Dunker AK (2005) Comparing and combining predictors of mostly disordered proteins. Biochemistry 44:1989–2000
47. Patil A, Kinoshita K, Nakamura H (2010) Domain distribution and intrinsic disorder in hubs in the human protein–protein interaction network. Protein Sci 19:1461–1468
48. Patil A, Kinoshita K, Nakamura H (2010) Hub promiscuity in protein-protein interaction networks. Int J Mol Sci 11:1930–1943
49. Radhakrishnan I, Pérez-Alvarado GC, Dyson HJ, Wright PE (1998) Conformational preferences in the ser133-phosphorylated and non-phosphorylated forms of the kinase inducible transactivation domain of CREB. FEBS Lett 430:317–322
50. Radhakrishnan I, Pérez-Alvarado GC, Parker D, Dyson HJ, Montminy MR, Wright PE (1997) Solution structure of the KIX domain of CBP bound to the transactivation domain of CREB: a model for activator:coactivator interactions. Cell 91:741–752
51. Sancho DD, Best RB (2012) Modulation of an IDP binding mechanism and rates by helix propensity and non-native interactions: association of HIF1α with CBP. Mol Biosyst 8: 256–267
52. Shoemaker BA, Portman JJ, Wolynes PG (2000) Speeding molecular recognition by using the folding funnel: the fly-casting mechanism. Proc Natl Acad Sci USA 97:8868–8873
53. Sickmeier M, Hamilton JA, LeGall T, Vacic V, Cortese MS, Tantos A, Szabo B, Tompa P, Chen J, Uversky VN, Obradovic Z, Dunker AK (2007) DisProt: the database of disordered proteins. Nucleic Acids Res 35:D786–D793
54. Spolar RS, Record MT Jr (1994) Coupling of local folding to site-specific binding of proteins to DNA. Science 263:777–784
55. Sugase K, Dyson HJ, Wright PE (2007) Mechanism of coupled folding and binding of an intrinsically disordered protein. Nature 447:1021–1025
56. Terakawa T, Takada S (2011) Multiscale ensemble modeling of intrinsically disordered proteins: p53 N-terminal domain. Biophys J 101:1450–1458
57. Tompa P, Fuxreiter M (2008) Fuzzy complexes: polymorphism and structural disorder in protein–protein interactions. Trends Biochem Sci 33:2–8
58. Umezawa K, Ikebe J, Takano M, Nakamura H, Higo J (2012) Conformational ensembles of an intrinsically disordered protein pKID with and without a KIX domain in explicit solvent investigated by all-Atom multicanonical molecular dynamics. Biomolecules 2:104–121
59. Wang J, Wang Y, Chu X, Hagen SJ, Han W, Wang E (2011) Multi-scaled explorations of binding-induced folding of intrinsically disordered protein inhibitor IA3 to its target enzyme. PLoS Comput Biol 7:e1001118
60. Ward JJ, Sodhi JS, McGuffin LJ, Buxton BF, Jones DT (2004) Prediction and functional analysis of native disorder in proteins from the three kingdoms of life. J Mol Biol 337:635–645
61. Wostenberg C, Kumar S, Noid WG, Showalter SA (2011) Atomistic simulations reveal structural disorder in the RAP74-FCP1 complex. J Phys Chem B 115:13731–13739
62. Wright PE, Dyson HJ (1999) Intrinsically unstructured proteins: re-assessing the protein structure-function paradigm. J Mol Biol 293:321–331
63. Yamane T, Okamura H, Nishimura Y, Kidera A, Ikeguchi M (2010) Side-chain conformational changes of transcription factor PhoB upon DNA binding: a population-shift mechanism. J Am Chem Soc 132:12653–12659
64. Zor T, Mayr BM, Dyson HJ, Montminy MR, Wright PE (2002) Roles of phosphorylation and helix propensity in the binding of the KIX domain of CREB-binding protein by constitutive (c-myb) and inducible (CREB) activators. J Biol Chem 277:42241–42248

Chapter 15
Coordination and Control Inside Simple Biomolecular Machines

Jin Yu

Abstract Biomolecular machines can achieve physiological functions precisely and efficiently, though they always operate under fluctuations and noises. We review two types of simple machinery that we have recently studied. The machinery can be regarded as molecular motors. They transform chemical free energy from NTP hydrolysis to mechanical work. One type belongs to small monomeric helicases that move directionally along single-stranded nucleic acid, and may further unwind the duplex part for gene replication or repair. The other type belongs to ring-shaped NTPase motors that also move or transport nucleic acid or protein substrate in a directional manner, such as for genome packaging or protein degradation. The central issue in this review is on how the machinery coordinates essential degrees of freedom during the mechanochemical coupling process. Further concerns include how the coordination and control are manifested in experiments, and how they can be captured well in modeling and computational research. We employed atomistic molecular dynamics simulations, coarse-grained analyses, and stochastic modeling techniques to examine the molecular machines at multiple resolutions and timescales. Detailed descriptions on how the protein interacts with its substrate at interface, as well as how multiple protein subunits are coordinated are summarized.

Keywords Biomolecular machine • Molecular motor • NTP hydrolysis • Mechanochemistry • Nucleic acid motor • MD simulation • Stochastical modeling • Single molecule experiment

---

J. Yu (✉)
Beijing Computational Science Research Center, No 3 Heqing Road, Haidian District, Beijing 100084, China
e-mail: jinyu@csrc.ac.cn

K. Han et al. (eds.), *Protein Conformational Dynamics*, Advances in Experimental Medicine and Biology 805, DOI 10.1007/978-3-319-02970-2_15,

15.1 Introduction

Biomolecular machines are microscopic counterparts of human-made machines that play diverse functions in living organisms. They are indispensible to genetic control, molecular transport, cell movements and a variety of metabolisms. Due to their small sizes and ambient solution conditions, the physics that governs the operations of the molecular machines appears quite different from that of the macroscopic machines [1, 2]. Essentially, viscosity forces and accompanied thermal fluctuations are significantly high in the small systems, such that stochasticity becomes critical for the operation. Viewed from atomistic details, the molecular machines are extremely complex entities that organize a large number of degrees of freedom. How to manipulate so many degrees in coordination to achieve precise control, in the presences of fluctuations and noises, is fundament to understand the operation mechanisms of the systems.

In this review, we focus on two types of simple machinery, the monomeric and ring-shaped NTPase motors that we have recently studied [3–6]. The molecular motors consume free energy from NTP hydrolysis to achieve directional movements along molecular tracks. They are self-sustaining *mechanochemical* systems. The monomeric ATPase motors can be regarded as smallest molecular machines [7]. The single peptide chain ATPase can fold into several structural subdomains, among which there locates a single chemical site for ATP binding and hydrolysis. The protein coordination, therefore, targets solely on linking the chemical site signal with subdomain conformational changes. The ring-shaped NTPase motor [8], however, consists of several protein subunits around the ring, with each of them a single polypeptide chain. The chemical sites are located at the interfaces between two neighboring subunits, and there are several of the sites. The multiple subunits along with the chemical sites need substantial coordination around the ring to ensure they work cooperatively.

From comparative genomic studies, the NTPase motors are affiliated to ASCE division of P-loop NTPases [9, 10], characterized by conserved NTP binding motif (Walker A), Mg^{2+} binding motif (Walker B), and some additional conserved residues in Walker B and related region. This division of proteins includes many helicases, packaging ATPases or terminases, pilus retraction motor proteins, ABC transporters, etc. Among them we have studied monomeric helicases, as well as viral DNA packaging motors that are multimeric and ring-shaped. The helicases and packaging motors use DNA or RNA as their molecular tracks to translocate linearly. It is interesting to notice that some evolutionary links may exist between the linear motors and an exemplary rotary motor, F- or V-type ATPase [11], which forms a trimer-of-dimer ring and encloses a protein substrate/track (γ subunit) for torque generation.

Experimental studies have improved substantially our knowledge on these motor proteins [7, 8, 12, 13]. In particular, single molecule experimental technologies have been implemented to monitor the functional motions of the stepping motors, one molecule at time, as well as to detect force responses of the molecules at real time [13, 14]. The information on dynamics of the motor, extracted from the experimental measurements, is ready to be incorporated into computational work to reveal mechanisms underlying. On the other hand, molecular dynamics (MD) simulations based on high-resolution structures can zoom into atomistic details of the system [15]. Nevertheless, as even the smallest molecular machines solved within water amount to ~100,000 atoms, MD simulations can barely reach microsecond (μs) time scale, far below the millisecond (ms) time resolution of single molecule experiments. Indeed, molecular machines usually work across time scale of several orders of magnitude, in particular, mechanochemical coupling involves degrees of freedom from supporting fast catalysis (femtosecond to picosecond) to propagating slow elastic deformations (μs to ms). Hence, it is expected that computational techniques should also accommodate different resolution and time scales when studying molecular machines. In this review, we summarize mainly our computational studies on some of nucleic acid motors, combining atomistic MD simulations, coarse-grained analyses, kinetic or stochastic modeling, etc. We want to demonstrate how to use a variety of computational techniques to fully take advantage of the experimentally detected information on structure and dynamics, and to discover how the protein machines deal with internal complexities amongst intrinsic and environmental noises.

15.2 Monomeric Helicases – Directionality Control and Active Unwinding

In this section we discuss how small helicase motor proteins move directionally along single-stranded (ss) DNA or RNA as well as how they further unwind the double-stranded (ds) DNA/RNA [7]. Basically, the motor proteins use free energy from ATP hydrolysis to support translocation as well as duplex unwinding. Helicases are involved in almost all aspects of DNA/RNA metabolism, including transcription, replication, and recombination. Defects in helicase function in humans can lead to genomic instability and pre-disposition to cancer [16].

We will introduce systematic studies on translocation of PcrA helicase from electronic to functional level. PcrA from bacteria is a monomeric ATPase that belongs to helicase superfamily 1 (SF1) [12]. As one of the smallest molecular motors, PcrA helicase is weighted about 80 kDa, and folds into four structural subdomains (1A, 2A, 1B and 2B). ATP binds into a cleft between the two RecA-like subdomains 1A and 2A, and modulates the alternative domain movements. We start by comparing the ATP binding site and hydrolysis characters of PcrA with that of F1-ATPase. Then we will focus more on discussing how the domain movements are coordinated for directional translocation, using classical MD, stochastic modeling, along with coarse graining and coevolutionary statistical analyses.

Following the studies of helicase translocation, we then investigated duplex unwinding of nucleic acids using a similar small helicase NS3, which belongs to helicase superfamily 2 (SF2). NS3 had been particularly studied by single molecule experiments on its unwinding properties [17–19]. We hence modeled NS3 unwinding mainly based on experimental data.

15.2.1 PcrA Helicase – Alternative Subdomain Stepping Coordinated by ATP Cycle

PcrA (plasmid copy reduced) is essential for plasmid rolling circle replication and repair of DNA damage in Gram-positive bacteria. It preferably binds to ssDNA containing 3′ overhangs, and moves directionally from 3′ to 5′ of the DNA strand. In this section, we illustrate how this directionality is maintained in molecular details. PcrA helicase had been crystalized and solved with high-resolution structures [20], in both a substrate (ATP bound) and a product (without ATP/ADP) state. The structure of the substrate state, in complex with a duplex DNA flanked by a piece of 3′ ssDNA, is shown in Fig. 15.1. Comparing to the product structure in the absence of ATP or ADP, the subdomains 1A and 2A come close to each other when ATP is bound in between.

15.2.1.1 ATP Hydrolysis

It has been noticed that high structural similarity exists between ATP binding or catalytic sites of PcrA and F1-ATPase. Both systems are RecA-like ATPases. The structural alignment between 1A and 1B subdomains of PcrA and β_{TP} domain of F1-ATPase, taken from [21], is shown in Fig. 15.1. Both catalytic sites have a RecA-like fold consisting of a central β sheet adjoined by α helices on both sides. The bound nucleotide is located at the interface between 1A and 2A subdomains in PcrA, while in F1-ATPase, the interface is provided by adjacent α and β subunits. Sequence alignment based on the structural fitting is also shown, with the upper panel the Walker A motif, the lower panel the Walker B motif. It is remarkable that PcrA and F1-ATPase show high degree of structural similarity though their sequence homology is fairly low.

In simulating the substrate state of PcrA using QM/MM, it was found that ATP hydrolysis reaction proceeds under a proton reply mechanism and is endothermic, with a product state energy of ~10 kcal/mol and a moderate transition state barrier of ~20 kcal/mol [21]. Similar mechanism and endothermicity was found in the F1-ATPase β_{TP} site that binds tightly with ATP [22, 23]. Simulations of F1 suggested that movement of the arginine finger residue αR373 toward the γ-phosphate group is necessary to convert the endothermic reaction to one that efficiently hydrolyses ATP with an equilibrium constant of $K \sim 1$. In PcrA, the arginine finger residue R287 seems properly positioned, yet another close-by arginine residue R601 was suspected to pose an unfavorable configuration. A slightly more closing between subdomains 1A and 2A seems to be required to bring R601 closer to ATP, and

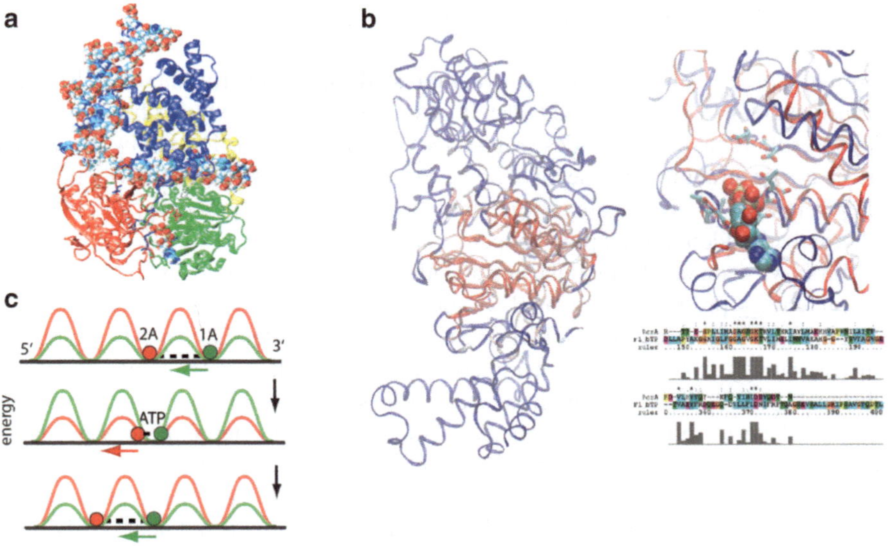

Fig. 15.1 PcrA helicase, its structural alignment with F1-ATPase, and the translocation model. (**a**) The structure of PcrA in complex with DNA and ATP [20]. The image is from [3]. The protein domains are in cartoon presentation (*red*, 2A domain; *green*, 1A; *blue*, 2B; *yellow*, 1B), along with DNA (van der Waals, vdW presentation: *red*, oxygen; *cyan*, carbon; *blue*, nitrogen; *tan*, phosphorus; *white*, hydrogen); the duplex DNA is bound to the *top left* of PcrA and is flanked by a 3′ ssDNA that crosses through the *middle* of PcrA from *left* to *right*. The ATP (analog) is bound in between domains 1A and 2A. (**b**) Structural alignment of PcrA and F1-ATPase. The images and captions are adopted from [21]. Depicted on the *left* are the aligned structures of the 1A and 1B domains of PcrA and the β_{TP} subunit of F1-ATPase (the color scheme from *red* to *white* to *blue* indicates a structural alignment quality ranging from good to weak). Shown on the *top right* is a close-up view of the aligned catalytic sites of PcrA (*red*) and F1-ATPase (*blue*) with bound ADP in vdW representation; conserved residues (Walker motifs A and B) are highlighted in licorice representation. Shown on the *bottom right* are the two parts of the sequence alignment that contain a stretch of two or more consecutive conserved residues; the *upper panel* corresponds to the Walker A or P loop motif, and the *lower panel* corresponds to the Walker B motif. (**c**) The simplified unidirectional translocation model of PcrA helicase [3, 4]. The image is from [4]. The *green/red bead* represents domain 1A/2A. The link between 1A and 2A hints for their association (e.g. an elastic spring). When there is no ATP, the two domains are separated. ATP binding draws the two domains close to each other. The *green/red curve* shows the potential of mean force of domain 1A/2A moving along the ssDNA, with periodicity in 1-nt distance

hence, to a more efficient ATP hydrolysis [21]. It was further proposed that there is a glutamine residue Q254 in the catalytic site of PcrA playing a role of ‘ssDNA sensor’. As mutations of Q254 significantly change the reaction profiles of ATP hydrolysis [21], Q254 was considered to be involved in the coupling of conformational changes induced by ssDNA base flipping into the Y257/F64 pocket to ATP hydrolysis in the catalytic site. The base flipping is an essential element in ssDNA translocation. The study gives a sneak peek on how chemical catalysis is coupled to amino acids movements.

15.2.1.2 Alternating Domain Mobilities

From the previous high-resolution structural studies, 'inchworm' model of helicase translocation had been proposed [20]. It was suggested that the two subdomains (1A and 2A) alternate their affinities with ssDNA in translocation. The inchworm model gives a nice structural perspective, however, it was lack of energetic evidence. We therefore tired to investigate energetically how 'inchworm' proceeds, i.e., how the alternative subdomain stepping happens in PcrA [3].

First, we decipher the model using a mathematical presentation, with two translocation energy profiles, or potentials of mean force (PMF) for both subdomains 1A and 2A (green and red in Fig. 15.1), respectively. The potential is a 1-D projection of free energy from high-dimension coordinate space. It is with 1-nucleotide (nt) distance periodicity as the helicase translocates, presumably, 1-nt at a time and repeats. Due to intrinsic stochasticity of the system, the stepping does not have to be fully synchronized with the ATP cycle. There can be 'diffusional' stepping without ATP participation, and there can also be futile ATP hydrolysis cycles without stepping. That says, the mechanochemical coupling is not necessarily tight or perfect. Besides, we assume that the potential inside 1-nt period is symmetric, or say, moving forward 1/2 nt is identically easy as moving backward 1/2 nt, though it does not have to be this way. Periodic yet asymmetric saw-tooth potentials have been widely used to describe ratcheting molecular motors [24]. The asymmetry in those potentials is essential in conducting the directionality, as moving forward is preferred to backward in transiting from a smooth potential to the saw-tooth potential. In our alternative domain stepping model, however, the directionality is not convoyed through any asymmetry in individual domain potentials (though the potential for the two subdomains combined does show an asymmetric shape and biases forward rather than backward, see figure 6 in reference [3]). As we illustrate below, the directionality in current model comes from alternating mobilities of respective domains coordinated by ATP hydrolysis cycle.

The mobilities of 1A and 2A are measured according to the potential or barrier height for each of subdomain to move 1-nt along ssDNA. The symmetry makes the barrier to be located at the center of the potential. The higher the barrier, the less mobile the (sub)domain. Hence, the mathematical presentation demonstrated in Fig. 15.1 delivers this scenario: When there is no ATP, 1A (green bead) and 2A (red bead) are fairly separated and 1A has a higher mobility (green potential lower than red one). As ATP binds in between 1A and 2A, it draws two subdomains closer by either having 1A moving forward 1-nt or 2A backward 1-nt. Since 1A has higher mobility than 2A at this moment, 1A moving forward dominates 2A moving backward. The domain mobilities, however, switch after ATP binds in the substrate state such that 2A becomes more mobile (red potential lower than the green one). The property persists through ATP hydrolysis to the moment that ADP/Pi product releases. The release relaxes the two subdomains so they separate again from each other, by either 2A moving forward or 1A moving backward. Since now 2A has a higher mobility, it moves forward faster and dominates to 1A backward. In this fashion, the helicase alternatively moves 1A and 2A forward during ATP binding and product release phases. The ATP hydrolysis reaction seems to play little role energetically to assist directional translocation. This seems to be consistent with a view that ATP hydrolysis reaction itself hardly releases much free energy (an equilibrium constant $K \sim 1$).

Then comes the next question, how to calculate the domain mobility potentials for 1A and 2A respectively? And what is the relationship between the mobility and the affinity of the domain with ssDNA? Intuitively, higher mobility would relate to lower affinity. We explored structural and energetic details behind this using atomistic simulations, summarized as below.

15.2.1.3 Determining Domain Mobility via Binding Affinities of Individual Nucleotides

The domain mobility is defined as the helicase moves *along* ssDNA, while the affinity between the protein domain and ssDNA is defined as protein approaches 'vertically' to bind to the DNA strand. Hence, the two quantities are considered through two 'orthogonal' directions, tangential to the DNA strand and perpendicular to the strand. Our calculations using MD were based on the idea that the protein domain mobility along the ssDNA can be measured through the change of protein-DNA affinities along the DNA strand. As the translocation is periodic, we only consider how the affinities change within 1-nt distance.

The protein-DNA affinity is measured as the binding free energy between the protein domain and a stretch of bound ssDNA, which consists of 5–6 individual nucleotides. If one knows a 'continuous' binding free energy curve along the ssDNA, as if a single nucleotide associates with the protein at different locations along the DNA strand, one can then calculate the domain translocation barrier, or the mobility: It can calculated as the sum of *changes* of binding free energies for all the 5 or 6 nucleotides, as the nucleotides move collectively along the strand through the protein [3].

However, the continuous binding free energy does not exist, as to obtain it one has to simulate the full translocation process or drag one nucleotide slowly enough along the DNA strand across the helicase. Since the translocation takes place in milliseconds, while atomistic MD simulations are limited by nanoseconds, direct simulation is impossible. Though steered simulation can accelerate the process, it is extremely hard to steer and coordinate several slow degrees of freedom involved in the helicase translocation, without artificially distorting the structure or energetics. Hence, what we really did was to estimate the binding free energy curves to come up with the domain mobility measurements: We calculated relative binding free energies of the 5–6 individual nucleotides along the DNA strand with the protein at one equilibrium configuration; then we 'guessed' the full binding free energy curve using data interpolation, with a free parameter determining the 'roughness' of the binding free energy curve. As we assumed that the translocation potential bears a symmetric shape, what we calculated are essentially the heights of translocation barriers for subdomain 1A and 2A [3].

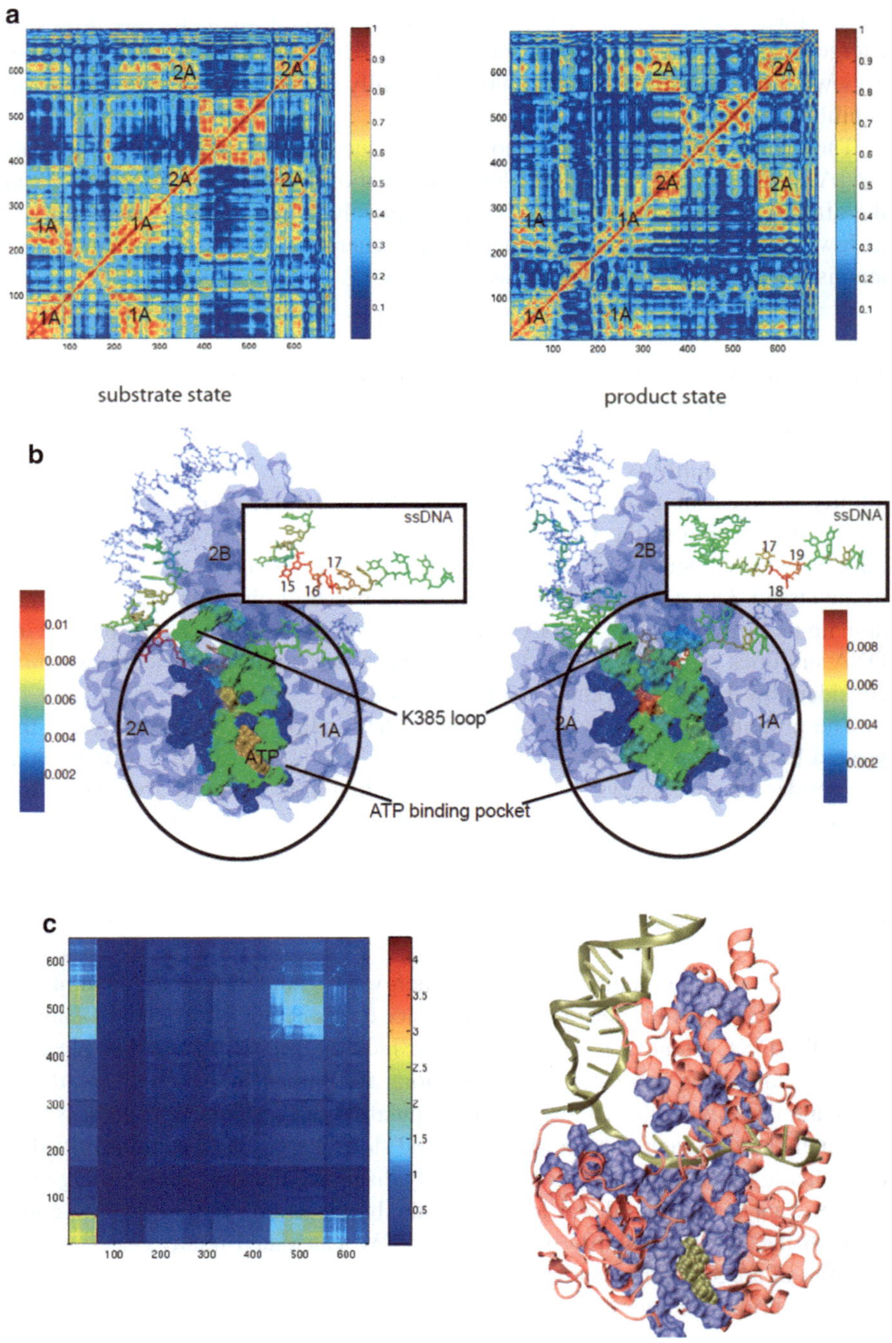
a
2A
1A
substrate state
product state
b
ssDNA
15
16
17
18
19
2B
2A
1A
ATP
K385 loop
ATP binding pocket
c

The calculations were conducted through the atomistic MD simulations, with PcrA helicase solved with explicit water molecules and ions (~110,000 atoms). The resulted two potential profiles for 1A and 2A, at both the substrate and the product states, are shown already in Fig. 15.1. From the calculations, we estimated the height of translocation barrier, subjecting to the undetermined interpolation parameter. Experimentally, it is known that the PcrA translocates at about 50 nt/s [25]. If the translocation is rate-limited by domain movements, one can fit the parameter by the translocation speed. However, it is not quite clear if chemical transitions such as ATP binding, hydrolysis, or product releases actually limit the speed. Hence, our calculations were still semi-quantitative. Nevertheless, the key results from the calculations are: First, we demonstrated the alternating 'asymmetry' in domain mobilities of 1A and 2A, with 1A more mobile in the product state, and 2A more mobile in the substrate state. Second, we identified essential amino acids that contribute most to the alternating domain mobilities thus the directionality. In brief, the studies provide energetic as well as structural details for directional mechanisms of a prototypical stepping motor [3].

15.2.1.4 Correlations Inside Protein and at Protein-DNA Interface

Based on MD simulation results, we can directly monitor how different parts of protein-DNA complex are correlated during the simulation period. The correlations are key in coordinating different essential degrees of freedom to achieve functional control in the molecular machine. First, we analyzed cross-correlations between any pair of residues using trajectories of Cα atoms from protein and P atoms from

Fig. 15.2 Correlation and coupling analyses from MD simulation, the elastic network model, and multi-sequence alignments. (**a**) Cross-correlation maps calculated from MD simulations of PcrA helicase complex, in both the substrate state with ATP bound (*left*) and the product state without ATP/ADP bound (*right*). The images are adopted from [3]. The maps are colored according to the amplitude of the cross-correlation matrix elements. (**b**) The coupling analyses based on the elastic network model of the PcrA-DNA complex, for both the substrate state (*left*) and the product state (*right*). The images are adopted from [3]. The complexes are colored according to the dynamical coupling of residues to the fluctuations of the ATP binding pocket. The dynamic coupling is probed through perturbation of one residue's spring constant and monitoring the ensuing effect on the vibrational fluctuation of the other interested site [26]. The protein, DNA, and ATP are shown in surface, licorice, and vdW presentations, respectively. (**c**) Co-evolutionary analyses for pair-wise mutational correlations between residues based on multi-sequence alignments of PcrA-related helicases. The images are adopted from [4]. The co-evolutionary statistical analyses, developed by [31, 32], were performed to sequences of over 800 proteins related to PcrA. The map on the *left* is colored according to the correlation matrix elements that describe the mutational coupling strength between two residues in the sequence alignment. This correlation map has been rearranged employing a procedure that clusters highly correlated core residues (see the *brightest square* region on *bottom left* of the map). The core residues are illustrated in *blue* surface representation in the PcrA-DNA complex on the *right* nucleic acids. The correlation map showing the amplitudes (absolute values) of the cross-correlation of the protein-DNA complex are displayed in Fig. 15.2, for both the substrate and product states. We see that in the product state (*right*), when subdomain 1A is more mobile than 2A, the correlation inside 1A is smaller than that inside 2A. While in the substrate state (*left*), when 1A is less mobile, the correlation inside 1A becomes larger than that inside 2A. The asymmetry from the internal correlation map suggests that when a structural domain is fairly 'rigid', i.e., with strong correlations or high collectivity inside the domain, the domain is also held tight by the DNA and is not very mobile. Vice versa, a mobile domain seems to have fairly low collectivity inside. Nevertheless, the mobility character would be considered more relevant at the interface between the motor protein and the DNA track. Our further cross-correlation analyses focusing on protein-DNA interaction did show that when 1A/2A is less mobile (in the substrate/product state), its correlation with the DNA segment on top of it is fairly high. The correlation is particularly for movements parallel to the protein-DNA interface rather than perpendicular to the interface [4]. So one sees that domain mobility is tuned through protein-DNA interactions at atomic level and links allosterically to internal domain collectivity. One also sees that the asymmetrical correlation pattern has already been manifested at nanoseconds time scale as our simulation conducted.

Next we tried to explore correlation properties of protein-DNA at longer time scales, as real translocation step takes milliseconds to happen. We implemented a method developed by [26] that analyzes 'dynamical coupling' in an elastic network model (ENM) of the protein-DNA complex. The ENM is a highly simplified model that has been widely used to describe large and slow conformational changes of protein. The model basically uses one node to represent each residue, and connects any pair of nodes within some cutoff distance by an elastic spring of the same strength [27]. The simple model can nevertheless describe well some experimental observables, such as B-factors. Combining with normal model analysis (NMA), ENM can predict large conformational changes of protein that overlap fairly well with real changes in many cases. The dynamical coupling is probed through perturbing a residue's spring constant in ENM and monitoring the ensuring effect on vibrational fluctuation of some other residue(s) [26]. The larger the fluctuation effect the higher the coupling is between the two parts. In particular, shown in Fig. 15.2, is the coupling pattern between ATP binding pocket and any residue in the protein-DNA complex, in both the substrate and product states. The purpose is to find out which part of the protein-DNA complex couples tightly to the chemical catalytic site. From analyses above, we know that in the substrate state, subdomain 2A is more mobile. We notice that in the dynamical coupling analyses, the DNA segment on top of 2A strongly couples with the catalytic site bound with ATP. While in the product state, 1A is more mobile, and the DNA segment on top of 1A strongly couples to the catalytic site, while there is no ATP/ADP bound. Hence, one infers that the domain mobility is somehow controlled by elastic couplings of the DNA segments to the chemical catalytic site. Without ATP, the catalytic site 'holds' the DNA nucleotides that are in contact with 1A such that the protein-ssDNA affinities are fairly low there, and therefore, 1A is quite mobile. When ATP binds, it changes the coupling pattern through allosteric interactions, and the DNA nucleotides in contact with 2A are 'held' by the catalytic site, 2A thus becomes fairly mobile.

Hence, we see that a highly coarse-grained network model is able to catch some mechanochemical coupling properties, which are crucial for directional control. Interestingly, we noticed that the above coupling pattern disappears when $3'$–$5'$ polarity of the strand is artificially switched in the simulation. Indeed, PcrA helicase does not bind well to a $5'$ piece of ssDNA. The coupling pattern keeps robust, however, when only the DNA sequences are changed (from poly-T to poly-C) [4]. The observations support the idea that it is mainly the protein-DNA backbone interactions that direct the coupling asymmetries, while ATP hydrolysis cycle alters the asymmetries in between the two subdomains, thus driving the directional domain stepping.

15.2.1.5 Co-evolutionary Coupling from Protein Sequences

Besides studying molecular dynamics, we tried to explore residue-residue couplings from protein sequence alignments. The related proteins share essential information about function and stability deposited into their protein sequences. Using BLAST (Basic Local Alignment Search Tool) [28], we collected hundreds of sequences from SF1 and SF2 proteins that are closely related to PcrA helicase. In particular, these proteins share similar 1A and 2A subdomains. We next made multi-sequence alignments using ClustalW [29] as well as structural alignment [30]. Then we implemented a 'co-evolutionary' statistical analysis (SCA) method developed by [31, 32]. The method detects how each pair of residues are coupled in protein sequence mutations. If two residues are highly correlated maintaining protein stability or function, the mutation of one would likely cause the mutation of the other. Hence, by counting how mutational events are correlated between two positions in a multi-sequence alignment, one obtains the coupling strength between each pair of residues in the related proteins. In this way we identified a 'co-evolutionary core region' of the PcrA-alike helicases, highlighted in the mutational correlation map and shown in blue in the helicase structure in Fig. 15.2. One sees that the core region links the ATP binding site to the protein-ssDNA interface, and to regions close to the protein-dsDNA interface. One would expect that the core region consists of most essential residues collectively maintaining structure and function throughout the evolutionary history.

In summary, for PcrA helicase, we have studied its directional translocation in structural and energetic details. We focused on the alternating domain stepping mechanisms of the motor, probed coordinated protein-ssDNA couplings during ATP hydrolysis cycle, from atomistic to residue and informatic level. There are similar SF1 helicases Rep, UvrD, RecB and RecD [7], which all contain similar helicase motifs and are likely use similar stepping mechanisms.

There is, however, another proposed 'ratcheting' model of translocation, which also requires the helicase switches between two states, weakly and strongly bound states with the ss [33]. It is proposed that, e.g., ATP binding loosens the grip of the

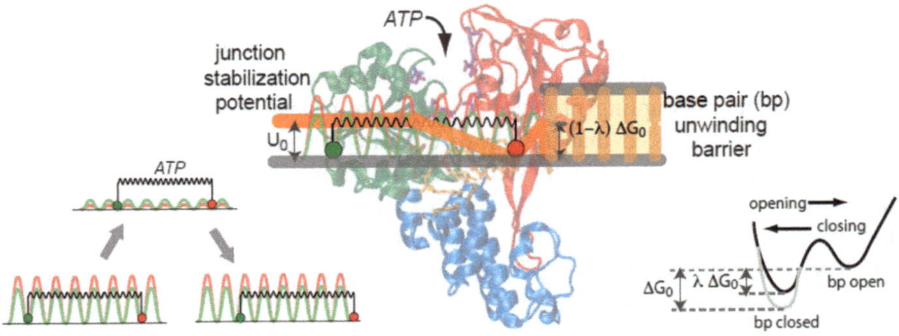

Fig. 15.3 Coupling translocation with nucleic acid unwinding in NS3 helicase. The images are adopted from the graphic abstract of [5]. Shown on the *bottom left* is the translocation model of NS3 helicase, similar to that of PcrA helicase [3, 4]. The *green/red bead* represents domain 1/2 in NS3, while the *green/red curve* represents the translocation potential of the domain 1/2 along the single strand. In particular, the potential barriers are low in the ATP bound state as the affinity between NS3 helicase and the single strand is low [47]. The structure of NS3 helicase is shown in the *middle* [35]. ATP is supposed to bind in between domain 1 (*green*) and 2 (*red*). At the junction formed by duplex DNA/RNA and a single stranded 3′ tail, there are two interaction potentials that affect the helicase action. One is the base par unwinding potential/barrier that hinders the helicase to move forward. The other is the junction stabilization potential, which prevents the helicase from moving away from the junction. On the *bottom right*, a two-state double well potential is shown for the opening and closing equilibrium of the end base pair on the duplex. Active helicase unwinding reduces the free energy difference from ΔG_0 to $\lambda\ \Delta G_0$, with $0 < \lambda < 1$

helicase on the ss so that the helicase may transiently diffuse along the ss (weakly bound); ATP hydrolysis and product release re-induce tight binding of the helicase to the nucleic acid (strongly bound), resulting in a biased forward movement that leads to directional translocation. The directional bias comes from asymmetric saw-tooth potentials between the protein and ss in the strongly bound state. As a matter of fact, the ‘ratcheting’ model can be accommodated well in the two-domain stepping framework we implemented above. Essentially, if one lowers the potential barriers for both subdomains 1A and 2A, e.g., in the ATP bound substrate state, one gets the weakly bound state that allows helicase diffusion to happen soon enough. While raising both domain barriers sufficiently high in the product state, one gets the strongly bound state. Instead of depicting potentials for individual subdomains, one can draw the potential for the motor as the sum of the individual domain potentials, and then obtain the asymmetric sawtooth-like potentials for ‘ratcheting’ (see reference [3] figure 6). We took this perspective to model another monomeric helicase NS3 (see Fig. 15.3 left), which ratchets or steps along ssRNA/DNA and is able to unwind dsRNA/DNA [5]. Our focus, however, would be then on helicase unwinding that is coupled to the ss translocation, rather than on the ss translocation alone.

15.2.2 NS3 Helicase – Coupling ss Translocation with ds Unwinding

NS3 from hepatitis C virus (HCV) is a non-structural protein with its C terminal part folds into a SF2 helicase. Identifying inhibitors that target HCV-NS3 helicase would help develop anti-HCV drugs [34]. In its monomeric form, NS3 helicase (NS3h) can translocate along the ss nucleic acid in the 3′–5′ direction, unwind the duplex region, and displace other bound proteins on the nucleic acid. The helicase shares with other monomeric helicases or translocases the two RecA-like domains [1, 2], with an ATP binding site located in between the two domains (see Fig. 15.3 middle). Additionally, NS3h has a domain 3, positioned similarly as the subdomain 1B (and part of 2B) in PcrA helicase. The high-resolution structure of NS3h was obtained early without ATP/ADP bound [35]. The substrate structure with ATP analog bound was discovered recently [36]. However, these structures contain only ss nucleic acids bound with the protein, while the binding configuration of NS3h to the duplex part of the nucleic acids is still missing. This causes difficulties studying protein-ds RNA/DNA interactions and hence, the duplex unwinding of NS3h.

On the other hand, experimental studies on NS3, including biochemical and single molecule measurements, have been abundant in recent years [17–19, 37–43]. In particular, single molecule optical tweezers monitored monomeric NS3h unwinding activities in real time [17, 18], providing substantial data to build a stochastic model of the helicase unwinding [5]. Below we illustrate essential ideas in the model, focusing first on how ss translocation is coupled to ds unwinding, and then on how the duplex unwinding is connected with junction stabilization of the helicase. Besides, we also summarize sequence effects on helicase unwinding, as well as on fluctuation properties of the helicase motor that have been caught and can be further probed.

15.2.2.1 Coupling Translocation with Unwinding

The helicase motor protein uses free energy from ATP hydrolysis cycle to assist its unidirectional translocation. When the helicase moves to the junction formed by the ds RNA/DNA and the ss tail, it can possibly move further and have the duplex region unwound. Basically, the unwinding is treated as simple as separation or opening of the base pair (bp) at the end of the duplex, breaking the hydrogen bonding interactions therein. Normally, the hydrogen bonds quickly form and break, at a time scale (μs) much faster than that of the helicase translocation (ms). Hence, the bp opening and closing are supposed to maintain a thermal equilibrium, with bp closing more favored. Separation of a stretch of bps altogether would be possible but was not considered yet.

For helicases that can unwind the duplex nucleic acids, there are two basic mechanisms proposed [44–46]: passive and active unwinding. Under the passive mechanism, the helicase does not interfere with the bp opening and closing equilibrium; it only takes advantage of spontaneous opening of the bp, makes a move 1-bp forward fast enough so that the bp cannot be closed back again. Indeed, even the bp is closed, there are still slight probabilities that the helicase can thermally force opening the bp (without affecting the opening-closing equilibrium) if the corresponding activation energy is not too high [5]. Anyhow, under this scenario, the helicase does not use its translocation free energy (from ATP hydrolysis cycle) to assist unwinding. In contrast, helicase works under the active unwinding mechanism does 'tilt' the bp opening and closing equilibrium toward the opening direction. We thus utilized a parameter λ ($0 \leq \lambda \leq 1$) to characterize how 'active' the helicase is (see Fig. 15.3 middle and right). The bp opening and closing free energy difference ΔG_0 (>0) is reduced to $\lambda\ \Delta G_0$ upon the presence of the helicase at the end of the duplex. The closer λ is toward 1, the less active the helicase is ($\lambda = 1$ for the passive case). The energy to 'tilt' the bp opening comes from the free energy being utilized for translocation. Consequently, the ratio between forward and backward rates of helicase translocation is reduced during the active unwinding. However, it is not clear yet if the forward or backward moves are limited by chemical transitions (such as ATP/ADP binding/unbinding, hydrolysis etc.), or by the protein domain movements. In practice, we assume that the forward and backward moves of the helicase are characterized by the domain stepping, either diffusional (e.g. in ATP bound state) or coupled to ATP binding or product release.

By fitting with single molecule experimental data on how fast the helicase unwinds under different ATP concentrations, we estimated the value of $\lambda < 0.5$ [5]. This suggests that energetically NS3h is a quite active helicase, as it reduces more than half of the bp opening-closing free energy difference on average during unwinding. Then comes the next question, how exactly the protein unwinds the dsDNA, kinetically, energetically, and structurally?

In studying PcrA helicase, we noticed that the duplex part of the DNA is significantly distorted by 2B subdomain of PcrA in the product state in the absence of ATP/ADP. The role of 2B itself is still unclear, however, as removal of it in some similar helicases (such as Rep) actually enhances unwinding activities of the helicase. Nevertheless, from a structural perspective, some regions in NS3 (domain 2 or 3) should interact closely or even distort the duplex, likely in the product state. Biochemistry measurements had indicated that NS3 shows low binding affinity with ssDNA in the ATP bound state [33, 47]. For the fluctuation properties detected in single molecule experiments, we suggested that the fluctuation is due to diffusional characters of the helicase in the low ss affinity ATP bound state (see later discussions). Indeed, stepping upon ATP binding (by domain 1) is hardly able to 'push' further for unwinding, as the protein transits to the low affinity state to ss (see Fig. 15.3 left). On the other hand, some evidence showed that Pi release generates power stroke for the duplex unwinding [42]. That is, after ATP hydrolysis, the stepping motions by domain 2 upon product release can 'push' further on the duplex region as the protein-ss affinity increases, making the end bp more likely to be open that it originally is, and hence actively assists ds unwinding.

15.2.2.2 Duplex Unwinding and Helicase Stabilization at Junction

Now we consider how unwinding proceeds at the junction. The junction is formed by the end of the duplex region and a piece of 3′ ss tail. Experiments had noticed that the helicase shows a higher affinity to the junction than to a piece of pure ss of the same tail length (7–10 nt) [33]. Thus one can estimate an association strength U_0 between the helicase and the duplex region of the junction at ~4 k_BT, in the *apo* state ($U_0 \sim 2$ k_BT in the ATP bound state as estimated further in [5]). Experiments had also discovered that unwinding could happen when NS3h binds the junction of a tail length <7 nt [33]. Combining the evidences, we inferred that in the stabilized binding configuration, NS3h covers 10-nt length on the junction, with a front 3-nt length bound with the duplex region, while the left 7-nt distance associated with the ss tail (see Fig. 15.3 middle). The association strength U_0 between the NS3h and the duplex, mainly through the front 3-nt length interface, would be responsible for preventing backward diffusion or dissociation of the helicase during unwinding. This could be important for NS3h as its diffusion is significant and the affinity to ss is low in the ATP bound state. Without strong enough ds association, the helicase may move away from the duplex region before it unwinds the bp.

Hence, one can imagine that in order to successfully unwind the bp at the junction, the helicase needs sufficiently high affinity with both the ss and ds part. For some of monomeric helicases, even they can translocate well along ss, they cannot unwind the ds part [7]. The possible reasons are (1) they may not be able to grab ssDNA/RNA strongly enough to make a power stroke for active unwinding; (2) they do not associate with dsDNA/RNA tightly enough and escape away from the ds most of time, failing even for the passive unwinding.

15.2.2.3 Sequence Effects of Unwinding and Diffusional Fluctuations

Through active bp unwinding at the duplex end, the helicase has its translocation velocity reduced compared to that during ss translocation, as the forward stepping rate of the front domain is inhibited. The average stepping efficiency, defined as the average number of steps advanced for each ATP consumed, is also reduced below 1 nt/ATP. Both of the reductions depend on the DNA/RNA sequence encountered by the front domain, or say, by the value of ΔG_0 that measures the sequence stability (~3 k_BT for GC and ~1.5 k_BT for AU on average). Even the dissociation rate of the helicase from the junction can be sequence dependent, as the helicase-ds interaction during unwinding bears the sequence effect. We have quantified the unwinding velocity, stepping efficiency, and dissociation rate in kinetic model. The resulted sequence dependencies match well with experimental measurements: with increasing sequence stability, the unwinding velocity and average stepping efficiency decrease, while the dissociation rate increases. As a result, the processivity length, defined as the average distance the helicase travels before its first dissociation (velocity over dissociation rate), decreases with the increasing sequence stability.

These experimental observables, hard to examine from detailed simulations, are easily trackable in kinetic or stochastic modeling.

An important character of the NS3h in this model is that its translocational diffusion is not negligible when the helicase is bound with ATP. The diffusion affects fluctuation properties of the helicase during ss translocation, though without impacting the *average* translocation velocity or the stepping efficiency (1 nt/ATP at average). During unwinding, however, the diffusive character does affect both the fluctuational and the average properties, such as the velocity and the stepping/unwinding efficiency of the helicase. In the absence of diffusion, the unwinding efficiency is already lower than 1 bp/ATP due to occasional futile ATP cycles upon the sequence barrier. The diffusion further lowers the average unwinding efficiency. The larger the diffusion rate, the more significant the sequence barrier decreases the efficiency, and this is achieved by increasing the frequency of backward jumps (-1 nt at a time) during unwinding. The interplay between the diffusion rate and sequence effects is also displayed in helicase dissociation during active unwinding. According to the numerical results, the sequence dependence of the dissociation rate is detectable only in the presence of diffusion; the larger the diffusion rate, the more pronounced the sequence dependency.

To quantify the diffusive character of the NS3 helicase, we estimated the diffusion rate to an order of 10 $nt^2\ s^{-1}$ based on the unwinding velocity fluctuations (standard deviation) measured from single-molecule experiments [17, 18]. This shows that a significant advantage of the single molecule experiments is to measure not only the average quantities, but also their fluctuation properties. More straightforwardly, one could obtain the diffusion rate from measuring velocity fluctuation or the effective diffusion rate during ss translocation of the helicase in the experiments. Further efforts are needed to link these 'long time' (ms) experimental observables to 'short time' (ns–μs) molecular dynamic quantities as well as structural properties of the molecular motor.

15.3 Multimeric Ring NTPase – DNA Packaging Motor and a Couple of Others

In this section we discuss how the ring-shaped multimeric NTPase or translocase moves along the molecular track, coordinating its individual subunits. The molecular track can be DNA, RNA or protein. The chemical or catalytic sites are distributed around the ring, and are located at the interfaces formed by any two neighboring subunits. We focus on our previously studied system, a viral DNA packaging motor from bacteriophage ϕ29 [6]. We want to highlight two key issues, first, how the motor interacts with the molecular track or the substrate, and second, how the motor subunits are coordinated around the ring.

Indeed, multimeric NTPases or translocases are widely distributed [8, 48–52]. For example, in helicases super-families other than SF1 and SF2, most of them are hexameric rings. NTPs bind in between two neighboring RecA-like domains on the ring. The binding site consists of Walker A and B motifs from one domain, and arginine finger from the other. In comparison with the DNA packaging motor, we briefly review studies on hexameric T7 helicase and ClpX motor as well. T7 helicase moves along ssDNA and unwinds the duplex DNA [48]. It is supposed to use arginine finger to coordinate around the ring. ClpX motor translocates and unfolds protein strand [53]. The subunits around the ClpX ring are not coordinated but rather 'fire' randomly. Hence, the protein- substrate interactions in T7 helicase and ClpX are different from that in the dsDNA packaging motor. The inter-subunit coordination, however, can be similar for T7 helicase and the packaging motor, while ClpX provides a special case.

15.3.1 DNA Packaging Motor – 'Push and Roll' and Coordination Around the Ring

Packaging the genome into the viral capsid is a key event in the life cycle of viruses. The bacteriophage ϕ29 is one of the simplest and most intensively studied systems [54]. Its genome is made of a linear ds DNA of about 19 kb, encoding 20 proteins. Packaging the long piece of dsDNA into a near- crystalline state inside a virus capsid ~50 nm in diameter generates a high internal pressure, due to entropic barriers, electrostatic repulsions and bending energies of DNA. The pressure can be utilized later on to eject the genome into the host cell during viral infection. The packaging is done by a powerful ATPase motor gp16. Figure 15.4 shows an image of the phi29 DNA packaging system obtained from the cryo-electron microscope (cryo-EM) density map [55]. It consists of three multimeric rings: an ATPase (gp16), a 174 base RNA (pRNA), and a dodecameric portal connector (gp10). The apparatus are located at a unique five-fold vertex of the icosahedral capsid (the prohead). It has shown that the ATPase and pRNA form pentamers [55]. A similar organization of the ATPase has been found in the DNA packaging motor from bacteriophage T4 [56].

In our modeling work, we studied ϕ29 DNA packaging ATPase, which has been investigated intensively in experiments using single-molecule manipulation techniques [51, 57–60]. The experiments indicated that DNA translocation is likely associated with Pi release after ATP hydrolysis, and the motor affinity for the DNA is high in the ATP-bound state (**T**) but low in the ADP-bound (**D**) or *apo* state (**E**) [58]. Further high-resolution optical tweezer measurements showed that the packaging proceeds in bursts of 10 bp, with each composed of four 2.5 bp substeps (Fig. 15.4) [59]. Following the burst phase is a dwell phase, composed of four ATP-binding events and several non-ATP binding events. Accordingly, we constructed a mechanochemical model based on experimental knowledge, homolog structural information, and a few necessary, but generic, assumptions [6]. The model provides a physical picture of how this multimeric motor translocates along dsDNA. We summarize the main features of the model as below.

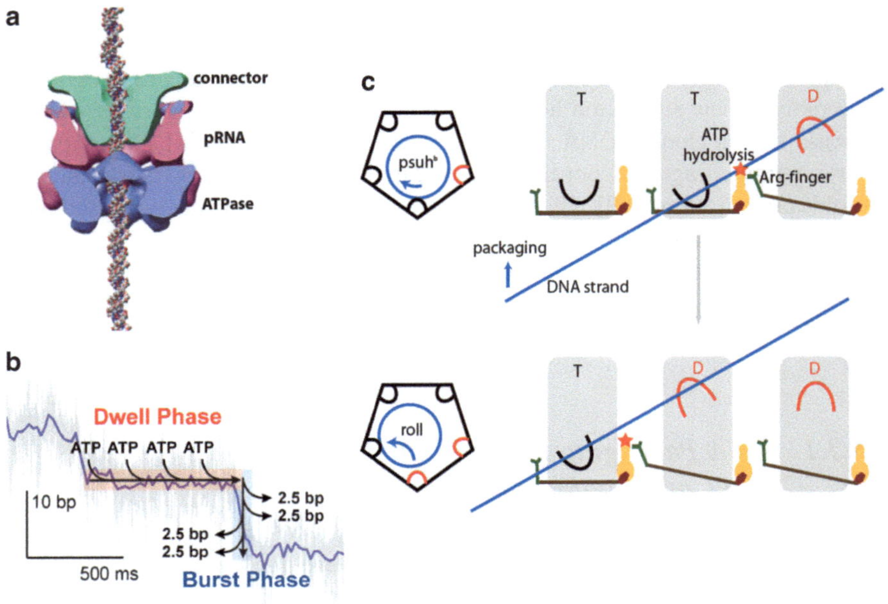

Fig. 15.4 A viral DNA packaging motor from bacteriophage ϕ29 and the push-and-roll model. The images are adopted from [6]. (**a**) Images (courtesy of M. Morais) from cryo-EM studies of the packaging system [55]: The dodecameric connector (gp10, *green*), pRNA (*magenta*) and ATPase pentamer (gp16, *blue*) with the DNA modeled for visualization. (**b**) The essential experimental results from high-resolution optical tweezer measurements [59]. DNA translocation proceeds in bursts of four power strokes of 2.5 bp, separated by dwell phases wherein four ATPs are loaded into four catalytic sites. There are also multiple slow events in the dwell phase that do not involve ATP binding. (**c**) The push-and-roll of DNA packaging and cooperative ATP hydrolysis mechanism proposed for ϕ29 DNA packaging [6]. The mechanism along with the cartoon is similar to that used for the ϕ12 packaging motor [63]. The molecular lever is down in the **T** (ATP-bound, color in *black*) state and up in the **D** (ADP-bound, color in *red*) state, while the DNA affinity of the lever/subunit is high in the **T** state but low in the **D** state. As the lever moves up during the power stroke (**T** $\rightarrow$ **D**), the DNA (*blue*) is pushed up by $\sim$2.5 bp, rotates ($\sim$$-30°$), and rolls to the next subunit. In the top view, the packaging up direction points toward the reader. Besides, ATP hydrolysis/Pi release in the current subunit triggers the Arg finger insertion into the next catalytic site, accelerating the (otherwise slow) ATP hydrolysis

15.3.1.1 Motor-DNA Interaction in 'Push and Roll'

The DNA packaging had been measured along with DNA rotation. We developed a 'push and roll' model to describe how the motor and DNA interact during the packaging cycle. The 'push' of the motor subunit onto the DNA is presumably conducted by a lever structure. Lack of the high-resolution structure of the packaging ATPase in ϕ29, we borrowed some structural features from another packaging motor P4 in ϕ12 [61–63]. These are basically the luminal loops that emanate from the central β-sheet, which also emanates the P-loop to grasp ATP. The loops are used as molecular levers by the motor subunits, to move up and down in cycles, to drive the translocation of the DNA. The packaging force generation can then be regarded as the consequence of conformational couplings between the loop and the ATP binding/catalytic site. The 'firing', or the power stroke, is likely generated upon the release of Pi, in the transition from the high lever-DNA affinity state **T** to the low affinity state **D** [58].

Intuitively, one expects the motor lever to push perpendicularly on the DNA helical strand (see Fig. 15.4), which is right-handed and tilted about 30° above the horizontal direction. The push, consequently, leads to movements of the DNA along two orthogonal directions, one 'vertically' up for the packaging, and one 'horizontally' toward left for rotating (CW in top view, with the packaging direction toward the reader, or '-'along the *under-winding* direction). The push leads to about $-30°$ of DNA rotation for every 2.5 bp distance of DNA packaging.

Energetically, the push comes from steric interactionis. Experiments had shown that the packaging could happen for the neutralized DNA, as long as the neutralized segment is less than 30 bp in length [60]. This suggests that the steric interaction is crucial for the packaging strokes. However, the electrostatic interactions are also important for 'steering' the lever toward the DNA backbone, as well as for 'holding' the DNA without slipping during the dwell phase. There can be a positively charged residue located at the tip of the lever (as that in ϕ12-P4 packaging motor). When the lever approaches to the DNA strand for push, the positively charged residue will grab to the nearest negatively charged phosphate group on the DNA backbone, so that the steric push can effectively happen. The electrostatic association is indispensible, in particular, when there is no steric push during the dwell phase. If the neutralized DNA region is too long ($>$30 bp), it poses too big an energy barrier even for four packaging strokes in a row (during the burst phase) to carry the DNA across (see supplementary information in ref [6]); as the motor subunit cannot grab on or associate with the neutralized DNA segment, slipping between the motor and the DNA will happen, and abolish the packaging.

Besides, DNA can roll around the internal ring from one subunit toward the next for each packaging substep. The cross-sections of DNA and the motor ring are circles. Rolling is the basic movement between two circular surfaces without slipping. If rolling of the DNA is CCW around the ring, a coupled rotation of DNA should be CW (-), leading to an even larger DNA rotation toward the under-winding direction ($<-30°$) for each packaging substep (2.5 bp). Preliminary (unpublished) measurements, however, indicated fairly small *negative* values of DNA rotations per step ($>-30°$). Hence, we inferred that the rolling happens in CW direction around the ring, leading to a positive (CCW) DNA rotation (Fig. 15.4). The amount of rotation depends actually on the relative size of the DNA cross-section to that of the motor ring [6]. A larger radius of the motor ring leads to a larger positive DNA rotation during the rolling. Hence, by measuring the exact amount of DNA rotation during each packaging substep, one can estimate the radius of the motor ring.

The energetic driving force for rolling can be electrostatic. We had shown that energetically, the rolling configurations of the DNA (attaching to the peripheral or internal surface of the ring) are more stabilized than hanging around the center of the ring (see supplementary information in ref [6]). The conclusion, however, was based on screened coulomb interactions between protein and DNA in a highly simplified model, with five positive lever charges equally distributed around the ring, and with long strands of negative charges representing the phosphate groups on the DNA. Further, we had assumed that the high protein-DNA affinity in the **T** state is largely due to the lever charge grabbing on the DNA backbone; after the power stroke (**T** $\rightarrow$ **D**) or steric push, the lever withdraws from the DNA and loosen its grip, hence, leading to a low protein-DNA affinity in the **D** state. As a result, the electrostatic attraction between the DNA and the *neighboring* subunit at high affinity **T** state makes the DNA to roll toward it.

15.3.1.2 Which Strand and Where on the DNA the Motor Pushes?

Based on the 'push and roll' model, the motor lever should push on whatever it can push on the DNA, without differentiating which DNA strand or the exact spot, though the positively charged residue likely grabs on the nearest phosphate group. However, experimental measurements did bring up some 'puzzles'. First, the motor seems to react differently to neutralizing different strands [60]: neutralizing the $5'$–$3'$ strand ($\sim$30 bp) abolished the packaging, while neutralizing the same length on the $3'$–$5'$ strand does not affect the packaging much. Second, the packaging step takes a fractional substep of 2.5 bp [59], why this happens, and would it bring *out of registry* trouble as the motor subunit 'sees' different spots for different pushes?

Indeed, there are 'asymmetries' between the two DNA strands, as the backbone of the $5'$–$3'$ ($3'$–$5'$) strand forms the upper edge of the DNA major (minor) groove. For the first 'puzzle', we inferred that during the power stroke the motor lever pushes more 'effectively' on the upper edge of the major groove ($\sim$12 Å wide in B-form DNA) than on the minor groove ($\sim$6 Å wide). Due to steric hindrances or some entropic barrier, the minor groove might be too narrow for the lever to produce an effective steric push. Neutralizing the $5'$–$3'$ strand would energetically lead the lever to approach the upper edge of the minor groove ($3'$–$5'$ strand) and push, which might not generate sufficient force to sustain packaging. Nevertheless, when the motor is packaging a normal DNA substrate, there is no such a large electrostatic energy barrier as that exists in packaging the neutralized DNA. Thus, the lever can approach to either strand and push on whatever steric elements it encounters, although we still expect that the packaging is more powerful or effective when the motor pushes on the $5'$–$3'$ strand than on the $3'$–$5'$ strand.

For the 2.5-bp packaging substep measured experimentally, we took it for granted in the model, without explaining why it is generated in the first place. A likely reason behind it is that the lever (a loop structure) lengthens or simply moves up about $2.5 \times 3.4 \sim 8.5$ Å for each power stroke. As such, will there be out of registry trouble every other 2.5-bp, when the lever tip stays in between two phosphate groups rather than on one of them? The trouble comes from an implicit assumption that the lever is a rigid device, and the motor requires grabbing exactly the same spot each time to make the push. The resolution can be, first, the lever is not rigid but flexible, it deforms from time to time to grab on the *nearest* phosphate group; second, as the push is *steric*, the pushing spot does not necessarily coincide with the electrostatic grabbing site. The electrostatic association is there only for holding the DNA, assisting packaging without slipping.

Hence, the above 'puzzles' could be artificial. Without direct measurements, one should not expect that the motor 'choses' some sophisticated way of packaging. The movements of the motor are guided by protein-DNA associations, tuned by solution condition and counterion effects. Lack of high-resolution structures of the packaging motor, as well as limited by the computational power, unbiased atomistic simulations of the full packaging process is not feasible in near future. On the other hand, toy models that catch some essential features of the system may provide certain insight into the functional mechanisms.

15.3.1.3 Coordination Mechanisms Around the Motor Ring

CryoEM studies had suggested that the ϕ29 packaging motor has a pentameric ring structure [55]. The high-resolution optical tweezer measurements, however, showed that in each 10-bp packaging burst, there are four instead of five substeps (2.5-bp each) [59]. It was not clear why the five motor subunits package four substeps in a row. Besides, it was not clear what else happen aside of four ATP loadings in the dwell phase. To provide feasible answers to these questions from a modeling point of view, we suggested a dominant packaging scenario (Fig. 15.5). The essential experimental clues, aside from that mentioned above (Fig. 15.4 left), include also some previous findings: **T** (**D**, **E**)-state motor subunits have high (low) DNA affinity, and that **T** $\rightarrow$ **D** transition (Pi release) likely delivers the packaging stroke [58]. The key properties in the suggested scenario are summarized as ***I*** to ***III*** below. They answer why that the packaging strokes happen in a row (***I***), why each packaging burst stops at the fourth rather than the fifth step (***II***), and what else happen besides the ATP binding during the dwell phase and how (***III***).

In Fig. 15.5, we provide the schematics of the dominant packaging events for the five motor subunits in one full packaging cycle. We start with a configuration in which all five motor subunits are loaded already with ATP molecules, hence in all **T** states. The state of the subunit is defined as the catalytic site on the subunit (e.g. formed with the subunit proceeding to it) is bound with ATP (**T**), ADP (**D**) or none (**E**). **T** $\rightarrow$ **D** $\rightarrow$ **E** $\rightarrow$ **T** transitions (reversible but with forward free energy bias) happen sequentially for each subunit and alternatively around the ring for the five.

(I) Arginine finger insertion from one subunit to the next accelerates the ATP hydrolysis, leading to packaging strokes/substeps happening in a row.

The arginine finger is supposed to locate at the interface of two neighboring subunits, inserting from the first subunit into the catalytic site formed by the two subunits (see Figs. 15.4 and 15.5 center). The knowledge that the arginine finger insertion can reduce activation barrier in the ATP hydrolysis and thus accelerate the process had been known [64, 65]. More specifically for the ring motor, it was proposed that the arginine finger insertion couples ATP hydrolysis events

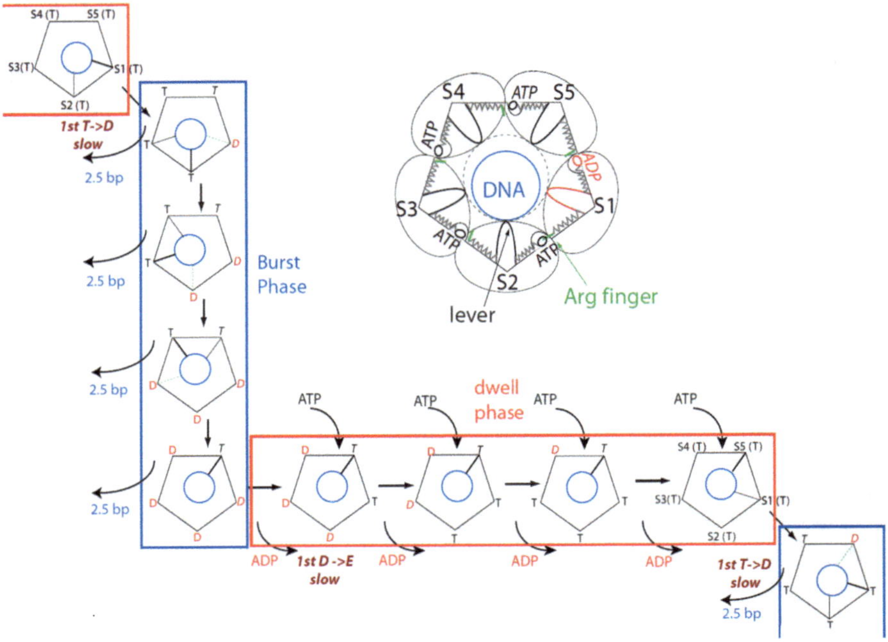

Fig. 15.5 A dominant mechanochemical scheme of ϕ29 DNA packaging motor. The image is adopted from [6]. The reaction cycle is divided into two phases. The burst phase contains four sequential power strokes at four consecutive catalytic sites. Each power stroke generates a 2.5-bp substep [59]. The dwell phase contains four consecutive ATP loadings and several non-ATP-binding events [59]. Each power stroke commences upon the **T** → **D** transition (Pi release) when the DNA is initially attached to the subunit. Each power stroke requires the next subunit to be in the **T** state to receive the DNA as the DNA rolls toward the subunit, following the completion of current power stroke (see text for property ***II***). After four contiguous power strokes in the burst phase, the motor pauses because the next subunit has been left in the low DNA affinity D state, and the system enters the dwell phase. During the dwell phase, the 1st ADP release is slow, but the following ADP releases (2nd to 4th **D** → **E** transition) proceed faster as ATP binds quickly (at high [ATP]) and accelerates ADP release at the next site (see text for property ***III***). The waiting time for the 1st power stroke is another rate-limiting non-ATP-binding event during the dwell phase after the four ATPs are loaded. The ensuing power strokes (2nd to 4th **T** → **D**) happen very quickly in the next burst phase. A related hydrolysis cooperative mechanism is the insertion of an Arg finger from the preceding subunit, driven by the hydrolysis/power stroke in that subunit (see text for property ***I***)

for neighboring subunits in ϕ12-P4 packaging motor [61–63], and we borrow this property. The arginine finger residue for ϕ29 packaging motor had also been identified from genomics studies [9]. Accordingly, we suggested that without arginine finger insertion from previous subunit, the first ATP hydrolysis during the 1st T → D happens spontaneously and slow, which partially limits the starting of the burst phase. However, once the 1st **T** → **D** happens, conformational changes trigger the Arg-finger insertion into the next catalytic site and accelerate the corresponding

$\mathbf{T} \rightarrow \mathbf{D}$ transition, hence 2nd to 4th packaging strokes or substeps happen fast and in a row.

(II) A packaging stroke requires the next subunit to be in high DNA-affinity **T** state to prevent slipping, and hence can bring to a stop at the 4th stroke.

To obtain four substeps out of five presumably identical subunits, one simple scenario is that each two neighboring subunits are coupled for one packaging substep. More specifically, it enforces both subunits to be in **T** state to have the first subunit pushes on the DNA and rolls the DNA toward the second subunit. The scenario is physical since the **T** state subunit has high DNA-affinity to 'attract' or hold the DNA, while the subunit in the low DNA-affinity state (**D** and **E**) does not. Starting from the configuration $\{\mathbf{T}_1\mathbf{T}_2\mathbf{T}_3\mathbf{T}_4\mathbf{T}_5\}$ (with the number index labeling subunits from 1 to 5 around the ring), the 4th packaging stroke ends up with a configuration $\{\mathbf{D}_1\mathbf{D}_2\mathbf{D}_3\mathbf{D}_4\mathbf{T}_5\}$. The configuration cannot support the fifth packaging stroke due to the 'end pair' $\mathbf{T}_5\mathbf{D}_1$ has subunit 1 in low DNA-affinity **D** state, which is not yet ready to 'accept' the DNA. If subunit 5 fires and has the DNA roll toward to subunit 1 at this moment, the DNA likely slips. Hence, until the ADPs are replaced by new ATPs, the motor stays in a dwell phase. One has to be aware, therefore, that the presence of the dwell phase also requires the ADP release to be sufficiently slow (discussed below). In brief, we see that $\mathbf{T}_i\mathbf{T}_{i+1}$ neighbor coupling for one packaging stroke is necessary for the packaging stops at every 4th stroke, though it is not sufficient.

(III) ADP releases slowly, sequentially, alternating and in coordination with ATP loading around the ring.

As mentioned above, slow ADP release is also essential to bring the packaging motor to the dwell phase. The motor relinquishes the reaction products and loads new fuels during this period. At the saturating ATP concentration, the dwell time statistics fits to a gamma distribution, which suggests multiple events and further gives an effective number of slow events in between 3 and 4 [59]. At this condition, ATP binding is very fast, so possible rate limiting events include four ADP releases and waiting for the 1st ATP hydrolysis. If the four ADP releases are equally slow, the effective number of events would likely be ~4 more. Hence, correlations likely exist among the four ADP releases. We suggested that the first ADP release is very slow, while the ADP release later on is accelerated by the ATP binding at the preceding site. The scenario seemed to work in F1-ATPase system [66]. For example, in the depicted scheme in Fig. 15.5, $\mathbf{D}_1 \rightarrow \mathbf{E}_1$ happens slowly after the 4th packaging stroke; at a high ATP concentration, $\mathbf{E}_1 \rightarrow \mathbf{T}_1$ happens immediately; the ATP then shrinks the binding pocket stronger than ADP does, and helps to open the binding pocket for subunit 2, thus accelerates $\mathbf{D}_2 \rightarrow \mathbf{E}_2$. Besides, the ADP releases need to go sequentially around the ring, as our numerical test on random ADP release turned out in conflict with experimental data.

Under the above scenario, we numerically solved Fokker-Planck equations describing both mechanical packaging and chemical transitions, and generated packaging trajectories and statistics. The results fit well with experimental data. More recently, further high-resolution optical tweezer measurements [51] confirmed that the ADP releases alternate with ATP loadings during the dwell phase, as suggested by (***III***) above. The measurements, however, suggested that a certain subunit plays a regulatory role without participating in ATP hydrolysis during each packaging cycle [51]. In the depicted scheme in Fig. 15.5, it implies that after the current packaging burst from subunit 1 to 4, the next cycle will repeat exactly (for subunit 1–4) and end up with the same configuration $\{\mathbf{D}_1\mathbf{D}_2\mathbf{D}_3\mathbf{D}_4\mathbf{T}_5\}$. In this case, subunit 5 appears to be the special 'regulatory' subunit that has ATP binding but no hydrolysis. Our previous suggestion, however, is that the five subunits likely play equal roles: If current cycle starts packaging with subunit 1, then next cycle will start with subunit 5 and ends up with $\{\mathbf{D}_1\mathbf{D}_2\mathbf{D}_3\mathbf{T}_4\mathbf{D}_5\}$ after the 4th stroke. Anyhow, the suggestion is irrelevant to properties (***I***) to (***III***) discussed above. For five identical subunits, our previous suggestion simply assumes regular and cyclic repeats around these subunits. However, it is possible that due to some structural differences, or more subtly, due to some dynamic reasons, there comes a 'special' subunit. For example, the DNA, held by subunit 5 ($\mathbf{T}_5$) after the 4th stroke, may dissociate from subunit 5 during the long dwell phase, and roll toward subunit 1 as $\mathbf{D}_1$ is replaced by $\mathbf{T}_1$. Without the DNA binding, subunit 5 cannot sustain ATP hydrolysis and thus is 'left out' for each cycle. Further studies are expected to investigate this issue.

We also notice that there is some additional or alternative mechanism suggested from literature [67], which may work together or replace the Arg-finger mechanism (***I***) to explain the continuous packaging strokes during the burst phase. The mechanism suggests that ATP binding to one subunit inhibits ATP hydrolysis by the neighboring (next) subunit, leading to coordinated rather than stochastic ATP hydrolysis within the ring. Under this mechanism, ATP hydrolysis can hardly happen for configurations with all or partial subunits bound with ATP; hydrolysis of one subunit ($\mathbf{T} \rightarrow \mathbf{D}$), however, releases the inhibition to the next subunit (**T**). The inter-subunit coordination or signaling is supposed to be conducted by pore loops in the AAA proteases [67], corresponding to the levers in the packaging motor. The overall effect is that the first ATP hydrolysis takes long time to happen in the ring, but once it takes place, the rest ones follow quickly. To find out the exact mechanism, further studies are needed.

15.3.2 Hexameric Helicase T7 – Mechanochemical Coupling and Unwinding

Bacteriophage T7 gp4 gene codes for a prototypical ring helicase. Like the translocation domain in the small helicases PcrA and NS3, the protein fold in individual subunits of T7 helicase is also RecA-like [68]. In the presence of dTTP and DNA, the T7 helicase can assemble into a multimeric ring of six-fold symmetry, with the DNA threading through its central channel. The ring is flexible and can deviate from the six-fold symmetry.

The essential issue for studying the ring helicase is to understand how the unwinding force is generated, and how motor subunits work together. Substantial experimental progress has been made on T7 helicase translocation and unwinding [69–71]. These studies provide information on pre-steady-state kinetics and single molecule dynamics of the system. Interestingly, the multi-meric T7 helicase had been examined comparatively in experiments with the monomeric NS3 helicase in nucleic acid unwinding [70]. Below, we first compare the subunit coordination around the ring in T7 helicase with that in ϕ29 DNA packaging motor. Then we discuss its unwinding activities, in comparison with NS3 helicase.

15.3.2.1 Mechanochemical Coupling as a Multimeric Ring Motor

By capturing the crystal structure of an active hexameric fragment of the T7 gp4 helicase [68], a sequential four-site 'binding change' mechanism was proposed. It explains how cooperative binding and hydrolysis of nucleotides are coupled to conformational changes in the helicase ring. Later experimental [72] and computational studies [73] supported the idea, and ssDNA is envisioned to be transferred from one subunit to the next, sequentially, around the helicase ring.

Indeed there are significant similarities between the T7 helicase and the ϕ29 packaging motor [6]. First, in the T7 helicase structure, luminal loops on the inner surface of the hexameric ring seem to provide binding sites for the ssDNA [68], analogous to molecular levers envisioned for the packaging motor. Second, the NTP-bound state of the motor has a high affinity to ssDNA, while the affinity is low in other chemical states – the same affinity trend found in between the ϕ29 packaging motor and DNA. This property is important for the packaging as illustrated in (***II***) in the previous section. Third, recent experimental analyses indicated that Pi release may also serve to trigger a power stroke [71], similar to that in the ϕ29 packaging motor. The ϕ29 packaging motor currently has no crystal structure available, so in modeling we borrowed features from the ϕ12 RNA packaging motor P4. These included the molecular lever and the arginine finger hydrolysis coupling mechanism (***I***). Indeed, the ϕ12 packaging motor P4 is a hexameric ATPase that is closely related to the helicase superfamily 4, to which the T7 gp4 helicase belongs [9].

Still, there are essential differences between these multimeric motors. Most importantly, the substrate of the T7 helicase translocation is ssDNA, while duplex DNA is involved only during helicase unwinding. As ϕ29 packaging motor pushes on dsDNA, the force generation mechanism cannot be simply transferred to the T7 helicase system. Instead, the essential properties that the ssDNA is polarized and highly flexible need to be taken into account if one is to study protein-substrate interactions in T7 helicase.

15.3.2.2 Active DNA Unwinding Helicase

T7 helicase is able to translocate unidirectionally along ssDNA [74]. The average translocation rate is ~130 bps in the 5′–3′ direction. The processivity of T7 helicase is fairly high: It can travel along ssDNA ~75 kb before dissociation. However, during unwinding *in* vitro [75], the processive rate of the helicase drops to about ten times slower than its translocation rate, and the processivity also decreases significantly. Recent single molecule studies have detected both ssDNA translocation and dsDNA unwinding activities of T7 helicase [69]. These studies clearly indicated that strand separation is a major barrier to unwinding, and suggested an active unwinding mechanism in the T7 helicase. The sequence dependence of the helicase unwinding has also been demonstrated in a recent experimental work [71]. The study showed that T7 helicase adjusts both its unwinding rate and coupling ratio (bp/dTTP) in response to different sequences: It unwinds the dsDNA slower with a lower coupling ratio when it encounters GC vs. AT bps.

From single molecule measurements [18] and our studies on NS3 helicase [5], we notice that NS3 has similar unwinding behaviors. Both helicases seem to be 'active' rather than 'passive', the unwinding rate and NTP coupling ratio display similar trends of sequence dependence. Although the two helicases have quite different molecular geometries, the unwinding force generation step in both systems seems to take place upon NTP hydrolysis or Pi release [33, 71].

In addition, T7 helicase unwinding has been studied in the vicinity of the replication fork where the T7 replisome is assembled [45, 48, 76]. The assembly includes T7 helicase and primase as well as DNA polymerases on both leading and lagging strand, along with ss binding proteins. Interestingly, the in *vivo* unwinding rate of the T7 helicase increases significantly compared to that in *vitro*, reaching about the same as its ssDNA translocation rate. This suggests that T7 helicase becomes more efficient in unwinding when assisted by other proteins.

15.3.3 ClpX – A Multimeric Ring Dismantling Protein Fold

ClpX is the ATPase component of the ClpXP protease in prokaryotes [77, 78]. It denatures native proteins and transports the denatured peptide into ClpP for degradation. ClpX belongs to AAA+ family (ATPases associated with various cellular activities) molecular machines that use energy from ATP hydrolysis to remodel a variety of molecular assemblies in the cell [79]. In *E. coli*, ClpX assembles into a hexameric ring docking onto the heptameric ClpP. Two ClpP rings stack back-to-back, creating the degradation chamber that aligns with the central channel of ClpX [77]. Protein substrates for denaturation and degradation are recognized by ClpXP via short peptide sequences [80].

Unlike the sequential and coordinated ATP binding and hydrolysis around the motor ring identified for the packaging motor and helicase systems, the six subunits in ClpX appear to fire independently and in random order [81]. That is, any subunit in the motor ring positioned best with the protein substrate can hydrolyze ATP, incrementally unfold the protein, and translocate the substrate. This is an intriguing property that appears to be uncommon amongst currently known ring ATPases, including the above mentioned T7 helicase, ϕ29 and ϕ12 packaging motor, as well as F1-ATPase, Rho and E1 helicase [59, 62, 68, 72, 82–84].

One also expects a somewhat different substrate translocation mechanism in ClpX from other motors working with nucleic acid substrate. In our model of ϕ29 DNA packaging motor, we have proposed a 'push-and-roll' mechanism for how the DNA moves through the ring lumen. In particular, the mechanism depends on DNA's helical structure and regular charge distribution. A protein substrate as in ClpX case, however, does not have this periodic structure: the charge distribution is not stereotyped, and the chain is much more flexible than dsDNA on the length scale of the motor dimension. On the other hand, the essential motor forces that directly transport the substrate appear to be steric in both ClpX and the packaging motor systems [49, 60, 85], providing some common theme in the force generation.

15.3.3.1 Random Firing and Structural Basis

Experimental studies linking covalently active and inactive subunits in ClpX showed that different geometric arrangements support the enzymatic unfolding of protein substrates and translocation of the denatured polypeptides into ClpP [81]. The studies indicate that the ClpX power stroke is generated autonomously in each subunit. They have also ruled out concerted and sequential hydrolysis, while suggesting a probabilistic sequence of hydrolyses around the ClpX ring.

High resolution structure of single subunit and the hexameric structure of ClpX reveal striking asymmetry due to large differences in rotation between large and small domains within individual subunits [53]. These differences in subunit rotation prevent nucleotide binding to two subunits and generate a staggered arrangement of ClpX subunits and pore loops around the hexameric ring. This could provide a mechanism for coupling conformational changes caused by ATP binding or hydrolysis in one subunit to flexing motions of the entire ring. It is possible that ATP binding, hydrolysis, and Pi or ADP release will alter the rotation between the large and small domains in the corresponding subunit. However, it is not clear how to link the structural characters to the mechanochemistry of the motor, so as to distinguish structural basis between random and coordinated firing.

15.3.3.2 Protein Unfolding and Translocation

Before the high resolution structure of the hexameric ClpX became available, it was shown in mutation experiments that a tyrosine residue in a pore loop of the ATPase links the hydrolysis to mechanical work by gripping substrates during unfolding and translocation [85]. The results support a model in which nucleotide-dependent conformational changes in the pore loops drive the substrate translocation and unfolding. Were the hexameric ClpX symmetric with a not-too-narrow central pore, the above suggestions would have provided a straightforward picture on how the protein substrate is driven through the motor, i.e., by each loop dragging one residue at a time.

However, significant asymmetries in the hexameric ClpX have been identified from the crystal structure [53], in which the tyrosine bearing loops occupy different axial positions. It is possible that ATP binding and/or hydrolysis could cause the loop to move downward as a consequence of the rigid-body movement of the large domain in an individual subunit. In addition, there are other types of pore loops that may participate in substrate interactions, though these extra loops are less conserved in AAA+ proteases. Anyhow, the pore loops fill most of the space in the pore such that it appears unlikely that even a single translocating polypeptide with bulky side chains could fit without some structural rearrangements.

Hence, the pore must be highly elastic in order to accommodate various sizes of protein substrates. One picture is that the protein substrate is 'swallowed' into the pore (as in the jaws of a snake), and 'squeezed' to drive unfolding, whereupon denatured polypeptides are translocated further by the loops. In order to describe this process, one should examine the size of the pore, the movement of the large domain, and the dominant loop configuration. These degrees of freedom are coordinated, loosely or tightly, as each subunit proceeds through its ATP hydrolysis cycles.

Indeed, it had been discovered early that the rate of ATP turnover was several fold slower during denaturation than translocation [86]. During the denaturation, the ATP turnover rate remains constant for substrates of different stabilities, but total ATP consumption increases with substrate stability. Hence, the effective unfolding force seems to be uniform and implemented iteratively. In addition, the protein unfolding process in ClpX has to be more or less coupled to the translocation of the unfolded polypeptide. In this aspect, it is similar to the helicase that couples ss translocation with duplex unwinding.

15.4 Conclusions

We have reviewed how simple biomolecular machinery, such as NTPase molecular motors, coordinate and control their internal degrees of freedom to achieve functional specificities. The mechanisms revealed from our computational work were based on recent experimental discoveries, in particular, from high-resolution structural studies and single molecule measurements. These NTPase motors share nucleotide-binding motifs in their catalytic sites, while developing variable types of architectures and functional modules. For the smallest helicase motors, the two RecA-like domains alternate their affinities to the nucleic acid strand as directed by ATP binding and product release, hence, leading to directional movements. For the ring-shaped NTPase motors, the inter-subunit coordination, though not indispensible in every case, is crucial as its presence ensures concerted motor firings around the ring. When the translocation of the motor is further coupled to activities such as DNA unwinding, protein displacement or unfolding, part of chemical energy utilized for translocation would be transferred to the activities.

For the P-loop NTPase motors of different types, there are some essential structural elements that play relatively conserved roles. For example, the arginine finger, present in both monomeric and ring-shaped systems, can stabilize some catalytic intermediate to accelerate the hydrolysis; the molecular levers in the cavity of the ring-shaped motors can push or pull on the substrate; the central beta-sheet, from which emanates both the P-loop (for nucleotide during) and the lever-like structure (for substrate contact), couple chemical transitions to mechanical force generation. It is interesting to consider how these molecular motors are evolutionarily connected, not only from a structural perspective, but also from their mechanochemic properties involving energy coupling and force generation.

References

1. Mavroidis C, Dubey A, Yarmush ML (2004) Molecular machines. Annu Rev Biomed Eng 6:363–395
2. Bustamante C, Liphardt J, Ritort F (2005) The nonequilibrium thermodynamics of small systems. Phys Today 58(7):43
3. Yu J, Ha T, Schulten K (2006) Structure-based model of the stepping motor of PcrA helicase. Biophys J 91(6):2097–2114
4. Yu J, Ha T, Schulten K (2007) How directional translocation is regulated in a DNA helicase motor. Biophys J 93(11):3783–3797
5. Yu J, Cheng W, Bustamante C, Oster G (2010) Coupling translocation with nucleic acid unwinding by NS3 helicase. J Mol Biol 404:439–455
6. Yu J, Moffitt J, Hetherington CL, Bustamante C, Oster G (2010) Mechanochemistry of a viral DNA packaging motor. J Mol Biol 400(2):186–203
7. Lohman TM, Tomko EJ, Wu CG (2008) Non-hexameric DNA helicases and translocases: mechanisms and regulation. Nat Rev Mol Cell Biol 9(5):391–401
8. Patel SS, Picha KM (2000) Structure and function of hexameric helicases. Annu Rev Biochem 69(1):651–697
9. Burroughs A, Iyer L, Aravind L (2007) Comparative genomics and evolutionary trajectories of viral ATP dependent DNA-packaging systems. In: Volff J-N (ed) Gene and protein evolution, vol 3, Genome dynamics. Karger, Basel, pp 48–65
10. Saraste M, Sibbald PR, Wittinghofer A (1990) The P-loop – a common motif in ATP- and GTP-binding proteins. Trends Biochem Sci 15:430–434
11. Mulkidjanian AY, Makarova KS, Galperin MY, Koonin EV (2007) Inventing the dynamo machine: the evolution of the F–type and V–type ATPases. Nat Rev Microbiol 5(11):892–899
12. Singleton MR, Dillingham MS, Wigley DB (2007) Structure and mechanism of helicases and nucleic acid translocases. Annu Rev Biochem 76(1):23–50
13. Bustamante C, Cheng W, Mejia YX (2011) Revisiting the central dogma one molecule at a time. Cell 144(4):480–497
14. Myong S, Ha T (2010) Stepwise translocation of nucleic acid motors. Curr Opin Struct Biol 20:121–127
15. Karplus M, McCammon A (2002) Molecular dynamics simulations of biomolecules. Nat Struct Biol 9:646–652
16. van Brabant A, Stan R, Ellis NA (2000) DNA helicases, genomic instability, and human genetic disease. Annu Rev Genomics Hum Genet 1:409–459

17. Dumont S et al (2006) RNA translocation and unwinding mechanism of HCV NS3 helicase and its coordination by ATP. Nature 439(7072):105–108
18. Cheng W, Dumont S, Tinoco I, Bustamante C (2007) NS3 helicase actively separates RNA strands and senses sequence barriers ahead of the opening fork. Proc Natl Acad Sci USA 104:13954–13959
19. Myong S, Bruno MM, Pyle AM, Ha T (2007) Spring-loaded mechanism of DNA unwinding by hepatitis C virus NS3 helicase. Science 317(5837):513–516
20. Velankar SS, Soultanas P, Dillingham MS, Subramanya HS, Wigley DB (1999) Crystal structures of complexes of PcrA DNA helicase with a DNA substrate indicate an inchworm mechanism. Cell 97(1):75–84
21. Dittrich M, Schulten K (2006) PcrA helicase, a prototype ATP-driven molecular motor. Structure 14(9):1345–1353
22. Dittrich M, Hayashi S, Schulten K (2003) On the mechanism of ATP hydrolysis in F1-ATPase. Biophys J 85:2253–2266
23. Dittrich M, Hayashi S, Schulten K (2004) ATP hydrolysis in the betaTP and betaDP catalytic sites of F1-ATPase. Biophys J 87:2954–2967
24. Astumian RD, Hanggi P (2002) Brownian motors. Phys Today 55(11):33–39
25. Dillingham MS, Wigley DB, Webb MR (1999) Demonstration of unidirectional single-stranded DNA translocation by PcrA helicase: measurement of step size and translocation speed. Biochemistry 39(1):205–212
26. Zheng W, Brooks B (2005) Identification of dynamical correlations within the myosin motor domain by the normal mode analysis of an elastic network model. J Mol Biol 346:745–759
27. Yang L-W, Chng C-P (2008) Coarse-grained models reveal functional dynamics – I. Elastic network models – theories, comparisons and perspectives. Bioinf Biol Insights 2:25–45
28. Altschul SF et al (1997) Gapped BLAST and PSI-BLAST: a new generation of protein database search programs. Nucleic Acids Res 25(17):3389–3402
29. Thompson JD, Higgins DG, Gibson TJ (1994) CLUSTAL W: improving the sensitivity of progressive multiple sequence alignment through sequence weighting, position-specific gap penalties and weight matrix choice. Nucleic Acids Res 22(22):4673–4680
30. Eargle J, Wright D, Luthey-Schulten Z (2006) Multiple alignment of protein structures and sequences for VMD. Bioinformatics 22(4):504–506
31. Lockless SW, Ranganathan R (1999) Evolutionarily conserved pathways of energetic connectivity in protein families. Science 286(5438):295–299
32. Suel GM, Lockless SW, Wall MA, Ranganathan R (2003) Evolutionarily conserved networks of residues mediate allosteric communication in proteins. Nat Struct Mol Biol 10(1):59–69
33. Levin MK, Gurjar M, Patel SS (2005) A Brownian motor mechanism of translocation and strand separation by hepatitis C virus helicase. Nat Struct Mol Biol 12(5):429–435
34. Frick DN (2007) The hepatitis C virus NS3 protein: a model RNA helicase and potential drug target. Curr Issues Mol Biol 9:1–20
35. Kim JL et al (1998) Hepatitis C virus NS3 RNA helicase domain with a bound oligonucleotide: the crystal structure provides insights into the mode of unwinding. Structure 6(1):89–100
36. Gu M, Rice CM (2010) Three conformational snapshots of the hepatitis C virus NS3 helicase reveal a ratchet translocation mechanism. Proc Natl Acad Sci USA 107:521–528
37. Jennings TA et al (2009) NS3 helicase from the hepatitis C virus can function as a monomer or oligomer depending on enzyme and substrate concentrations. J Biol Chem 284(8):4806–4814
38. Serebrov V, Beran RKF, Pyle AM (2009) Establishing a mechanistic basis for the large kinetic steps of the NS3 helicase. J Biol Chem 284(4):2512–2521
39. Matlock DL et al (2010) Investigation of translocation, DNA unwinding, and protein displacement by NS3h, the helicase domain from the hepatitis C virus helicase. Biochemistry 49(10):2097–2109
40. Rajagopal V, Gurjar M, Levin MK, Patel SS (2010) The protease domain increases the translocation stepping efficiency of the hepatitis C virus NS3-4A helicase. J Biol Chem 285(23):17821–17832

41. Khaki AR et al (2010) The macroscopic rate of nucleic acid translocation by hepatitis C virus helicase NS3h is dependent on both sugar and base moieties. J Mol Biol 400(3):354–378
42. Wang Q, Arnold JJ, Uchida A, Raney KD, Cameron CE (2010) Phosphate release contributes to the rate-limiting step for unwinding by an RNA helicase. Nucleic Acids Res 38(4):1312–1324
43. Cheng W, Arunajadai SG, Moffitt JR, Tinoco I, Bustamante C (2011) Single-base pair unwinding and asynchronous RNA release by the hepatitis C virus NS3 helicase. Science 333(6050):1746–1749
44. Lohman TM, Bjornson KP (1996) Mechanisms of helicase-catalyzed DNA unwinding. Annu Rev Biochem 65:169–214
45. von Hippel PH, Delagoutte E (2001) A general model for nucleic acid helicases and their 'coupling' within macromolecular machines. Cell 104:177–190
46. Betterton MD, Jülicher F (2005) Opening of nucleic-acid double strands by helicases: active versus passive opening. Phys Rev E 71(1):011904
47. Levin MK, Gurjar MM, Patel SS (2003) ATP binding modulates the nucleic acid affinity of hepatitis C virus helicase. J Biol Chem 278(26):23311–23316
48. Donmez I, Patel SS (2006) Mechanisms of a ring shaped helicase. Nucleic Acids Res 34(15):4216–4224
49. Barkow SR, Levchenko I, Baker TA, Sauer RT (2009) Polypeptide translocation by the AAA+ ClpXP protease machine. Chem Biol 26:605–612
50. Lyubimov AY, Strycharska M, Berger JM (2011) The nuts and bolts of ring-translocase structure and mechanism. Curr Opin Struct Biol 21(2):240–248
51. Chistol G et al (2012) High degree of coordination and division of labor among subunits in a homomeric ring ATPase. Cell 151(5):1017–1028
52. Iino R, Noji H (2013) Intersubunit coordination and cooperativity in ring-shaped NTPases. Curr Opin Struct Biol 23(2):229–234
53. Glynn SE, Martin A, Nager AR, Baker TA, Sauer RT (2009) Structures of asymmetric ClpX hexamers reveal nucleotide-dependent motions in a AAA+ protein-unfolding machine. Cell 139:744–756
54. Rao VB, Feiss M (2008) The bacteriophage DNA packaging motor. Annu Rev Genet 42:19.11–19.35
55. Morais M et al (2008) Defining molecular and domain boundaries in the bacteriophage phi29 DNA packaging motor. Structure 16:1267–1274
56. Sun S et al (2008) The structure of the phage T4 DNA packaging motor suggests a mechanism dependent on electrostatic forces. Cell 135:1251–1262
57. Smith D et al (2001) The bacteriophage phi29 portal motor can package DNA against a large internal force. Nature 413:748–752
58. Chemla Y et al (2005) Mechanism of force generation of a viral DNA packaging motor. Cell 122:683–692
59. Moffitt J et al (2009) Intersubunit coordination in a homomeric ring ATPase. Nature 457:446–450
60. Aathavan K et al (2009) Substrate interactions and promiscuity in a viral DNA packaging motor. Nature 461:669–673
61. Mancini E et al (2004) Atomic snapshots of an RNA packaging motor reveal conformational changes linking ATP hydrolysis to RNA translocation. Cell 118:743–755
62. Lisal J, Tuma R (2005) Cooperative mechanism of RNA packaging motor. J Biol Chem 280:23157–23164
63. Kainov D et al (2008) Structural basis of mechanochemical coupling in a hexameric molecular motor. J Biol Chem 283:3607–3617
64. Ahmadian MR, Stege P, Scheffzek K, Wittinghofer A (1997) Confirmation of the arginine-finger hypothesis for the GAP-stimulated GTP-hydrolysis reaction of Ras. Nat Struct Biol 4(9):686–689
65. Komoriya Y, Ariga T, Iino R, Noji H (2012) Principal role of the arginine finger in rotary catalysis of F1-ATPase. J Biol Chem 287(18):15134–15142

66. Wang H, Oster G (1998) Energy transduction in the F1 motor of ATP synthase. Nature 396:279–282
67. Augustin S et al (2009) An intersubunit signaling network coordinates ATP hydrolysis by m-AAA proteases. Mol Cell 35:574–585
68. Singleton MR, Sawaya MR, Ellenberger T, Wigley DB (2000) Crystal structure of T7 gene 4 ring helicase indicates a mechanism for sequential hydrolysis of nucleotides. Cell 101:589–600
69. Johnson DS, Bai L, Smith BY, Patel SS, Wang MD (2007) Single-molecule studies reveal dynamics of DNA unwinding by the ring-shaped T7 helicase. Cell 129:1299–1309
70. Donmez I, Rajagopal V, Jeong Y-J, Patel SS (2007) Nucleic acid unwinding by hepatitis C virus and bacteriophage T7 helicases is sensitive to base pair stability. J Phys Chem 282:21116–21123
71. Donmez I, Patel SS (2008) Coupling of DNA unwinding to nucleotide hydrolysis in a ring-shaped helicase. EMBO J 27(12):1718–1726
72. Crampton DJ, Mukherjee S, Richardson CC (2006) DNA-induced switch from independent to sequential dTTP hydrolysis in the bacteriophage T7 DNA helicase. Mol Cell 21:165–174
73. Liao J-C, Jeoong Y-J, Kim D-E, Patel S, Oster G (2005) Mechanochemistry of T7 helicase. J Mol Biol 350:452–475
74. Kim D-E, Narayan M, Patel SS (2002) T7 DNA helicase: a molecular motor that processively and unidirectionally translocates along single-stranded DNA. J Mol Biol 321:807–819
75. Jeong Y-J, Levin MK, Patel SS (2004) The DNA-unwinding mechanism of the ring helicase of bacteriophage T7. Proc Natl Acad Sci USA 101:7264–7269
76. Stano NM et al (2005) DNA synthesis provides the driving force to accelerate DNA unwinding by a helicase. Nature 435:370–373
77. Sauer RT et al (2004) Sculpting the proteome with AAA+ proteases and disassembly machines. Cell 119(1):9–18
78. Zolkiewski M (2006) A camel passes through the eye of a needle: protein unfolding activity of Clp ATPases. Mol Microbiol 61:1094–1100
79. Iyer LM, Leipe DD, Koonin EV, Aravind L (2004) Evolutionary history and higher order classification of AAA+ ATPases. J Struct Biol 146(1–2):11–31
80. Flynn JM, Neher SB, Kim Y-I, Sauer RT, Baker TA (2003) Proteomic discovery of cellular substrates of the ClpXP protease reveals five classes of ClpX-recognition signals. Mol Cell 11(3):671–683
81. Martin A, Baker TA, Sauer RT (2005) Rebuilt AAA+ motors reveal operating principles for ATP-fuelled machines. Nature 437:1115–1120
82. Kinosita K, Adachi K, Itoh H (2004) Rotation of F1-ATPase: how an ATP-driven molecular machine may work. Annu Rev Biophys Biomol Struct 33:245–268
83. Skordalakes E, Berger J (2003) Structure of the Rho transcription terminator: mechanism of mRNA recognition and helicase loading. Cell 114:135–146
84. Enemark E, Joshua-Tor L (2006) Mechanism of DNA translocation in a replicative hexameric helicase. Nature 442:270–275
85. Martin A, Baker TA, Sauer RT (2008) Pore loops of the AAA+ ClpX machine grip substrates to drive translocation and unfolding. Nat Struct Mol Biol 15:1147–1151
86. Kenniston JA, Baker TA, Fernandez JM, Sauer RT (2003) Linkage between ATP consumption and mechanical unfolding during the protein processing reactions of an AAA+ degradation machine. Cell 114:511–520

Chapter 16
Multi-state Targeting Machinery Govern the Fidelity and Efficiency of Protein Localization

Mingjun Yang, Xueqin Pang, and Keli Han

Abstract Proper localization of newly synthesized proteins is essential to cellular function. Among different protein localization modes, the signal recognition particle (SRP) and SRP receptor (SR) constitute the conserved targeting machinery in all three life kingdoms and mediate about one third of the protein targeting reactions. Based on experimental and computational studies, a detailed molecular model is proposed to explain how this molecular machinery governs the efficiency and fidelity of protein localizations. In this targeting machinery, two distinct SRP GTPases are contained into the SRP and SR that are responsible to the interactions between SRP and SR. These two GTPases can interact with one another through a series of sequential and discrete interaction states that are the early intermediate formation, stable complex association, and GTPase activation. In contrast to canonical GTPases, a floppy and open conformation adopted in free SRP GTPases can facilitate efficient GTP/GDP exchange without the aid of any external factors. As the apo-form free SRP GTPases can adopt the conformational states of GDP- or GTP-bound form, the binding of GTP/GDP follows a mechanism of conformational selection. In the first step of complex formation, the two SRP GTPases can rapidly assemble into an unstable early intermediate by selecting and stabilizing one another's primed states from the equilibrium conformational ensemble. Subsequently, extensive inter- and intra-domain changes rearrange the early complex into a tight and closed state of stable complex through induced fit mechanism. Upon stable complex association, further tune of several important interaction networks activates the SRP GTPase for GTP hydrolysis. These different conformational states are coupled to corresponding protein targeting events, in which the complex formation deliveries the translating ribosome to the target membrane and the GTPase activation couples to the cargo release from SRP-SR

M. Yang • X. Pang • K. Han (✉)
Dalian Institute of Chemical Physics, Chinese Academy of Sciences, Group 1101, Zhongshan Road 457, Dalian City, Liaoning Province 116023, China
e-mail: klhan@dicp.ac.cn

K. Han et al. (eds.), *Protein Conformational Dynamics*, Advances in Experimental Medicine and Biology 805, DOI 10.1007/978-3-319-02970-2_16,

machinery to the translocation channel. It is thus suggested that the SRP GTPases constitute a self-sufficient system to execute exquisite spatial and temporal control of the complex targeting process. The working mechanism of the SRP and SR provides a novel paradigm of how the protein machinery functions in controlling diverse biological processes efficiently and faithfully.

Keywords Signal recognition particle • Protein targeting • Protein localization • Molecular machinery • Protein machinery • Protein conformational dynamics • Principal component analysis • Targeted molecular dynamics

Abbreviations

cpSRP	chloroplast SRP
cpSR	chloroplast SR
Ffh	SRP54 homologous protein in bacteria and archaea
FtsY	SRα homologous protein in bacteria and archera
GDP	guanosine diphosphate
GTP	guanosine triphosphate
IBD	insertion box domain
RNC	ribosome nascent chain complex
SRP	signal recognition particle
SR	SRP receptor
T.aq.	Thermus aquaticus

16.1 Introduction

Proteins play a variety of fascinating roles in virtually all intra- and inter-cellular events. The functions of proteins are exerted through their unique three-dimensional structures dictated by the amino acid sequence. However, the protein structures are not rigid and can adopt a series of different functional conformations in response to biological cues, including interactions with other molecules and exterior environmental changes (light, heat, pressure, and concentration etc.) [1, 2]. Many studies have shown that the transitions between different functional states are intrinsically hierarchical in both time and space, from the local changes of side chain rotamers with typically picoseconds timescale to the collective inter-domain motions within micro- or milliseconds. Thus the dynamic personalities of specific proteins potentially govern the temporal and spatial precisions of diverse cellular processes [3].

In light of the central role of protein conformational dynamics in cellular functions, this question has attracted extensive experimental and computational studies in the past several decades [4]. In experiments, the studies from X-ray crystallography [5], nuclear magnetic resonance spectroscopy [6, 7], time-resolved X-ray scattering [8, 9], single particle cryo-electron microscopy [10, 11], fluorescence-based ensemble/single-molecule techniques [12–14], infrared spectroscopy [15], and neutron scattering [16] have offered insightful information of the protein structures and conformational dynamics. From the computational aspects, the persistent developments of novel algorithms and simulation skills have provided powerful tools in dissecting the detailed correlation between protein conformational dynamics and functions. These computational methods include different levels of system representations (from quantum mechanics [17–23], molecular mechanics [24, 25], to coarse grained models [26, 27]), the enhanced conformational sampling [28–31] or accelerated exploration [32], the analysis algorithms to extract useful information from simulated trajectories [33, 34], and the flexible fitting methods between different resolutions of structures [35, 36] etc. These experimental and computational methods have shown distinct advantages and efficiencies in demonstrating diverse aspects of protein functioning mechanisms. Combining the advantages of both types of methods, people have attained significant insights into the role of the protein conformational dynamics in a number of different biological processes, e.g. enzyme catalysis [3, 37], protein folding [38], molecular transport [39, 40], and signal transduction [41, 42]. In this chapter, we review the working mechanism of the protein targeting machinery consisting of signal recognition particle (SRP) and its receptor (SR) [43, 44]. In this molecular machinery, two SRP GTPases interact with each other through a series of discrete and sequential conformational switches, each of which can be utilized by external cues in controlling the efficiency and fidelity of the targeting events. Studies of this system from both experiments and computations have provided a novel paradigm of the structure-dynamics-function relation in the protein targeting process.

Proteins are synthesized by ribosomes in cytosol and roughly one third of the newly synthesized proteins are destined as secretive or membrane proteins. Of these proteins, the first translated 15–30 amino acid residues compose the signal peptide, which determines their cellular localization. According to properties of different signal sequence, the corresponding molecular machinery are utilized to mediate correct localization of the proteins, among which the SRP and SR together form an evolutionarily conserved co-translational protein targeting machinery in all three life kingdoms [45–47]. The SRP and SR target the translating ribosome-nascent chain complex (RNC cargo) to a protein translocation channel in the endoplasmic reticulum membrane in eukaryotes or the plasma membrane in prokaryotes. In this targeting process, the SRP firstly recognizes and binds the newly synthesized signal peptide and then load the RNC cargo to membrane translocon through a series of discrete and sequential interaction states with SR associated to the membrane [43, 44].

16.2 Bacteria Preserve a Minimal Functional Core of the SRP and SR Targeting Machinery

There are two distinct SRP pathways existing in cytosol and chloroplast to mediate the co-translational or post-translational protein targeting, respectively [48–50]. The composition of SRP and SR in cytosol varies widely among different organisms.

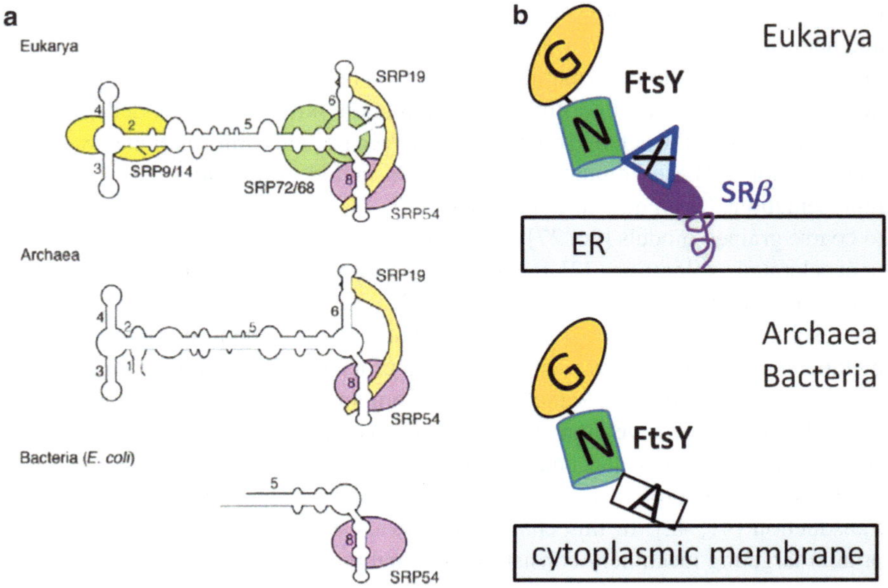

Fig. 16.1 The composition of SRP and SR in eukarya, archaea, and bacteria. (**a**) Components of SRP (The figure is used under permission from the original journal [55]). (**b**) Components of SR

In higher eukaryotes, SRP is a nucleoprotein particle consisting of one 7S RNA and six proteins including SRP9/14, SRP72/68, SRP19, and SRP54 [51]. This complex can be classified into the ALU domain and the S domain that are structurally and functionally independent to each other. The ALU domain includes SRP9/14 and helices 1–4 of SRP RNA and the S domain includes SRP72/68, SRP19, SRP54, and helices 6–8 of SRP RNA [52] (Fig. 16.1a). Studies have shown that the ALU domain plays a role in arresting protein translation on ribosome and the S domain is mainly responsible to signal peptide recognition and SRP-SR interaction [53, 54]. The SR is a heterodimer composed of two subunits SRα and SRβ (Fig. 16.1b). The SRβ is a transmembrane protein and anchors SRα onto the endoplasmic reticulum membrane through its interaction with the N-terminal X domain of SRα. The SRα shares high sequence and structure similarity with SRP54 and is responsible to the interaction with SRP during protein targeting. In archaels, only one 7S SRP RNA and the SRP54 and SRP19 homologues are included in SRP and only SRα homologue in SR [55]. Bacteria occupy a set of minimal components of the targeting machinery, in which only one 4.5S SRP RNA and the SRP54 homolog are included in SRP and the SRα homologous protein in SR [47, 56]. In both archaea and bacteria, the SRP54 and SRα homologous proteins are termed as Ffh and FtsY, respectively (Fig. 16.1b) [57]. The 4.5S RNA in bacterial SRP shares a conserved identity to helix 8 of the 7S RNA in both eukaryotes and archaels. Despite the simple constitution of the targeting machinery in bacteria, it carries the same functional role as in other complicated organisms and thus most experiments have been carried out using this reduced system.

Besides the canonical composition of SRP and SR in cytosol, another distinct SRP targeting system exists in the chloroplast of higher plants, in which the conserved cytosolic SRP RNA is absent and only one SRP54 homologue (cpSRP54) and one chloroplast specific protein SRP43 (cpSRP43) compose the chloroplast SRP (cpSRP). In chloroplast SR, only one SRα homologous protein (cpFtsY) is included. In fact, the cpSRP RNA are also found in the chloroplast of some photosynthetic organisms, which can be the evolutionary intermediate state of this machinery from bacteria to higher plants [48].

16.3 Structural Characteristics of the Functional Core

X-ray crystallographic studies have provided insightful understandings of the SRP and SR structures [58–63]. Because a minimal composition of SRP system can have the full function in protein targeting and is conserved among other complicated organisms, we will take these essential components as an example to address their structure features. In bacteria and archaea, the homologous proteins of SRP54 and SRα are termed as Ffh and FtsY, respectively. Both Ffh and FtsY contain two universally conserved N and GTPase (G) domain [62, 63] (Fig. 16.2a). The N domain is composed of four helices packing into a bundle and opens at one end to accommodate the hydrophobic core of the G domain, which results in a structurally and functionally coupled NG unit (Fig. 16.2b). The G domain is a unique GTPase domain, which shares high structural similarity to the GTP-binding domain of RAS-like GTPases. Distinct from RAS-like GTPases, the SRP GTPases include an additional insertion-box domain (IBD). A number of conserved motifs, including the motifs I-V, DGQ, DARGG in the G domain and the ALLEADV in the N domain, are responsible to GTP binding and hydrolysis and comprised of the main interaction interface between SRP and SR [60, 61]. Besides the NG domain, there is an additional M domain linked to the C-terminus of G domain through a flexible helix [64, 65]. The M domain is responsible to the signal peptide recognition through a hydrophobic binding groove and forms primary contacts with the SRP RNA [66] (Fig. 16.2c). In bacteria and archaea where no SRβ component exists, an additional A domain at the N-terminus of FtsY attaches itself onto the plasma membrane. However, the A domain is absent in FtsY from Thermus aquaticus (*T.aq.*). Although the SRP RNA is much variable, the 4.5S RNA preserves a minimal functional core, in which two internal loops (symmetric and asymmetric) interact with the M domain of SRP54. A tetraloop at the proximal end plays a critical role in regulation of protein translocation [67–70].

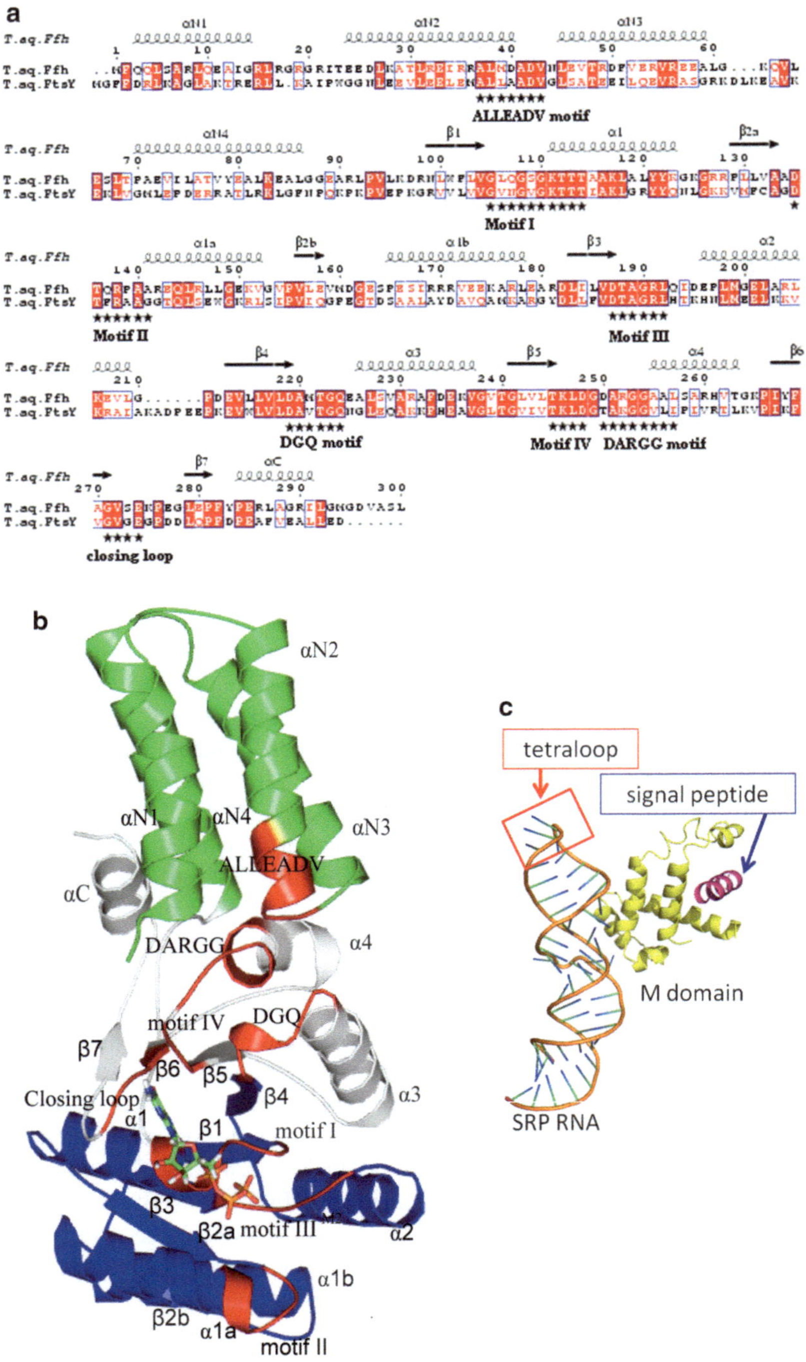
a
T.aq.Ffh
T.aq.FtsY
ALLEADV motif
Motif I
Motif II
Motif III
DGQ motif
Motif IV
DARGG motif
closing loop
b
αN2
αN1
αN4
αN3
ALLEADV
αC
DARGG
α4
motif IV
DGQ
β7
β6
β5
β4
α3
Closing loop
α1
β1
motif I
β3
β2a
motif III
α2
α1b
β2b
α1a
motif II
c
tetraloop
signal peptide
M domain
SRP RNA

Previous studies have shown that the SRP GTPases in both SRP54 and SRα undergo a series of discrete and sequential conformational states in the process of protein targeting [43, 69, 71] (Fig. 16.3). These conformations respond to external cues and provide different regulation points of the targeting events. In this targeting pathway, the SRP recognizes and binds to the RNC cargo and then loads it to the membrane translocon through interaction with SR. The SRP and SRα interaction begins with fast assembly of an unstable early complex and then extensive conformational rearrangements lead to a stable complex formation. Following the stable SRP-SR complex, the SRP GTPases are activated and GTP hydrolysis drives dissociation of the complex to produce free SRP and SR for the next round of protein targeting.

16.4 Free SRP GTPases: Floppy Open Conformation in apo-, GDP-, or GTP-Binding States

Unlike the canonical RAS-like GTPases that require the guanine exchange factor (GEF) to switch from GDP-bound to GTP-bound states, the intrinsic nucleotide exchange in free (monomeric) SRP GTPases does not require the aid of any external factors and has a rate of 2–4 orders faster than that of canonical GTPases [72, 73]. However, the basal GTPase activity in free SRP GTPases is very low compared to canonical GTPases (Fig. 16.4a). Biophysical studies based on fluorescent techniques have determined similar binding affinity between GDP and GTP to free SRP GTPases, with a dissociation constant of 0.2–2.0 μM. However, considering the different concentrations of GTP (900 μM) and GDP (100 μM) in physiological conditions, the GTP-bound form is predominant in vivo [72]. Structural studies by X-ray diffraction indicate that free SRP GTPases bind GDP or GTP in a floppy and open conformation, which is very similar to that of apo-form state (Fig. 16.4b, c) [74–78]. Computational simulations show that the conformation of free SRP GTPases is flexible in both the conserved motifs and inter-domain orientations, which can accommodate the variations in different crystal structures [79, 80]. It is thus suggested that GDP or GTP binding to free SRP GTPases utilizes the conformational selection rather than the induced-fit mechanism, in which the ligand selectively binds to a specific conformational ensemble of the apo-receptor and stabilizes the conformation after formation of the receptor-ligand complex [41].

Fig. 16.2 Sequence alignment and structural features of the NG domain and signal peptide-M domain-RNA interaction. (**a**) Sequence alignment between the NG domains of Ffh (*T.aq.*) and FtsY (*T.aq.*). (**b**) Cartoon representation of the NG domain of Ffh. The N domain is colored *green*, the conserved sequence motifs in *red* and the core region of the G domain in *blue*. (**c**) The signal peptide-M domain-RNA interaction model was constructed using resolved crystal structures (PDBID: 1DUL [66] and 3NDB [59]) (Figures (**a**) and (**b**) are used under permission from original journal [79])

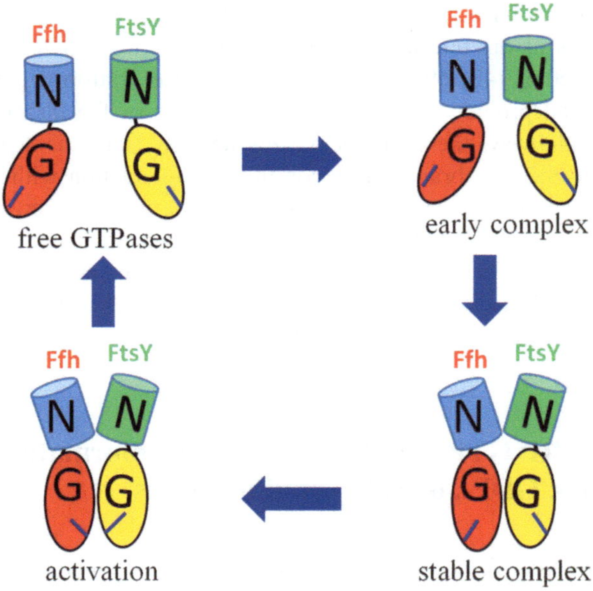

Fig. 16.3 Schematic draw of the discrete and sequential interaction states between Ffh and FtsY in the SRP mediated protein targeting process

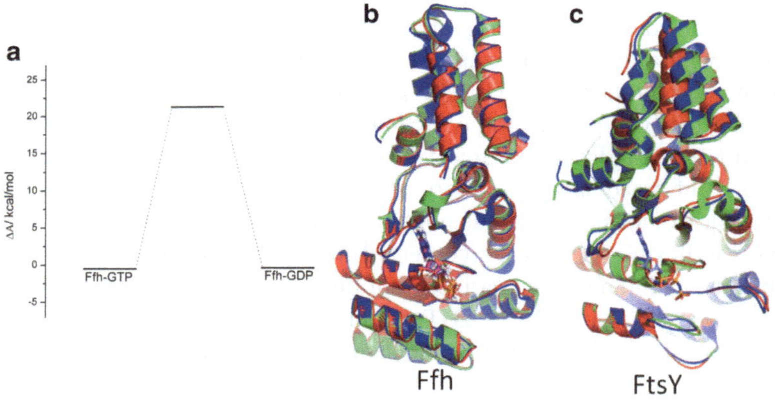

Fig. 16.4 Free energy profile for GTP hydrolysis in free SRP GTPases and structural superposition over the apo-, GDP-, and GTP-analog bounded Ffh and FtsY. (**a**) The free energy profile is computed using rate formula $k = \frac{k_B T}{h} e^{-\beta \Delta G}$, where k, ΔG, T, k_B, h, and β are the rate constant determined in experiment [73], free energy barrier of the reaction, experimental temperature, Boltzmann constant, Planck constant, and inverse temperature, respectively. (**b**) Structural superposition of *T.aq.* Ffh (PDBID: 1LS1 [75], 2C03 [78], and 2C04 [78]). (**c**) Structural superposition of *T.aq.* FtsY (PDBID: 2Q9A [77], 2Q9C [77], and 2IYL [76])

16.5 Unstable Early Complex Formation Between Ffh and FtsY: A Conformational Selection Mechanism

In the process of SRP-mediated protein targeting, the SRP interacts with SR through a series of discrete and sequential conformational states, which begins from the fast early-complex formation between Ffh and FtsY [43]. This early complex is an on-pathway transient intermediate and can be stalled by GDP-bound or apo-form SRP GTPases [81]. Biophysical studies using electron paramagnetic resonance and time-resolved fluorescent resonance energy transfer methods reveal that the early complex can occupy a broad conformational ensemble with relatively closer N domain-N domain contacts and further separated G domain-G domain interactions. According to charge distribution in the two proteins, the formation of early complex is mediated mainly by different electrostatic properties in the N domains of the two SRP GTPases [69].

The conformation rearrangements are believed to be small in formation of the transient intermediate since a fast kinetic rate was determined experimentally [81]. Intriguingly, computational simulations show that a cooperative inter- and intra-domain motion embedded into the equilibrium fluctuation of free SRP GTPases can be functionally relevant to the formation of Ffh-FtsY complex (Fig. 16.5) [79]. However, the magnitude of functional fluctuation is far from that of the conformational changes required for stable Ffh-FtsY complex formation. It is suggested that these functional relevant motions can be utilized by external interactions to shift the equilibrium towards the association of the heterodimeric complex. In this context, the fast assembly of early intermediate can be regarded as the first stage of stable complex formation, in which free SRP GTPases can select and stabilize one another's primed state from the equilibrium conformational ensembles [82].

16.6 Stable Ffh-FtsY Complex Association: Extensive Inter- and Intra-domain Rearrangements from the Induced-Fit Interaction

In the presence of GTP or GTP analogs, the early intermediate proceeds to a stable Ffh-FtsY complex. Based on crystal structures resolved for the heterodimeric complex and the free SRP GTPases, two significant changes are observed: the N/G inter-domain orientation and the position of conserved motifs (Fig. 16.5b) [54, 58, 60, 83–85]. Along with stable complex association, the floppy open conformation in free SRP GTPases evolves into a tightly closed state in the heterodimeric complex in which extensive interaction interface are formed between the conserved motifs in the two proteins. Structural analyses show that the core region of the G domain is composed of several alternatively connected α helices and β sheet, and this region changes as a rigid body with respect to the N domain. Computational studies using targeted molecular dynamics suggest that the relative

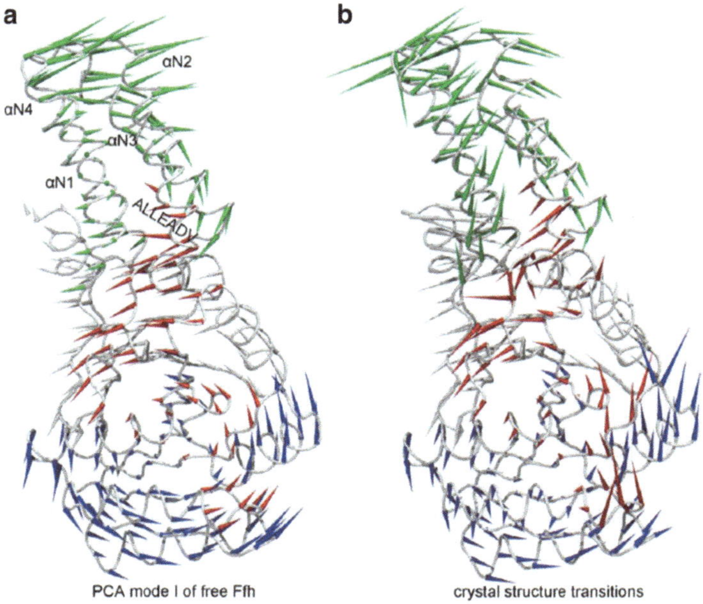

Fig. 16.5 Functional relevant motions in free *T.aq.* Ffh. The protein structures are shown in tube, with a cone attached to each CA atom indicating the direction of displacement. The length of the cone represents twice the magnitude of the displacement for clarity. The N domain is colored in *green*, the conserved sequence motifs in *red* and the core region of the G domain in *blue*. (**a**) The first mode of principal component analysis of the equilibrium fluctuation of free Ffh. (**b**) Crystal structural transitions from free Ffh (PDBID: 1JPJ [74]) to Ffh-FtsY complex (PDBID: 1RJ9 [61]) (This figure is used under permission from original journal [79])

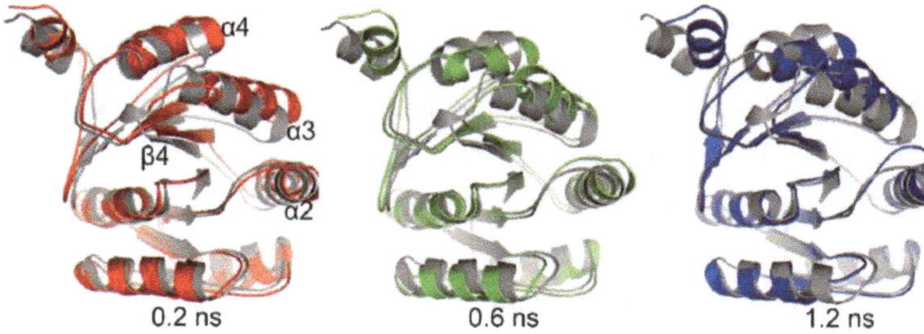

Fig. 16.6 Structural rearrangements of α3 and α4 helices from free Ffh to heterodimeric Ffh in targeted molecular dynamics simulations. The initial conformation is colored *gray*. Only the G domain is shown for clarity (This figure is used under permission from original journal [79])

position of two α helices (α3 and α4) packing at the domain interface can move flexibly to serve as a bridge between the N domain and the core region of the G domain to accommodate significant inter-domain reorientations in the process of complex formation (Fig. 16.6) [79]. However, these extensive inter- and intra-domain changes are only observed in the reciprocal interaction between GTP-bound Ffh and FtsY, which can be assigned to the induced-fit mechanism from early intermediate to stable complex [82].

16.7 SRP GTPase Activation: Fine Tune of the Positions of Several Important Residues from Stable Complex

Upon the formation of the heterodimeric complex between Ffh and FtsY, a composite active chamber is formed between the *cis/trans* packed GTP molecules and the conserved motifs in the two proteins (Fig. 16.7a) [60, 61]. Distinct from canonical RAS-like GTPases, which rely on external GTPase activating protein (GAP) to hydrolyze GTP, the SRP GTPases can reciprocally stimulate one another's enzymatic activity to catalyze GTP hydrolysis upon complex association. However, in all the available structures of the stable heterodimeric complex, some residues are not arranged properly to explain their role in GTPase activation observed from mutational studies [60, 61, 71]. These discrepancies from structural and biochemical observations suggest that additional tune of the stable complex conformation is required to achieve GTPase activation.

Computational studies of the *T.aq.* Ffh-FtsY complex show that three key interaction networks contribute to the GTPase activation via well-tuned conformational changes [86]. The first network involves the conserved Ffh:R191 residue (or its homolog FtsY:R195), the mutation of which diminishes GTPase activation in both Ffh and FtsY (Fig. 16.7b) [71, 87]. Analyses of the crystal structures in complex with GTP analogs show that the Ffh:L198 residue (FtsY:L202) hinders further readjustment of Ffh:R191 (FtsY:R195). Molecular simulations with GTP bound in both active sites and in solvent environment suggest that flip of the loop connecting Ffh:R191 and Ffh:M199 (FtsY:R195 and FtsY:M203) can remove the hindrance and accommodate further rearrangements of Ffh:R191 (FtsY:R195) side chain. In several trajectories under different simulation conditions, the side chain of Ffh:R191 (FtsY:R195) can rotate across the Ffh-FtsY interface to form additional interactions with FtsY:E284 (Ffh:E274). In this new model of the arginine residues, mutation of Ffh:R191 near the Ffh active site disrupts its interaction with FtsY:E284 and results in deficiency of both active sites. In the second network, two positively charged residues Ffh:R138 and FtsY:R142 position in the center of the composite active chamber and play critical role in GTP hydrolysis (Fig. 16.7c). A rotation of FtsY:R142 side chain is reproduced in several different simulation trajectories when replacing the GTP analogs in crystal structures by GTP molecules in the computational models. The new conformation of FtsY:R142 can generate stronger interaction with the leaving phosphate group of GTP, which is suggested to maintain better electrostatic balance of the active chamber and stabilize the transition state in GTP hydrolysis. In the third interaction network, one crystal water molecule is suggested to attack the leaving phosphate group in GTP hydrolysis and is stabilized

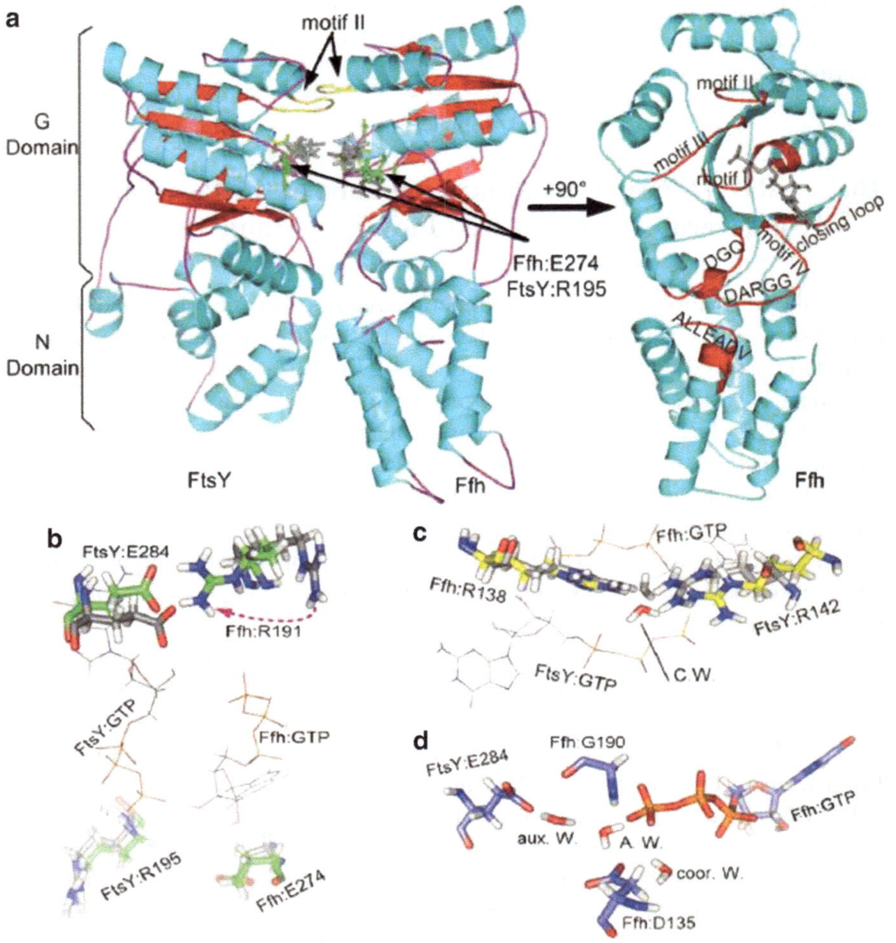

Fig. 16.7 The activation model of SRP GTPases in Ffh-FtsY complex. (**a**) Cartoon representation of the heterodimeric complex. FtsY:R195 and Ffh:E274, mirrored by Ffh:R191 and FtsY:E284 on the other side of the complex were shown in *green sticks* (*left*). Positions of conserved motifs were labeled and colored in *red* (*right*). The bound GTP molecules were represented in *gray sticks*. (**b**) Conformational rearrangement of the pending residues Ffh:R191 and FtsY:R195 before (*gray*) and after (*green*) the formation of salt bridge. (**c**) Rotation of FtsY:R142 side chain from bended to extended conformation in MD simulations. This interaction network involving FtsY:R142, Ffh:R138, the central water (C.W.) and two GTP molecules. Initial conformation was in *gray* and the rotated conformation in *yellow*. The central water shifted from the initial position (*gray*) to a new position (*red*) toward FtsY:GTP. (**d**) Structure model of the interaction network including Ffh:D135 (FtsY:D139), nucleophilic (attacking) water (A.W.), auxiliary water (aux. W.), coordinating water (coor. W.), Ffh:G190 (FtsY:G194), Ffh:GTP (FtsY:GTP), and FtsY:E284 (Ffh:274) (The figure is used under permission from original journal [86])

by an auxiliary water (Fig. 16.7d) [83]. However, the auxiliary water is too mobile to maintain the interaction network for attacking water stabilization. Instead, a stable hydrogen bonding interaction between the conserved residue Ffh:G190 or FtsY:G194 and the attacking water is observed in the simulations. Therefore, it is the glycine residue rather than the auxiliary water molecule suggested from crystal structures is responsible to stabilize the attacking water in GTP hydrolysis.

Taken together, these conformational rearrangements from stable Ffh-FtsY complex present a more favorable interaction model for SRP GTPase activation in comparison to the conformational states resolved from crystallographic studies. These findings provide a connection between the structural characteristics of the key interaction networks and their functional roles for understanding of the distinct GTPase activation mode by hetero- or homo-dimerization [88].

16.8 GTP-Binding Primes the cpFtsY Conformation for Efficient Ffh-FtsY Complex Association

Besides the cytosolic SRP pathway in which the SRP RNA plays an indispensible role in protein targeting, there is another distinct SRP system in chloroplast of higher plants [48, 49, 89]. The chloroplast SRP mediates protein localization to thylakoid membrane through both post-translational and co-translational modes. In the co-translational pathway, some chloroplast encoded proteins are transported when translating on the ribosome; whereas the post-translational pathway mainly target the nucleus-encoded light harvesting chlorophyll a/b binding proteins after finishing translation. In comparison to cytosolic SRP, three distinct features are found in the cpSRP pathway: (1) in the absence of SRP RNA, cpSRP54 can interact with cpFtsY (the SRα homolog in chloroplast) with a similar efficiency as their bacterial homologs [90, 91]; (2) in contrast to bacterial FtsY that exhibits low discrimination between cognate and noncognate nucleotide in its free form, free cpFtsY displays substantial GTP specificity [90]; (3) in comparison to free cytosolic FtsY, the N/G inter-domain orientation in cpFtsY is much closer to that of the stable *T.aq.* FtsY-Ffh complex [48, 92, 93]. It is thus interesting to address why cpSRP pathway can bypass the requirement of SRP RNA to achieve efficient protein localization.

Based on structural analyses and kinetic rate measurements, it is suggested that free cpFtsY is preorganized into a closed state that allows an optimal interaction with cpSRP54 [48, 90–93]. However, all the crystal structures for cpFtsY are in apo form, which cannot provide direct evidence to support the biochemical hypothesis [48, 92, 93]. In this case, computational simulation is a convenient tool to virtually construct the GTP-bound cpFtsY and investigate the conformational dynamics induced by GTP molecule. In a well-designed computational study of both apo-cpFtsY and GTP-cpFtsY, GTP binding induces important inter-domain

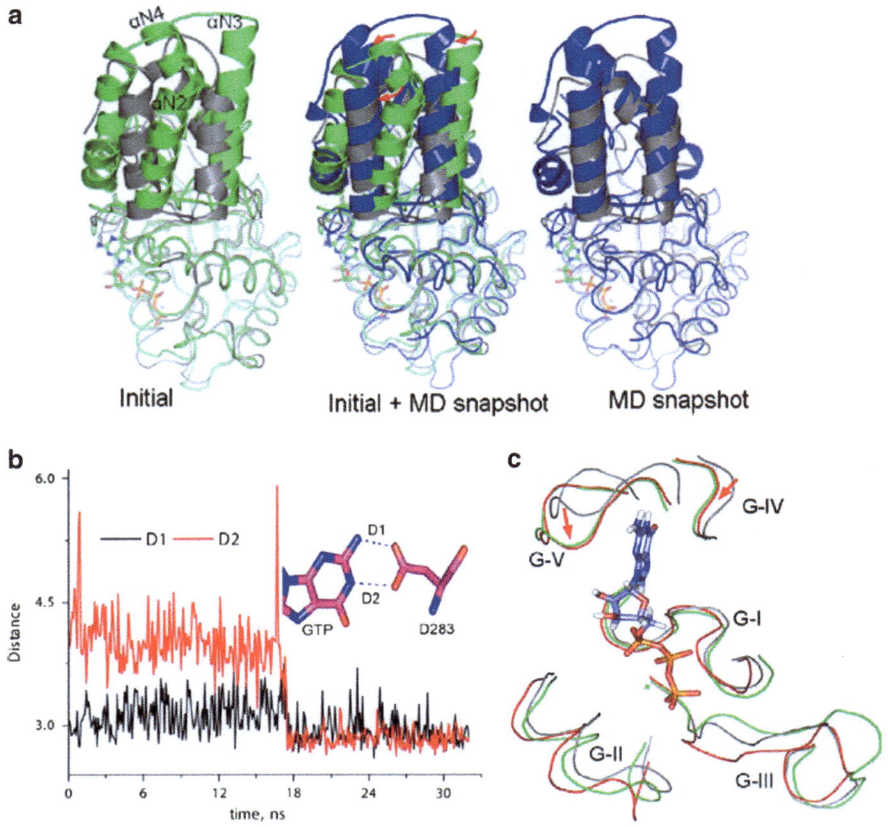

Fig. 16.8 GTP induced inter- and intra-domain rearrangements in free cpFtsY. (**a**) Superposition among crystal structure of the cpFtsY (PDB code 2OG2 [93], *green*), structure of the *T. aq.* FtsY in the stable complex (PDB code 1OKK [60], *gray*), and a representative MD snapshot (*blue*) at 19 ns in the simulation of the GTP–cpFtsY that shared the same N–G domain orientation with the *T. aq.* FtsY. *Red arrows* (*middle panel*) indicate movement of the N domain from the initial cpFtsY structure. (**b**) Distance variation between the nucleotide specificity determinant cpFtsY:D283 and the guanine base of GTP molecule. (**c**) Conformational rearrangements of motifs G-I–G-V. The crystal structure of the cpFtsY (PDB code 2OG2 [93], *gray*), the structure of the T. aq. FtsY in the Ffh–FtsY complex (PDB code 1OKK [60], *green*), and one representative MD snapshot (*red*) at 19 ns of GTP–cpFtsY simulation were superimposed onto the core region of the G domain. *Red arrows* indicated movements towards the conformation of Ffh–FtsY complex

reorientation towards the structure observed in the Ffh-FtsY complex (Fig. 16.8a) [80]. This inter-domain rearrangement is strictly GTP dependent where no similar dynamic changes were observed in a longer simulation time of the apo-cpFtsY model. Along with the inter-domain change, a GTP specificity determinant residue in the conserved motif V of the G domain is brought closer to the nucleotide ring and produces a tight hydrogen bonding interaction as observed in the stable Ffh-FtsY complex, explaining why free cpFtsY exhibits substantial discrimination over cognate and noncognate nucleotides (Fig. 16.8b). In addition, a number of conserved motifs in the G domain that are responsible to GTP binding and interaction interface formation between cpSRP54 and cpFtsY also rearrange into a closed conformation (Fig. 16.8c). Further analyses using the binding energy decomposition show that the GTP binding in cpFtsY follows the same principle as in the bacterial homolog because the same set of residues contribute remarkably to the binding affinity. These observations suggest that the conformation of GTP-cpFtsY could access to the preorganized structure for formation of a stable complex under the equilibrium condition. In this case, large energy penalty is not paid for significant conformational rearrangement from free cpFtsY to stable cpFtsY-cpSRP54 complex, rationalizing why the cpSRP system can bypass the requirement of SRP RNA in the targeting process.

16.9 SRP RNA Regulates the Kinetic and Thermodynamic Properties of the Interaction Between Ffh and FtsY

In the cytolic SRP pathway, the SRP RNA is an indispensible component. It profoundly influences the kinetic and thermodynamic properties of Ffh and FtsY interaction; the SRP RNA stabilizes the early complex by 50-fold and accelerates assembly of the stable complex and the followed GTPase activation by 200–400 and 6-fold, respectively [73, 81, 94, 95]. However, the stabilization of stable complex is not affected, suggesting a typical catalytic role of SRP RNA in the stable Ffh-FtsY complex formation. Mutational studies indicate that the tetraloop region in the proximal end of SRP RNA plays a critical role in acceleration of Ffh-FtsY complex association and the distal end specifically stimulate the subsequent SRP GTPase activation [69, 81, 85]. Structural studies by both X-ray crystallography and cryo-EM show that the SRP RNA tetraloop binds to the G domains in the Ffh-FtsY early conformation and the distal end binds to the G domains in stable Ffh-FtsY complex (Fig. 16.9) [85, 96]. These results suggest that the interaction of Ffh and FtsY takes place initially at the tetraloop end of SRP RNA and then is transferred to the distal end. Based on observation of a large-scale movement of the relative position between SRP RNA and Ffh-FtsY complex from single-molecule fluorescence microscopy, Shan and co-workers suggest that the SRP RNA can act as a scaffold to handover the RNC cargo onto membrane translocation channel [70].

16.10 Conformational Rearrangements in the Interaction Process of Ffh and FtsY Provide Key Regulation Points for External Cues

In the SRP mediated protein targeting pathway, the interactions between SRP and SR undergo a series of discrete and sequential conformational states. These conformational rearrangements provide important regulation points by external

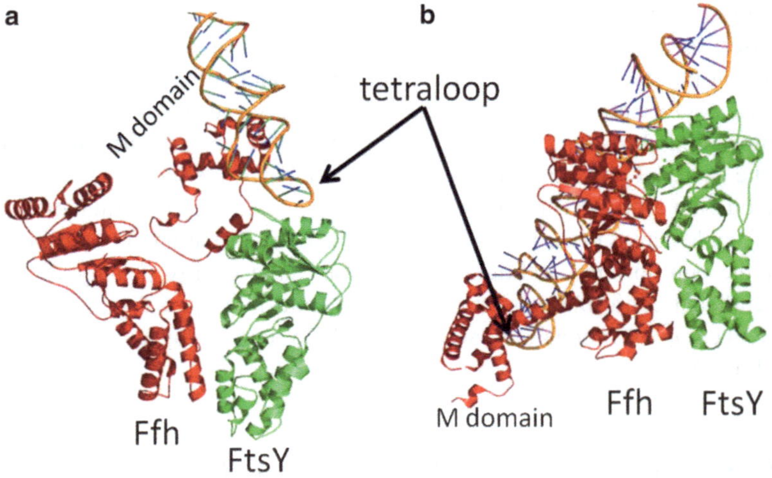

Fig. 16.9 The relative position between SRP RNA and the NG domains of Ffh and FtsY in different stages of protein targeting. (**a**) In early complex (PDBID: 3ZN8 [58]), the tetraloop of SRP RNA contacts with the G domain of FtsY. (**b**) In the stable complex after cargo release (PDBID: 2XXA [85]), the distal end of SRP RNA contacts with the G domains of Ffh-FtsY

cues, e.g. bindings to cargo, membrane, and membrane translocon, to control the spatial and temporal accuracy of the targeting reaction (Fig. 16.10). It has been shown that the RNC binding can speed up the association rate of the closed SRP-SR complex by 100-fold [97, 98]. Simultaneously, the closed complex is destabilized and the subsequent SRP GTPase activation is delayed by tenfold. Besides the RNC cargo, the anionic lipid membrane can also accelerate the stable complex formation by 100-fold and stabilize the activation state by 40-fold [99, 100]. These results suggest that the RNC cargo can be efficiently delivered to the target membrane in the process of stable SRP-SR complex formation. In addition, the membrane translocon can re-activate the SRP GTPases in the heterodimer for cargo releasing from SRP to the translocation channel [101].

16.11 How Do the SRP and SR Targeting Machinery Govern the Efficiency and Fidelity of Protein Localization?

Based on previous studies from both experiments and computations, a detailed molecular model is proposed to explain how the SRP and SR govern the efficiency and fidelity of protein localizations [43, 102]. In minimal composition of this targeting machinery in bacteria, the SRP consists of one 4.5S RNA and one SRP54 homologous protein Ffh and only the SRα homolog FtsY is contained into SR. Both Ffh and FtsY interact with one another using their NG domains and proceed the early intermediate formation, stable complex association, and GTPase activation

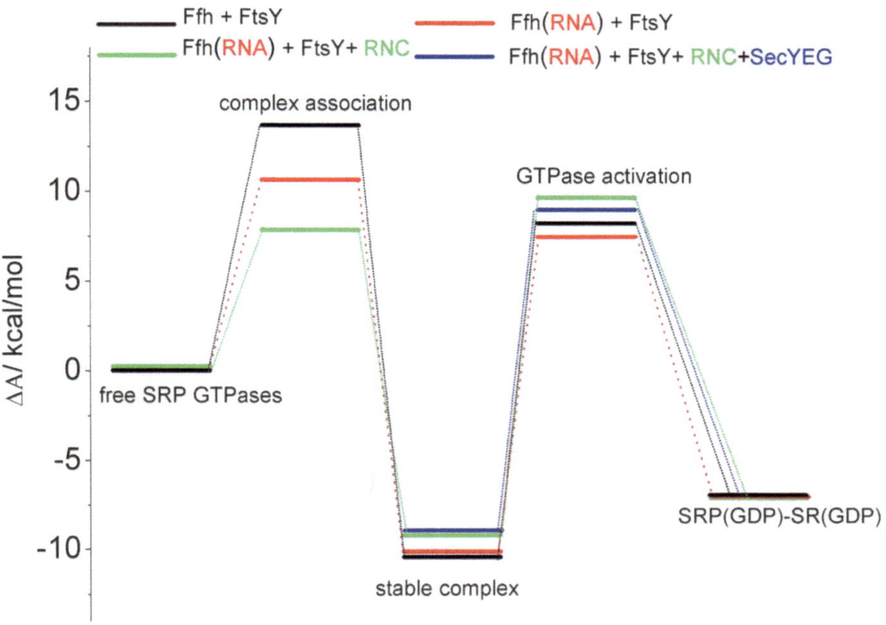

Fig. 16.10 Free energy profiles of stable SRP-SR complex association and subsequent SRP GTPases activation. The free energy profile is computed using rate formula $k = \frac{k_B T}{h} e^{-\beta \Delta G}$, where k, ΔG, T, k_B, h, and β are the rate constant determined in experiment, free energy barrier of the reaction, experimental temperature, Boltzmann constant, Planck constant, and inverse temperature, respectively. The rate constants determined for Ffh(±RNA) and FtsY interactions [72, 79, 92], Ffh(+RNA), FtsY, and RNC interactions [97, 98], Ffh(+RNA), FtsY, RNC, and SecYEG interactions [101] are used to compute the free energy differences

for GTP hydrolysis. These different conformational states of Ffh and FtsY are coupled to corresponding protein targeting events. The stable complex deliveries the translating ribosome exposing the signal peptide to the plasma membrane; whereas the GTPase activation releases the cargo from SRP-SR complex to the translocation channel. Thus the SRP GTPases constitute a self-sufficient system to execute exquisite spatial and temporal control of the complex targeting process [97]. By summarizing all these observations of this system together, a fascinating picture of the SRP mediated targeting process can be depicted as follows (Fig. 16.11):

(1) The SRP recognizes and binds to the signal peptide on the translating ribosome. The ribosome binding with SRP positions the tetraloop of SRP RNA in a suitable orientation relative to the NG domain of Ffh and selects out the functional relevant motion in the equilibrium fluctuation of free Ffh [58, 79, 96]. These conformational and dynamic pre-organizations prime SRP for subsequent interaction with SR attached on membrane surface. However, in this step, only the binding affinity difference between SRP and diverse signal peptide cannot provide sufficient discrimination against incorrect cargos and

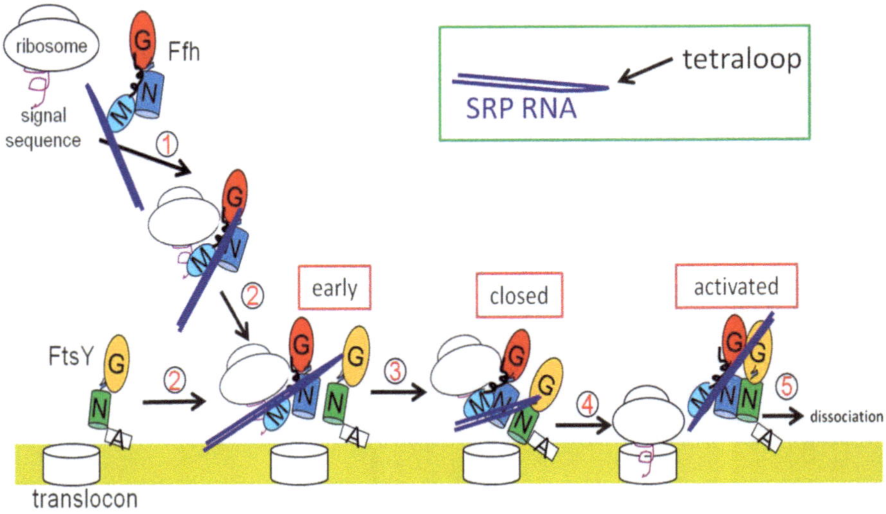

Fig. 16.11 Schematic draw of the SRP mediated protein targeting process

other kinetic checkpoints are required in following events to guarantee the fidelity of protein targeting [97].

(2–3) The SRP interacts with membrane anchored SR to transport the RNC cargo to the target membrane. In this process, the SRP and SR firstly assemble into an unstable on-pathway early intermediate and then evolve into the stable SRP-SR heterodimer via extensive inter- and intra-domain rearrangements [79, 81]. Formation of the early complex is supposed to adopt the conformational selection mechanism in which the pre-organized conformations in both free SRP GTPases are selected and stabilized by one another. In addition, the electrostatic interactions between the N domains of the two GTPases are suggested to be the main driving force in this early intermediate association [69]. From the early complex, significant inter-domain reorientation and extensive adjustments of conserved motifs turn the floppy open state in free SRP GTPases to the closed conformation in stable SRP-SR complex [60, 61]. Two α helices (α3 and α4) at the domain interface adjust their relative positions flexibly to accommodate the inter-domain rearrangements between the N domain and the core region of the G domain [79]. In the formation of stable Ffh-FtsY complex, the SRP RNA and lipid membrane can accelerate the rate by 200–400 and 100-fold, respectively [73, 94, 99, 100]. The RNC cargo can also speed up the stable SRP-SR association by 100-fold but delay the subsequent GTP activation rate by 10-fold [98]. In addition, there is also a ~1,000-fold kinetic discrimination between the correct and incorrect cargos in the stable complex assembly, which offers additionally checkpoint for the fidelity of protein targeting [97]. The delay of GTPase activation by RNC binding can provide longer time window for efficient cargo delivery to target membrane, which can compete with the translating rate of the nascent chain on ribosome to regulate the temporal accuracy of the targeting events. In addition, the incorrect state of SRP in step (1), e.g. the apo- or GDP-bound SRP, can be aborted from this pathway at early complex formation.

(4) SRP GTPases activation following SRP-SR complex formation unloads the cargo from the targeting machinery to the membrane translocon. Upon stable association of SRP-SR complex, further rearrangement of a number of critical residues in conserved motifs II and III are required to fine tune the interaction network for GTPase activation around the active chamber [71, 86, 87]. In this step, the NG domains of Ffh-FtsY complex is relocalized from the tetraloop end to the distal end of the SRP RNA, which release the ribosome exit site for initial binding of the membrane translocation channel. This large-scale movement of the SRP GTPase domains along the RNA scaffold is negatively regulated by RNC cargo and positively regulated by the translocation machinery [70]. In addition, it is shown that the GTP hydrolysis competes with cargo unloading, implying that the targeting reaction would be aborted if GTP hydrolysis is too fast in the heterodimeric complex. Correspondingly, the RNC can stabilize the activation state by tenfold, which provide extra point for fidelity check of the targeting [97]. Therefore, only at this point after handover of the RNC cargo to membrane translocation complex, the SRP GTPases are reactivated for GTP hydrolysis, which are regulated by the cargo and translocon interaction to govern the spatial and temporal fidelity of the targeting reactions [101].

(5) GTP hydrolysis in the SRP-SR heterodimer drives dissociation of the complex to recover free SRP and SR states for next round of protein targeting [71].

16.12 Perspective

Although the critical functioning mechanism of the SRP-SR targeting machinery has been presented through several decades of persistent efforts, further studies are required to dissect the details of how this machinery work, e.g. how the composite active chamber in the heterodimer catalyzes *cis/trans* bound GTP hydrolysis [60, 61], how the signal peptide binding to SRP cross talks to SRP RNA to exert their role in a synergistic way [103, 104], how the SRP RNA transmits the information of cargo and translocon binding to regulate the stable complex formation and GTPase activation [97, 98, 101]. However, due to the complexity of this system and the large-scale temporal and spatial changes spanned by the targeting process, great challenges still exist in both experimental and computational methods. Experimentally, the newly developing time-resolved techniques with higher resolution are promising for presenting profound insight into the conformational dynamics of complicated biomolecular systems [5, 9, 105]. Computationally, consistent multiscale modeling algorithms in combination with robust enhanced sampling techniques will be more powerful in elucidating the mechanism of large-scale conformational dynamics of various macromolecular systems [26, 28]. Furthermore, it is of most importance to actively combine the advantages of both experimental and computational methods in one study, which will provide us great opportunities to tackle more difficult and important problems.

Acknowledgement We thank Dr. Xin Zhang for his insightful comments to this manuscript. This work was supported by the National Basic Research Program of China (2013CB834604).

References

1. Tokuriki N, Tawfik DS (2009) Protein dynamism and evolvability. Science 324(5924): 203–207
2. James LC, Tawfik DS (2003) Conformational diversity and protein evolution – a 60-year-old hypothesis revisited. Trends Biochem Sci 28(7):361–368
3. Henzler-Wildman K, Kern D (2007) Dynamic personalities of proteins. Nature 450:964–972
4. Austin RH, Beeson KW, Eisenstein L, Frauenfelder H, Gunsalus IC (1975) Dynamics of ligand-binding to myoglobin. Biochemistry 14(24):5355–5373
5. Spence JCH, Weierstall U, Chapman HN (2012) X-ray lasers for structural and dynamic biology. Rep Prog Phys 75(10):102601
6. Kay LE (2005) NMR studies of protein structure and dynamics. J Magn Reson 173(2): 193–207
7. Ishima R, Torchia DA (2000) Protein dynamics from NMR. Nat Struct Biol 7(9):740–743
8. Neutze R, Moffat K (2012) Time-resolved structural studies at synchrotrons and X-ray free electron lasers: opportunities and challenges. Curr Opin Struct Biol 22(5):651–659
9. Aquila A, Hunter MS, Doak RB, Kirian RA, Fromme P, White TA, Andreasson J, Arnlund D, Bajt S, Barends TRM, Barthelmess M, Bogan MJ, Bostedt C, Bottin H, Bozek JD, Caleman C, Coppola N, Davidsson J, DePonte DP, Elser V, Epp SW, Erk B, Fleckenstein H, Foucar L, Frank M, Fromme R, Graafsma H, Grotjohann I, Gumprecht L, Hajdu J, Hampton CY, Hartmann A, Hartmann R, Hauriege S, Hauser G, Hirsemann H, Holl P, Holton JM, Hoemke A, Johansson L, Kimmel N, Kassemeyer S, Krasniqi F, Kuehnel K, Liang M, Lomb L, Malmerberg E, Marchesini S, Martin AV, Maia FRNC, Messerschmidt M, Nass K, Reich C, Neutze R, Rolles D, Rudek B, Rudenko A, Schlichting I, Schmidt C, Schmidt KE, Schulz J, Seibert MM, Shoeman RL, Sierra R, Soltau H, Starodub D, Stellato F, Stern S, Strueder L, Timneanu N, Ullrich J, Wang X, Williams GJ, Weidenspointner G, Weierstall U, Wunderer C, Barty A, Spence JCH, Chapman HN (2012) Time-resolved protein nanocrystallography using an X-ray free-electron laser. Opt Express 20(3):2706–2716
10. Elmlund H, Baraznenok V, Linder T, Szilagyi Z, Rofougaran R, Hofer A, Hebert H, Lindahl M, Gustafsson CM (2009) Cryo-EM reveals promoter DNA binding and conformational flexibility of the general transcription factor TFIID. Structure 17(11):1442–1452
11. Heymann JB, Conway JF, Steven AC (2004) Molecular dynamics of protein complexes from four-dimensional cryo-electron microscopy. J Struct Biol 147(3):291–301
12. Torres T, Levitus M (2007) Measuring conformational dynamics: a new FCS-FRET approach. J Phys Chem B 111(25):7392–7400
13. Weiss S (2000) Measuring conformational dynamics of biomolecules by single molecule fluorescence spectroscopy. Nat Struct Biol 7(9):724–729
14. Weiss S (1999) Fluorescence spectroscopy of single biomolecules. Science 283(5408):1676–1683
15. Barth A (2007) Infrared spectroscopy of proteins. Biochim Biophys Acta Bioenerg 1767(9):1073–1101
16. Gabel F, Bicout D, Lehnert U, Tehei M, Weik M, Zaccai G (2002) Protein dynamics studied by neutron scattering. Q Rev Biophys 35(4):327–367

17. Yang WT, Lee TS (1995) A density-matrix divide-and-conquer approach for electronic-structure calculations of large molecules. J Chem Phys 103(13):5674–5678
18. Kitaura K, Ikeo E, Asada T, Nakano T, Uebayasi M (1999) Fragment molecular orbital method: an approximate computational method for large molecules. Chem Phys Lett 313(3–4):701–706
19. Nakano T, Kaminuma T, Sato T, Akiyama Y, Uebayasi M, Kitaura K (2000) Fragment molecular orbital method: application to polypeptides. Chem Phys Lett 318(6):614–618
20. Zhang DW, Zhang JZH (2003) Molecular fractionation with conjugate caps for full quantum mechanical calculation of protein-molecule interaction energy. J Chem Phys 119(7):3599–3605
21. Xie W, Gao J (2007) Design of a next generation force field: the X-Pol potential. J Chem Theory Comput 3(6):1890–1900
22. Hu H, Yang W (2008) Free energies of chemical reactions in solution and in enzymes with Ab initio quantum mechanics/molecular mechanics methods. Annu Rev Phys Chem 59:573–601
23. Song L, Han J, Lin YL, Xie W, Gao J (2009) Explicit polarization (X-Pol) potential using Ab initio molecular orbital theory and density functional theory. J Phys Chem A 113(43):11656–11664
24. Lindorff-Larsen K, Piana S, Palmo K, Maragakis P, Klepeis JL, Dror RO, Shaw DE (2010) Improved side-chain torsion potentials for the Amber Ff99sb protein force field. Proteins Struct Funct Bioinform 78(8):1950–1958
25. MacKerell AD, Bashford D, Bellott M, Dunbrack RL, Evanseck JD, Field MJ, Fischer S, Gao J, Guo H, Ha S, Joseph-McCarthy D, Kuchnir L, Kuczera K, Lau FTK, Mattos C, Michnick S, Ngo T, Nguyen DT, Prodhom B, Reiher WE, Roux B, Schlenkrich M, Smith JC, Stote R, Straub J, Watanabe M, Wiorkiewicz-Kuczera J, Yin D, Karplus M (1998) All-atom empirical potential for molecular modeling and dynamics studies of proteins. J Phys Chem B 102(18):3586–3616
26. Takada S (2012) Coarse-grained molecular simulations of large biomolecules. Curr Opin Struct Biol 22(2):130–137
27. Zhang Z, Pfaendtner J, Grafmueller A, Voth GA (2009) Defining coarse-grained representations of large biomolecules and biomolecular complexes from elastic network models. Biophys J 97(8):2327–2337
28. Yang L, Shao Q, Gao Y (2012) Enhanced sampling method in molecular simulations. Prog Chem 24(6):1199–1213
29. Hu Y, Hong W, Shi Y, Liu H (2012) Temperature-accelerated sampling and amplified collective motion with adiabatic reweighting to obtain canonical distributions and ensemble averages. J Chem Theory Comput 8(10):3777–3792
30. Kaestner J (2011) Umbrella sampling. Wiley Interdiscip Rev Comput Mol Sci 1(6):932–942
31. Fiore CE, da Luz MGE (2010) Comparing parallel- and simulated-tempering-enhanced sampling algorithms at phase-transition regimes. Phys Rev E 82(3):031104-1–031104-11
32. Markwick PRL, McCammon JA (2011) Studying functional dynamics in bio-molecules using accelerated molecular dynamics. Phys Chem Chem Phys 13(45):20053–20065
33. Haider S, Parkinson GN, Neidle S (2008) Molecular dynamics and principal components analysis of human telomeric quadruplex multimers. Biophys J 95(1):296–311
34. Chodera JD, Swope WC, Pitera JW, Seok C, Dill KA (2007) Use of the weighted histogram analysis method for the analysis of simulated and parallel tempering simulations. J Chem Theory Comput 3(1):26–41
35. Chan KY, Trabuco LG, Schreiner E, Schulten K (2012) Cryo-electron microscopy modeling by the molecular dynamics flexible fitting method. Biopolymers 97(9):678–686
36. Zheng W, Tekpinar M (2011) Accurate flexible fitting of high-resolution protein structures to small-angle X-ray scattering data using a coarse-grained model with implicit hydration shell. Biophys J 101(12):2981–2991
37. Torbeev VY, Raghuraman H, Hamelberg D, Tonelli M, Westler WM, Perozo E, Kent SBH (2011) Protein conformational dynamics in the mechanism of HIV-1 protease catalysis. Proc Natl Acad Sci USA 108(52):20982–20987

38. Zhuang W, Sgourakis NG, Li Z, Garcia AE, Mukamel S (2010) Discriminating early stage a beta 42 monomer structures using chirality-induced 2DIR spectroscopy in a simulation study. Proc Natl Acad Sci USA 107(36):15687–15692
39. Zhang B, Miller TF III (2010) Hydrophobically stabilized open state for the lateral gate of the Sec translocon. Proc Natl Acad Sci USA 107(12):5399–5404
40. Silva JR, Pan H, Wu D, Nekouzadeh A, Decker KF, Cui J, Baker NA, Sept D, Rudy Y (2009) A multiscale model linking ion-channel molecular dynamics and electrostatics to the cardiac action potential. Proc Natl Acad Sci USA 106(27):11102–11106
41. Nussinov R, Ma B (2012) Protein dynamics and conformational selection in bidirectional signal transduction. BMC Biol 10:2
42. Boehr DD, Nussinov R, Wright PE (2009) The role of dynamic conformational ensembles in biomolecular recognition. Nat Chem Biol 5(11):789–796
43. Saraogi I, Shan SO (2011) Molecular mechanism of co-translational protein targeting by the signal recognition particle. Traffic 12(5):535–542
44. Pool MR (2005) Signal recognition particles in chloroplasts, bacteria, yeast and mammals (review). Mol Membr Biol 22(1–2):3–15
45. Park E, Rapoport TA (2012) Mechanisms of Sec61/SecY-mediated protein translocation across membranes. In: Rees DC (ed) Annual review of biophysics, vol 41, Annual Rieviews, Palo Alto California, pp 21–40
46. Cheng ZL (2010) Protein translocation through the Sec61/SecY channel. Biosci Rep 30(3):201–207
47. Driessen AJM, Nouwen N (2008) Protein translocation across the bacterial cytoplasmic membrane. Annu Rev Biochem 77:643–667
48. Traeger C, Rosenblad MA, Ziehe D, Garcia-Petit C, Schrader L, Kock K, Richter CV, Klinkert B, Narberhaus F, Herrmann C, Hofmann E, Aronsson H, Schuenemann D (2012) Evolution from the prokaryotic to the higher plant chloroplast signal recognition particle: the signal recognition particle RNA is conserved in plastids of a wide range of photosynthetic organisms. Plant Cell 24(12):4819–4836
49. Richter CV, Bals T, Schunemann D (2010) Component interactions, regulation and mechanisms of chloroplast signal recognition particle-dependent protein transport. Eur J Cell Biol 89(12):965–973
50. Rosenblad MA, Gorodkin J, Knudsen B, Zwieb C, Samuelsson T (2003) SRPDB: signal recognition particle database. Nucleic Acids Res 31(1):363–364
51. Rapoport TA (2007) Protein translocation across the eukaryotic endoplasmic reticulum and bacterial plasma membranes. Nature 450(7170):663–669
52. Zwieb C, Van Nues RW, Rosenblad MA, Brown JD, Samuelsson T (2005) A nomenclature for all signal recognition particle RNAs. RNA-A Publ RNA Soc 11(1):7–13
53. Zhang DW, Shan SO (2012) Translation elongation regulates substrate selection by the signal recognition particle. J Biol Chem 287(10):7652–7660
54. Halic M, Becker T, Pool MR, Spahn CMT, Grassucci RA, Frank J, Beckmann R (2004) Structure of the signal recognition particle interacting with the elongation-arrested ribosome. Nature 427(6977):808–814
55. Calo D, Eichler J (2011) Crossing the membrane in archaea, the third domain of life. Biochim Biophys Acta Biomembr 1808(3):885–891
56. Dalbey RE, Kuhn A (2012) Protein traffic in gram-negative bacteria – how exported and secreted proteins find their way. Fems Microbiol Rev 36(6):1023–1045
57. Verstraeten N, Fauvart M, Versees W, Michiels J (2011) The universally conserved prokaryotic GTPases. Microbiol Mol Biol Rev 75(3):507–542
58. von Loeffelholz O, Knoops K, Ariosa A, Zhang X, Karuppasamy M, Huard K, Schoehn G, Berger I, Shan SO, Schaffitzel C (2013) Structural basis of signal sequence surveillance and selection by the SRP–FTSY complex. Nat Struct Mol Biol 5:604–610
59. Hainzl T, Huang SH, Merilainen G, Brannstrom K, Sauer-Eriksson AE (2011) Structural basis of signal-sequence recognition by the signal recognition particle. Nat Struct Mol Biol 18(3):389–391

60. Focia PJ, Shepotinovskaya IV, Seidler JA, Freymann DM (2004) Heterodimeric GTPase core of the SRP targeting complex. Science 303(5656):373–377
61. Egea PF, Shan SO, Napetschnig J, Savage DF, Walter P, Stroud RM (2004) Substrate twinning activates the signal recognition particle and its receptor. Nature 427(6971):215–221
62. Montoya G, Svensson C, Luirink J, Sinning I (1997) Crystal structure of the NG domain from the signal-recognition particle receptor FtsY. Nature 385(6614):365–368
63. Freymann DM, Keenan RJ, Stroud RM, Walter P (1997) Structure of the conserved GTPase domain of the signal recognition particle. Nature 385(6614):361–364
64. Hainzl T, Huang SH, Sauer-Eriksson AE (2007) Interaction of signal-recognition particle 54 GTPase domain and signal-recognition particle RNA in the free signal-recognition particle. Proc Natl Acad Sci USA 104:14911–14916
65. Rosendal KR, Wild K, Montoya G, Sinning L (2003) Crystal structure of the complete core of archaeal signal recognition particle and implications for interdomain communication. Proc Natl Acad Sci USA 100(25):14701–14706
66. Batey RT, Rambo RP, Lucast L, Rha B, Doudna JA (2000) Crystal structure of the ribonucleoprotein core of the signal recognition particle. Science 287(5456):1232–1239
67. Neher SB, Bradshaw N, Floor SN, Gross JD, Walter P (2008) SRP RNA controls a conformational switch regulating the SRP-SRP receptor interaction. Nat Struct Mol Biol 15(9):916–923
68. Siu FY, Spanggord RJ, Doudna JA (2007) SRP RNA provides the physiologically essential GTPase activation function in cotranslational protein targeting. RNA-A Publ RNA Soc 13(2):240–250
69. Zhang X, Lam VQ, Mou Y, Kimura T, Chung J, Chandrasekar S, Winkler JR, Mayo SL, Shan SO (2011) Direct visualization reveals dynamics of a transient intermediate during protein assembly. Proc Natl Acad Sci USA 108(16):6450–6455
70. Shen K, Arslan S, Akopian D, Ha T, Shan SO (2012) Activated GTPase movement on an RNA scaffold drives co-translational protein targeting. Nature 492(7428):271–275
71. Shan SO, Walter P (2005) Co-translational protein targeting by the signal recognition particle. FEBS Lett 579(4):921–926
72. Jagath JR, Rodnina MV, Lentzen G, Wintermeyer W (1998) Interaction of guanine nucleotides with the signal recognition particle from *Escherichia coli*. Biochemistry 37(44):15408–15413
73. Peluso P, Shan SO, Nock S, Herschlag D, Walter P (2001) Role of SRP RNA in the GTPase cycles of Ffh and FtsY. Biochemistry 40(50):15224–15233
74. Padmanabhan S, Freymann DM (2001) The conformation of bound GMPPNP suggests a mechanism for gating the active site of the SRP GTPase. Structure 9(9):859–867
75. Ramirez UD, Minasov G, Focia PJ, Stroud RM, Walter P, Kuhn P, Freymann DM (2002) Structural basis for mobility in the 1.1 Angstrom crystal structure of the NG domain of *Thermus aquaticus* Ffh. J Mol Biol 320(4):783–799
76. Gawronski-Salerno J, Coon JSV, Focia PJ, Freymann DM (2007) X-ray structure of the T-aquaticus FtsY: GDP complex suggests functional roles for the C-terminal helix of the SRP GTPases. Proteins Struct Funct Bioinform 66(4):984–995
77. Reyes CL, Rutenber E, Walter P, Stroud RM (2007) X-ray structures of the signal recognition particle receptor reveal targeting cycle intermediates. PLoS One 2(7):e607
78. Ramirez UD, Focia PJ, Freymann DM (2008) Nucleotide-binding flexibility in ultrahigh-resolution structures of the SRP GTPase Ffh. Acta Crystallogr Sect D Biol Crystallogr 64:1043–1053
79. Yang MJ, Zhang X, Han KL (2010) Molecular dynamics simulation of SRP GTPases: towards an understanding of the complex formation from equilibrium fluctuations. Proteins Struct Funct Bioinform 78(10):2222–2237
80. Yang MJ, Pang XQ, Zhang X, Han KL (2011) Molecular dynamics simulation reveals preorganization of the chloroplast FtsY towards complex formation induced by GTP binding. J Struct Biol 173(1):57–66

81. Zhang X, Kung S, Shan SO (2008) Demonstration of a multistep mechanism for assembly of the SRP·SRP receptor complex: implications for the catalytic role of SRP RNA. J Mol Biol 381(3):581–593
82. Boehr DD, Wright PE (2008) How do proteins interact? Science 320(5882):1429–1430
83. Focia PJ, Gawronski-Salerno J, Coon JSV, Freymann DM (2006) Structure of a GDP:ALF4 complex of the SRP GTPases Ffh and FtsY, and identification of a peripheral nucleotide interaction site. J Mol Biol 360(3):631–643
84. Gawronski-Salerno J, Freymann DM (2007) Structure of the GMPPNP-stabilized NG domain complex of the SRP GTPases Ffh and FtsY. J Struct Biol 158(1):122–128
85. Ataide SF, Schmitz N, Shen K, Ke A, Shan SO, Doudna JA, Ban N (2011) The crystal structure of the signal recognition particle in complex with its receptor. Science 331(6019):881–886
86. Yang MJ, Zhang X (2011) Molecular dynamics simulations reveal structural coordination of Ffh-FtsY heterodimer toward GTPase activation. Proteins Struct Funct Bioinform 79(6):1774–1785
87. Shan SO, Stroud RM, Walter P (2004) Mechanism of association and reciprocal activation of two GTPases. PLoS Biol 2(10):1572–1581
88. Gasper R, Meyer S, Gotthardt K, Sirajuddin M, Wittinghofer A (2009) It takes two to tango: regulation of G proteins by dimerization. Nat Rev Mol Cell Biol 10(6):423–429
89. Celedon JM, Cline K (2013) Intra-plastid protein trafficking: how plant cells adapted prokaryotic mechanisms to the eukaryotic condition. Biochim Biophys Acta Mol Cell Res 1833(2):341–351
90. Jaru-Ampornpan P, Chandrasekar S, Shan SO (2007) Efficient interaction between two GTPases allows the chloroplast SRP pathway to bypass the requirement for an SRP RNA. Mol Biol Cell 18(7):2636–2645
91. Nguyen TX, Chandrasekar S, Neher S, Walter P, Shan SO (2011) Concerted complex assembly and GTPase activation in the chloroplast signal recognition particle. Biochemistry 50(33):7208–7217
92. Stengel KF, Holdermann I, Wild K, Sinning I (2007) The structure of the chloroplast signal recognition particle (SRP) receptor reveals mechanistic details of SRP GTPase activation and a conserved membrane targeting site. FEBS Lett 581(29):5671–5676
93. Chandrasekar S, Chartron J, Jaru-Ampornpan P, Shan SO (2008) Structure of the chloroplast signal recognition particle (SRP) receptor: domain arrangement modulates SRP-receptor interaction. J Mol Biol 375(2):425–436
94. Peluso P, Herschlag D, Nock S, Freymann DM, Johnson AE, Walter P (2000) Role of 4.5s RNA in assembly of the bacterial signal recognition particle with its receptor. Science 288(5471):1640–1643
95. Bradshaw N, Walter P (2007) The signal recognition particle (SRP) RNA links conformational changes in the SRP to protein targeting. Mol Biol Cell 18(7):2728–2734
96. Estrozi LF, Boehringer D, Shan SO, Ban N, Schaffitzel C (2011) Cryo-EM structure of the *E. coli* translating ribosome in complex with SRP and its receptor. Nat Struct Mol Biol 18(1):88–90
97. Zhang X, Rashid R, Wang K, Shan SO (2010) Sequential checkpoints govern substrate selection during cotranslational protein targeting. Science 328(5979):757–760
98. Zhang X, Schaffitzel C, Ban N, Shan SO (2009) Multiple conformational switches in a GTPase complex control co-translational protein targeting. Proc Natl Acad Sci USA 106(6):1754–1759
99. de Leeuw E, Kaat KT, Moser C, Menestrina G, Demel R, de Kruijff B, Oudega B, Luirink J, Sinning I (2000) Anionic phospholipids are involved in membrane association of FtsY and stimulate its GTPase activity. EMBO J 19(4):531–541
100. Lam VQ, Akopian D, Rome M, Henningsen D, Shan SO (2010) Lipid activation of the signal recognition particle receptor provides spatial coordination of protein targeting. J Cell Biol 190(4):623–635

101. Akopian D, Dalal K, Shen K, Duong F, Shan SO (2013) SecYEG activates GTPases to drive the completion of cotranslational protein targeting. J Cell Biol 200(4):397–405
102. Shan SO, Schmid SL, Zhang X (2009) Signal recognition particle (SRP) and SRP receptor: a new paradigm for multistate regulatory GTPases. Biochemistry 48(29):6696–6704
103. Bradshaw N, Neher SB, Booth DS, Walter P (2009) Signal sequences activate the catalytic switch of SRP RNA. Science 323(5910):127–130
104. Shen K, Zhang X, Shan SO (2011) Synergistic actions between the SRP RNA and translating ribosome allow efficient delivery of the correct cargos during cotranslational protein targeting. RNA-A Publ RNA Soc 17(5):892–902
105. Schotte F, Cho HS, Kaila VRI, Kamikubo H, Dashdorj N, Henry ER, Graber TJ, Henning R, Wulff M, Hummer G, Kataoka M, Anfinrud PA (2012) Watching a signaling protein function in real time via 100-ps time-resolved Laue crystallography. Proc Natl Acad Sci USA 109(47):19256–19261

Chapter 17
Molecular Dynamics Simulations of F_1-ATPase

Yuko Ito and Mitsunori Ikeguchi

Abstract F_1-ATPase is a rotary motor enzyme. Despite many theoretical and experimental studies, the molecular mechanism of the motor rotation is still not fully understood. However, plenty of available data provide a clue as to how this molecular motor rotates: with nucleotide perturbations, the catalytically active β subunit propagates its structural changes to the entire $\alpha_3\beta_3$ complex via both sides of the subunits, resulting that asymmetry is created in the $\alpha_3\beta_3$ hexamer ring. In the sequential reaction step, the structure of the asymmetrical $\alpha_3\beta_3$ complex changes from one state to the other due to the nucleotide perturbations, and the γ subunit axis follows the sequentially changing $\alpha_3\beta_3$ structure. Therefore, there are mainly two essential elements for motor rotation: the conformational change of the β subunit and the asymmetrical structure of the $\alpha_3\beta_3$ subunit complex. Therefore, this chapter reports a series of studies focused on these two elements via combinational approaches of molecular dynamics (MD) simulations and experimental or other theoretical studies. In addition to the motor rotation factors, the combined study also revealed other important elements of F_1-ATPase, such as torque transmission and the chemical reaction pathway, which is described in the later part of this chapter. All of these results provide insight into the rotational mechanism and deepen the understanding of this molecular motor.

17.1 F_1-ATPase

F_1-ATPase is an ATP-driven rotary motor enzyme [1–11]. This enzyme can perform ATP synthesis/hydrolysis using reversible motor rotation (Fig. 17.1, inside the rectangular box) [13]. ATP synthesis/hydrolysis via a rotation of the γ subunit

Y. Ito (✉) • M. Ikeguchi (✉)
Department of Medical Life Science, Yokohama City University, 1-7-29, Suehiro-cho, Tsurumi-ku, Yokohama 230-0045, Japan
e-mail: yukoito@tsurumi.yokohama-cu.ac.jp; ike@tsurumi.yokohama-cu.ac.jp

K. Han et al. (eds.), *Protein Conformational Dynamics*, Advances in Experimental Medicine and Biology 805, DOI 10.1007/978-3-319-02970-2_17,

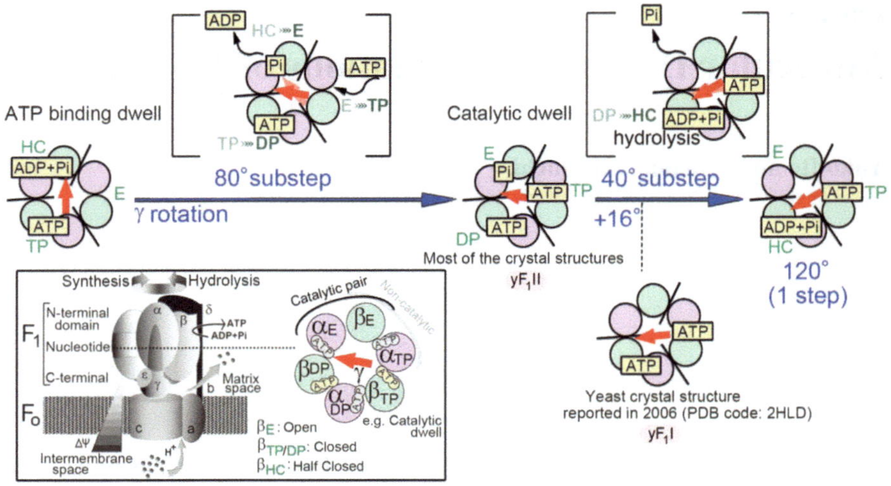

Fig. 17.1 Reaction scheme for the 120° rotation of the γ subunit. The figures inside the *brackets* indicate each intermediate for the 80° and 40° substeps. The *bottom left* figure inside the *rectangular box* indicates the general information for F_1-ATPase: structure of the F_0F_1-ATPase complex and the cross-sectional arrangement of the subunits in the F_1 moiety, viewed from the side of the C-terminal domains. In a yeast crystal structure reported in 2006 [12], the central stalk is rotated +16° in the hydrolysis direction from the catalytic dwell state, and Pi molecule is released from the β_E subunit, which is mentioned in the latter part of this article

in F_1-ATPase (the F_1 moiety) is coupled to an electrochemical diffusion gradient across a membrane-embedded F_0 unit [14]. The three-dimensional structure of F_1-ATPase was determined for the first time in 1994 [15]. The $\alpha_3\beta_3$ subunits are arranged hexagonally around the central γ-subunit stalk. Only the β subunit is catalytically active and changes its conformation during nucleotide binding/release and ATP hydrolysis. In most of the crystal structures, the three β subunits are in different states: two closed states (β_{DP} and β_{TP}) and one open state (β_E) (Fig. 17.1). The hydrolysis reaction of F_1-ATPase occurs during the 120° rotation step of the γ subunit [16], which can be further divided into 80° and 40° substeps, as described below [17]. The crystal structure corresponds to the structure after the 80° rotation [18].

A single molecule experiment has revealed the sequential conformational changes of the β subunit along with the 120° rotation of the γ subunit (Fig. 17.1) [19]. According to the study, before the 80° rotation, the three β subunits in the F_1-ATPase complex adopt the closed (β'_{TP}), open (β'_E), and half-closed (β_{HC}) conformations, where β_{HC} is the "half-closed" structure, and the apostrophe (') is used to distinguish the β subunit structures from those found in the crystal structure. This structure corresponds to the ATP-binding dwell state [19]. The 80° rotation is then induced by the ATP binding and the ADP release, and the structure of two different β subunits changes: $\beta'_E \rightarrow \beta_{TP}$ and $\beta_{HC} \rightarrow \beta_E$ [19]. After the 80° rotation, the F_1-ATPase complex corresponding to the crystal structure is at the catalytic dwell state [18]. At this state, ATP hydrolysis occurs for 1 ms [20]. Subsequently, the 40° rotation is induced by the β subunit conformational change of $\beta_{DP} \rightarrow \beta_{HC}$ due to the ATP hydrolysis [19] and the Pi release from the β_E subunit [21].

Only the β subunits change conformations during nucleotide events (substrate binding/release or ATP hydrolysis). However, the conformational changes of the β subunits propagate to the entire $\alpha_3\beta_3$ complex via both sides of the α subunits and create the asymmetrical $\alpha_3\beta_3$ ring structure. In the sequential reaction step, the structure of the asymmetrical $\alpha_3\beta_3$ complex changes from one state to the other due to the nucleotide perturbations, and the γ axis follows the sequentially changing $\alpha_3\beta_3$ structure. It is the so-called "γ subunit rotation". Accordingly, there are mainly two essential elements for motor rotation: the conformational change of the β subunit and the asymmetry of the $\alpha_3\beta_3$ subunit complex. Therefore, a series of studies have focused on these two elements using combinations of molecular dynamics (MD) simulations with experimental and theoretical approaches.

17.2 The Main "Engine" of the F_1-ATPase Motor: The β Subunit

17.2.1 Structural Properties of the β Subunit Obtained from Thermal Fluctuations at Equilibrium

When all-atom MD simulations are conducted for hundreds of nanoseconds, proteins only fluctuate around their starting structures (which are generally crystal structures). However, the thermal fluctuations observed at equilibrium are a good indicator of the intrinsic flexibility of a protein, and this flexibility is closely related to the large conformational change during biological functional process. According to theoretical methods, such as the application of the fluctuation dissipation theorem that includes linear response theory, there is a relation between the fluctuation properties of a system at thermal equilibrium and the response of the system to an applied perturbation. This relation can be applied to protein dynamics [22]. Protein flexibility, which is linked to functional activity, is an intrinsic property of the protein and is encoded in its folding. Therefore, analyzing structural fluctuations at equilibrium to investigate and discuss protein dynamics on a slow time scale is reasonable and has been established as a common method [23–26]. We used this technique on isolated β subunits and conducted equilibrium MD simulations for 100 ns [27]. In fact, before this study, Böckmann et al. performed MD simulations on the isolated β subunit [28], and the open β subunit tended to close during the 12-ns simulation without nucleotide binding. However, this observation is

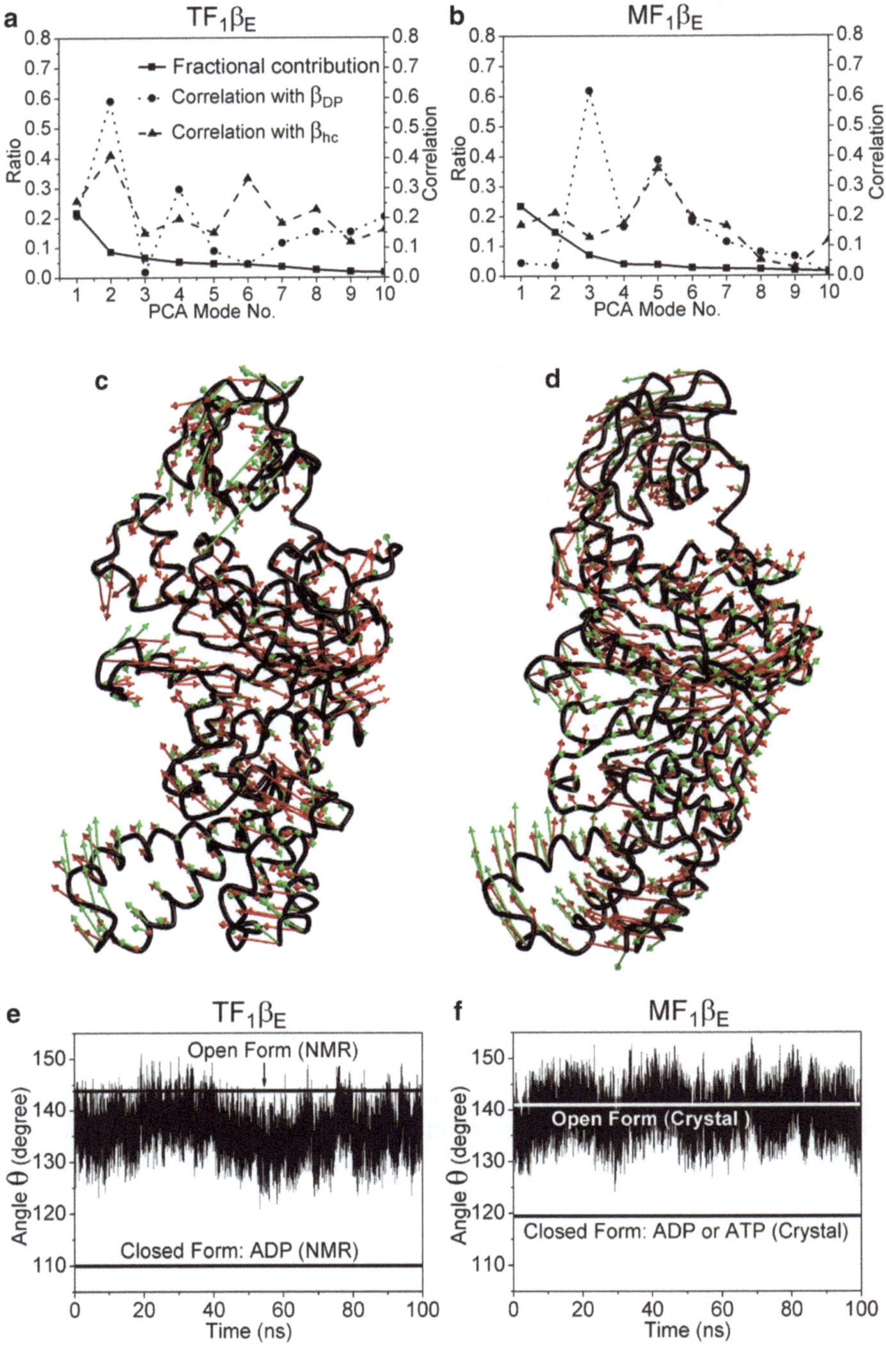
a
$TF_1\beta_E$
Fractional contribution
Correlation with β_{DP}
Correlation with β_{hc}
Ratio
Correlation
PCA Mode No.
b
$MF_1\beta_E$
Ratio
Correlation
PCA Mode No.
c
d
e
$TF_1\beta_E$
Open Form (NMR)
Closed Form: ADP (NMR)
Angle θ (degree)
Time (ns)
f
$MF_1\beta_E$
Open Form (Crystal)
Closed Form: ADP or ATP (Crystal)
Angle θ (degree)
Time (ns)

inconsistent with NMR results in which the isolated β subunit remains open, and only nucleotide binding is able to induce the changes in conformation to reach the closed form [29]. To clarify this inconsistency, all-atom MD simulations of isolated open β subunits were performed on a longer time scale (100 ns) using two different species of the isolated β subunit, MF_1 (bovine mitochondrial F_1-ATPase) and TF_1 (thermophilic Bacillus PS3 F_1-ATPase). These species were used because the cause of the inconsistency between the previous MD simulation [28] and the NMR experiment [29] might have been the difference in species (the MD simulation used MF_1, whereas the NMR experiment used TF_1).

First, to investigate the structural flexibility of the β subunit, principal component analysis (PCA) was used. PCA can decompose the overall protein motion over the entire simulation time into a set of modes that can be ranked by the size of their contribution (from largest to smallest) to the protein root mean square fluctuation, RMSF. The RMSF is the fluctuation around the time-averaged structure and is defined as $RMSF(i) = \sqrt{\left\langle (\mathbf{R}_i - \langle \mathbf{R}_i \rangle)^2 \right\rangle}$, where $\mathbf{R}_i$ is the position vector of atom i and the chevron brackets represent the time average over the entire trajectory. The PCA results show that the motions in the low-frequency mode are well correlated with the functionally important structural transition of the β subunit from the open to the closed conformation in both species (Fig. 17.2a–d). In the TF_1 simulation, the motions in the second dominant mode (PC2) are correlated with the β_E-$\beta_{DP/TP}$ structural transition (Fig. 17.2a, c). In the MF_1 simulation, the third dominant mode (PC3) corresponds to the structural transition direction (Fig. 17.2b, d).

Subsequently, the fluctuations of the structural transition direction were analyzed using the angle θ (measured between an axis connecting the centers of mass of the N- and C-domains and the axis of a helix:Glu399-Lys409 [29]) representing the β_E-$\beta_{DP/TP}$ structural transition (Fig. 17.3b, shown in red). This analysis was also applied to the two species (Fig. 17.2e, f). Although 12-ns MD simulations of the isolated β subunit previously showed that the open β subunit tended to close [28], the longer time scale simulations clarify that the β subunits in both the TF_1 and MF_1 simulation fluctuates only around the open form. The fluctuation is insufficient to reach the fully closed conformation.

The results of the fluctuation analysis using PCA and the angle θ indicate that, regardless of the species, the β subunit remains in the open form unless a nucleotide binds, which is consistent with the NMR experimental data [24]. Only nucleotide

Fig. 17.2 PCA and oscillating direction of the isolated β subunit. PCA for (**a**) TF_1 and (**b**) MF_1. The *solid line* indicates the fractional contribution in PCA, while the *dotted and dashed lines* indicate the correlation with the β_{DP} and β_{HC} subunit, respectively. The average structure of (**c**) PC2 in TF_1 and (**d**) PC3 in MF_1. The *green arrows* indicate the strongest correlation mode. The *red arrows* indicate the structural transitional direction to the closed form (β_{DP}). Angle θ versus time for the isolated β_E subunits of (**e**) TF_1 and (**f**) MF_1. The *upper and lower lines* in (**e**) and (**f**) represent the open and closed forms, respectively. (**e**) TF_1 in NMR [30] and (**f**) MF_1 in the 1E79 crystal structure [31]

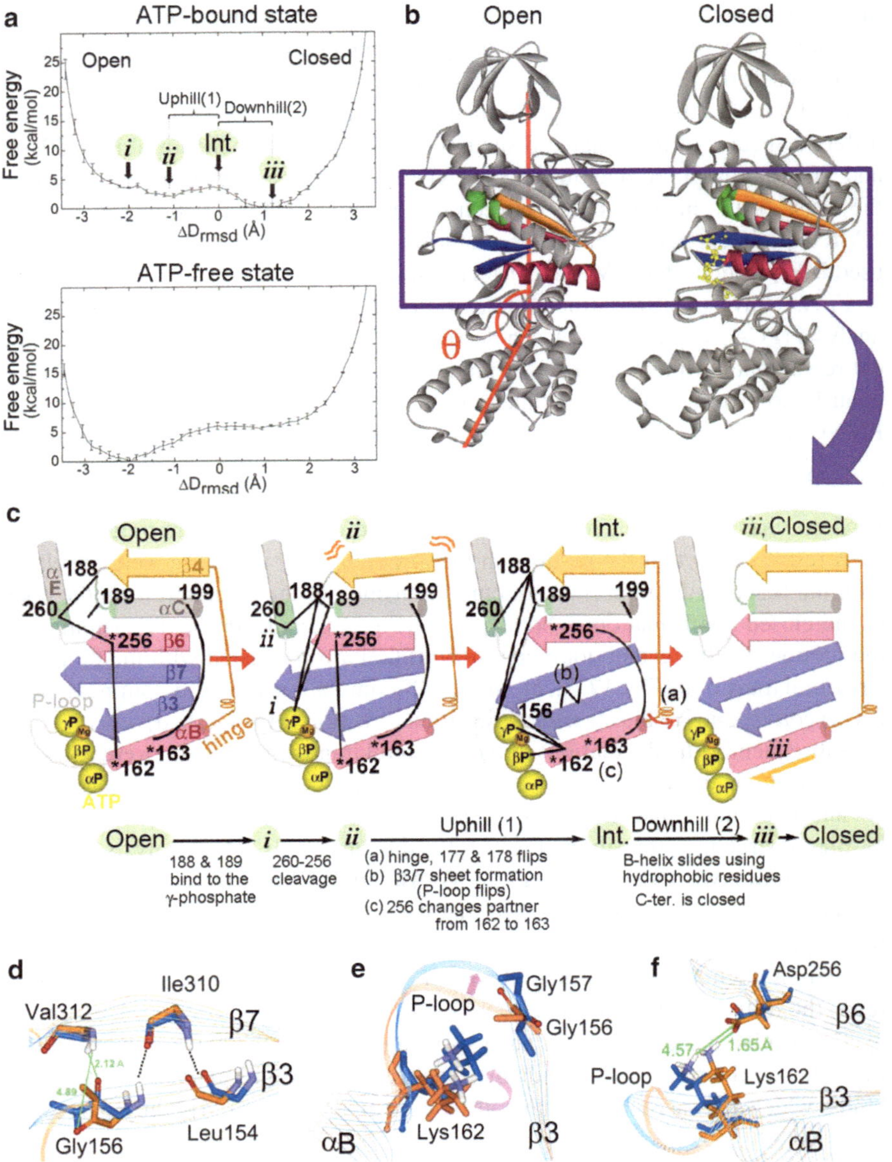

Fig. 17.3 (**a**) Free-energy profiles associated with the conformational transitions of the isolated β subunit. The one-dimensional energy profile along the ΔD_{rmsd} reaction coordinate in the ATP-bound and ATP-free pathway. The error bars represent the standard error of the mean energy values determined from the trajectory, which was divided into three phases. In the ATP-bound pathway, the *arrows* indicate minima *i*, *ii*, and *iii*. The main barrier between minima *ii* and *iii* is divided into Uphill (1) and Downhill (2). The ΔD_{rmsd} value of the open (β_E) and closed (β_{TP}) forms in the 2JDI crystal structure is −3.92 and 3.92, respectively. (**b**) Isolated β subunits in the open (β_E) and closed states (β_{TP}) taken from a crystal structure (PDB code: 2JDI). *Colored parts* show key regions of the conformational change, which were identified using experimental studies.

binding can induce the conformational change from the open to the closed form in the β subunit. In short, this NMR experiment [29] and the MD simulations [27] suggest that although flexibility in the direction of the structural transition is an intrinsic structural feature of the β subunit, there is an energy barrier in the pathway between the open and closed states and that this barrier cannot be overcome without nucleotide binding.

17.2.2 Mechanism of the Conformational Change of the β Subunit Revealed by Free-Energy Simulations

17.2.2.1 Computational Methods

To confirm the structural dynamics of the β subunit obtained from the equilibrium MD simulations [27], free-energy simulations were performed to determine the mechanism of the structural change from the open to the closed state in the isolated β subunit [33]. Generally, for the temporal and spatial scales, this open-closed conformational change is too broad to simulate directly using equilibrium all-atom MD simulations. Therefore, to bridge the sampling between the open and closed states, many alternative methods have been developed. Targeted MD (TMD) and steered MD (SMD) are used to investigate the conformational change of proteins using external perturbations along a conformational progress variable to guide the transition into a predefined direction in conformational space. These simulation methods have previously been used to show the conformational change of F_1-ATPase [34, 35]. However, the structural transitions generated by TMD or SMD appear to depend strongly on the external forces, and the path obtained between the initial and final states appears to be affected by the starting MD structures. In our experience, for example, an NMR study [32] identified that switching an important salt bridge induces an overall structural change of the β subunit. However, even after the overall structural change was completed in the TMD simulation, the salt bridge remained until the last moment of the simulation. Therefore, to overcome these

Fig. 17.3 (continued) *Pink, blue, green, and orange* indicate the αB-helix and β6 strand, the β3/β7 sheet, binding residues for the γ phosphate of ATP, and the β4 strand and hinge, respectively. The angle θ represents the open/closed structural transition. (**c**) A schematic model for the conversion from the open to the closed form of the β subunit. Color coding is the same as in (**b**). *Numerals* indicate the residue numbers. "*" indicates essential residues for the open/closed conversion, identified by NMR [32]. Roman numerals correspond to the minima in the energy profile of the ATP-bound state in (**a**). Int. represents the intermediate structure before Downhill (2) starts. (**d**) Superimposition of the structures of the β3 and β7 strands before (*orange*) and after (*blue*) the dihedral angles at Gly156 and Gly157 rotate. (**e**) Superimposition of the structure around the P-loop and the (**f**) Walker A and B motifs before (*orange*) and after (*blue*) the dihedral angles of the Gly156 main chain and the Lys162 side-chain flip, corresponding to the structure at $\Delta D_{rmsd} = -0.5$ and 0.0 Å, respectively problems, a different sampling method, a combination [36, 37] of nudged elastic band (NEB) [38, 39] and umbrella sampling MD simulations [40], was used to calculate the potential of mean force (PMF) [41]. For the atomistic conformational change of the biomolecules, various other simulation methods, such as the "string method" and "transition path sampling" have been developed [42–56]. However, considering the size of the simulation system of the β subunit, NEB is a less computationally demanding approach to obtaining converged energy profiles. For the reaction coordinate in the simulations, a difference in the root-mean-square deviation between the open (β_E) and closed states (β_{TP}) (ΔD_{rmsd}) of the β subunit was chosen such that the difference was able to successfully characterize the transition pathways in various protein and DNA molecules [36, 37, 57, 58]. The NEB method was able to find a minimum energy path between the open and closed structures: first, dozens of structures were generated via linear interpolation between the open and closed structures, and subsequently these structures were minimized using the adopted-basis Newton–Raphson method. When the initial path was obtained, the resulting structures were subjected to umbrella sampling MD simulations with the restraint w_j on the ΔD_{rmsd} order parameter. $w_j = K_{rmsd}(\Delta D_{rmsd} - \Delta D_{min})^2$, where ΔD_{min} is the value around which ΔD_{rmsd} is restrained and K_{rmsd} is a force constant. The PMF can provide the behavior of the system between the open and closed forms with free-energy variation along the reaction coordinate.

17.2.2.2 Overview of the Obtained Energy Profiles

Figure 17.3a shows the free-energy profiles associated with the conformational transition pathway of the isolated β subunit with/without ATP along the ΔD_{rmsd} reaction coordinate. The profiles indicate that in the β subunit without ATP, the open state is favored by 6 kcal/mol (Fig. 17.3a, bottom). This finding is consistent with the results of the equilibrium MD simulations [27] and the NMR experiments [29]; the ATP-free β subunit fluctuates only around the open form, and ligand binding is required to attain the fully closed conformation of the β subunit. In contrast, in the β subunit with ATP, the closed state is favored (Fig. 17.3a, top). In the energy landscape of this ATP bound state, there is a metastable state between the open and closed state (Fig. 17.3a, *ii*), indicating that before the β subunit turns into the fully closed from, the intermediate structure, i.e., the open β subunit that binds ATP, is transiently stable.

17.2.2.3 Details of the Conformational Change of the β Subunit with ATP

To obtain a more-detailed mechanism of the conformational change of the ATP-bound pathway, various local structure changes along the ΔD_{rmsd} order parameter were computed. The results reveal that the β subunit conformational change is accomplished roughly in two characteristic steps: ***(A)*** changing of the hydrogen-bond network around ATP (Fig. 17.3c, from Open to Int.) and ***(B)*** dynamic movement of the C-terminal domain via sliding of the B-helix (Fig. 17.3c, from Int. to Closed). The details revealed by the simulation are described below and are classified using the ΔD_{rmsd} order parameter from the open to the closed state.

17.2.2.4 First Step, *(A)*

In the first stage of the stepwise conformational change of ***(A)*** (Fig. 17.3a, c, open → *i*), Glu188 and Arg189 form a salt bridge and subsequently bind to the γ phosphate of ATP, where Glu188 and Arg189 have been identified as the residues that contribute to the catalytic reaction (the residue numbering is based on bovine mitochondrial F_1-ATPase) [59–63]. Before these interactions are formed, the side chain of Arg189 is able to swing without any restrictions, and Glu188 is coordinated by a salt bridge to Arg260 (Fig. 17.3c, left).

In the second stage (from minimum *i* to *ii*), the hydrogen bond between Arg260 and Asp256 is broken. Asp260 has been reported to be an essential residue for recognizing P_i (the cleaved phosphate) [64]. Asp256 is recognized as the residue of the Walker B motif that is conserved in many ATPases, and this residue was also reported to be essential for the β-subunit conformational change induced by nucleotide binding [32]. In the open structure, Arg260 initially interacts with both Asp256 and Glu188. However, the newly formed conformation of Glu188 in the former step (open → *i*) intrudes into the Arg260-Asp256 space, breaking this salt bridge.

The third stage corresponds to Uphill (1) of the main barrier (Fig. 17.3a, c, *ii* → Int.) and includes three processes: (a) hinge flips (ϕ, Gly178; ψ, His177) [65], (b) formation of the $\beta 3/\beta 7$ sheet [32], and (c) switching of the partner of Asp256 (Walker B) from Lys162 to Thr163 (Walker A) [32], the details of which follow below in (a)–(c). These structural changes appear to be coupled with one another, leading to a small loss of free energy.

(a) As described above, in the open state, Arg260 initially forms a salt bridge with both Asp256 and Glu188. However, the Arg260-Asp256 salt bridge is broken in the second step (*i* → *ii*), but Glu188 still maintains interactions with Arg260. Therefore, along with the elongation of the distance between Asp256 and Arg260 due to the bond breaking, the entire Glu188 residue is dragged toward the direction of Arg260. The pulled Glu188 residue imposes stress on the hinge region (His177 and Gly178 through the $\beta 4$ sheet) (Fig. 17.3c, orange part). When this strain on the $\beta 4$ sheet exceeds the elasticity, the hinge (the backbone dihedral angles: ϕ, Gly178; ψ, His177) is flipped. The hinge residues have been identified by Masaike et al. as essential for the structural conversion [65].
(b) After (a), the $\beta 3$ and $\beta 7$ strands become closer (the reason is described in the original paper [33]). When the $\beta 3$ and $\beta 7$ strands become close enough to interact with each other, Gly156 on the P-loop, which resides next to the

β3 strand, rotates its backbone to form a hydrogen bond with Val312 on β7 (Fig. 17.3d). Once this additional hydrogen bond between the β3 and β7 strands is established, the formation of the β3/β7 sheet is completed. The formation of the β3/β7 sheet has also been determined by an NMR study to be the essential element for the structural conversion [32].

(c) During the β3/β7 sheet formation, the rotation of the backbone dihedral angles ψ at Gly156 and ϕ at Gly157 results in the structure of the P-loop backbone changing (Fig. 17.3e). Along with the structural change of the P-loop backbone, the side chain of Lys162 relocates by rotations of the $\chi 1$ and $\chi 4$ dihedral angles of Lys162 (data shown in the original paper [33]). This rotation of the alkyl side chain of Lys162 distances Lys162 from Asp256, which triggers the hydrogen bond of Asp256 to switch from Lys162 to Thr163 (Fig. 17.3f). The released side chain of Lys162 forms new hydrogen bonds with the backbone CO of Gly156 as well as with the β and γ phosphates of ATP. The NMR study [32] indicated that the switching of the hydrogen bonding partner is also essential for the conformational change.

17.2.2.5 The Second Step, *(B)*

These local structural changes of ***(A)*** increase the structural strains around the ATP-binding site. To alleviate this strain, the B-helix, which is located beside the P-loop, slides using its hydrophobic residues on the other hydrophobic surface (Fig. 17.3c, right). This sliding occurs during Downhill (2) (Fig. 17.3a). In fact, this process corresponds to ***(B)*** because this B-helix displacement is coordinated with the movement of the lower half of the β subunit (depicted with the light color in Fig. 17.4c). In this movement, the values of both the RMSD of the B helix (Fig. 17.4b) and the angle θ, which represents the structural transition (Fig. 17.4a), simultaneously change at ΔDrmsd = 0.0–1.5. The large open/closed motion of the C-terminal domain in the β subunit is responsible for changing the $\alpha_3\beta_3$ complex ring and eventually leads to the γ subunit rotation. Consequently, this torque generation is ascribed to the one-turn shift of the B helix.

The large closing motion of the C-terminal domain associated with the B-helix displacement leads to a stabilization energy of ~5.0 kcal/mol relative to that of the structure at the top of the peak (Fig. 17.3a, top). The stabilization occurs because the packing rearrangement of the hydrophobic interface of the B-helix is improved along with the B-helix sliding (Fig. 17.4e, f). After the completion of the B-helix displacement, the space between these interfaces is too small for even a single water molecule to occupy (Fig. 17.4d). Generally, water molecules inside hydrophobic surroundings prevent the formation of hydrophobic interactions. Therefore, after the water molecules are excluded, hydrophobic side chains make contacts, which lead to the lowest free-energy configuration. This rearrangement has often been postulated, such as in the process of protein folding [66–68].

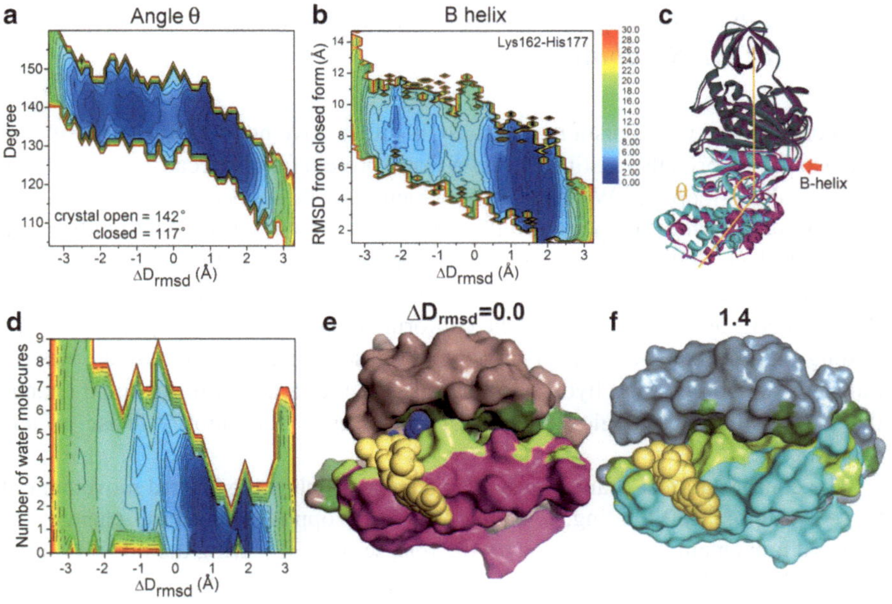

Fig. 17.4 Local conformational change associated with the structural transition of the β subunit in the ATP-bound pathway. Free-energy surface along the ΔD_{rmsd} reaction coordinate (horizontal axis) and various variations (vertical axis): the angle θ (**a**), the RMSD of residues of the B-helix from the closed β_{TP} structure (**b**), and the number of water molecules in the space formed by the hydrophobic surfaces of the B-helix and the other helices/sheets (i.e., the C-helix and β3-7 sheets) (**d**). Superposition of the structure before and after the C-terminal domain movement, corresponding to the structures at $\Delta D_{rmsd} = 0.0$ (*magenta*) and 1.4 Å (*cyan*), respectively. The fit is performed over the N-terminal domain. The parts exhibiting a small structural change (i.e., residues 9–123 and 178–329) and a large change (i.e., residues 124–177 and 330–474) in reaction coordinate from $\Delta D_{rmsd} = 0.0$ and 1.4 Å are depicted with dark and light colors, respectively. The B-helix is marked by a *red arrow* (**c**). Snapshots of the packing of the hydrophobic surface at $\Delta D_{rmsd} = 0.0$ (*magenta*) and 1.4 Å (*cyan*) are exhibited in (**e**) and (**f**), respectively. The hydrophobic parts of the interface are colored *green*. The trapped water molecules in the interspace are indicated in *blue*. Color coding for dark/light is matched to (**c**)

Here, the proposed pathway of the entire open/closed conversion for the β subunit agrees well with experimental data. Because the conformational change in the β subunit is responsible for the driving force of the rotation of the γ subunit, several essential regions for the structural change have been identified experimentally [32, 65]. However, it is still difficult to comprehend the entire picture of the consecutive conformational changes using only the experimental information. Therefore, simulations are a powerful tool to show the entire sequential process of protein conformational changes at the atomistic level.

17.2.3 Discussion of the Conformational Change of the β Subunit

Furthermore, the mechanism in this study is informative not only for the β subunit but also for the broadly existing P-loop ATPase protein family because the β subunit contains the P-loop and Walker motifs, which exist throughout the entire ATPase protein family. The P-loop and Walker motifs are important for ATP binding and ATP hydrolysis in the ATPase family. However, in addition to those functions, both experimental [32] and theoretical studies [33] have revealed that the P-loop and Walker motifs are also involved in the conformational change of the β subunit. This finding suggests that these universally conserved sequences play important roles not only in ATP binding and hydrolysis, but also in the large conformational change that occurs during biological functional processes in the entire ATPase protein family.

The conformational change investigated here, which is the inward movement of $\beta_E \rightarrow \beta_{TP}$ upon ATP binding, is coupled with the opposite structural change, the outward movement of $\beta_{HC} \rightarrow \beta_E$ with the release of ADP during the 80° γ rotation in the 120° cycle of the F_1-ATPase complex [16, 19, 20, 69]. As shown in the free-energy profile (Fig. 17.3a, top), the inward structural change yields energy. Although the free-energy profile of the outward movement has not been calculated, it is most likely endoergic, and we assumed that the exoergic conformational change ($\beta_E \rightarrow \beta_{TP}$) would cover the endoergic energy loss of the outward movement ($\beta_{HC} \rightarrow \beta_E$). Moreover, "reversibility" is one of the major distinctive features of this rotary motor enzyme. For the ATP synthesis direction using the obtained energy surface for the reverse mechanism, the process becomes endoergic and must be coupled with a structural change that yields energy. Combining different conformational changes with energy compensation also contributes to the reversibility of this motor. Such energy compensation is a prime reason for F_1-ATPase adopting the binding change mechanism [70].

Our results show that ATP binding is tightly coupled to the conformational change in the β subunit. The advantage of the tight coupling for the motor engine would be that the γ rotation can be strictly regulated via the β subunit, which can change its conformation only through nucleotide binding, thus avoiding unproductive rotations that might otherwise be caused by severe thermal fluctuations.

17.3 Packing Exchange Mechanism

In this section, we describe the other factor in the γ rotation: the asymmetry of the $\alpha_3\beta_3$ complex. The conformational changes of the β subunit propagate to the entire $\alpha_3\beta_3$ complex via both side of the α subunits and create the asymmetrical $\alpha_3\beta_3$ ring structure, and the constant changes of the $\alpha_3\beta_3$ asymmetrical ring with nucleotide perturbations allow the γ subunit to rotate. However, as described in the β subunit section, it is very difficult to simulate this series of events directly using all atom MD simulations due to both the total number of atoms (the F_1-ATPase complex is substantially larger than the β subunit) and the time scale of the movement (sub-millisecond). To overcome the limitation, coarse-grained (CG) simulations are used to simulate the motor rotation; in these simulations, small groups of atoms are treated collectively as large particles to reduce number of degrees of freedom, speeding-up the simulation by several orders of magnitude [71, 72]. In fact, the CG simulation performed by Koga and Takada [71] demonstrated that the conformational changes of three β subunits allow the γ subunit to rotate. From this finding, it was concluded that the crystal structures correspond to a snapshot of the catalytic dwell state rather than the ATP-binding dwell state, which was not obvious previously. Thus, CG simulations can handle large-scale conformational changes of large biomolecules.

In contrast to CG simulation, analyses at the atomistic level revealed that the intrinsic structural flexibility contributes to motor rotation [73–77]. Of these studies, all-atom MD simulations [76] and the statistical thermodynamics of molecular liquids [77] indicate that there are substantial differences in the interface configurations of each subunit in the $\alpha_3\beta_3$ complex, which allows us to deduce that these different interface configurations are cyclically exchanged during nucleotide perturbations, resulting in the γ subunit rotation. We refer to this proposed hypothesis as the "packing exchange mechanism".

17.3.1 Packing Exchange Mechanism: MD Simulations

First, to investigate the packing change mechanism, an equilibrium MD simulation was conducted for 30 ns [76]. The catalytic waiting dwell state [18] (PDB code: 2JDI [78]) was used for the initial structure of the simulation. The trajectory was analyzed in terms of the structural fluctuations and the subunit interface interactions using an RMSF calculation and contact analysis, respectively. The RMSF indicates the intensity of the structural fluctuation of each residue. In contact analysis [79], residue pairs that maintained their inter-subunit interactions within a certain distance over a certain percentage of the MD trajectory were selected. In this study, the following basic threshold values were used: <4.5 Å for the interatomic distance and more than 98 % of the MD trajectory. Satisfying these threshold values indicates that although that residue pair resides on different neighboring subunits, the residues stay close to each other during the entire MD simulation. In other words, the subunit interface around the selected residue pairs is tightly packed. Hereinafter, the contacts between residue pairs are referred to as "stable contacts". The subunit interface interactions identified by contact analysis affect the magnitude of the structural fluctuation (e.g., a tightly packed interface suppresses structural fluctuation, whereas a loose interface allows significant structural fluctuations) [76]. In particular, distinct differences in both the RMSF value and stable contacts for the subunits are observed in the C-terminal domain, because the most distinct structural differences are found

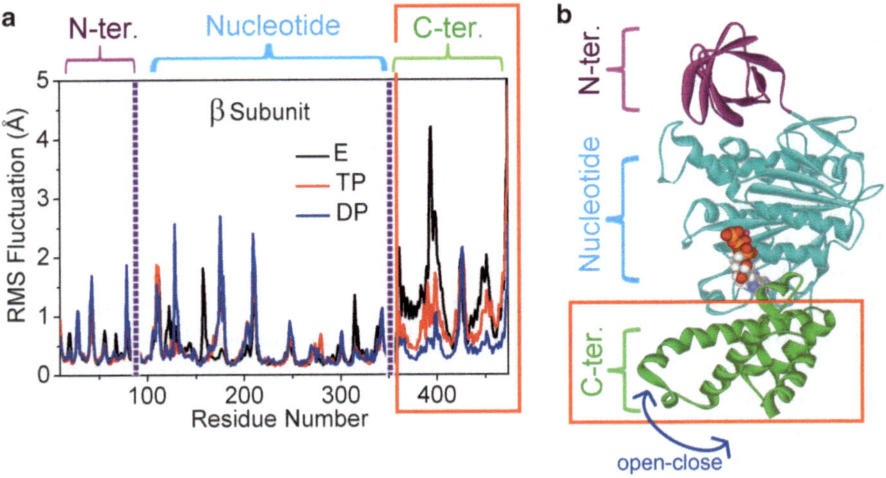

Fig. 17.5 (**a**) RMSF as a function of the residue number for the β subunits of the simulated catalytic dwell structure. The data are partitioned by domains (the N-terminal, nucleotide binding, and C-terminal). (**b**) The three domains of the β subunit are color-coded. The parts enclosed by *red rectangles* indicate the C-terminal domain, which has relatively large fluctuations

in the C-terminal domain in the β subunit for some nucleotide states (Fig. 17.5). Hence, the results using the contact analysis and the RMSF calculation are focused on the C-terminal domain.

The MD simulations show that stable contacts appear in all β_{DP} subunit interfaces (both sides of the α subunits and the γ subunit), indicating that the β_{DP} subunit interacts intensively with all neighboring subunits (Fig. 17.6, left). Because these tight interface interactions restrict fluctuations, the β_{DP} subunit has the smallest fluctuation magnitude of the three β subunits (Fig. 17.5a, blue line). The conformation of the β_{TP} subunit is fairly similar to that of the β_{DP} subunit, and β_{TP} binds ATP in the same manner as the β_{DP} subunit (the Cα-RMSD between β_{TP} and β_{DP} in 2JDI is 0.67 Å). However, the β_{TP} subunit has stable contacts with only the γ subunit. Therefore, the β_{TP} subunit fluctuates to some extent (Fig. 17.5a, red line). Compared with the β_{TP} and β_{DP} subunits, the β_E subunit has few stable contacts, resulting in large fluctuations. In this MD simulation, three different interface configurations (tight, moderate, and loose) are revealed in the catalytic dwell state of the F_1-ATPase complex (Fig. 17.6, left).

17.3.2 Packing Exchange Mechanism: Water-Entropy Effects

In addition to the MD simulation, we analyzed the characteristics of the asymmetric packing of the same F_1-ATPase structure in terms of the water-entropy effect using the statistical thermodynamics of molecular liquids [77]. The asymmetric interface

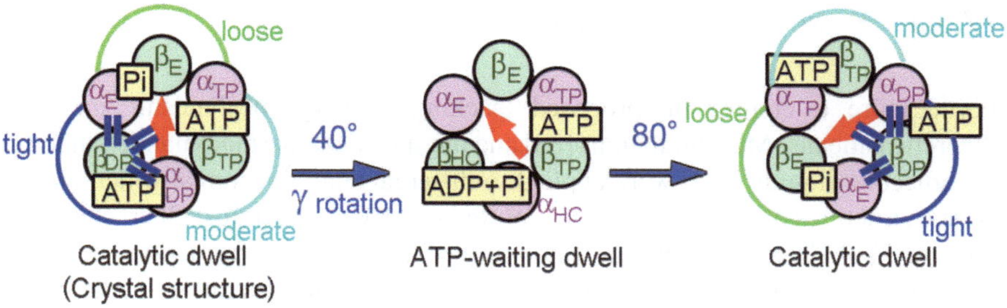

Fig. 17.6 Schematic representation of the subunit interface configuration of F_1-ATPase during the 120° rotation of the γ subunit, obtained from MD simulations and the statistical thermodynamics of molecular liquids. The *blue double lines* indicate the tightly packed interface configurations. When the αβα subunit sub-complex (the β subunit is in the center, and both sides of the α subunits are included) is considered, the three different types of interface configurations are indicated by *circular arcs*

interactions observed in the MD simulations were successfully reproduced using a theoretical method focused on the water-entropy effect. We have asserted that this type of asymmetric packing is driven by the water-entropy effect. In general, when overall, impartial tight packing is not achievable in a protein, the portions that can be tightly packed are chosen for preferential tight packing to maximize the water entropy. When a tightly packed portion is perturbed, the structure with the maximum water entropy is recovered by forming tight packing in the other portion. This asymmetry of the catalytic dwell state of the F_1-ATPase structure found in these studies suggests that after this state, the tightly interacting interfaces around the β_{DP} subunit are loosened due to perturbations, such as ATP hydrolysis and Pi release. Furthermore, instead of the β_{DP} subunit interfaces, the other subunit interfaces reach tighter packing than that in the catalytic dwell state. These subunit rearrangements within the $\alpha_3\beta_3$ complex eventually induce the γ subunit rotation. This effect can be expanded to the following complete (80° + 40°) rotational mechanism: a nucleotide event occurs; the asymmetric $\alpha_3\beta_3$ complex structure is perturbed, leading to a decrease in the water entropy; and tightly or weakly interacting interfaces are reorganized within the $\alpha_3\beta_3$ complex ring, allowing the γ subunit to rotate and resulting in maximization of the water entropy of the system. This description is a view of the packing exchange mechanism in terms of the water-entropy effects. In this mechanism, the complex always tries to form three regions (tightly packed, moderately packed, and loosely packed), and these regions are cyclically exchanged (Fig. 17.6). In fact, the importance of rearrangement in the $\alpha_3\beta_3$ complex is supported by experiments. In single molecules studies, the γ subunit still rotated in the correct direction during ATP hydrolysis, even with most of the γ axle truncated [80, 81]. Even without the γ subunit, unidirectional β subunit conformational changes in the presence of ATP were observed in an atomic force microscopy (AFM) study [82]. These data clearly indicate that the sequential packing exchange in the $\alpha_3\beta_3$ asymmetric ring is primarily responsible for the γ subunit rotation.

17.3.3 *Yeast F_1-ATPase Case*

Furthermore, to confirm the proposed "packing exchange mechanism", the same theoretical means (MD simulations [83] and the statistical thermodynamics analyses of molecular liquids [84]) were applied to different state structures, which are the crystal structures of yeast F_1-ATPase reported in 2006 [12]. Almost all crystal structures of F_1-ATPase represent the catalytic waiting dwell conformation. However, the yeast crystallographic unit contains two different state structures. One structure corresponds to the catalytic dwell state with a structure similar to other existing crystal structures, and this structure is named "yF_1II". In the other structure, the central stalk is rotated +16° in the hydrolysis direction, and this structure liberates Pi from the β_E subunit, which is "yF_1I" (Fig. 17.1). Because single molecule experiments demonstrated that Pi release occurs at the β_E subunit and that the γ subunit is rotated after ATP hydrolysis and Pi release [21, 85], these yeast structures are supposed to represent snapshots of before and after the Pi release in the 40° substep. Accordingly, these structures are suitable for studying the sequentially changing structures and for characterizing the rotational mechanism.

17.3.3.1 The $\alpha_{DP}\beta_{DP}$ and $\alpha_{TP}\beta_{TP}$ Subunit Interfaces in yF_1II and yF_1I

First, the $\alpha_3\beta_3$ ring structures of yF_1II and yF_1I were analyzed using the structural fluctuations and the subunit interface interactions. The RMSF magnitudes are summarized in Fig. 17.7a. In the same figure, the total numbers of stable contacts between the neighboring subunits are also shown. The data clearly show that the fluctuation tendency and the interface configuration of each subunit differ between yF_1II and yF_1I. As shown in the left part of Fig. 17.7a, in yF_1II, the smallest fluctuation is found in the β_{DP} subunit (the order is $\beta_{DP} < \beta_{TP} < \beta_E$ [83]), and this order is equivalent to that of the catalytic dwell structure determined in bovine F_1-ATPase [76] shown in Fig. 17.5a. This RMSF magnitude is interpreted using the subunit interface interactions derived from the contact analysis. The β_{DP} subunit has stable contacts in all β_{DP} subunit interfaces (both sides of the α subunits and the γ subunit), indicating that the β_{DP} subunit interacts intensively with all neighboring subunits. Because these tight interface interactions restrict fluctuations, the β_{DP} subunit has the smallest fluctuation magnitude of the three β subunits. In contrast, for yF_1I, the order of the magnitude of the structural fluctuations is $\beta_{TP} < \beta_{DP} < \beta_E$ (Fig. 17.7a, right), differing from that determined for yF_1II. This fluctuation order can also be explained by subunit interface interactions. Since the β_{TP} subunit fluctuates with the smallest magnitude in yF_1I, more contacts are found in the β_{TP} subunit of yF_1I than yF_1II (particularly on the α_{TP} subunit side). These gained contacts are depicted as a contiguous surface in Fig. 17.7b, appearing as red marks in the C-terminal domain in those subunits (Fig. 17.7b, panel 3). In contrast to the β_{TP} subunit, the β_{DP} subunit, whose fluctuation is not the smallest in yF_1I, loses contacts with neighboring subunits (Fig. 17.7a). As shown in Fig. 17.7b, panel 5, the stable contacts found in the C-terminal domain of the $\alpha_{DP}\beta_{DP}$ subunit interface for yF_1II are completely absent.

17.3.3.2 The $\alpha_E\beta_E$ Subunit Interface in yF_1II and yF_1I

We further investigated the $\alpha_E\beta_E$ subunit interface of yF_1I and yF_1II; this subunit interface is directly associated with the Pi release. However, as shown in Fig. 17.7a, b, panel 1, distinct differences between the two structures are not found because the threshold of the contact analysis for obtaining permanent interactions (<4.5 Å interatomic distance and more than 98 % of the MD trajectory) is too high to examine the differences of the $\alpha_E\beta_E$ subunit interfaces. Therefore, for this interface, the contacting percentage of the MD trajectory (the latter threshold value) was gradually reduced. The reduction in the value from 98 to 70 % (70 % means interacting with a moderate frequency) yields different results for the $\alpha_E\beta_E$ subunit interface of yF_1II and yF_1I. The overall $\alpha_E\beta_E$ subunit interface in yF_1I, particularly the side facing the γ subunit, loses contacts, suggesting that after the Pi release, that part of the interface becomes looser and more flexible (Fig. 17.7c).

17.3.3.3 Position of the γ Subunit

Finally, the positions of the γ subunit relative to the different $\alpha_3\beta_3$ complexes in yF_1II and yF_1I were investigated to characterize the induction of the 16° rotation of the γ subunit via the subunit rearrangements of the $\alpha_3\beta_3$ complex due to the Pi release. According to the contiguous surface figures (Fig. 17.7d), the additional α_{TP} and β_{TP} contacts with the γ subunit appear in the C-terminal domain after the Pi release. In contrast, the α_{DP} and β_{DP} subunits do not show distinctive differences before and after the Pi release (Fig. 17.7d). However, when the contacted surface areas of the α/β subunit with the γ subunit were calculated, the value between the β_{DP} and γ subunits was significantly reduced (data shown in the original paper [83]). These observations: the increased number of contacts for the α_{TP} and β_{TP} subunits not only with the $\alpha_3\beta_3$ subunit complex but also with the γ subunit, suggest that after the Pi release, the tightly packed interface regions are reorganized from the interfaces around the β_{DP} to those around the β_{TP} subunit.

17.3.3.4 Overall Structural Change due to the 16° Rotation of the γ Subunit

The results of the entire subunit rearrangements are visualized in the three-dimensional model in Fig. 17.8a and are summarized in Fig. 17.8b. With the Pi release from the $\alpha_E\beta_E$ subunit interface, the interface becomes looser and more

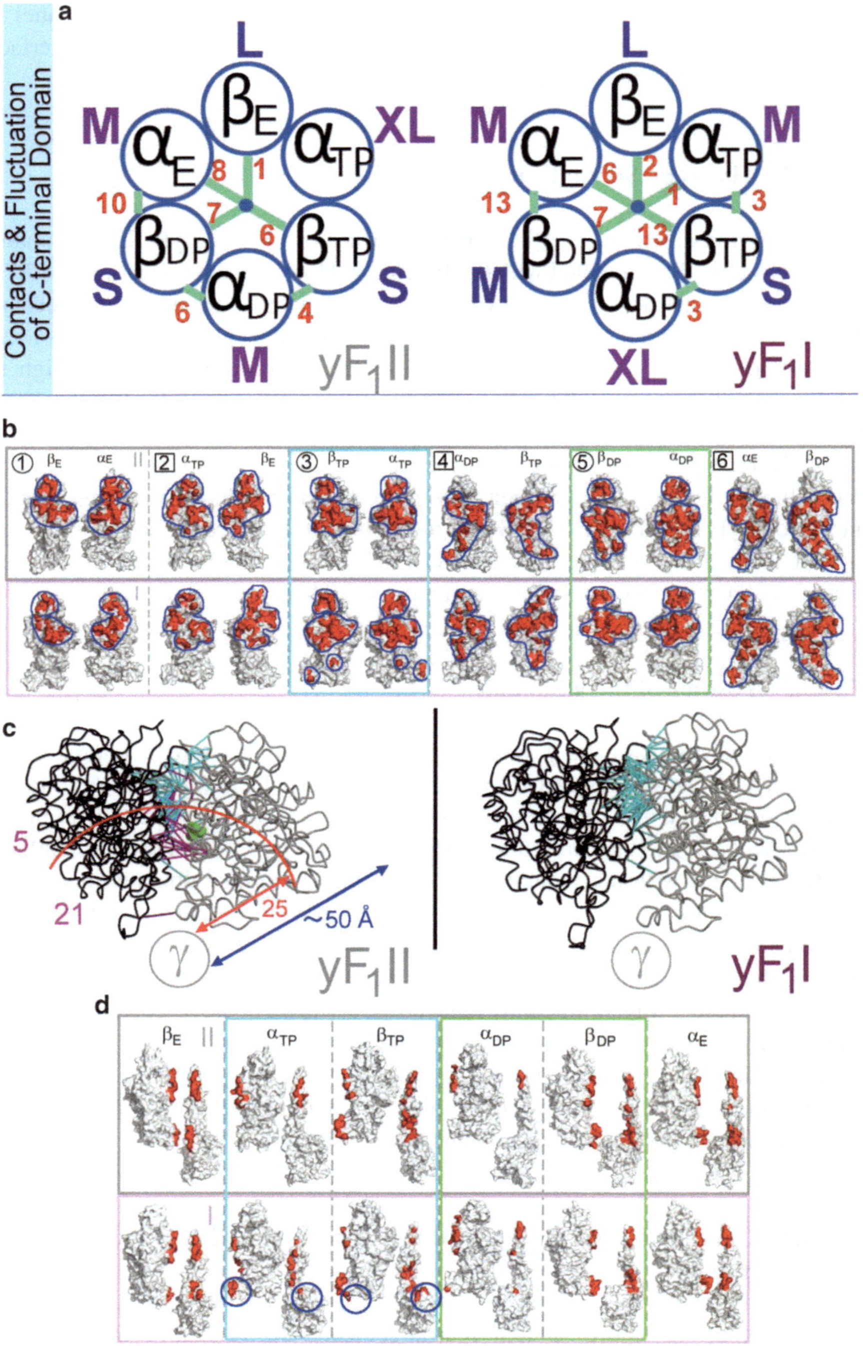
a
Contacts & Fluctuation of C-terminal Domain
L
M
XL
S
M
S
βE
αE
αTP
βTP
αDP
βDP
8
1
7
6
10
6
4
yF1II
L
M
M
M
S
XL
βE
αE
αTP
βTP
αDP
βDP
6
2
1
7
13
13
3
3
yF1I
b
① βE αE II
2 αTP βE
③ βTP αTP
4 αDP βTP
⑤ βDP αDP
6 αE βDP
I
c
5
21
25
~50 Å
γ
yF1II
γ
yF1I
d
βE II
αTP
βTP
αDP
βDP
αE
I

flexible, allosterically causing the $\alpha_{DP}\beta_{DP}$ subunit interface to loosen as well. This structural communication between these two interfaces occurs through tightening of the $\alpha_{TP}\beta_{TP}$ subunit interface because the $\alpha_E\beta_{DP}$ interface hardly changes (data shown in the original paper [83]). The interactions of the γ subunit with the α_{DP} and β_{DP} subunits weaken, whereas the interactions with the α_{TP} and β_{TP} subunits strengthen. With this displacement of the γ subunit, the γ axis leans into the loosened $\alpha_E\beta_E$ subunit interface (data shown in the original paper [83]). After the Pi release, the tightly packed interfaces are reorganized from the interfaces around the β_{DP} to those around the β_{TP} subunit, inducing the 16° rotation. This view of the 16° rotation is consistent with the results suggested from an analysis of the water-entropy effect [84]. These results are also consistent with our proposed "packing exchange mechanism" [77].

17.3.3.5 Proposed Mechanism for the Structural Change in the 40° Rotation

The results of this simulation show the details of the structural displacement after the 16° rotation of the γ subunit, which corresponds to the state after the Pi release but before the ATP hydrolysis. In fact, the 40° rotation of the γ subunit is induced by both the ATP hydrolysis and the Pi release. On the basis of our series of studies [76, 77, 83, 84] combined with past theoretical and experimental insights, a view of the structural changes occurring during the 40° rotation of F_1-ATPase can be deduced.

Fig. 17.7 (**a**) Summary of the fluctuation magnitude and the number of the interface contacts for the C-terminal domain in yF_1II and yF_1I. "*S*", "*M*", "*L*", and "*XL*" indicate small, medium, large and extra large, respectively, for the RMSF of three α or β subunits. These indicators are defined by the average of the RMSF value in the C-terminal domain: "S" < 1.0 Å; 1.0 Å ≤ "M" < 1.3 Å; 1.3 ≤ "L" < 1.5; and 1.5 ≤ "XL". The numbers outside and inside the F_1 complex model indicate the net number of stable contacts in the interface of the C-terminal domain between the α and β subunits and those between the α/β and γ subunits, respectively. Further data on the RMSF and the stable contacts are in the original paper [83]. (**b**) Stable contacts (*in red*) that maintain the interatomic distance at <4.5 Å for more than 98 % of MD trajectory The *top and bottom* panels indicate yF_1II and yF_1I, respectively. The $\beta_{TP}\alpha_{TP}$ and $\beta_{DP}\alpha_{DP}$ subunit interfaces, which show noticeable change in the contiguous surface between yF_1II and yF_1I, are enclosed by *cyan and green rectangles*, respectively. (**c**) Contact analysis result for the $\alpha_E\beta_E$ subunit interfaces, which are viewed from the C-terminal domain. *Colored lines* indicate stable contacts that maintain the interatomic distances at <4.5 Å for more than 70 % of the MD trajectory. The residue pairs found only in yF_1II (i.e., after the Pi release, these residue combinations no longer interact even with a moderate frequency) are indicated by the *pink color*. The *pink lines* appear mainly on the side facing the γ subunit. In the figure for yF_1II, to emphasize the localization of the *pink lines*, the $\alpha_E\beta_E$ interface is divided into the γ subunit side and the outside interfaces, using a half radius of 25 Å (from the γ subunit to the edge of the α/β subunit is ~50 Å). The *pink lines* in each section are counted, and the total numbers are indicated. (**d**) *Red* indicates the stable contacts between the γ and α/β subunits that maintain the interatomic distance at <4.5 Å for more than 70 % of the MD trajectory; the low-threshold value for the contacts that maintain the percentage of the MD trajectory is used to obtain the difference between yF_1II and yF_1I

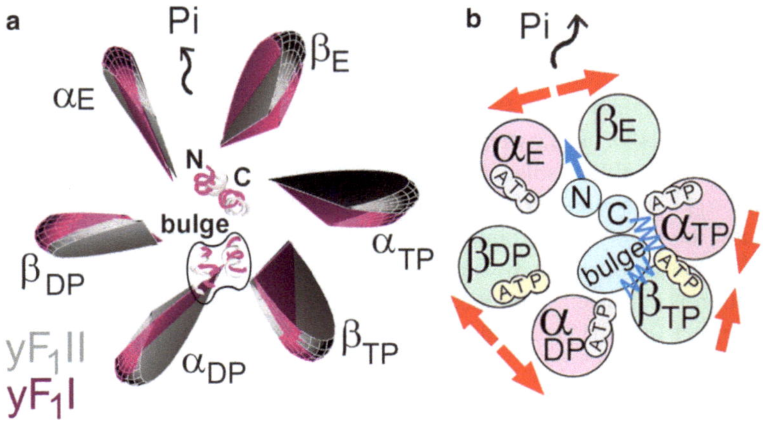

Fig. 17.8 Displacements of the $\alpha_3\beta_3$ complex before (*gray*) and after (*magenta*) Pi release as viewed from the C-terminal domain side; the fit was performed over the N-terminal domain of all the subunits (**a**). The geometric arrangements of the average domain centroids of the $\alpha_3\beta_3$ subunits are described by a three-dimensional model using cut-orange-like objects. "*N*" and "*C*" indicate the N- and C-terminal α helix of the γ subunit, respectively. The portions for the N- and C-terminal α helices are Lys18-Ile25 and Ala236-Asn243, respectively. To indicate the bulge position of the γ subunit, two α helices (residues from Leu91 to His98 and from Lys113 to Arg120) are also presented. Summary of the subunit rearrangements after the Pi release (**b**). *Cyan color* indicates the γ subunit. "*N*" and "*C*" indicate the N- and C-terminal α helices of the γ subunit, respectively

ATP hydrolysis depends weakly on the angle of the γ subunit [86]. Our simulation results are consistent with this weak dependence. The local structures of the ATP-binding site remain similar before and after the 16° rotation of the γ subunit (data shown in the original paper [83]), suggesting that the local structures of the ATP-binding site are insusceptible to the rotation angle of the γ subunit in the range of $\sim$10°. In contrast, Pi release is strongly dependent on the angle of the γ subunit [21, 85]. Our simulation results are also consistent with the strong angle dependence. With the Pi release, the $\alpha_E\beta_E$ interface becomes looser at both the local and global levels, contributing to the γ subunit rotation. The loosening of the $\alpha_E\beta_E$ subunit interface allosterically makes the $\alpha_{DP}\beta_{DP}$ subunit interface looser through a tightening of the $\alpha_{TP}\beta_{TP}$ subunit interface and the γ subunit. Once the tightly packed $\alpha_{DP}\beta_{DP}$ subunit interface is allosterically perturbed, both the loosened interfaces and the ATP hydrolysis concertedly allow conformational changes in the β_{DP} subunit from the closed to the half-closed form ($\beta_{DP} \rightarrow \beta_{HC}$). These conformational changes of the $\alpha_3\beta_3$ complex drive the 40° rotation. Consequently, it is not until both the Pi release and the ATP hydrolysis are accomplished that the F_1-ATPase completes the 40° rotation of the γ subunit.

MD simulations and the statistical thermodynamics analysis of molecular liquids identify the consistent structural characteristics via structural dynamics and thermodynamics properties, respectively. These features indicate that the structural fluctuations at these portions are highly suppressed with the formation of tightly packed configurations due to the water-entropy effect. These different theoretical approaches are fundamentally related to each other.

17.4 Role of the DELSEED Loop Revealed by Single Molecule Experiment and MD Simulations

In this section, we briefly introduce an example of combining single molecule experiments and MD simulations. The structural information derived from MD simulations facilitates a deep understanding of the facts observed in single molecule experiments. This combined study revealed the key factor for torque transmission [87]. As described so far, the β subunit undergoes a conformational change using nucleotide events, and the largest conformational change appears in the C-terminal domain (Fig. 17.5). Therefore, the definitive asymmetry of the $\alpha_3\beta_3$ complex ring also appears in part of the C-terminal domain. Because the γ subunit (axis) follows the sequentially created asymmetrical $\alpha_3\beta_3$ ring, the C-terminal domain has the primary responsibility of transmitting the conformational change of the β subunit to the γ axis. In particular, the DELSEED loop (residue number: 386–394 in thermophilic *Bacillus* PS3, TF_1) of the C-terminal domain plays a critical role in the transmission of torque. This loop comprises a strongly conserved sequence of the amino acids "DELSEED", the loop forms a helix-turn-helix motif that bulges toward the γ subunit, and the γ subunit mostly contacts this loop during the rotation. The aim of this combined study was to elucidate the factor of the DELSEED loop that was crucial for torque transmission.

In the experiments, all of the residues in the DELSEED loop were substituted either with alanine or glycine. The purpose of using the alanine mutant is to diminish the specific interaction between the DELSEED loop and the γ axis; the purpose of the glycine mutant is to disrupt the loop structure because the structure of poly-glycine is much more flexible than that of poly-alanine. The resulting glycine substitution mutants generate half the torque of the wild-type, whereas the alanine substitution mutants generate comparable torque. The MD simulations show that the DELSEED loop is disordered by the glycine substitution, whereas the loop forms the original secondary structure of the alanine mutant (Fig. 17.9). This result is reflected in the magnitude of the structural fluctuations, and the RMSD value of the glycine mutant is much larger than that of the alanine mutant. Consequently, the combination of experimental approaches and MD simulations emphasize the importance of loop rigidity for the efficient transmission of torque.

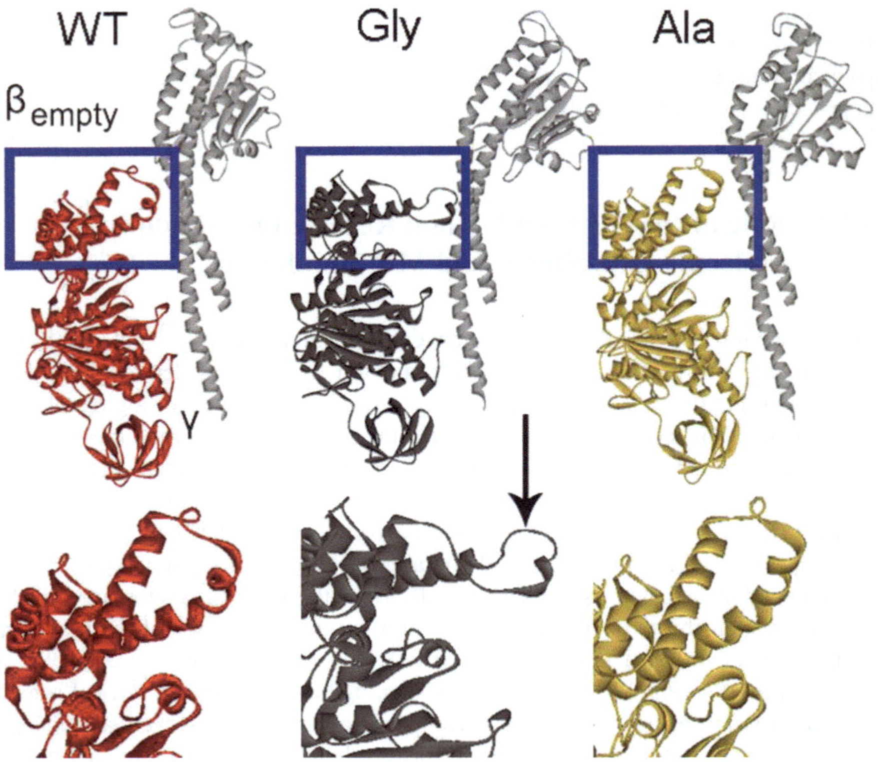

Fig. 17.9 The average structures of the β_E subunit during the last 15 ns of the simulations are shown. The structures of the β_E subunit in the wild-type, glycine mutant, and alanine mutant of F_1-ATPase are shown in *red*, *black*, and *orange*, respectively. Expanded views of the structure of the DELSEED loop regions are shown in the *bottom* of the figure (This figure was taken from the original article [87])

17.5 Chemical Reaction Mechanism of ATP Hydrolysis in the F_1-ATPase

Chemical–mechanical energy conversion is also important for this rotary molecular motor. To elucidate the energy conversion, the chemical reaction (the ATP hydrolysis) during the structural change (motor rotation) was elucidated [63, 88–92]. The hybrid quantum-mechanical/molecular-mechanical (QM/MM) methodology was developed to treat the chemical reaction involving the formation/cleavage of chemical bonds in a biosystem. The QM/MM methodology allows us to calculate a chemical reaction involving a substrate in a binding pocket by taking the molecular interactions with the surrounding protein environment into account.

Providing a quantum-chemical description of the highly polar electric nature of the ATP substrate and its binding pocket is difficult. Therefore, the reaction mechanisms proposed for F_1-ATPase to date are diverse and depend on the simulation protocols and computational accuracies [63, 88, 89]. Therefore, Hayashi et al. [92] conducted a combination of molecular simulations and single molecule experiments; thus, the reaction mechanisms predicted by the simulations underwent a solid verification process using single molecule experiments.

In the predicted reaction mechanism (Fig. 17.10b), the P_γ-O_β bond dissociates first. Subsequently, a proton is transferred from the lytic water molecule, which is strongly activated by a metaphosphate that is generated by the preceding P_γ-O_β bond dissociation. The latter proton transfer is the rate-determining step. The activation energy of the transition state for the proton transfer is computed to be 71.1 kJ/mol. The overall reaction is calculated to be slightly exothermic by −6.7 kJ/mol (Fig. 17.10a).

As described above, this mechanism was verified using single molecule experiments, in which the ATP hydrolysis reaction rates in D_2O and with a substrate analogue, ATPγS, were measured. Considering the reaction order of each substance (the proton transfer or the P_γ-O_β dissociation) and the possible cases for the rate-determining step for either the proton transfer or the P_γ-O_β dissociation, there are three other possible reaction pathways in addition to the path predicted by the molecular simulations (Fig. 17.11). The D_2O solvent should remarkably reduce the reaction rate if the rate-determining step is the proton transfer because of the kinetic isotope effect (cases a and c). In contrast, the hydrolysis reaction should be slowed with ATPγS if the rate-determining step is the P_γ-O_β bond dissociation because the P_γ-O_β dissociation is affected by the substitution of S for O_γ, which is adjacent to the dissociating P-O bond (cases b and d). The reaction profile proposed by the calculation predicts that the reaction rate should decrease when both the D_2O solvent and the ATPγS substrate are used (case c). Single-molecule experiments show positive rate sensitivity when using both D_2O and ATPγS, proving that the reaction mechanism predicted by the molecular simulations is reasonable.

The computed small reaction energy in this QM/MM study suggests that the energy of the hydrolysis reaction would be utilized to regulate the unidirectional rotation by rectifying the thermal equilibrium among the ligand binding and unbinding states, rather than the actual motor function. The major driving forces of the rotational torque should be derived from the ligand binding/release. This explanation is consistent with the mechanism proposed by single molecule experiments [21, 85, 86] and the MD simulation [83].

Despite the fact that the rate-determining step is the latter proton transfer, the first P_γ-O_β bond cleavage is fulfilled by hydrogen bonds between the Walker A motif and an arginine finger. These residues commonly exist in many NTPases and trigger the chain activation of the proton transfer. This finding indicates that the overall activity for the function via both the P_γ-O_β bond dissociation and the proton transfer steps is regulated by the protein system.

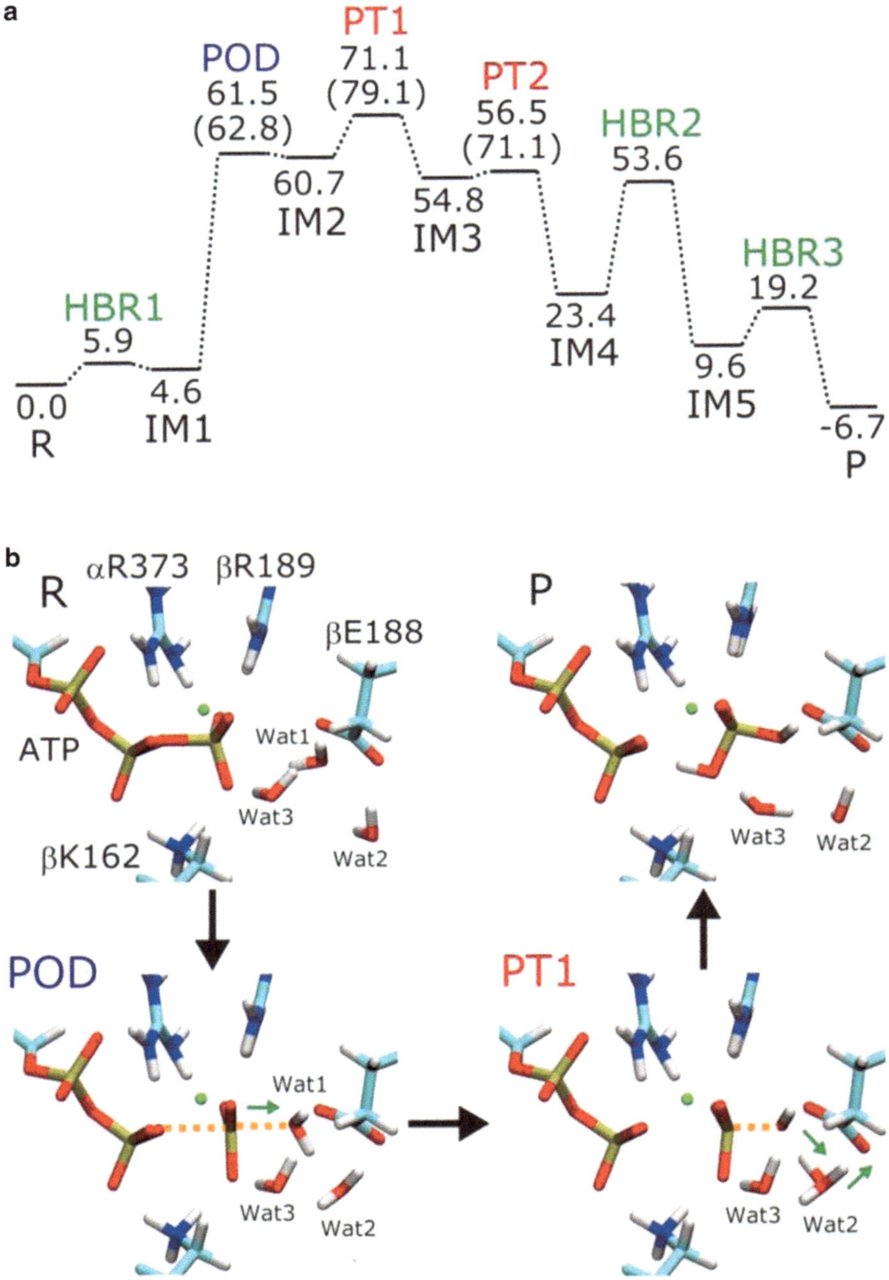

Fig. 17.10 Reaction profile determined using the QM/MM calculations. (**a**) Energy diagram of the overall reaction path from reactant (R) to product (P). (**b**) Structural changes at important states along the reaction path. *POD* and *PT* indicate P_{γ}-O_{β} bond dissociation and proton transfer, respectively (This figure was taken from the original article [92])

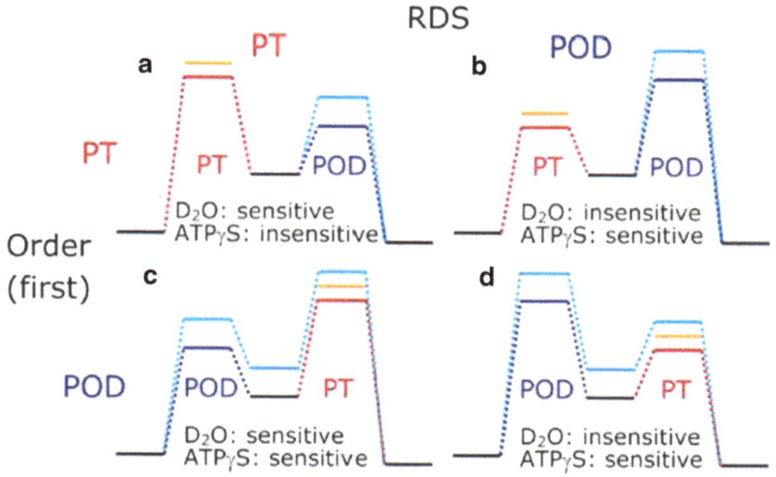

Fig. 17.11 Schematic energy diagrams of possible reaction mechanisms. Energy profiles of the possible reaction paths for the native substrate in H_2O solvent are drawn with *black line* for the reactant, intermediate, and product states, *red lines* for PT (proton transfer), and *blue lines* for POD (P-O bond dissociation). Two possibilities each for the RDS (rate-determining step) and the sequential order produce four possible reaction schemes (**a–d**). The *yellow lines* at the activation barriers of the PT steps indicate energy increases in D_2O solvent. Energy profiles for ATPγS are represented with cyan lines (This figure was taken from the original article [92])

17.6 Summary

Structural fluctuation analysis suggests that only nucleotide binding can change the conformation of the β subunit from the open to the closed form. The subsequent free-energy simulations confirm this suggestion and further demonstrated the details of the β subunit conformational change. This change is accomplished in roughly two characteristic steps: the change of the hydrogen-bond network around ATP and the subsequent dynamic movement of the C-terminal domain via sliding of the B-helix.

Studying the entire F_1-ATPase complex reveals that the complex structure is asymmetric. The subunits in the $\alpha_3\beta_3$ subunit complex form three different interface configurations (tight, moderate, and loose). This heterogeneity of the subunit interfaces indicates that the F_1-ATPase structure is frustrated, in terms of free-energy. In general, the frustrated structure has greater variability and can alter its structure to other free-energy minimum states. The F_1-ATPase can switch among the three different configurations of the packing interface in the $\alpha_3\beta_3$ subunit complex. Therefore, perturbations (substrate binding/release or ATP hydrolysis) allow this type of frustrated F_1-ATPase to change from one to the other sequentially with reorganization of the subunit interfaces, resulting in the γ subunit (axis) rotation.

Interestingly, the asymmetry observed using structural fluctuations via MD simulations are consistent with the results suggested by considering the water-entropy effect. In general, a phenomenon in which a portion of a protein forms a tightly packed configuration can be explained by the water-entropy effect. Accordingly, all of our results suggest that the water-entropy effect plays an important role in the creation of the asymmetrical (frustrated) F_1-ATPase structure that leads to motor rotation. Therefore, this effect fundamentally underlies the rotational mechanism.

The asymmetry of the $\alpha_3\beta_3$ complex structure is created for the rotation of the γ subunit axis. The definitive asymmetry of the $\alpha_3\beta_3$ complex ring appears in part of the C-terminal domain. Therefore, the C-terminal domain, in particular the DELSEED loop, has the primary responsibility of transmitting the conformational change of the β subunit to the γ axis. The factor of the DELSEED loop that is crucial for torque transmission was investigated via a combined study using MD simulations and single molecule experiments. This study shows that loop rigidity, rather than the specific residual interactions between the γ subunit and DELSEED, is important for the efficient transmission of torque.

Finally, the chemical–mechanical energy conversion, another important aspect of this rotary molecular motor, was also studied. To elucidate the chemical reaction of the ATP hydrolysis, QM/MM calculations were employed, and the predicted reaction mechanism was subsequently verified using single molecule experiments. This combination study identifies the sequential order of two elementary processes (the first is the P_γ-O_β bond dissociation, and latter, the proton transfer) and the rate-determining step (the proton transfer) in the ATP hydrolysis reaction. The small reaction energy (−6.7 kJ) that is obtained suggests that the role of ATP hydrolysis is to regulate the unidirectional rotation by rectifying the thermal equilibrium among the ligand binding and unbinding states, rather than regulating the actual motor function. The generation of the rotational torque and the major driving forces of the γ rotation would be derived from the ligand binding/release.

Acknowledgments The authors are grateful to all of our collaborators for their cooperation and helpful discussions. This work was supported by the following: a Grant-in-Aid for the Japan Society for the Promotion of Science (JSPS) fellows; Grants-in-Aid for Scientific Research on Innovative Areas from the Ministry of Education, Culture, Sports, Science and Technology of Japan (MEXT); Grants-in-Aid for Scientific Research (B); the Grand Challenges in Next-Generation Integrated Simulation of Living Matter, a part of the Development and Use of the Next-Generation Supercomputer Project of MEXT; the Platform for Drug Design, Informatics and Structural Life Sciences (MEXT); and the X-ray Free Electron Laser Priority Strategy Program (MEXT).

References

1. Futai M, Kanazawa H (1983) Structure and function of proton-translocating adenosine triphosphatase (F_0F_1): biochemical and molecular biological approaches. Microbiol Rev 47:285–312
2. Futai M, Noumi T, Maeda M (1989) ATP synthase (H^+-ATPase): results by combined biochemical and molecular biological approaches. Annu Rev Biochem 58:111–136
3. Senior AE (1990) The proton-translocating ATPase of *Escherichia coli*. Annu Rev Biophys Biophys Chem 19:7–41

4. Pedersen PL, Amzel LM (1993) ATP synthases. Structure, reaction center, mechanism, and regulation of one of nature's most unique machines. J Biol Chem 268:9937–9940
5. Boyer PD (1997) The ATP synthase – a splendid molecular machine. Annu Rev Biochem 66:717–749
6. Walker JE (1998) ATP synthesis by rotary catalysis (Nobel lecture). Angew Chem Int Ed 37:2308–2319
7. Weber J, Senior AE (2000) ATP synthase: what we know about ATP hydrolysis and what we do not know about ATP synthesis. Biochim Biophys Acta 1458:300–309
8. Kinosita K Jr, Yasuda R, Noji H, Ishiwata S, Yoshida M (1998) F_1-ATPase: a rotary motor made of a single molecule. Cell 93:21–24
9. Gao YQ, Yang W, Karplus M (2005) A structure-based model for the synthesis and hydrolysis of ATP by F_1-ATPase. Cell 123:195–205
10. Karplus M, Gao YQ (2004) Biomolecular motors: the F_1-ATPase paradigm. Curr Opin Struct Biol 14:250–259
11. Noji H, Yasuda R, Yoshida M, Kinosita K Jr (1997) Direct observation of the rotation of F_1-ATPase. Nature 386:299–302
12. Kabaleeswaran V, Puri N, Walker JE, Leslie AGW, Mueller DM (2006) Novel features of the rotary catalytic mechanism revealed in the structure of yeast F_1-ATPase. EMBO J 25:5433–5442
13. Itoh H, Takahashi A, Adachi K, Noji H, Yasuda R, Yoshida M, Kinosita K Jr (2004) Mechanically driven ATP synthesis by F_1-ATPase. Nature 427:465–468
14. Rastogi VK, Girvin ME (1999) Structural changes linked to proton translocation by subunit c of the ATP synthase. Nature 402:263–268
15. Abrahams JP, Leslie AG, Lutter R, Walker JE (1994) Structure at 2.8 Å resolution of F_1-ATPase from bovine heart mitochondria. Nature 370:621–628
16. Yasuda R, Noji H, Kinosita K Jr, Yoshida M (1998) F_1-ATPase is a highly efficient motor that rotates with discrete 120-degree steps. Cell 93:1117–1124
17. Yasuda R, Noji H, Yoshida M, Kinosita K Jr, Itoh H (2001) Resolution of distinct rotational substeps by submillisecond kinetic analysis of F_1-ATPase. Nature 410:898–904
18. Okuno D, Fujisawa R, Iino R, Hirono-Hara Y, Imamura H, Noji H (2008) Correlation between the conformational states of F_1-ATPase as determined from its crystal structure and single-molecule rotation. Proc Natl Acad Sci USA 105:20722–20727
19. Masaike T, Koyama-Horibe F, Oiwa K, Yoshida M, Nishizaka T (2008) Cooperative three-step motions in catalytic subunits of F_1-ATPase correlate with 80° and 40° substep rotations. Nat Struct Mol Biol 15:1326–1333
20. Shimabukuro K, Yasuda R, Muneyuki E, Hara KY, Kinosita K Jr, Yoshida M (2003) Catalysis and rotation of F_1 motor: cleavage of ATP at the catalytic site occurs in 1 ms before 40° substep rotation. Proc Natl Acad Sci USA 100:14731–14736
21. Adachi K, Oiwa K, Nishizaka T, Furuike S, Noji H, Itoh H, Yoshida M, Kinosita K Jr (2007) Coupling of rotation and catalysis in F_1-ATPase revealed by single-molecule imaging and manipulation. Cell 130:309–321
22. Ikeguchi M, Ueno J, Sato M, Kidera A (2005) Protein structural change upon ligand binding: linear response theory. Phys Rev Lett 94:078102
23. Verma CS, Caves LS, Hubbard RE, Roberts GCK (1997) Domain motions in dihydrofolate reductase: a molecular dynamics study. J Mol Biol 266:776–796
24. Radkiewicz JL, Brooks CL III (2000) Protein dynamics in enzymatic catalysis: exploration of dihydrofolate reductase. J Am Chem Soc 122:225–231
25. Rod TH, Radkiewicz JL, Brooks CL III (2003) Correlated motion and the effect of distal mutations in dihydrofolate reductase. Proc Natl Acad Sci USA 100:6980–6985
26. Thorpe IF, Brooks CL III (2004) The coupling of structural fluctuations to hydride transfer in dihydrofolate reductase. Proteins 57:444–457
27. Ito Y, Ikeguchi M (2010) Molecular dynamics simulations of the isolated β subunit of F_1-ATPase. Chem Phys Lett 490:80–83

28. Böckmann RA, Grubmüller H (2003) Conformational dynamics of the F_1-ATPase β-subunit: a molecular dynamics study. Biophys J 85:1482–1491
29. Yagi H, Tsujimoto T, Yamazaki T, Yoshida M, Akutsu H (2004) Conformational change of H^+-ATPase β monomer revealed on segmental isotope labeling NMR spectroscopy. J Am Chem Soc 126:16632–16638
30. Shirakihara Y, Leslie AG, Abrahams JP, Walker JE, Ueda T, Sekimoto Y, Kambara M, Saika K, Kagawa Y, Yoshida M (1997) The crystal structure of the nucleotide-free α3β3 subcomplex of F_1-ATPase from the thermophilic Bacillus PS3 is a symmetric trimer. Structure 5:825–836
31. Gibbons C, Montgomery MG, Leslie AG, Walker JE (2000) The structure of the central stalk in bovine F_1-ATPase at 2.4 Å resolution. Nat Struct Biol 7:1055–1061
32. Yagi H, Kajiwara N, Iwabuchi T, Izumi K, Yoshida M, Akutsu H (2009) Stepwise propagation of the ATP-induced conformational change of the F_1-ATPase β subunit revealed by NMR. J Biol Chem 284:2374–2382
33. Ito Y, Oroguchi T, Ikeguchi M (2011) Mechanism of the conformational change of the F_1-ATPase β subunit revealed by free-energy simulations. J Am Chem Soc 133:3372–3380
34. Ma J, Flynn TC, Cui Q, Leslie AG, Walker JE, Karplus M (2002) A dynamic analysis of the rotation mechanism for conformational change in F_1-ATPase. Structure 10:921–931
35. Böckmann R, Grubmüller H (2002) Nanoseconds molecular dynamics simulation of primary mechanical energy transfer steps in F_1-ATP synthase. Nat Struct Biol 9:198–202
36. Arora K, Brooks CL III (2007) Large-scale allosteric conformational transitions of adenylate kinase appear to involve a population-shift mechanism. Proc Natl Acad Sci USA 104:18496–18501
37. Arora K, Brooks CL III (2009) Functionally important conformations of the Met20 loop in dihydrofolate reductase are populated by rapid thermal fluctuations. J Am Chem Soc 131:5642–5647
38. Jonsson H, Mills G, Jacobsen KW (1998) Nudged elastic band method for finding minimum energy paths of transitions. In: Berne BJ, Cicotti G, Coker DF (eds) Classical and quantum dynamics in condensed phase simulations. World Scientific, Rivers Edge, pp 385–404
39. Chu JW, Trout BL, Brooks BR (2003) A super-linear minimization scheme for the nudged elastic band method. J Chem Phys 119:12708–12717
40. Torrie GM, Valleau JP (1974) Monte Carlo free energy estimates using non-Boltzmann sampling: application to the sub-critical Lennard-Jones fluid. Chem Phys Lett 28:578–581
41. Kirkwood JG (1935) Statistical mechanics of fluid mixtures. J Chem Phys 3:300–313
42. Weinan E, Ren W, Vanden-Eijnden E (2002) String method for the study of rare events. Phys Rev B 66:052301–052304
43. Weinan E, Ren W, Vanden-Eijnden E (2007) Simplified and improved string method for computing the minimum energy paths in barrier-crossing events. J Chem Phys 126:164103–164108
44. Weinan E, Ren W, Vanden-Eijnden E (2005) Finite temperature string method for the study of rare events. J Phys Chem B 109:6688–6693
45. Vanden-Eijnden E, Venturoli M (2009) Revisiting the finite temperature string method for the calculation of reaction tubes and free energies. J Chem Phys 130:194103–194117
46. Maragliano L, Fischer A, Vanden-Eijnden E, Ciccotti G (2006) String method in collective variables: minimum free energy paths and isocommittor surfaces. J Chem Phys 125:024106–024115
47. Maragliano L, Vanden-Eijnden E (2007) On-the-fly string method for minimum free energy paths calculation. Chem Phys Lett 446:182–190
48. Pan AC, Sezer D, Roux B (2008) Finding transition pathways using the string method with swarms of trajectories. J Phys Chem B 112:3432–3440
49. Gan W, Yang S, Roux B (2009) Atomistic view of the conformational activation of Src kinase using the string method with swarms-of-trajectories. Biophys J 97:L8–L10
50. Dellago C, Bolhuis PG, Geissler PL (2002) Transition path sampling. Adv Chem Phys 123:1–78
51. Dellago C, Bolhuis PG (2007) Transition path sampling simulations of biological systems. Top Curr Chem 268:291–317

52. Hagan MF, Dinner AR, Chandler D, Chakraborty AK (2003) Atomistic understanding of kinetic pathways for single base-pair binding and unbinding in DNA. Proc Natl Acad Sci USA 100:13922–13927
53. Juraszek J, Bolhuis PG (2006) Sampling multiple folding pathways of Trp-cage mini-protein in explicit solvent. Proc Natl Acad Sci USA 103:15859–15864
54. Radhakrishnan R, Schlick T (2004) Orchestration of cooperative events in DNA synthesis and repair mechanism unraveled by transition path sampling of DNA polymerase β's closing. Proc Natl Acad Sci USA 101:5970–5975
55. Martí J, Csajka FS (2004) Transition path sampling study of flip-flop transitions in model lipid bilayer membranes. Phys Rev E 69:061918
56. Quaytman SL, Schwartz SD (2007) The reaction coordinate of an enzymatic reaction: TPS studies of lactate dehydrogenase. Proc Natl Acad Sci USA 104:12253–12258
57. Banavali NK, Roux B (2005) Free energy landscape of a-DNA to B-DNA conversion in aqueous solution. J Am Chem Soc 127:6866–6876
58. Banavali NK, Roux B (2005) The N-terminal end of the catalytic domain of Src kinase Hck is a conformational switch implicated in long-range allosteric regulation. Structure 13:1715–1723
59. Amano T, Tozawa K, Yoshida M, Murakami H (1994) Spatial precision of a catalytic carboxylate of F_1-ATPase β subunit probed by introducing different carboxylate-containing side chains. FEBS Lett 348:93–98
60. Löbau S, Weber J, Wilke-Mounts S, Senior AE (1997) F_1-ATPase: roles of three catalytic site residues. J Biol Chem 272:3648–3656
61. Ariga T, Muneyuki E, Yoshida M (2007) F_1-ATPase rotates by an asymmetric, sequential mechanism using all three catalytic subunits. Nat Struct Mol Biol 14:841–846
62. Nadanaciva S, Weber J, Senior AE (1999) The role of β-Arg-182, an essential catalytic site residue in Escherichia coli F_1-ATPase. Biochemistry 38:7670–7677
63. Dittrich M, Hayashi S, Schulten K (2004) ATP hydrolysis in the β_{TP} and β_{DP} catalytic sites of F_1-ATPase. Biophys J 87:2954–2967
64. Ahmad Z, Senior AE (2004) Mutagenesis of residue β-Arg-246 in the phosphate-binding subdomain of catalytic sites of Escherichia coli F_1-ATPase. J Biol Chem 279:31505–31513
65. Masaike T, Mitome N, Noji H, Muneyuki E, Yasuda R, Kinosita K Jr, Yoshida M (2000) Rotation of F_1-ATPase and the hinge residues of the β subunit. J Exp Biol 203:1–8
66. Yoshidome T, Kinoshita M, Hirota S, Baden N, Terazima M (2008) Thermodynamics of apoplastocyanin folding: comparison between experimental and theoretical results. J Chem Phys 128:225104(1–9)
67. Kauzmann W (1959) Some factors in the interpretation of protein denaturation. Adv Protein Chem 14:1–63
68. Gerstman BS, Chapagain PP (2005) Self-organization in protein folding and the hydrophobic interaction. J Chem Phys 123:054901(1–6)
69. Watanabe R, Iino R, Shimabukuro K, Yoshida M, Noji H (2008) Temperature-sensitive reaction intermediate of F_1-ATPase. EMBO Rep 9:84–90
70. Boyer PD (1993) The binding change mechanism for ATP synthase–some probabilities and possibilities. Biochim Biophys Acta 1140:215–250
71. Koga N, Takada S (2006) Folding-based molecular simulations reveal mechanisms of the rotary motor F_1-ATPase. Proc Natl Acad Sci USA 103:5367–5372
72. Pu J, Karplus M (2008) How subunit coupling produces the γ-subunit rotary motion in F_1-ATPase. Proc Natl Acad Sci USA 105:1192–1197
73. Cui Q, Li GH, Ma JP, Karplus M (2004) A normal mode analysis of structural plasticity in the biomolecular motor F_1-ATPase. J Mol Biol 340:345–372
74. Czub J, Grubmüller H (2011) Torsional elasticity and energetics of F_1-ATPase. Proc Natl Acad Sci USA 108:7408–7413
75. Okazaki K, Takada S (2011) Structural comparison of F_1-ATPase: interplay among enzyme structures, catalysis, and rotations. Structure 19:588–598
76. Ito Y, Ikeguchi M (2010) Structural fluctuation and concerted motions in F_1-ATPase: a molecular dynamics study. J Comput Chem 31:2175–2185

77. Yoshidome T, Ito Y, Ikeguchi M, Kinoshita M (2011) On the rotation mechanism of F_1-ATPase: crucial importance of water-entropy effect. J Am Chem Soc 133:4030–4039
78. Bowler MW, Montgomery MG, Leslie AGW, Walker JE (2007) Ground state structure of F_1-ATPase from bovine heart mitochondria at 1.9 Å resolution. J Biol Chem 282:14238–14242
79. Oroguchi T, Hashimoto H, Shimizu T, Sato M, Ikeguchi M (2009) Intrinsic dynamics of restriction endonuclease EcoO109I studied by molecular dynamics simulations and X-ray scattering data analysis. Biophys J 96:2808–2822
80. Furuike S, Hossain MD, Maki Y, Adachi K, Suzuki T, Kohori A, Itoh H, Yoshida M, Kinosita K Jr (2008) Axle-less F_1-ATPase rotates in the correct direction. Science 319:955–958
81. Hossain MD, Furuike S, Maki Y, Adachi K, Suzuki T, Kohori A, Itoh H, Yoshida M, Kinosita K Jr (2008) Neither helix in the coiled coil region of the axle of F_1-ATPase plays a significant role in torque production. Biophys J 95:4837–4844
82. Uchihashi T, Iino R, Ando T, Noji H (2011) High-speed atomic force microscopy reveals rotary catalysis of rotorless F_1-ATPase. Science 333:755–758
83. Ito Y, Yoshidome T, Matubayasi N, Kinoshita M, Ikeguchi M (2013) Molecular dynamics simulations of yeast F_1-ATPase before and after 16° rotation of the γ subunit. J Phys Chem B 117:3298–3307
84. Yoshidome T, Ito Y, Matubayasi N, Ikeguchi M, Kinoshita M (2012) Structural characteristics of yeast F_1-ATPase before and after 16-degree rotation of the γ subunit: theoretical analysis focused on the water-entropy effect. J Chem Phys 137:035102(1–8)
85. Watanabe R, Iino R, Noji H (2010) Phosphate release in F_1-ATPase catalytic cycle follows ADP release. Nat Chem Biol 6:814–820
86. Watanabe R, Okuno D, Sakakihara S, Shimabukuro K, Iino R, Yoshida M, Noji H (2012) Mechanical modulation of catalytic power on F_1-ATPase. Nat Chem Biol 8:86–92
87. Tanigawara M, Tabata KV, Ito Y, Ito J, Watanabe R, Ueno H, Ikeguchi M, Noji H (2012) Role of the DELSEED loop in torque transmission of F_1-ATPase. Biophys J 103:970–978
88. Beke-Somfai T, Lincoln P, Nordén B (2011) Double-lock ratchet mechanism revealing the role of αSER-344 in F_oF_1 ATP synthase. Proc Natl Acad Sci USA 108:4828–4833
89. Dittrich M, Hayashi S, Schulten K (2003) On the mechanism of ATP hydrolysis in F_1-ATPase. Biophys J 85:2253–2266
90. Yang W, Gao YQ, Cui Q, Ma J, Karplus M (2003) The missing link between thermodynamics and structure in F_1-ATPase. Proc Natl Acad Sci USA 100:874–879
91. Gao YQ, Yang W, Marcus RA, Karplus M (2003) A model for the cooperative free energy transduction and kinetics of ATP hydrolysis by F_1-ATPase. Proc Natl Acad Sci USA 100:11339–11344
92. Hayashi S, Ueno H, Shaikh AR, Umemura M, Kamiya M, Ito Y, Ikeguchi M, Komoriya Y, Iino R, Noji H (2012) Molecular mechanism of ATP hydrolysis in F_1-ATPase revealed by molecular simulations and single-molecule observations. J Am Chem Soc 134:8447–8454

Chapter 18
Chemosensorial G-proteins-Coupled Receptors: A Perspective from Computational Methods

Francesco Musiani*, Giulia Rossetti*, Alejandro Giorgetti, and Paolo Carloni

Abstract G-protein coupled receptors (GPCRs) constitute the targets of about 40 % of all the pharmaceutical drugs in the market and, among other functions, a large portion of the family detects odorants and a variety of tastant molecules. Computational techniques are instrumental to understand structure, dynamics and function of the cascades triggered by these receptors. As an example, here we

---

*Author contributed equally with all other contributors.

Supported by Programma Operativo del Fondo Sociale Europeo 2007/2013 of Regione Autonoma Friuli Venezia Giulia.

F. Musiani
Scuola Internazionale Superiore di Studi Avanzati (SISSA/ISAS), Trieste, Italy

G. Rossetti
Institute for Research in Biomedicine (IRB Barcelona), Barcelona, Spain

Joint IRB-BSC Program in Computational Biology, Barcelona, Spain

Jülich Supercomputing Centre (JSC), Forschungszentrum Jülich, Jülich, Germany

Computational Biophysics, German Research School for Simulation Sciences, Jülich, Germany

Institute for Advanced Simulation, Forschungszentrum Jülich, Jülich, Germany

A. Giorgetti (✉)
Computational Biophysics, German Research School for Simulation Sciences, Jülich, Germany

Department of Biotechnology, University of Verona, Ca' Vignal 1, Strada le Grazie 15, Verona 37134, Italy

Institute for Advanced Simulation, Forschungszentrum Jülich, Jülich, Germany
e-mail: alejandro.giorgetti@univr.it

P. Carloni
Computational Biophysics, German Research School for Simulation Sciences, Jülich, Germany

Institute for Advanced Simulation, Forschungszentrum Jülich, Jülich, Germany

K. Han et al. (eds.), *Protein Conformational Dynamics*, Advances in Experimental Medicine and Biology 805, DOI 10.1007/978-3-319-02970-2_18,

report our own computational work aimed to dissect GPCR molecular mechanisms for chemical senses. The implications of our work for systems biology and for pharmacology are discussed.

Keywords G-protein coupled receptors • Multi-scale modeling • Odor and bitter taste perception • Bioinformatics • Molecular dynamics

18.1 Introduction

G-proteins-coupledreceptors (GPCRs) belong to the largest membrane-bound receptor family expressed by mammalians (encompassing ca. 4 % of the protein-coding human genome) [1] and are of paramount importance for pharmaceutical intervention (ca. 40 % of currently marketed drugs target GPCRs) [2]. Structurally, GPCRs are characterized by an extracellular N-terminus, followed by seven transmembrane (7-TM) α-helices (TM-1 to TM-7) connected by three intracellular (IL-1 to IL-3) and three extracellular loops (EL-1 to EL-3), and finally by an intracellular C-terminus (Fig. 18.1a). GPCRs' tertiary structure resembles a barrel, with the seven transmembrane helices forming a cavity within the plasma membrane that serves as ligand-binding domain, often covered by EL-2.

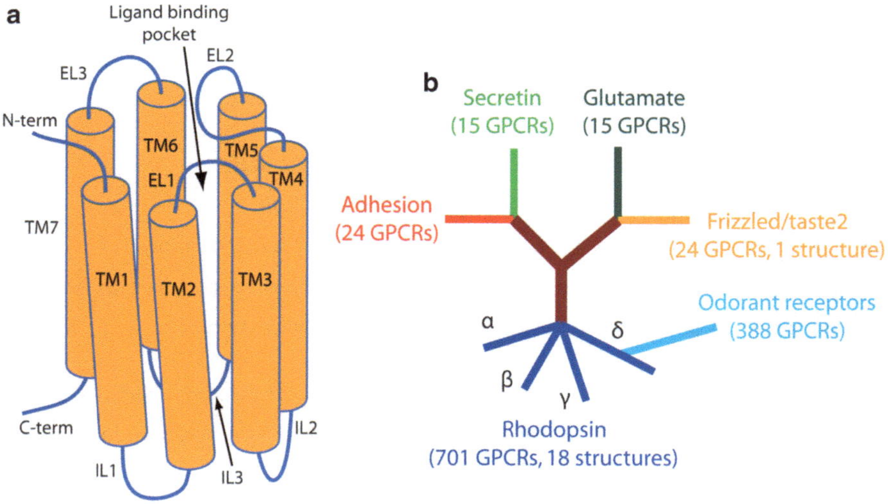

Fig. 18.1 (**a**) Schematic representation of GPCR fold. Trans-membrane helices (TM) are depicted as *orange cylinders*. The positions of intracellular (IL) and extracellular (EL) loops are indicated. (**b**) Phylogenetic tree of human GPCRs according to the GRAFS system. Human GPCRs sequences can be divided in five phylogenetic families following the so called GRAFS classification (glutamate, rhodopsin, adhesion, frizzled/taste2, secretin) [105]. The rhodopsin family can be further divided in four subfamilies (named α-, β-, γ-, and δ-branches). This classification excludes the olfactory receptors (that form a separate sub-branch of 388 receptors in the rhodopsin δ-branch) and pheromone receptors of type 1

The first solved structure of a GPCR was that of bovine rhodopsin, back in 2000 [3]. In rhodopsin, the ligand is covalently bound to the protein. Seven years were needed to get the high-resolution structure of the human β_2-adrenergic receptor (β2AR) [4, 5], the first example of a GPCR with a non-covalently bound ligand. That structure was followed by other GPCRs in the rhodopsin family (See Fig. 18.1b), including β1AR, A_2A adenosine (A2AAR), chemokine CXCR4, dopamine D_3, histamine H1, opioid receptors, sphingosine 1-phosphate receptor, 5-HT_{1B} and 5-HT_{2B} serotonin receptors [9] and most recently by the structure of the smoothened receptor belonging to the frizzled/taste2 family [10]. The explosion of crystallography of GPCRs, while confirming the common seven transmembrane GPCR fold, provided the first insights of structural diversity in GPCRs at various levels of evolution. Thus, with the exception of the smoothened receptor, the structures represent closely related GPCR subtypes, different subfamilies within the aminergic family, different sub-branches within the same major α-branch, as well as different major α-, β- and γ-branches of GPCRs belonging to the rhodopsin family. The ligand binding pocket, which is similar across the GPCR X-ray structures [11], is located in the extracellular side of the TM bundle. Because of the high sequence diversity of the protein [11], in particular of the N-term [12] and of the extracellular loops [13], they bind ligand molecules of diverse shapes, sizes and chemical properties. These molecules range from metal ions [14] to small molecules [11] and to short peptides [15].

The GPCR signaling cascade is initiated by the binding of a ligand to the GPCR in the extracellular binding pocket (or by the interaction with the electromagnetic field in the case of vision). GPCRs in the resting state are usually bound to their cognate heterotrimeric guanine nucleotide-binding proteins (G-protein) (Fig. 18.2a). G-proteins function as molecular switches [16]. In their inactive form, they are trimers, formed by subunits Gα (in complex with GDP), Gβ and Gγ [17].[1] When the ligand is bound to the GPCR, the activated receptor undergoes a conformational change, causing a rearrangement of its cognate G-protein, which then exchanges GDP for GTP [16, 18]. This triggers the dissociation (or the weakening of interactions) between the Gα•GTP subunit and the Gβ•Gγ dimer (Gβγ hereafter). Gα•GTP and Gβγ activate a downstream cascade of events by binding to specific target proteins. The G-protein may also rearrange itself and/or bind to molecules involved in the downstream pathway [19, 20].

[1]Gα subunits are divided in classes. In mammals, these are: the stimulatory $G\alpha_s$ family (which comprises $G\alpha_s$ and $G\alpha_{olf}$), the inhibitory $G\alpha_i$ family (which includes $G\alpha_{i/o}$, $G\alpha_t$, $G\alpha_{gus}$, and $G\alpha_z$), $G\alpha_q$, $G\alpha_{12/13}$ and rod transducin $G\alpha_{t1}$. They feature from 35 to 95 % sequence identity (SI) among each other [6]. Gβ In mammals, isoforms ($G\beta_{1-5}$), with splice variants of $G\beta_3$ ($G\beta_{3s}$, $G\beta_{3s2}$ and $G\beta_{3v}$) and of $G\beta_5$ ($G\beta_{5L}$), feature 50–90 % SI [7]. In mammals, 12 isoforms of Gγ ($G\gamma_{T1}$, $G\gamma_{T2}$, $G\gamma_{2\text{-}4}$, $G\gamma_5$, $G\gamma_{5ps}$, $G\gamma_{7,8,10\text{–}12}$) share 31–77 % SI [7].

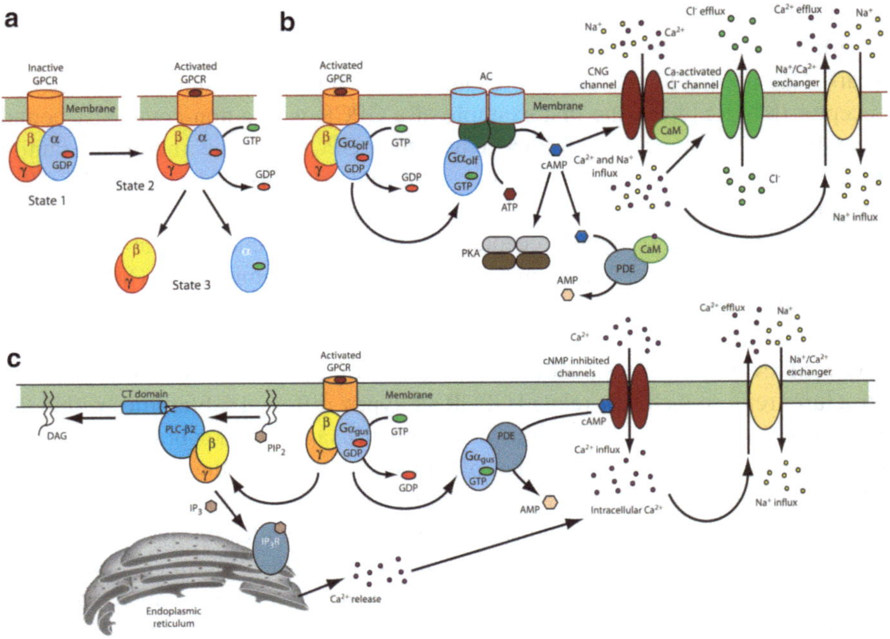

Fig. 18.2 (**a**) Schematic representation of GPCR/G-protein interactions during signalling in eukaryotes. State 1: G-protein inactive state. State 2: GDP/GTP exchange, first step of activation. State 3: dissociation of Gα•GTP from the tightly bound Gβ•Gγ dimer. Initial phases of the olfactory (**b**) and of the bitter taste (**c**) pathways. In the olfactory pathway, $G\alpha_{olf}$•GTP stimulates the transmembrane AC enzyme, which catalyzes the formation of cAMP from ATP. cAMP in turns activates ion channels as well as activating protein kinase A (PKA) enzyme. The latter in turn activates a variety of downstream processes including the activations of CNG channels. This process causes Ca^{2+} and Na^{+} inflow from the extracellular to the intracellular side of the membrane and Cl^{-} efflux. The signal is quenched by Ca^{2+} binding to calmodulin (CaM), which stimulates the activity of a phosphodiesterase (PDE) that converts cAMP into AMP and closes the CNG channel. In the bitter taste pathway, $G\alpha_{gus}$•GTP stimulates PDE activity, thus reducing intracellular cAMP concentration and opening the cNMP inhibited channels. Concomitantly, Gβγ subunits stimulate the PLC-β2 enzyme, which catalyzed the formation of DAG and IP_3 from PIP_2. PIP2 in turns stimulates Ca^{2+} release from the endoplasmatic reticulum. Thus, both $G\alpha_{gus}$•GTP and Gβγ concur in the increase of intracellular Ca^{2+} concentration

18.2 Chemosensorial GPCR'S and Their Cascades

More than half of the GPCRs encoded in mammalian genomes are olfactory receptors (ORs) [21]. The second largest sensorial GPCR subfamily is the bitter-taste receptor family, formed by about 30 members in the human genome [22].

18.2.1 Olfactory Signaling Pathway

In the cilia of olfactory sensory neurons, volatile odorant molecules binding to ORs cause conformation changes, with the consequence of activating its cognate G-protein (G_{olf} and/or the G_s isoforms; Fig. 18.2b) [23]. The $G\alpha$ subunit cellular partner is isoform 3 of adenylate cyclase (AC3) enzyme, which converts adenosine-5′-triphosphate (ATP) into cyclic adenosine monophosphate (cAMP), triggered by the binding of $G\alpha_{olf}$•GTP. cAMP acts as "second messenger", activating the protein kinase A (PKA), which phosphorylates downstream targets [24]. cAMP also binds to and opens the cytoplasmic domains of the olfactory cyclic nucleotide gated (CNG) ion channels. This allows Na^+ and Ca^{2+} cations to flow along their electrochemical gradients from the extracellular to the intracellular side of the membrane [25, 26]. The increased Ca^{2+} concentration in the cilia causes the opening of Ca^{2+}-activated Cl^- channels and the subsequent Cl^- efflux, which further depolarizes the cell [27–29]. Thus, the chemical interactions of ORs with volatile molecules lead ultimately to the production of action potentials that will carry information about the external world to the brain [30, 31]. The axons of the olfactory sensory neurons from the nasal cavity send information to second-order neurons in the olfactory bulb, which in turn project to the olfactory cortex and then to other brain areas. The increase of Ca^{2+} concentration has an inhibitory effect, which eventually terminates the signal, as obviously required for the function of this apparatus. This is achieved by Ca^{2+} binding to calmodulin (CaM), which also stimulates the activity of a phosphodiesterase (PDE). Ca^{2+} is then extruded by a Na^+/Ca^{2+} exchanger (Fig. 18.2b).

18.2.2 Bitter Taste Perception

Bitter taste perception discourages humans and other mammals from ingesting bitter, possibly toxic, substances. The perception stems from the binding of bitter molecules to ca. 25 specific GPCRs referred to as taste 2 receptors (TAS2Rs) [32, 33]. TAS2Rs are located in special subsets of taste receptor cells [32–35]. They are able to detect multiple and diverse natural and synthetic organic molecules [34]. Single nucleotide polymorphisms can cause "blindness" to its agonists. This is the case, for instance of the phenylthiocarbamide (PTC) and propylthiouracil (PROP) [36] agonists of the TAS2R38 receptor [37]. Indeed, normal populations can be divided in two different phenotypes, i.e. those subjects that perceive phenylthiocarbamide (PTC) and its related compounds and subjects that do not. The cognate G-protein of TAS2Rs is the heterotrimeric gustducin protein (G_{gus}, a transducin-like G-protein selectively expressed in ca. 25–30 % of taste receptors cells, Fig. 18.2c) [38]. $G\alpha_{gus}$•GTP activate the phosphodiesterase (PDE) enzyme [38], which in turns reduces the concentration of cAMP in the cell and thus decreases cAMP inhibition of the cNMP-inhibited channels, causing an increase of Ca^{2+} concentration in the cell [38]. At the same time, the Gβγ subunit interacts with isoform 2 of 1-phosphatidylinositol-4,5-bisphosphate phosphodiesterase beta (PLC-β2) [38]. PLC-β2 catalyzes the hydrolysis of phosphatidylinositol 4,5-bisphosphate (PIP_2) into two second-messenger molecules: 1,4,5-trisphosphate (IP_3) and diacylglycerol (DAG). IP_3 then stimulates the release of Ca^{2+} in the cytoplasm [39]. Both DAG and calcium are essential second messengers which activate downstream signaling components such as protein kinase C and calmodulin dependent kinase [39].

In spite of GPCRs widespread presence in humans (as well as that in other animals [40, 41]) experimental structural information at atomic level is so far lacking. Hence, computer methods are the method of choice to investigate structure, dynamics and function of these GPCRs [42–44]. Here we review some of our computational work in the olfactory and the bitter taste receptor pathways.

18.3 Olfactory Receptor Cascade Proteins: From Structural Predictions to Large Scale Motions

18.3.1 Olfactory Receptors

Structural predictions of the ORs have been carried out by several groups [45–48]. The observation that structural features are well conserved across GPCRs belonging to the rhodopsin family,[2] has led to the suggestion that template-based structural predictions, together with the use of restraints extracted from point mutagenesis experiments, may lead to the prediction of fairly reliable models [52, 53]. This is confirmed by the observation that GPCR X-ray structures exhibit a large predictability and can thus be used as templates for structural models of GPCRs sharing similar sequences [52, 53]. Nine crucial amino acids involved in ligand binding and selectivity on the helices TM3, TM5, and TM6[3] of ORs set the stage for structural predictions [45]. Our model of one of the 29 ORs for which ligand binding data are available – the MOR174-9 – were validated against experimental information [55]. This model was then used to predict which amino acids could be involved in the ligand binding.

18.3.2 G-proteins

Upon ligand binding, the active GPCR-bound heterotrimeric G-protein, bound to GDP (Fig. 18.3a) undergoes a structural reorganization, including a rearrangement

---

[2]It includes ORs and rhodopsin [49], along with the structures of other GPCRs (such as the human β2AR [4], the turkey β1AR [50] and the human A2AAR [51]).

[3]The identified residues were 3.40, 5.45, 5.46, 5.50, 5.51, 6.44, 6.47, 6.48, and 6.51; following the numbering of Ballesteros et al. [54].

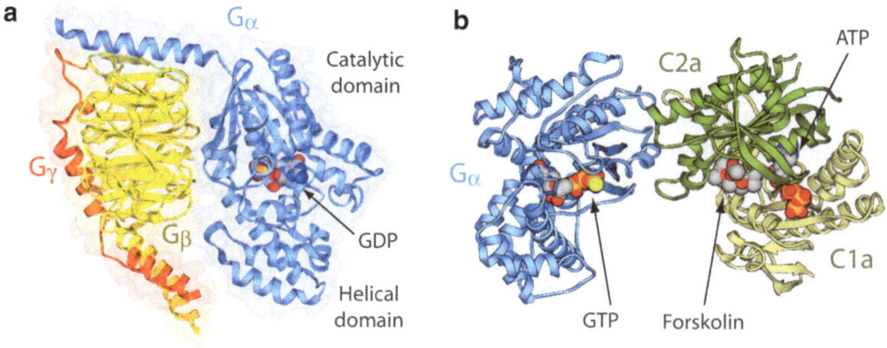

Fig. 18.3 (**a**) Structure of a mammalian heterotrimeric G protein (PDB code: 1GP2 [8]). The Gα subunit is formed by a helical domain composed of six helices and by a catalytic domain (also referred as RAS-like or GTPase domain) (Fig. 18.3a). The two domains are connected by two flexible linkers and their interface hosts the nucleotide binding pocket. Gβ consists of an N-terminal α-helix followed by a β-propeller domain, formed by seven "WD" repeat motifs, each made by approximately 43 amino acids. Its overall fold is completed by the interactions of strands from WD1 and WD7. The N-terminal helix of Gα interacts with WD1 of Gβ, while β2 strand, α2 helix and β3/α2 loop of Gα interact with six out of seven WD repeats of Gβ (WD1-WD5, WD7). Gγ, which is smaller than the other two subunits, consists of two helices connected by a loop. The N-terminal helix of Gγ forms a tight coiled-coil interaction with the N-terminal helix of Gβ. (**b**) Crystal structure of the dimeric catalytic extra membrane domain of trans membrane adenylyl cyclase in complex with $G\alpha_s$•GTP, forskolin and ATP (PDB: 3C16) [50]. The monomers present a ferredoxin-like fold. In the structure, AC dimer is formed by C1a from AC isoform 5 (AC5) and C2a from isoform 2 (AC2). AC2/5 features 37–70 % SI with the other isoenzymes

of the α5 helix (Fig. 18.3a) [56, 57]. These conformational changes lead to GDP release, with formation of a transient and conformationally dynamic empty state [58]. GTP then replaces GDP, triggering a new structural change that causes the detachment of one of the three subunits (Gα) from the other two, the Gβ and Gγ. The Gα subunit then binds and activates several enzymes and effectors [59–61]. In the case of ORs, it activates the AC3 enzyme.

MD simulations on the inactive state have shown that large scale motions involve the relative movement of the helical and catalytic domains [45], as highlighted by previous observations [62, 63] suggesting that this conformational change may be essential for GDP release [57]. Notably, almost all of the residues relevant for the Gα/Gβ interface (identified by computational alanine scanning) are conserved or conservatively mutated [64]. Instead, when the G-proteins are found in their empty state, GDP removal causes instabilities in the β6-α5 region, consistently with the suggestion that α5 helix might be involved in the process [56, 62]. The latter rearrangement is assisted by the conformational flexibility in the Gly202-Gly203 region.

18.3.3 Adenylyl Cyclases and Their Complex with Gα Subunit

AC enzymes catalyze the synthesis of the universal second messenger cAMP from ATP. Depending on the AC isoform involved in the process, they can be activated or inhibited by binding with GTP-bound α-subunit of specific G-proteins or by Gβγ subunits [65]. cAMP, synthesized by the enzyme, activates target proteins such as protein kinases, ion channels, and transcription factors, firing up the cellular response to the primary external stimulus. These proteins contain two TM domains (M1 and M2), each crossing the membrane six times. The main functional parts are located in the cytoplasm and can be subdivided into the N-terminus, C1a, C1b, C2a, and C2b. The C1 region exists between TM helices 6 and 7 and the C2 region follows TM helix 12. The C1a and C2a domains form a catalytic dimer where ATP binds and is converted to cAMP (Fig. 18.3b). The structural determinants of the cytoplasmic domain of an enzyme involved in OR signaling (AC3) in complex with its cognate Gα-subunit ($G\alpha_{olf}$), in the presence of the essential Mg^{2+} ion and forskolin (MPFsk), was predicted by using homology modeling and using the MPFk-bound AC3•$G\alpha_s$•Mg^{2+}•2′-deoxy-3′-adenosine monophosphate complex (SI >50 %) as template. The model suggests that the active site residues binding to MPFsk are the same as in the template [66].

18.3.4 Cyclic Nucleotide Gated Channels

CNG ion channels are tetrameric proteins gated by cGMP and cAMP second messengers. They produce the electrical signal in response not only to odor stimulation but also to light in the vision process. CNG channels belong to the superfamily of tetrameric voltage-gated ion channels [67, 68]. Their structure consist of: (i) a transmembrane domain formed by six transmembrane helices (S1–S6) and a pore helix (P-helix); and (ii) a cytoplasmic domain formed by the cyclic nucleotide binding domain, which is linked to the transmembrane domain through the so-called C-linker region. The structure of the pore region of the CNG channels was predicted by a knowledge-based guided homology model protocol, in which sequence alignment and experimental constraints were used to provide a structural basis for these channels [69]. The experimental constraints were obtained by cysteine scanning mutagenesis of residues present principally along the channel axis. Mutated channels were then studied by measuring the differences of current blockage upon the introduction of metals, such as Cd^{2+}, and agents capable of interacting with cysteines in the solution [70]. A combination of experimental and theoretical studies lead to the suggestion that a rotational movement begins in the C-linker region. This rotational movement is then transmitted upwards, making the upper part of S6 rotate anticlockwise. Due to the direct interaction of S6 with the P-helix, this motion is transmitted to the latter, which rearranges itself so that its terminal Thr360 residues and, therefore, the lower part of the pore wall, lead to the opening of the pore lumen. Thus, the initial event of cyclic nucleotide binding is transmitted to the pore walls by a remarkable and sophisticated coupling of conformational changes spanning the entire cytoplasmic and transmembrane domains of the channel.

18.3.5 Chloride Channels

Anoctamin 2/TMEM16B is likely the major subunit of the Ca^{2+}-activated Cl^- channel (CaCC) of olfactory sensory neurons [71], although other subunits may also be involved. Previous studies proposed that a member of the bestrophin protein family [72] plays a role as CaCC [73, 74], but recent works appear to refute this hypothesis [75]. Bioinformatic studies suggest that the N- and C-terminal domains of bestrophins would be located at the intracellular side of the membrane and would be connected to four or five hydrophobic domains forming the channel [76, 77]. Computational molecular biology techniques are currently in use to identify aspartate and glutamate residues that bind Ca^{2+} and to predict the effects of their mutations to alanine. Selected mutations have been investigated by electrophysiological experiments [78, 79].

18.3.6 Calmodulin

Calmodulin is a calcium-binding protein found ubiquitously in eukaryotes. It is capable of regulating biological activities of several calcium-sensitive enzymes, ion channels, and other proteins by performing conformational changes upon binding to calcium. This events in turn enables the binding to cognate proteins that bring a specific response. In particular, they regulate the activities of the CNG channels in the olfactory signaling pathway by: (i) decreasing the probability of opening CNG channels by binding to the channel subunits CNGA4 and CNGB1b; (ii) binding and activating the CaM-dependent phosphodiesterase (PDE1C2) and therefore catalyzing the transformation of cAMP in AMP; and (iii) activating the CaM-dependent protein kinase II and thus inhibiting AC3 by Ser1076 phosphorylation [80]. Calmodulin is composed of two globular domains connected together by a flexible linker. Each end contains two EF-hand motifs, each of which can bind a calcium ion. Despite a large number of experimental and theoretical studies, the detailed mechanisms of CaM target peptide recognition are not fully understood. Metadynamics-based free energy simulations [81] were used to investigate the final steps of CaM-peptide complex formation. Due to the lack of structural information for CaM in complex with the olfactory CNG channel target segment, the complex between CaM and M13, a peptide which is part of the skeletal muscle myosin light chain kinase (skMLCK), was considered. This complex is experimentally well characterized and involves an important biological CaM partner in the muscle tissue. The calculations were validated with a comparison between calculated and NMR-derived structural and dynamical properties. The results of the calculations [80] provide novel insights into the mechanism of protein/peptide recognition: it was shown that the process is associated with a free energy gain similar to that experimentally measured for the CaM complex with the homologous smooth muscle MLCK peptide [82]. The simulations suggested that CaM binding is dominated by entropic effects, in agreement with previous proposals. Furthermore, it was demonstrated that the large flexibility of the conserved methionine side chains play a key role in the binding mechanism. Finally, a rationale is provided for the experimental observation that in all CaM complexes the C-terminal domain seems to be hierarchically more important in establishing the interaction. The metadynamics simulation in this work has provided a first step toward predicting the complete energetics of the molecular recognition of CaM and CNG channels.

18.4 Bitter Taste Receptors: Conformational Fluctuations of Agonist Binding by Multiscale Simulations

Functional assay-validated bioinformatics approaches, complemented with molecular docking, have recently provided structural insights on agonist/bitter receptor interactions [34, 83, 84]. In particular, for the TAS2R38 receptor – one of the most widely characterized bitter receptors at the genetic level [36, 85] – the responses of the different receptor mutants have been measured upon application of increasing concentrations of agonists such as PTC and PROP (Fig. 18.4a). These pieces of information were included in the generation of the model of the ligand/receptor adduct, providing insights into structure/function relationships [34]. The modeling/blind-docking procedure allowed us to capture just one of the residues involved in binding [34] and the use of a knowledge-guided docking potential, as that of Haddock [86, 87], improved the description of the binding cavity. To improve the accuracy of our predictions, recently we have developed an hybrid Molecular Mechanics/Coarse-Grained (MM/CG) approach tailored for GPCRs (see Appendix and Fig. 18.4b, c) [88]. One-microsecond long MM/CG simulations allowed for conformational fluctuations of the complexes. These fluctuations eventually lead to poses consistent with most of the experiments carried out for this research. These consisted in functional calcium-imaging experiments [34, 88], in which the responses of the different proposed receptor mutants were measured upon application of increasing concentrations of agonists. Hence, the MM/CG approach allowed a description of the system with an unprecedented level of detail for a low-sequence identity homology model. In perspective, the protocol described in Ref. [88], that includes extensive MM/CG simulations on homology models combined with site-directed mutagenesis experiments, could be applied to different members of the bitter taste receptors as well as other receptors from the GPCR superfamily.

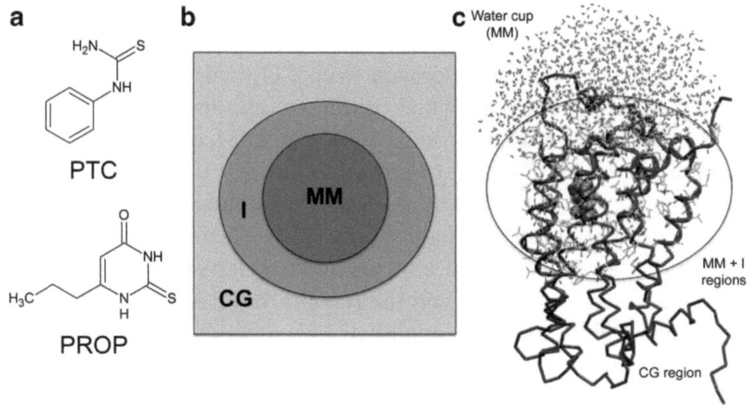

Fig. 18.4 (**a**) Schematic structure of PTC and PROP, agonists of the hTAS2R38 receptor. (**b**, **c**) Molecular Mechanics/Coarse-grained system set-up. (**b**) Schematic representation of the regions defined in the MM/CG model. The MM, I and CG regions are colored in *dark gey*, *grey*, and *light grey*, respectively. (**c**) MM/CG representation of the hTAS2R38 receptor in complex with PTC. Water molecules and residues belonging to the MM and I regions are represented as *lines*. The agonist atoms are represented as *grey spheres*. The protein Cα atoms are represented in order to show the protein backbone (Adapted from Ref. [88])

18.5 Conclusions and Perspective

Our computing studies on the bitter taste and olfactory receptors have been presented here with a twofold scope. First, we show, as done by many other groups (see Ref. [89] for a recent review and references within it), that even without experimental structural information, one can obtain good accuracy in predicting binding poses of GPCR by combining available biological data with computational techniques. Second, we have described our first attempt at investigating pathways at the molecular level. This is a necessary and crucial step towards understanding the behavior of biological systems [90]. Most importantly, multi-scale approaches will strongly impact on pharmaceutical sciences and toxicology, elucidating the drugs' effect on entire pathways, rather than on single biomolecules [91–93].

Appendix: Molecular Mechanics/Coarse-Grained Hybrid Approach

Coarse-grained (CG)-based MD approaches allow the study of longer timescales than all-atom force field simulations [94, 95]. The reduction of the number of degrees of freedom makes the model computationally very efficient, allowing a reduction of the simulation time by ca. 2–3 orders of magnitude compared to full atom force fields [96]. Unfortunately, without a detailed description of the side-chains, as in cases such as GPCRs, these approaches cannot describe in detail the intermolecular ligand/protein interactions. Thus a possible solution to this problem may be to combine atomistic with CG modeling [97–101]. Indeed, a hybrid/multi-scale approach in which different representations of the system are modeled concurrently was proposed, i.e. Molecular Mechanics/Coarse-Grained (MM/CG) simulations. In this kind of procedure, a coupling scheme is needed to connect the boundary of the different models. This approach has been developed for proteins by several groups, including ours [99–102]. In our scheme, a region of interest (i.e. the active site of an enzyme, MM region) is treated at molecular level using an atomistic force field and the protein frame is described at CG level using a Go-like model. Recently we modified and extended the use of the method, previously developed for soluble enzymes, to the case of GPCRs [103] in which the presence of the lipid bilayer must be imposed. In addition, one has to avoid that water from the binding site diffuses into the hydrophobic regions of the lipid bilayer. The accuracy of the new version of our MM/CG method was established by comparing MM/CG simulations with all-atom MD calculations on the human β2AR (hβ2AR) [103], in complex with two different ligands: the co-crystallized ligand and inverse agonist S-Carazolol (S-Car) [5] and its agonist R-Isoprenaline (R-ISO). The MM region consisted of 476 and 486 atoms, while the overall system was made of only 4,597 and 4,587 atoms, for the hβ2AR/S-Car and hβ2AR/R-ISO complexes, respectively. This allowed us to simulate more than 70 ns/day on 16 CPUs, which is a speed up of 15 times compared to the MD simulations of the same system. The trajectory obtained with our MM/CG scheme are able to reproduce the key structural features of the active site found in the MD simulations [104]. With these results, due to both its low cost and high reliability, using the MM/CG methodology emerges as a useful approach to study the ligand cavity of these proteins [103], indeed we have extensively used it for characterizing the binding cavity of the human TAS2R38 bitter taste receptor (see above).

References

1. Schoneberg T, Schulz A, Biebermann H, Hermsdorf T, Rompler H et al (2004) Mutant G-protein-coupled receptors as a cause of human diseases. Pharmacol Ther 104:173–206
2. Overington JP, Al-Lazikani B, Hopkins AL (2006) How many drug targets are there? Nat Rev Drug Discov 5:993–996
3. Palczewski K, Kumasaka T, Hori T, Behnke CA, Motoshima H et al (2000) Crystal structure of rhodopsin: a G protein-coupled receptor. Science 289:739–745
4. Cherezov V, Rosenbaum DM, Hanson MA, Rasmussen SGF, Thian FS et al (2007) High-resolution crystal structure of an engineered human β2-adrenergic G protein–coupled receptor. Science 318:1258–1265
5. Rasmussen SG, Choi HJ, Rosenbaum DM, Kobilka TS, Thian FS et al (2007) Crystal structure of the human beta2 adrenergic G-protein-coupled receptor. Nature 450:383–387
6. Downes GB, Gautam N (1999) The G protein subunit gene families. Genomics 62:544–552
7. McIntire WE (2009) Structural determinants involved in the formation and activation of G protein betagamma dimers. Neurosignals 17:82–99

8. Wall MA, Coleman DE, Lee E, Iñiguez-Lluhi JA, Posner BA et al (1995) The structure of the G protein heterotrimer Gi alpha 1 beta 1 gamma 2. Cell 83:1047–1058
9. Zhao Q, B-l W (2012) Ice breaking in GPCR structural biology. Acta Pharmacol Sin 33:324–334
10. Wang C, Wu H, Katritch V, Han GW, Huang XP et al (2013) Structure of the human smoothened receptor bound to an antitumour agent. Nature. doi:10.1038/nature12167
11. Venkatakrishnan AJ, Deupi X, Lebon G, Tate CG, Schertler GF et al (2013) Molecular signatures of G-protein-coupled receptors. Nature 494:185–194
12. Lagerstrom MC, Schioth HB (2008) Structural diversity of G protein-coupled receptors and significance for drug discovery. Nat Rev Drug Discov 7:339–357
13. Unal H, Karnik SS (2012) Domain coupling in GPCRs: the engine for induced conformational changes. Trends Pharmacol Sci 33:79–88
14. Depoortere I (2012) GI functions of GPR39: novel biology. Curr Opin Pharmacol 12:647–652
15. White JF, Noinaj N, Shibata Y, Love J, Kloss B et al (2012) Structure of the agonist-bound neurotensin receptor. Nature 490:508–513
16. Sprang S (1997) G protein mechanisms: insights from structural analysis. Annu Rev Biochem 66:639–678
17. Tesmer JJ (2010) The quest to understand heterotrimeric G protein signaling. Nat Struct Mol Biol 17:650–652
18. Bos JL, Rehmann H, Wittinghofer A (2007) GEFs and GAPs: critical elements in the control of small G proteins. Cell 129:865–877
19. Chung KY, Rasmussen SG, Liu T, Li S, DeVree BT et al (2011) Conformational changes in the G protein Gs induced by the β2 adrenergic receptor. Nature 477:611–615
20. Westfield GH, Rasmussen SG, Su M, Dutta S, DeVree BT et al (2011) Structural flexibility of the G alpha s alpha-helical domain in the beta2-adrenoceptor Gs complex. Proc Natl Acad Sci USA 108:16086–16091
21. Gaillard I, Rouquier S, Giorgi D (2004) Olfactory receptors. Cell Mol Life Sci 61:456–469
22. Chandrashekar J, Hoon MA, Ryba NJ, Zuker CS (2006) The receptors and cells for mammalian taste. Nature 444:288–294
23. Lupieri P, Nguyen CH, Bafghi ZG, Giorgetti A, Carloni P (2009) Computational molecular biology approaches to ligand-target interactions. HFSP J 3:228–239
24. Oldham W (2008) Heterotrimeric G protein activation by G-protein-coupled receptors. Nat Rev Mol Cell Biol 9:60–71
25. Kaupp UB, Seifert R (2002) Cyclic nucleotide-gated ion channels. Physiol Rev 82:769–824
26. Pifferi S, Boccaccio A, Menini A (2006) Cyclic nucleotide-gated ion channels in sensory transduction. FEBS Lett 580:2853–2859
27. Kleene SJ (1993) The cyclic nucleotide-activated conductance in olfactory cilia: effects of cytoplasmic Mg^{2+} and Ca^{2+}. J Membr Biol 131:237–243
28. Lowe G, Gold GH (1993) Nonlinear amplification by calcium-dependent chloride channels in olfactory receptor cells. Nature 366:283–286
29. Frings S, Hackos DH, Dzeja C, Ohyama T, Hagen V et al (2000) Determination of fractional calcium ion current in cyclic nucleotide-gated channels. Methods Enzymol 315:797–817
30. Menini A (1999) Calcium signalling and regulation in olfactory neurons. Curr Opin Neurobiol 9:419–426
31. Firestein S (2001) How the olfactory system makes sense of scents. Nature 413:211–218
32. Nadler W, Brunger AT, Schulten K, Karplus M (1987) Molecular and stochastic dynamics of proteins. Proc Natl Acad Sci USA 84:7933–7937
33. Matsunami H, Montmayeur J-P, Buck LB (2000) A family of candidate taste receptors in human and mouse. Nature 404:601–604
34. Biarnes X, Marchiori A, Giorgetti A, Lanzara C, Gasparini P et al (2010) Insights into the binding of phenyltiocarbamide (PTC) agonist to its target human TAS2R38 bitter receptor. PLoS One 5:e12394
35. Shi P, Zhang J (2006) Contrasting modes of evolution between vertebrate sweet/umami receptor genes and bitter receptor genes. Mol Biol Evol 23:292–300

36. Bufe B, Breslin PA, Kuhn C, Reed DR, Tharp CD et al (2005) The molecular basis of individual differences in phenylthiocarbamide and propylthiouracil bitterness perception. Curr Biol 15:322–327
37. Lee JH, Lee IH, Choe YJ, Kang S, Kim HY et al (2009) Real-time analysis of amyloid fibril formation of alpha-synuclein using a fibrillation-state-specific fluorescent probe of JC-1. Biochem J 418:311–323
38. Margolskee RF (2002) Molecular mechanisms of bitter and sweet taste transduction. J Biol Chem 277:1–4
39. Berridge MJ, Lipp P, Bootman MD (2000) The versatility and universality of calcium signalling. Nat Rev Mol Cell Biol 1:11–21
40. Nei M, Niimura Y, Nozawa M (2008) The evolution of animal chemosensory receptor gene repertoires: roles of chance and necessity. Nat Rev Genet 9:951–963
41. Dong D, Jin K, Wu X, Zhong Y (2012) CRDB: database of chemosensory receptor gene families in vertebrate. PLoS One 7:e31540
42. Grossfield A (2011) Recent progress in the study of G protein-coupled receptors with molecular dynamics computer simulations. Biochim Biophys Acta 1808:1868–1878
43. Johnston JM, Filizola M (2011) Showcasing modern molecular dynamics simulations of membrane proteins through G protein-coupled receptors. Curr Opin Struct Biol 21:552–558
44. Bruno A, Costantino G (2012) Molecular dynamics simulations of G protein-coupled receptors. Mol Inform 31:222–230
45. Khafizov K, Lattanzi G, Carloni P (2009) G protein inactive and active forms investigated by simulation methods. Proteins 75:919–930
46. Singer MS (2000) Analysis of the molecular basis for octanal interactions in the expressed rat 17 olfactory receptor. Chem Senses 25:155–165
47. Floriano WB, Vaidehi N, Goddard WA 3rd (2004) Making sense of olfaction through predictions of the 3-D structure and function of olfactory receptors. Chem Senses 29:269–290
48. Goddard WA 3rd, Abrol R (2007) 3-Dimensional structures of G protein-coupled receptors and binding sites of agonists and antagonists. J Nutr 137:1528S–1538S; discussion 1548S
49. Kolakowski LF Jr (1994) GCRDb: a G-protein-coupled receptor database. Receptors Channels 2:1–7
50. Warne T, Serrano-Vega MJ, Baker JG, Moukhametzianov R, Edwards PC et al (2008) Structure of a beta1-adrenergic G-protein-coupled receptor. Nature 454:486–491
51. Jaakola VP, Griffith MT, Hanson MA, Cherezov V, Chien EY et al (2008) The 2.6 angstrom crystal structure of a human A2A adenosine receptor bound to an antagonist. Science 322:1211–1217
52. Costanzi S (2008) On the applicability of GPCR homology models to computer-aided drug discovery: a comparison between in silico and crystal structures of the beta2-adrenergic receptor. J Med Chem 51:2907–2914
53. Jorgensen AM, Tagmose L, Jorgensen AM, Topiol S, Sabio M et al (2007) Homology modeling of the serotonin transporter: insights into the primary escitalopram-binding site. ChemMedChem 2:815–826
54. Ballesteros JA, Weinstein H, Stuart CS (1995) Integrated methods for the construction of three-dimensional models and computational probing of structure-function relations in G protein-coupled receptors. In: Methods in neurosciences. Academic Press, San Diego, pp 366–428
55. Katada S, Hirokawa T, Oka Y, Suwa M, Touhara K (2005) Structural basis for a broad but selective ligand spectrum of a mouse olfactory receptor: mapping the odorant-binding site. J Neurosci 25:1806–1815
56. Oldham WM, Van Eps N, Preininger AM, Hubbell WL, Hamm HE (2006) Mechanism of the receptor-catalyzed activation of heterotrimeric G proteins. Nat Struct Mol Biol 13:772–777
57. Gales C, Van Durm JJ, Schaak S, Pontier S, Percherancier Y et al (2006) Probing the activation-promoted structural rearrangements in preassembled receptor-G protein complexes. Nat Struct Mol Biol 13:778–786

58. Abdulaev NG, Ngo T, Ramon E, Brabazon DM, Marino JP et al (2006) The receptor-bound "empty pocket" state of the heterotrimeric G-protein alpha-subunit is conformationally dynamic. Biochemistry 45:12986–12997
59. Kristiansen K (2004) Molecular mechanisms of ligand binding, signaling, and regulation within the superfamily of G-protein-coupled receptors: molecular modeling and mutagenesis approaches to receptor structure and function. Pharmacol Ther 103:21–80
60. Oldham WM, Hamm HE (2006) Structural basis of function in heterotrimeric G proteins. Q Rev Biophys 39:117–166
61. Johnston CA, Siderovski DP (2007) Receptor-mediated activation of heterotrimeric G-proteins: current structural insights. Mol Pharmacol 72:219–230
62. Ceruso MA, Periole X, Weinstein H (2004) Molecular dynamics simulations of transducin: interdomain and front to back communication in activation and nucleotide exchange. J Mol Biol 338:469–481
63. Mello LV, van Aalten DM, Findlay JB (1998) Dynamic properties of the guanine nucleotide binding protein alpha subunit and comparison of its guanosine triphosphate hydrolase domain with that of ras p21. Biochemistry 37:3137–3142
64. Wall MA, Posner BA, Sprang SR (1998) Structural basis of activity and subunit recognition in G protein heterotrimers. Structure 6:1169–1183
65. Hanoune J, Defer N (2001) Regulation and role of adenylyl cyclase isoforms. Annu Rev Pharmacol Toxicol 41:145–174
66. Tesmer JJ, Sunahara RK, Gilman AG, Sprang SR (1997) Crystal structure of the catalytic domains of adenylyl cyclase in a complex with Gsalpha. GTPgammaS. Science 278:1907–1916
67. Biel M, Michalakis S (2007) Function and dysfunction of CNG channels: insights from channelopathies and mouse models. Mol Neurobiol 35:266–277
68. Anselmi C, Carloni P, Torre V (2007) Origin of functional diversity among tetrameric voltage-gated channels. Proteins 66:136–146
69. Giorgetti A, Nair AV, Codega P, Torre V, Carloni P (2005) Structural basis of gating of CNG channels. FEBS Lett 579:1968–1972
70. Nair AV, Mazzolini M, Codega P, Giorgetti A, Torre V (2006) Locking CNGA1 channels in the open and closed state. Biophys J 90:3599–3607
71. Pifferi S, Cenedese V, Menini A (2012) Anoctamin 2/TMEM16B: a calcium-activated chloride channel in olfactory transduction. Exp Physiol 97:193–199
72. Hartzell HC, Qu Z, Yu K, Xiao Q, Chien LT (2008) Molecular physiology of bestrophins: multifunctional membrane proteins linked to best disease and other retinopathies. Physiol Rev 88:639–672
73. Pifferi S, Pascarella G, Boccaccio A, Mazzatenta A, Gustincich S et al (2006) Bestrophin-2 is a candidate calcium-activated chloride channel involved in olfactory transduction. Proc Natl Acad Sci USA 103:12929–12934
74. Boccaccio A, Menini A (2007) Temporal development of cyclic nucleotide-gated and Ca2+ -activated Cl− currents in isolated mouse olfactory sensory neurons. J Neurophysiol 98:153–160
75. Pifferi S, Dibattista M, Sagheddu C, Boccaccio A, Al Qteishat A et al (2009) Calcium-activated chloride currents in olfactory sensory neurons from mice lacking bestrophin-2. J Physiol 587:4265–4279
76. Jentsch TJ, Stein V, Weinreich F, Zdebik AA (2002) Molecular structure and physiological function of chloride channels. Physiol Rev 82:503–568
77. Loewen ME, Forsyth GW (2005) Structure and function of CLCA proteins. Physiol Rev 85:1061–1092
78. Xiao Q, Prussia A, Yu K, Cui YY, Hartzell HC (2008) Regulation of bestrophin Cl channels by calcium: role of the C terminus. J Gen Physiol 132:681–692
79. Kranjc A, Grillo FW, Rievaj J, Boccaccio A, Pietrucci F et al (2009) Regulation of bestrophins by Ca2+: a theoretical and experimental study. PLoS One 4:e4672

80. Fiorin G, Pastore A, Carloni P, Parrinello M (2006) Using metadynamics to understand the mechanism of calmodulin/target recognition at atomic detail. Biophys J 91:2768–2777
81. Laio A, Parrinello M (2002) Escaping free-energy minima. Proc Natl Acad Sci USA 99:12562–12566
82. Ehrhardt MR, Urbauer JL, Wand AJ (1995) The energetics and dynamics of molecular recognition by calmodulin. Biochemistry 34:2731–2738
83. Sakurai T, Misaka T, Ishiguro M, Masuda K, Sugawara T et al (2010) Characterization of the beta-D-glucopyranoside binding site of the human bitter taste receptor hTAS2R16. J Biol Chem 285:28373–28378
84. Singh UC, Kollman PA (1986) A combined ab initio quantum mechanical and molecular mechanical method for carrying out simulations on complex molecular systems: applications to the CH3Cl + Cl− exchange reaction and gas phase protonation of polyethers. J Comput Chem 7:718–730
85. Kim UK, Jorgenson E, Coon H, Leppert M, Risch N et al (2003) Positional cloning of the human quantitative trait locus underlying taste sensitivity to phenylthiocarbamide. Science 299:1221–1225
86. de Vries SJ, van Dijk AD, Krzeminski M, van Dijk M, Thureau A et al (2007) HADDOCK versus HADDOCK: new features and performance of HADDOCK2.0 on the CAPRI targets. Proteins 69:726–733
87. Dominguez C, Boelens R, Bonvin AM (2003) HADDOCK: a protein-protein docking approach based on biochemical or biophysical information. J Am Chem Soc 125:1731–1737
88. Marchiori A, Capece L, Giorgetti A, Gasparini P, Behrens M et al (2013) Coarse-grained/molecular mechanics of the TAS2R38 bitter taste receptor: experimentally-validated detailed structural prediction of agonist binding. PLoS One 8(5):e64675
89. Levit A, Barak D, Behrens M, Meyerhof W, Niv M (2012) Homology model-assisted elucidation of binding sites in GPCRs. In: Vaidehi N, Klein-Seetharaman J (eds) Membrane protein structure and dynamics. Humana Press, pp 179–205
90. Chang X, Xu T, Li Y, Wang K (2013) Dynamic modular architecture of protein-protein interaction networks beyond the dichotomy of 'date' and 'party' hubs. Sci Rep 3:1691
91. Pujol A, Mosca R, Farrés J, Aloy P (2010) Unveiling the role of network and systems biology in drug discovery. Trends Pharmacol Sci 31:115–123
92. Wass MN, Fuentes G, Pons C, Pazos F, Valencia A (2011) Towards the prediction of protein interaction partners using physical docking. Mol Syst Biol 7:1–8
93. Vidal M, Cusick ME, Barabási A-L (2011) Interactome networks and human disease. Cell 144:986–998
94. Ayton GS, Noid WG, Voth GA (2007) Multiscale modeling of biomolecular systems: in serial and in parallel. Curr Opin Struct Biol 17:192–198
95. Hyeon C, Thirumalai D (2011) Capturing the essence of folding and functions of biomolecules using coarse-grained models. Nat Commun 2:487
96. Monticelli L, Kandasamy SK, Periole X, Larson RG, Tieleman DP et al (2008) The MARTINI coarse-grained force field: extension to proteins. J Chem Theory Comput 4:819–834
97. Kalli AC, Campbell ID, Sansom MSP (2011) Multiscale simulations suggest a mechanism for integrin inside-out activation. Proc Natl Acad Sci USA 108:11890–11895
98. Messer BM, Roca M, Chu ZT, Vicatos S, Kilshtain AV et al (2010) Multiscale simulations of protein landscapes: using coarse-grained models as reference potentials to full explicit models. Proteins 78:1212–1227
99. Shi Q, Izvekov S, Voth GA (2006) Mixed atomistic and coarse-grained molecular dynamics: simulation of a membrane-bound ion channel. J Phys Chem B 110:15045–15048
100. Villa E, Balaeff A, Schulten K (2005) Structural dynamics of the lac repressor-DNA complex revealed by a multiscale simulation. Proc Natl Acad Sci USA 102:6783–6788
101. Neri M, Baaden M, Carnevale V, Anselmi C, Maritan A et al (2008) Microseconds dynamics simulations of the outer-membrane protease T. Biophys J 94:71–78
102. Neri M, Anselmi C, Cascella M, Maritan A, Carloni P (2005) Coarse-grained model of proteins incorporating atomistic detail of the active site. Phys Rev Lett 95:218102

103. Leguebe M, Nguyen C, Capece L, Hoang Z, Giorgetti A et al (2012) Hybrid molecular mechanics/coarse-grained simulations for structural prediction of G-protein coupled receptor/ligand complexes. PLoS One 7:e47332
104. Vanni S, Neri M, Tavernelli I, Rothlisberger U (2011) Predicting novel binding modes of agonists to B adrenergic receptors using all-atom molecular dynamics simulations. PLoS Comput Biol 7:e1001053
105. Bjarnadottir TK, Gloriam DE, Hellstrand SH, Kristiansson H, Fredriksson R et al (2006) Comprehensive repertoire and phylogenetic analysis of the G protein-coupled receptors in human and mouse. Genomics 88:263–273

Index

K. Han et al. (eds.), *Protein Conformational Dynamics*, Advances in Experimental Medicine and Biology 805, DOI 10.1007/978-3-319-02970-2,

K

S

MIX
Papier aus verantwortungsvollen Quellen
Paper from responsible sources
FSC® C105338

If you have any concerns about our products,
you can contact us on
ProductSafety@springernature.com

In case Publisher is established outside the EU,
the EU authorized representative is:
Springer Nature Customer Service Center GmbH
Europaplatz 3, 69115 Heidelberg, Germany

Printed by Libri Plureos GmbH
in Hamburg, Germany